The Enzymes

VOLUME XXVI

GLYCOSYLPHOSPHATIDYLINOSITOL (GPI) ANCHORING OF PROTEINS

THE ENZYMES

Edited by

Anant K. Menon

Department of Biochemistry
Weill Cornell Medical College
New York, NY 10065, USA

Peter Orlean

Department of Microbiology
University of Illinois at
Urbana-Champaign
Urbana, IL 61801, USA

Taroh Kinoshita

Research Institute for Microbial Diseases
Osaka University, 3-1 Yamada-oka
Suita, 565-0871 Osaka, Japan

Fuyuhiko Tamanoi

Department of Microbiology,
*   Immunology, and Molecular Genetics*
Molecular Biology Institute
University of California, Los Angeles
Los Angeles, CA 90095, USA

Volume XXVI

GLYCOSYLPHOSPHATIDYLINOSITOL (GPI) ANCHORING OF PROTEINS

AMSTERDAM • BOSTON • HEIDELBERG • LONDON
NEW YORK • OXFORD • PARIS • SAN DIEGO
SAN FRANCISCO • SINGAPORE • SYDNEY • TOKYO
Academic Press is an imprint of Elsevier

ELSEVIER

Academic Press is an imprint of Elsevier
32 Jamestown Road, London NW1 7BY, UK
Radarweg 29, PO Box 211, 1000 AE Amsterdam, The Netherlands
Linacre House, Jordan Hill, Oxford OX2 8DP, UK
30 Corporate Drive, Suite 400, Burlington, MA 01803, USA
525 B Street, Suite 1900, San Diego, CA 92101-4495, USA

First edition 2009

ISBN: 978-0-12-374963-5   10059566 62X
ISSN: 1874-6047

For information on all Academic Press publications
visit our website at elsevierdirect.com

Printed and bound in USA

09 10 11 12 13 10 9 8 7 6 5 4 3 2 1

Contents

Preface... xi

1. Overview of GPI Biosynthesis

    Taroh Kinoshita and Morihisa Fujita

    I.   Abstract..1
    II.   Introduction of GPI-Anchored Proteins.............................2
    III. Structure of GPI-APs..6
    IV.   Biosynthesis of GPI and Attachment to Proteins 7
    V.   Lipid Remodeling of GPI and GPI-APs..............................15
    VI.   Enzymes Involved in Modification of GPI Glycan in GPI-APs 20
          References..21

2. The N-Acetylglucosamine-PI Transfer Reaction, the GlcNAc-PI
    Transferase Complex, and Its Regulation

    David E. Levin and Ronald J. Stamper

    I.   Abstract...31
    II.   The N-Acetylglucosamine-PI Transfer Reaction......................32
    III. Biological Importance of the GPI-GlcNAc Transferase...............33
    IV.   The GPI-GlcNAc Transferase Complex................................33
    V.   Regulation of GPI-GlcNAc Transferase Activity 38
    VI.   Concluding Remarks 43
          Acknowledgment..43
          References..44

3. The GlcNAc-PI de-N-acetylase: Structure, Function, and Activity

    Michael D. Urbaniak and Michael A.J. Ferguson

    I.   Abstract...49
    II.   Introduction...50
    III. Protein Structure and Function 50
    IV.   Recombinant Protein Expression....................................55
    V.   Activity Assays...58
    VI.   Enzyme Substrate Specificity 61
          References..62

v

4. Inositol Acylation/Deacylation

TAKEHIKO YOKO-O AND YOSHIFUMI JIGAMI

I. Abstract 65
II. Introduction: PI-PLC-Resistant GPI Molecules 66
III. Variations in Inositol Acylation/Deacylation Among Organisms 67
IV. Enzymes Involved in Inositol Acylation/Deacylation and Their Functions 72
V. Perspective and Concluding Remarks 82
Acknowledgments 83
References 83

5. Mannosylation

YUSUKE MAEDA AND YASU S. MORITA

I. Abstract 91
II. Overview of Biosynthetic Pathway for GPI Mannosylation 92
III. Dol-P-Man as a Substrate for GPI-Man-Ts 92
IV. GPI-Mannosyltransferases (GPI-Man-Ts) 94
V. Structural Consideration of Glycosyltransferases that use
Dol-P-monosaccharides as Donor Substrates 97
VI. Substrate Specificities and Inhibitors for Mannosylations 102
VII. GPI Mannosylation-Related Diseases 103
VIII. Polyprenol-Phosphate-Mannose (PPM)-Dependent Mannosyltransferases
in Mycobacteria 104
IX. Mycobacteria Genes Homologous to Mammalian PIG-M 107
References 109

6. Phosphoethanolamine Addition to Glycosylphosphatidylinositols

PETER ORLEAN

I. Abstract 117
II. Sites of Etn-P Modification on Protein-Bound and Free GPIs 118
III. Phosphoethanolamine Donor 119
IV. Proteins Involved in Etn-P Addition 120
V. Concluding Remarks 127
References 128

7. Attachment of a GPI Anchor to Protein

AITA SIGNORELL AND ANANT K. MENON

I. Abstract 133
II. The GPI Signal Sequence 134
III. The Transamidation Reaction 136
IV. GPI Transamidase (GPIT) 136

    V. GPI8/Gpi8p...138
   VI. GAA1/Gaa1p 140
  VII. PIG-T/Gpi16p..141
 VIII. PIG-U/Gab1p..142
   IX. PIG-S/Gpi17 143
    X. TTA1 and TTA2...143
   XI. Final Remarks...144
  XII. Future Work 144
       Acknowledgements 145
       References..145

8. Split Topology of GPI Biosynthesis

ANANT K. MENON

    I. Abstract..151
   II. Synthesis and De-*N*-Acetylation of GlcNAc-PI Occur on the
       Cytoplasmic Face of the ER...151
  III. Inositol Acylation Probably Occurs in the ER Lumen.............................152
  IV. Beyond Inositol Acylation: Later Reactions of GPI Assembly (Mannosylation,
       Phosphoethanolamine Addition, and GPI Transfer to Protein)
       Occur in the ER Lumen 153
   V. GlcN-PI Flips Across the ER Membrane During GPI Biosynthesis 154
  VI. Conclusion...156
       Acknowledgments 156
       References..157

9. GPIs of Apicomplexan Protozoa

HOSAM SHAMS-ELDIN, FRANÇOISE DEBIERRE-GROCKIEGO,
JÜRGEN KIMMEL, AND RALPH T. SCHWARZ

    I. Abstract..159
   II. Introduction: Apicomplexan Parasites 160
  III. GPI Structures of *Plasmodium falciparum* and *Toxoplasma gondii*...................162
  IV. Immunological Functions of Protozoan GPIs 170
       References..176

10. Chemical Synthesis of Glycosylphosphatidylinositol (GPI) Anchors

RAM VISHWAKARMA AND DIPALI RUHELA

    I. Abstract..181
   II. Introduction...182
  III. Synthesis of GPI Anchor of VSG of *T. brucei*....................................183
  IV. Synthesis of GPI Anchor of *Saccharomyces cerevisiae* (Yeast).......................195
   V. Synthesis of GPI Anchor of Rat Brain Thy-1......................................200

VI.   Synthesis of GPI Anchor of *P. falciparum* 204
VII. Synthesis of GPI Anchors of *Trypanosoma cruzi* 209
VIII. Synthesis of GPI Anchor of CD52...216
IX.   Synthesis of GPI Anchor of Lipophosphoglycan (LPG) of *Leishmania* parasite 218
X.   Synthesis of GPI Anchor of GIPL of *T. cruzi* 221
XI.   Other Notable Contributions 223
XII. Conclusion...224
      References...224

11. GPI-Based Malarial Vaccine: Past, Present, and Future

XINYU LIU, DANIEL VARON SILVA, FAUSTIN KAMENA,
AND PETER H. SEEBERGER

I.   Abstract...229
II. Introduction to GPI in Malarial Pathogenesis 230
III. Synthetic GPI as Antitoxic Malarial Vaccine Candidate in a Rodent Model 233
IV. Synthetic GPI Microarray to Define Antimalarial Antibody Response................235
V.   Synthetic GPI as Tools to Study Malaria Associated Anemia.......................238
VI. Conclusion and Perspectives..240
     Acknowledgments 242
     References..242

12. Inhibitors of GPI Biosynthesis

TERRY K. SMITH

I.   Abstract...247
II.   Introduction..248
III.   GPI Biosynthesis 249
IV.   GlcNAc Transferase...250
V.   GlcNAc-PI De-*N*-Acetylase 250
VI.   Inositol Acyltransferase..253
VII.   Mannosylation of GPI Anchor Intermediates 254
VIII. Mannose Analogs...254
IX.   α1–4-Mannosyltransferase (MT-I)..255
X.   α1–6-Mannosyltransferase (MT-II) 256
XI.   α1–2-Mannosyltransferase (MT-III) 256
XII.   Ethanolamine Phosphate Transferases 257
XIII. Inositol Deacylation...257
XIV.   GPI Lipid Remodelling..258
XV.   GPI Transamidase 258
XVI.   Species-Specific Modifications to the Core GPI Structure.......................259
XVII. Perspectives...260
       Acknowledgments 260
       References..261

13. Transport of GPI-Anchored Proteins: Connections to Sphingolipid and Sterol Transport

GUILLAUME A. CASTILLON AND HOWARD RIEZMAN

I. Abstract 269
II. ER Exit of GPI-Anchored Proteins..270
III. Sorting of GPI-Anchored Proteins upon ER Exit................................273
IV. Defects in GPI-Anchored Protein Trafficking and Folding........................275
V. GPI-Anchored Protein and Lipid Traffic 278
VI. Parallels and Differences between Yeast and Mammalian Cells.....................283
    Acknowledgments 283
    References..283

14. Mechanisms of Polarized Sorting of GPI-anchored Proteins in Epithelial Cells

SIMONA PALADINO AND CHIARA ZURZOLO

I. Abstract 289
II. Introduction..290
III. Secretory Pathway and Polarized Sorting......................................290
IV. Site of Sorting and Routes to the Surface......................................300
V. Regulation of Membrane Traffic 306
VI. Conclusion/Perspectives 308
    References..309

15. GPI Proteins in Biogenesis and Structure of Yeast Cell Walls

MARLYN GONZALEZ, PETER N. LIPKE, AND RAFAEL OVALLE

I. Abstract 321
II. Fungal GPI-Anchored Proteins and the Cell Wall: General Introduction.............322
III. A Brief History of the Discovery of Fungal GPI Proteins in Yeast Cell Walls.........324
IV. Structure of Yeast Cell Walls 326
V. Ordered Cell Wall Assembly and Addition of GPI Proteins.......................329
VI. Phylogenetics of GPI-Cell Wall Transglycosylation 330
VII. Roles of GPIs in Biogenesis of GPI-Cell Wall Cross-Links........................332
VIII. Surface Display in Yeast 337
IX. Summary 345
    References..345

16. Inherited GPI Deficiency

ANTONIO ALMEIDA, MARK LAYTON, AND ANASTASIOS KARADIMITRIS

I. Abstract 357
II. Introduction..358
III. Disorders of GPI Deficiency...361

IV. A Tentative Model of Transcriptional Control of PIG-M by Sp1
     and its Dysregulation in IGD 367
 V. IGD: Questions and Perspectives. 369
     Acknowledgment 370
     References. 370

Author Index 375
Index 409

Preface

This volume is the third publication in this series dealing with posttranslational modification of proteins. Volume 21 dealt with protein lipidation, specifically protein prenylation, S-acylation, and N-myristoylation, while the topic of volume 24 was protein methyltransferases. The topic of this volume is "Glycosylphosphatidylinositol (GPI) anchoring of cell surface proteins." GPI anchors are glycolipids composed of a phosphatidylinositol whose headgroup is extended by a glucosamine and mannose-containing glycan bearing one or more phosphoethanolamine moieties. The distal phosphoethanolamine links the GPI structure to the C-terminal amino acid of a mature protein via an amide bond. The GPI's hydrophobic lipid moiety serves to anchor the protein to the membrane. GPI-anchored proteins exhibit a variety of functions. They play critical roles in receptor-mediated signal transduction pathways. They are markers of specialized plasma membrane domains and are important in apical protein positioning and in cell wall construction in fungi.

We start with an overview of GPI biosynthesis in Chapter 1. Chapters 2–8 discuss details of the steps involved in this series of biosynthesis reactions. Enzymes, substrates, and enzymatic mechanisms involved in these reactions are discussed. Chapter 9 has special emphasis on the GPIs of protozoa. Chapters 10 and 11 discuss chemical synthesis of GPI anchors and their use in vaccine development. Chapter 12 deals with chemical inhibitors of GPI biosynthesis. Chapters 13–15 discuss aspects of the cell biology of GPI-anchored proteins, including their transport, polarized sorting, and their involvement in cell wall synthesis. Finally, Chapter 16 discusses hemostatic and neurological problems due to GPI deficiency.

The idea for this volume was conceived at the 2006 FASEB meeting on Protein Lipidation, Signaling, and Membrane Domains in Indian Wells, California, where we had a long discussion about the content. We would like to thank the contributors for preparing their chapters in a timely fashion. We also thank Lisa Tickner for her expert help in organizing the publication. We are also grateful to Gloria Lee who helped edit chapters at UCLA.

Fuyuhiko Tamanoi
Anant K. Menon
Taroh Kinoshita
Peter Orlean
May 11, 2009

1

Overview of GPI Biosynthesis

TAROH KINOSHITA • MORIHISA FUJITA

Research Institute for Microbial Diseases
Osaka University
3-1 Yamada-oka Suita
565-0871 Osaka
Japan

I.   Abstract

Glycosylphosphatidylinositol (GPI) anchor is a form of posttranslational modification of many cell surface proteins common to all phyla of eukaryotes. GPI acts to anchor proteins on the outer leaflet of the plasma membrane. The lipid part is either phosphatidylinositol (PI) or inositol phosphoceramide. Due to saturated fatty chains in the lipid part of GPI, GPI-anchored proteins (GPI-APs) are mainly present in membrane microdomains of mammalian and yeast plasma membranes. The glycan part consists of a conserved core backbone and variable side branches. Some of the functions of the glycan part have been characterized in yeast and protozoan parasite trypanosome, while they are largely unclear in mammalian cells. The core backbone structure, EtNP-6Manα1–2Manα1–6Manα1–4GlcNα1–6*myo*Inositol-phospholipid (where EtNP is ethanolamine phosphate; Man is mannose; and GlcN is glucosamine), is common to all of them. The terminal EtNP is linked to the C-terminus of protein via an amide bond. Precursors of GPI-anchor are synthesized in the endoplasmic reticulum (ER) from PI through at least nine sequential reaction steps. The complete GPI precursor is attached to proteins bearing a C-terminal GPI attachment signal peptide by GPI transamidase (GPI-TA). Nascent GPI-APs are then transported via the secretory pathway through the Golgi to the plasma membrane. On the way of transportation, the lipid part of GPI is remodeled and the glycan part is modified. At least 23 gene products are involved in biosynthesis and attachment to proteins of GPI-anchor precursors. Several genes involved in lipid remodeling have been identified.

II.   Introduction of GPI-Anchored Proteins

Glycosylphosphatidylinositol (GPI) anchor is a form of posttranslational modification of many cell surface proteins common to all phyla of eukaryotes [1–5]. GPI acts to anchor proteins on the outer leaflet of the plasma membrane. Whereas two other major posttranslational modifications for membrane anchoring of proteins, namely acylation and prenylation, include only lipids, GPI consists of lipid and glycan parts. The lipid part is either phosphatidylinositol (PI) or inositol phosphoceramide. Due to saturated fatty chains in the lipid part of GPI, GPI-anchored proteins (GPI-APs) are mainly present in membrane microdomains (membrane rafts or lipid rafts) in mammalian and yeast plasma membranes [6]. The glycan part consists of the conserved core backbone and the variable side branches. Some of the functions of the glycan part have been characterized in yeast and protozoan parasite trypanosome, while they are largely unclear in mammalian cells.

About 150 human proteins with various functions are GPI anchored. GPI-APs include hydrolytic and other enzymes (alkaline phosphatase, 5′-nucleotidase/CD73, erythrocyte acetylcholinesterase, renal dipeptidase, and mono-ADP-ribosyltransferase ART), adhesion molecules (neural cell adhesion molecule 120, TAG1, and isoform of CD58), receptors (folate receptor, CD14, CD16b, uPA receptor/CD87, ciliary neurotrophic factor receptor α subunit, and glial-cell-derived neurotrophic factor receptor α subunit), complement regulatory proteins (CD55 and CD59), immunologically important proteins (CD24, CD48, CD52, and CD90/Thy-1), and other proteins (prion protein and glypicans). Complete deficiency of GPI causes early embryonic lethality due to malformation of the brain as shown by knockout mice with defective GPI biosynthesis [7]. Inherited partial deficiency causes inherited GPI deficiency, a disease characterized by hepatic and/or portal vein thromboses and seizures [8]. Acquired GPI deficiency due to a somatic mutation in the hematopoietic stem cell causes paroxysmal nocturnal hemoglobinuria characterized by complement-mediated hemolytic anemia, venous thrombosis, and bone marrow failure [9, 10].

It is estimated that about 60 out of 6000 proteins of yeast *Saccharomyces cerevisiae* are GPI anchored, based on the presence of GPI-attachment signal peptide at the C-terminus of the protein sequence predicted from genome data [2, 11]. Many, perhaps majority, of the GPI-APs are cell wall proteins rather than plasma membrane proteins. Cell wall localization of GPI-APs is achieved by a transglycosidation reaction between the glycan part of GPI and cell wall β-1,6 glucan [12]. GPI biosynthesis is essential for the growth of *S. cerevisiae* [2].

The plant Arabidopsis may have 248 GPI-APs as predicted in a similar way [4]. They include various enzymes and receptors. GPI-APs are required for root development, cell wall synthesis, pollen germination, and tube growth [13, 14].

GPI-APs are by far the most popular type of the cell surface proteins in protozoa, such as trypanosomes and malaria parasites [15, 16]. African trypanosome, *Trypanosoma brucei*, has two proliferative stages, a bloodstream form that grows in the blood plasma of mammalian hosts and a procyclic form that grows in the midgut of tsetse fly vector. Bloodstream form parasites have a dense cell surface coat consisting of 10 million molecules of a single GPI-AP, variant surface glycoprotein (VSG) [15]. GPI biosynthesis is essential for growth of the bloodstream form, being exploited as a target of antitrypanosomal drug development [17–20]. The surface of procyclic form parasites is coated by one million molecules of procyclins, GPI-AP with a large side branch of GPI-containing terminal sialic acids. GPI biosynthesis is not essential for growth of the procyclic form in *in vitro* culture [17], whereas it is critical for survival in tsetse fly [21, 22].

Major cell surface proteins of sporozoites and merozoites of malaria parasites are also GPI anchored. Sporozoites that invade hepatocytes after injection by mosquitos are covered by GPI-anchored circumsporozoite proteins [23]. Merozoites that invade erythrocytes have major GPI-APs, such as MSP1 and MSP2 [24].

Glycan and lipid structures have been determined for various GPI-APs from mammalian cells [5, 25–27], yeast (*S. cerevisiae*) [28], protozoan parasites (*T. brucei* and *Plasmodium falciparum*) [24, 29], and plant (*Pyrus communis*) [30] (Figure 1.1). The core backbone structure, EtNP-6Manα1–2Manα1–6Manα1–4GlcNα1–6*myo*Inositol-phospholipid (where EtNP is ethanolamine phosphate; Man is mannose; and GlcN is glucosamine), is common to all of them [1–3, 5]. The terminal "bridging" EtNP is linked to the C-terminus of the protein via an amide bond. Various side branches decorate the core backbone of GPI [1–3].

Precursors of the GPI anchor are synthesized in the ER from PI through sequential reaction steps, such as addition of monosaccharides, EtNP, and fatty acid, and removal of acetyl group from *N*-acetylGlcN (GlcNAc) and fatty acid [1–3, 5] (Figure 1.2). The complete GPI precursor is then attached

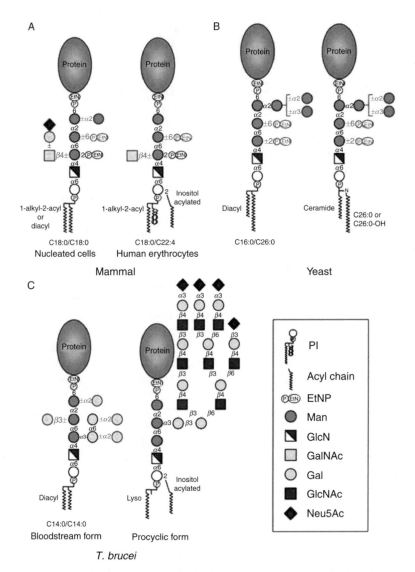

FIG. 1.1. Structures of GPI-APs in mammalian cells (A), yeast (B), and trypanosome (C). (A) Left, GPI-AP from nucleated cells; right, GPI-AP from human erythrocytes. (B) Left, diacylglycerol-type GPI-AP; right, ceramide-type GPI-AP in budding yeast, *S. cerevisiae*. (C) Left, GPI-AP from blood stream form of *T. brucei*; right, GPI-AP from procyclic form of *T. brucei* [5]. Symbol representations of monosaccharides are according to Ref. [148].

to proteins bearing a C-terminal GPI attachment signal peptide on the luminal side of the ER membrane [1]. The GPI attachment is mediated by a transamidase that cleaves the C-terminal signal peptide and replaces it

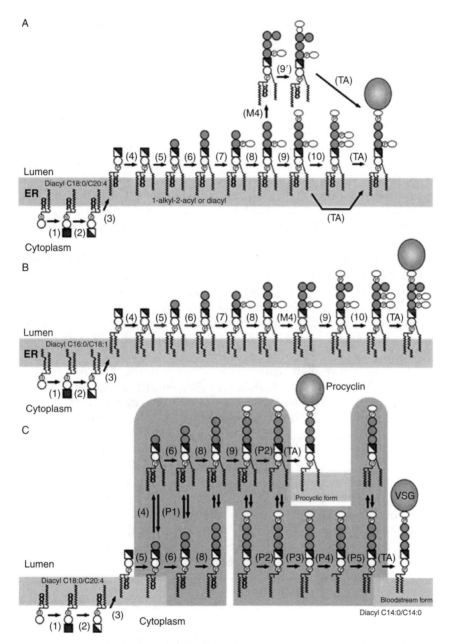

FIG. 1.2. Biosynthesis of GPI on the ER membrane in mammalian cells (A), yeast (B), and trypanosome (C). Steps 1-TA correspond to those in Tables 1.1 and 1.2. In C, reactions are not numbered in order, but those equivalent to mammalian steps (A) are given with the same numbers.

with GPI [1, 2]. Nascent GPI-APs are then transported via the secretory pathway through the Golgi to the plasma membrane. On the way of transportation, GPI is structurally remodeled and modified [1–3, 31].

III.   Structure of GPI-APs

A.   MAMMALIAN GPI-APS

All mammalian GPI-APs have an EtNP side branch linked to the 2-position of the first α1–4 linked Man (Man-1) of the core backbone [1, 3] (Figure 1.1A). Occasionally, the third EtNP is on the 6-position of the second Man (Man-2) [32, 33]. In some mammalian GPI-APs, the fourth Man is linked via an α1–2 bond to the third Man (Man-3) [34]. These modifications occur during biosynthesis of the GPI precursor before attachment to proteins [1, 3](Figure 1.2A). Heterogeneity in the glycan side-chain is found even in one GPI-AP. N-acetylgalactosamine (GalNAc) is linked via β1–4 bond to Man-1 in all rat brain Thy-1 and in a fraction of human erythrocyte CD59 [27, 29, 33] (Figure 1.1A). The GalNAc is occasionally modified by galactose with or without sialic acid. This GalNAc modification has never been found in GPI precursors, suggesting that it is added after attachment to proteins. There is a report that GPI-APs contain βGlcNAc phosphate linked to Man-2 [35].

The lipid part of mammalian GPI-AP is always PI that has two characteristics. First, most mammalian GPI-APs on nucleated cells have two saturated fatty chains in their PI moiety [36, 37] (Figure 1.1A). This unique fatty chain structure of GPI-APs is elaborated by fatty acid remodeling (see Section V.A for more detail). Second, in many GPI-APs, PI is a mixture of diacy and 1-alkyl, 2-acyl PI, the latter being usually a major form [36]. The biological significance of the alkyl chain is yet to be determined.

GPI-APs on human erythrocytes have an exceptional lipid structure [27, 38]. They have unsaturated fatty acids at the sn-2 position like endogenous free PI, suggesting that they are not subjected to fatty acid remodeling. In addition, they have a palmitoyl chain linked to the 2-position of the inositol ring, which is usually removed soon after the attachment of GPI to proteins (Figure 1.1A) [27, 38]. Therefore, GPI-APs on human erythrocytes have three fatty chains.

B.   YEAST GPI-APS

GPI-APs of *S. cerevisiae*, like mammalian GPI-APs, have an EtNP side branch linked to Man-1 [39] (Figure 1.1B). Man-2 in yeast GPI-anchor precursors has an EtNP side branch [40] (Section IV.B); however, it is unclear whether GPI-APs have it [39]. Yeast GPI-APs always have the α1–2 linked to the fourth Man (Man-4) [28]. These components are incorporated during biosynthesis of the GPI precursor (Figure 1.2B). Yeast GPI-APs may have more mannoses linked to Man-4, occurring after attachment of GPI to protein [41].

The lipid part of yeast GPI-AP is either PI or inositol phosphoceramide [28, 42–44] (Figure 1.1B). Diacylglycerol in the PI moiety contains palmitic acid (C16:0) in the sn-1 position and C26:0 fatty acid in the sn-2 position. Ceramide in the inositol phosphoceramide moiety has phytosphingosine and a C26:0 or a hydroxy-C26:0 fatty acid. The ceramide-type GPI-APs are generated from the diacylglycerol-type GPI-APs by lipid remodeling [1, 2, 31] (see Section V.A.4).

C. GPI-APs FROM *TRYPANOSOMA BRUCEI*

GPI-APs from *T. brucei* do not have an EtNP side branch [29]. VSG has side branches consisting of only galactoses and has strictly di-myristoyl PI [29, 45] (Figure 1.1C). The di-myristoyl PI moiety is generated by fatty acid remodeling occurring before attachment to proteins [46] (Section V.B.1). Procyclins have a Man-2-linked big side branch consisting of repeated and branched lactosamines. Sialic acids are linked to some of the galactoses in the side branch (Figure 1.1C). The lipid part of procyclins is unusual in that it has lyso PI with palmitoyl chain linked to the inositol ring [47].

IV. Biosynthesis of GPI and Attachment to Proteins

A. MAMMALIAN GPI BIOSYNTHETIC PATHWAY

The GPI-anchor precursors are synthesized in the ER from PI through at least nine sequential reaction steps [1, 3] (Figure 1.2A). The biosynthetic pathway is initiated on the cytoplasmic side of the ER membrane [48] by the transfer of GlcNAc to PI from UDP-GlcNAc, generating GlcNAc-PI (step 1). GlcNAc-PI is de-*N*-acetylated to generate GlcN-PI (step 2). It is most likely that GlcN-PI is flipped across the ER membrane into the luminal side (step 3). An acyl (usually palmitoyl) chain is added to the inositol ring from acyl-CoA to form GlcN-(acyl)PI (step 4) [49]. There is still an argument as to whether GlcN-(acyl)PI or GlcN-PI is flipped into the luminal side [50, 51]. The membrane orientations of GlcN-PI and the catalytic site of the acyltransferase should be definitively determined to resolve this issue.

Endogenous PI in mammalian cells is predominantly 1-stearoyl, 2-arachidonoyl PI (C18:0/C20:4), whereas the PI moiety in mammalian GPI-APs is more heterogeneous and 1-alkyl, 2-acyl PI is dominant. Recently, the fatty chain compositions of GlcNAc-PI, GlcN-PI, and GlcN-(acyl)PI accumulated in mutant Chinese hamster ovary (CHO) cells defective in PIG-L, PIG-W, and DPM2, respectively (see below and Table 1.1 for biosynthetic genes), were determined [52]. GlcNAc-PI and GlcN-PI had predominantly 1-stearoyl, 2-arachidonoyl PI like free PI, whereas the PI moiety in GlcN-(acyl)PI was a mixture of diacyl and 1-alkyl, 2-acyl PIs, the latter being a major form (Figure 1.2A). The alkyl chain in the sn-1 position was palmityl (C16:0), stearyl (C18:0), or a C18:1 (very likely an alkenyl form) chain, whereas the acyl chain in the sn-2 position was oleic (C18:1), arachidonic (C20:4), or docosatetraenoic (C22:4) acid. Therefore, the diacyl to 1-alkyl, 2-acyl change occurs in GlcN-(acyl)PI [52].

Man-1 and Man-2 are sequentially transferred to GlcN-(acyl)PI from dolichol-phosphate-mannose (Dol-P-Man) to generate Man-Man-GlcN-(acyl)PI (steps 5 and 6) [53]. The EtNP side branch is added to the 2-position of Man-1 from phosphatidylethanolamine (PE) generating Man-(EtNP)Man-GlcN-(acyl)PI (step 7). Man-3 is then transferred from Dol-P-Man (step 8). Finally, the "bridging" EtNP is added to Man-3 from PE (step 9), generating a form of mature GPI-anchor precursor, EtNP-Man-Man-(EtNP)Man-GlcN-(acyl)PI [1, 3].

The third EtNP is added to Man-2 as a side branch from PE to generate another form of mature GPI-anchor precursor (step 10) [54]. After step 8, the fourth Man (Man-4) can be transferred from Dol-P-Man (step M4) to generate an intermediate bearing four mannoses, which is then converted by EtNP addition to Man-3 (step 9′) into the third form of mature GPI-anchor precursor [34].

B.   GPI Biosynthetic Pathway in Yeast *S. cerevisiae*

Reaction steps 1 through 8 in yeast GPI biosynthesis are common with the mammalian pathway [2] (Figure 1.2B). In contrast to the mammalian pathway, the addition of Man-4 (step M4) is the essential step in yeast before the addition of "bridging" EtNP to Man-3 (step 9) [2]. The third EtNP is then added to Man-2 (step 10) [40]. Cellular free PI in yeast is predominantly 1-palmitoyl, 2-oleoyl PI (C16:0/C18:1). It is unknown whether PI moiety is modified during GPI biosynthesis and is assumed that GPI transferred to protein has the same chain composition. The chain composition of the diacylglycerol-type GPI-APs (C16:0/C26:0) is achieved by fatty acid remodeling (Section V.A).

TABLE 1.1

GENES INVOLVED IN GPI BIOSYNTHESIS AND REMODELING IN MAMMALIAN CELLS

Step	Enzyme	Donor Substrate	Gene	Function
1	GPI-GlcNAc transferase (GPI-GnT)	UDP-GlcNAc	*PIG-A*	Catalytic
			PIG-C	
			PIG-H	
			PIG-P	
			PIG-Q	
			PIG-Y	
			DPM2	
2	GlcNAc-PI de-*N*-acetylase		*PIG-L*	Catalytic
3	Flippase		Not	identified
4	Inositol acyltransferase	Palmitoyl-CoA	*PIG-W*	Catalytic
5	α1–4mannosyltransferase I (GPI-MT I)	Dol-P-Man	*PIG-M*	Catalytic
			PIG-X	Stabilization
6	α1–6mannosyltransferase II (GPI-MT II)	Dol-P-Man	*PIG-V*	Catalytic
7	EtNP transferase I (GPI-ET I)	PE	*PIG-N*	Catalytic
8	α1–2mannosyltransferase III (GPI-MT III)	Dol-P-Man	*PIG-B*	Catalytic
M4	α1–2mannosyltransferase IV (GPI-MT IV)	Dol-P-Man	*PIG-Z* (*SMP3*)	Catalytic
9	EtNP transferase III (GPI-ET III)	PE	*PIG-O*	Catalytic
			PIG-F	Stabilization
10	EtNP transferase II (GPI ET II)	PE	*PIG-G* (*GPI7*)	Catalytic
			PIG-F	Stabilization
TA	GPI transamidase (GPI-TA)		*PIG-K*	Catalytic
			GAA1	
			PIG-S	
			PIG-T	
			PIG-U	
P1	Inositol deacylase		*PGAP1*	Catalytic
P2	GPI-AP phospholipase A2		*PGAP3*	Catalytic
P3	Lyso-GPI acyltransferase	Stearyl-CoA	*PGAP2*	Noncatalytic

C. GPI BIOSYNTHETIC PATHWAY IN *T. BRUCEI*

Like mammalian cells and yeast, GPI biosynthesis in *T. brucei* is initiated by the transfer of GlcNAc on the cytoplasmic side of the ER [55] to generate GlcNAc-PI (step 1), which is then de-*N*-acetylated to generate

GlcN-PI (step 2) [56–58]. GlcN-PI must flip into the luminal side (step 3) because in *T. brucei* Man-1 is next added to generate Man-GlcN-PI (step 5) before inositol acylation occurs [59] (Figure 1.2C). Therefore, the third intermediate in the trypanosome pathway is Man-GlcN-PI, whereas that in mammalian and yeast pathways is GlcN-(acyl)PI. Man-GlcN-PI is then inositol acylated to generate Man-GlcN-(acyl)PI (step 4). Man-2 and 3 are added to generate Man-Man-Man-GlcN-(acyl)PI (steps 6 and 8 in the upper branch of the pathway). Man-2 and -3 can also be added to Man-GlcN-PI (steps 6 and 8 in the lower branch). Inositol acylation and deacylation can occur in the intermediates bearing one, two, or three mannoses [60]. The addition of the "bridging" EtNP occurs only to the inositol-acylated intermediate, generating EtNP-Man-Man-Man-GlcN-(acyl)PI (step 9). After step 9, the pathway is different between procyclic and bloodstream forms.

In procyclic form, fatty acid in the sn-2 position is removed to generate EtNP-Man-Man-Man-GlcN-(acyl)lysoPI, the mature form GPI that is attached to procyclins (step P2). In bloodstream form, the inositol-linked acyl chain is removed and the fatty acid remodeling of PI moiety occurs to replace both chains with myristic acid. For this, EtNP-Man-Man-Man-GlcN-(acyl)PI is subjected to either inositol deacylation and removal of the sn-2 chain or removal of the sn-2 chain and then inositol deacylation to generate EtNP-Man-Man-Man-GlcN-2-lysoPI. After the fatty acid remodeling (see Section V.B.1), GPI with di-miristoyl PI is attached to VSG.

D. POSTTRANSLATIONAL ATTACHMENT OF GPI ANCHORS TO PROTEINS

Proteins that are to be GPI anchored have a GPI-attachment signal peptide at the C-terminus. The signal peptide is recognized, cleaved, and replaced by preassembled GPI by the action of a GPI-TA residing in the ER (reviewed in [1, 2, 61]). The GPI-attachment signal peptides from various GPI-AP precursors do not contain any consensus sequence, but have several common features, i.e., (i) The ω-site amino acids to which the GPI anchor is amide bonded are those with small side-chains, namely Gly, Ala, Ser, Asn, Glu, and Cys; (ii) the ω + 2 amino acids are also those with a small side-chains, such as Gly, Ala, and Ser; (iii) a hydrophilic spacer sequence with six or more residues starting at the ω + 3 site; and (iv) a C-terminal hydrophobic sequence long enough to span the ER membrane. The GPI-attachment signal peptides contain all the information necessary and sufficient for GPI anchoring because they can convert non-GPI-APs to GPI-APs when attached to their C-termini.

E.   ENZYMES INVOLVED IN BIOSYNTHESIS OF GPI AND
     ATTACHMENT TO PROTEINS

1.   Mammalian Enzymes for GPI Biosynthesis

At least 18 gene products are involved in the biosynthesis of GPI-anchor precursors (Table 1.1) (reviewed in [1]). Step 1, generation of GlcNAc-PI, is mediated by GPI-GlcNAc transferase (GPI-GnT) that is an unusually complex glycosyltransferase, consisting of six core (PIG-A [62], PIG-C [63], PIG-H [64], PIG-P [65], PIG-Q [66], and PIG-Y [67]) subunits [1, 68]. In addition, a fraction of GPI-GnT contains one extra, the seventh, subunit DPM2 [65]. DPM2, which is also one of the three subunits of Dol-P-Man synthase [69], enhances GPI-GnT activity threefold [65]. GPI biosynthesis and Dol-P-Man biosynthesis may be coregulated by DPM2. PIG-A is the catalytic subunit [70] belonging to the glycosyltransferase family 4 (GT4) in the CAZy classification [71, 72]. The functions of the five other core subunits are not clear yet, but all of them are essential or nearly essential. Functionally important sites in some of these subunits, including the catalytic site in PIG-A, are on the cytoplasmic side of the ER [73, 74].

Step 2 is mediated by ER-membrane protein PIG-L having GlcNAc-de-*N*-acetylase activity [75]. Functionally important sites in PIG-L face the cytoplasm, indicating that GlcN-PI is generated on the cytoplasmic side [75].

A putative enzyme that mediates step 3, flipping of GlcN-PI into the luminal side, has not been identified [1]. Step 4, the transfer of the acyl chain to inositol, is mediated by multitransmembrane protein PIG-W having acyltransferase activity [76].

Step 5 is mediated by α1–4 mannosyltransferase, GPI-MTI, consisting of PIG-M and PIG-X [77, 78]. PIG-M is the catalytic subunit belonging to the GT50 family [72] and has 10 transmembrane domains and a functionally important DXD motif within the first luminal domain [77]. PIG-X, an ER transmembrane protein with a large luminal domain, is associated with PIG-M and is required for the stable expression of PIG-M [78]. Whether PIG-X has any role in mannose transfer is not clear. Step 6, the transfer of Man-2, requires PIG-V. Although its catalytic activity has not been demonstrated, PIG-V is most likely the GPI-MTII, α1–6 mannosyltransferase, because PIG-V has multiple (eight) transmembrane domains and a functionally important acidic amino acid (Asp) in the first luminal domain like other ER-resident, Dol-P-monosaccharide-utilizing glycosyltransferases [53]. PIG-V is classified into GT76 [1].

Step 7, EtNP transfer to Man-1, requires PIG-N, which is most likely the GPI-ethanolamine phosphate transferase I (GPI-ETI) itself [79]. PIG-N has three motifs conserved in phosphatases [80]. Man-3 is then added by

PIG-B, GPI-MTIII, and α1–2 mannosyltransferase (step 8) [81]. In step 9, the "bridging" EtNP is transferred by GPI-ETIII, a complex of PIG-O [82] and PIG-F [83]. PIG-O must be the catalytic subunit because it has three conserved motifs similar to PIG-N and forms a protein family with PIG-N and PIG-G (also called GPI7) [82]. PIG-F, a hydrophobic protein with two transmembrane domains, binds to and stabilizes PIG-O [82]. The addition of the EtNP side branch to Man-2 (step 10) is mediated by GPI-ETII consisting of PIG-G and PIG-F [84]. Similarly to GPI-ETIII, PIG-G is stabilized by PIG-F [84]. Transfer of Man-4 is mediated by PIG-Z (also called SMP3), an α1–2 mannosyltransferase [34]. PIG-B, PIG-Z, and ALG9, ALG12 (α mannosyltransferases involved in N-glycan synthesis) form a family that is classified as GT22 [72].

2. Yeast Enzymes for GPI Biosynthesis

The first enzyme GPI-GnT of yeast consists of six subunits homologous to mammalian counterparts; Gpi3p/PIG-A [85–87], Gpi2p/PIG-C [87], Gpi15p/PIG-H [88], Gpi19p/PIG-P [89], Gpi1p/PIG-Q [90], and Eri1p/ PIG-Y [91] (Table 1.2). Yeast does not have homologue of DPM2 that is an extra subunit in a fraction of mammalian GPI-GnT. DPM2 is an essential subunit of tri-molecular mammalian Dol-P-Man synthase, whereas yeast Dol-P-Man synthase (DPM1) is a single-protein enzyme. The second and the third enzymes are also homologous to mammalian counterparts, Gpi12p/PIG-L [75] and Gwt1p/PIG-W [92, 93].

It was reported recently that yeast Arv1p is required for the efficient delivery of GlcN-(acyl)PI to the first GPI-mannosyltransferase (GPI-MTI) and the possibility that Arv1p is the flippase of GlcN-(acyl)PI was discussed [50]. The latter point remains to be experimentally tested. Yeast GPI-MTI consists of catalytic Gpi14p homologous to PIG-M, and noncatalytic Pbn1p homologous to PIG-X [78]. Gpi18p, S. cerevisiae PIG-V homolog [94], requires another gene product Pga1p to act as GPI-MTII [95]. Mammalian genomes have no homolog of Pga1, suggesting that either PIG-V alone is sufficient or a cofactor protein not homologous to Pga1p is involved. Yeast GPI-ETI (Mcd4p) [96], MTIII (Gpi10p) [97], MTIV (Smp3p) [98], ETIII (requiring Gpi13p and Gpi11p) [99, 100], and -ETII (requiring Gpi7p and Gpi11p) [80, 100] are homologous to the mammalian counterparts.

3. Trypanosome Enzymes for GPI Biosynthesis

Trypanosome homologs of many of mammalian and yeast genes in the GPI biosynthetic pathway have been identified in the genome database (Table 1.2). The function of PIG-L/GPI12-homolog TbGPI12 was proven by gene disruption in the bloodstream form to be the gene required for

TABLE 1.2

GENE COMPARISON AMONG MAMMALS, YEAST, AND TRYPANOSOME

Step	Enzyme	Mammals	Yeast	*T. brucei*
1	GPI-GlcNAc transferase (GPI-GnT)	*PIG-A*	*GPI3*	*TbGPI3*
		PIG-C	*GPI2*	*TbGPI2*
		PIG-H	*GPI15*	*TbGPI15?*
		PIG-P	*GPI19*	*TbGPI19*
		PIG-Q	*GPI1*	*TbGPI1*
		PIG-Y	*ERI1*	–
		DPM2	–	–
2	GlcNAc-PI de-*N*-acetylase	*PIG-L*	*GPI12*	*TbGPI12*
3	Flippase	Not identified	*ARV1?*	Not identified
4	Inositol acyltransferase	*PIG-W*	*GWT1*	–
				Not identified[a]
5	α1–4mannosyltransferase I (GPI-MT I)	*PIG-M*	*GPI14*	*TbGPI14*
		PIG-X	*PBN1*	–
6	α1–6mannosyltransferase II (GPI-MT II)	*PIG-V*	*GPI18*	*TbGPI18*
		–	*PGA1*	–
7	EtNP transferase I (GPI-ET I)	*PIG-N*	*MCD4*	–
8	α1–2mannosyltransferase III (GPI-MT III)	*PIG-B*	*GPI10*	*TbGPI10*
M4	α1–2mannosyltransferase IV (GPI-MT IV)	*PIG-Z (SMP3)*	*SMP3*	–
9	EtNP transferase III (GPI-ET III)	*PIG-O*	*GPI13*	*TbGPI13*
		PIG-F	*GPI11*	*TbGPI11?*
10	EtNP transferase II (GPI ET II)	*PIG-G (GPI7)*	*GPI7*	–
		PIG-F	*GPI11*	*TbGPI11?*
TA	GPI transamidase (GPI-TA)	*PIG-K*	*GPI8*	*TbGPI8*
		GAA1	*GAA1*	*TbGAA1*
		PIG-S	*GPI17*	–
		PIG-T	*GPI16*	*TbGPI16*
		PIG-U	*GAB1*	–
		–	–	*TTA1[b]*
		–	–	*TTA2*
P1	Inositol deacylase	*PGAP1*	*BST1*	*GPIdeAc2 (GPIdeAc)*
P2	GPI phospholipase A2	*PGAP3*	*PER1*	–
				Not identified[a]
P3	Lyso-GPI acyltransferase I	*PGAP2*	*CWH43-N*	–
		Not identified	*GUP1*	*TbGUP1*
CR	Ceramide remodelase		*CWH43-C*	–

(*Continued*)

TABLE 1.2 (*Continued*)

Step	Enzyme	Mammals	Yeast	*T. brucei*
P4	GPI phospholipase A1	–	–	Not identified
P5	Lyso-GPI acyltransferase II	–	–	Not identified

[a]The reactions exist but the genes required are not found.
[b]TTA1 is remotely related to PIG-S.

GPI-GlcNAc de-*N*-acetylase [101, 102]. TbGPI10 complemented PIG-B-defective murine cells and gpi10-mutant yeast, showing that it is trypanosome GPI-MTIII [17]. Open reading frames with homology to PIG-H/GPI15 and PIG-F/GPI11 exist in the trypanosome genome; however, whether they are functional homologs is not conclusive and requires experimental evidence. Consistent with the facts that trypanosome GPI does not have EtNP side branches and Man-4, there are no homologs of PIG-N/MCD4, PIG-G/GPI7, and PIG-Z/SMP3. Homologs of PIG-Y/ERI1, DPM2, ARV1, PIG-X/PBN1, and PGA1 do not exist in the trypanosome genome. Regarding the gene for inositol acylation, apparent sequence homolog of PIG-W/GWT1 does not exist, suggesting that a nonhomologous GPI inositol acyltransferase exists. This seems reasonable because substrate specificities are different in that trypanosome enzyme can acylate GPI intermediates with one, two, or three mannoses [60]; whereas mammalian and yeast enzymes acylate before mannosylation [76, 93]. Two GPI inositol deacylases, GPIdeAc [103, 104] and GPIdeAc2 [105], have been cloned, characterized, and disrupted in the bloodstream form. GPIdeAc2 appeared to account for major inositoldeacylase activity [105].

4. GPI Transamidases

The GPI-TA is a membrane-bound multisubunit enzyme. Mammalian GPI-TA consists of five subunits; PIG-K (also called GPI8) [106], GAA1 [107, 108], PIG-S, PIG-T [109], and PIG-U [110] proteins (Table 1.1). In the complex, PIG-K is disulfide-bonded to PIG-T, whereas other subunits are associated noncovalently [111]. PIG-K/GPI8 is the catalytic subunit that has sequence homology to cysteine protease family members [112–114]. The other four subunits are also essential or nearly essential for GPI attachment, however their functions are not clear (see recent review by Orlean and Menon [1] for further discussion about GPI-TA). Yeast GPI-TA consists of five subunits homologous to its mammalian counterparts;

Gpi8p/PIG-K, Gaa1p/GAA1, Gpi17p/PIG-S, Gpi16p/PIG-T, and Gab1p/
PIG-U (Table 1.2) [115–118]. Trypanosome GPI-TA also consists of five
subunits. TbGPI8, TbGAA1, and TbGPI16 are homologous to PIG-K/
Gpi8p, GAA1/Gaa1p, and PIG-T/Gpi16p, respectively [21, 119]. Like
mammalian complex, TbGPI8 is disulfide-bonded to TbGPI16 [111]. Two
other subunits, TTA1 and TTA2, are unique to trypanosome. Their homologs are found only in trypanosomatid protozoa, such as *T. brucei*, *T. cruzi*,
and *Leishmania* species, hence termed Trypanosomatid Transamidase 1
and 2 [119]. They have similar membrane orientation to PIG-S and
PIG-U, respectively. It was found that TTA1 is remotely related to PIG-S.

When the nascent protein to be GPI anchored is transported into the ER
lumen via the translocon machinery, GPI-TA recognizes the C-terminal
GPI-attachment signal peptide, cleaves it between the ω and $\omega + 1$ sites,
and generates an enzyme-substrate intermediate. The ω-site amino acid is
linked to the catalytic cysteine in PIG-K/GPI8 via a thiol-ester bond, which
is then attacked by an amino group in the "bridging" EtNP, completing the
transamidation reaction [1].

V.   Lipid Remodeling of GPI and GPI-APs

A.   LIPID REMODELING OF GPI-APs IN MAMMALIAN CELLS AND YEAST

GPI anchors are synthesized from PI containing unsaturated fatty acid at
the sn-2 position. After attachment to proteins, lipid moieties of GPI
anchors are remodeled in both yeast and mammalian cells [3, 31]. These
lipid remodeling processes are essential for the association of GPI-APs with
specialized membrane domains, called membrane microdomains or membrane rafts. Recently, several genes required for the GPI lipid remodeling
were identified in yeast and mammalian cells. In this section, we describe
the genes involved in lipid remodeling of GPI anchors in mammals, yeast,
and trypanosomes.

1.   Inositol Deacylation

After GPI attachment to proteins, lipid moieties of GPI-APs are dynamically remodeled in the ER and the Golgi apparatus during their transport
to the plasma membrane in both yeast and mammalian cells (Figure 1.3).
First, the acyl group linked to the inositol residue of GPI anchor is elimi-
nated in the ER. Mammalian PGAP1 and yeast Bst1p mediate this deacylation reaction [120]. Deacylation from inositol is essential for the following
fatty acid remodeling [121, 122]. Molecular characterizations of PGAP1
and Bst1p are described in Chapter 4.

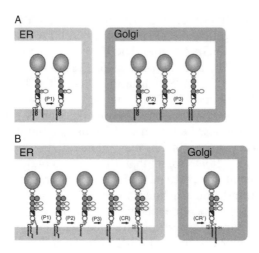

FIG. 1.3. Lipid remodeling of GPI-APs in mammalian cells (A) and yeast (B). Steps P1–P3 and CR correspond to those in Tables 1.1 and 1.2.

2. Removal of Unsaturated Fatty Acid from the sn-2 Position in GPI Anchor

The first step in the fatty acid remodeling is removal of unsaturated acyl chain from the sn-2 position to form lyso-GPI (Figure 1.3). Mammalian PGAP3 and yeast Per1p are required for the formation of lyso-GPI [121, 123]. Yeast *PER1* was found as a gene that showed genetic interactions similar to those of GPI inositol deacylase *BST1* and lyso-GPI acyltransferase *GUP1* in the comprehensive genetic interaction maps, called E-MAPs (epistatic miniarray profiles) [123, 124]. It was also isolated as a mutational suppressor of the *cdc1* mutant and as a gene involved in the unfolded protein response and protein folding [125, 126]. Per1p is a multi-spanning membrane protein localized in the ER. The Per1Δ cells showed calcofluor white (CFW) sensitivity and temperature sensitivity, suggesting a defect in the cell wall integrity. Mature form of Gas1p, a major GPI-AP, is greatly decreased in Per1Δ cells. Instead, a significant amount of Gas1p is released from the plasma membrane into the culture medium in Per1Δ cells. GPI-APs were not associated with the detergent-resistant membranes (DRMs) in Per1Δ cells. The mutant cells had a defect in the GPI lipid remodeling from regular PI to a C26 fatty-acid-containing PI. *In vitro* analysis showed that *PER1* is required for the production of lyso-GPI, suggesting that Per1p possesses or regulates the GPI-phospholipase A2 activity. While Per1p does not have any lipase-like motif, H177 and H326 in the regions conserved among Per1p homologs are essential for the Per1 function.

To identify the gene involved in fatty acid remodeling of GPI-APs in mammalian cells, a double mutant cell line, termed PGAP2&3, was first established from chemically mutagenized PGAP2-deficient cells [121]. PGAP2&3 cells restored the surface expression of GPI-APs in PGAP2 single mutant cells (see below PGAP2-deficient cells). PGAP2&3 double mutant cells possess an unsaturated fatty acid at the sn-2 position in GPI anchors, whereas the unsaturated chain is replaced by the saturated stearoyl chain in wild-type cells. In PGAP2&3 double mutant cells, GPI-APs could not associate with DRMs. PGAP3, the gene responsible for the second defect in the double mutant cells, was the mammalian homolog of yeast *PER1*. Transfection of yeast *PER1* also complemented the PGAP3 mutation. The predicted amino acid sequences of yeast *PER1* and human PGAP3 shared 28% identity. The PGAP3 gene also restored the phenotypes of yeast Per1Δ cells, clearly indicating that mammalian and yeast share the same fatty acid remodeling of GPI anchor.

In yeast, Per1p is mainly localized in the ER, whereas PGAP3 is localized mainly in the Golgi and weakly in the ER. From the localization of these proteins, it appeared that this reaction is carried out in the ER in yeast, but mainly in the Golgi in mammalian cells (Figure 1.3). It is consistent with the place where GPI-APs are incorporated into membrane rafts in these organisms. In yeast, GPI-APs are associated with raft-like domains in the ER, whereas mammalian GPI-APs are incorporated into rafts in the Golgi [127, 128]. In fact, incorporation of GPI-APs into raft-like domains is a prerequisite for the efficient transport from the ER to the Golgi in yeast, but not in mammalian cells [129, 130]. Yeast Per1p, but not mammalian PGAP3, is required for the efficient transport of GPI-APs.

3. Transfer of Saturated Fatty Acid to Lyso-GPI Anchor

The second step in fatty acid remodeling is transfer of saturated acyl chain to lyso-GPI (Figure 1.3). Saturated acyl chain is usually C26:0 in yeast and C18:0 stearic acid in mammalian cells, each donated from the corresponding acyl-CoA [43, 121]. The difference in the length of remodeled fatty acid between yeast and mammalian cells seems to be correlated with typical chain length of sphingolipids, which contain C26 fatty acid in yeast and C16~24 fatty acids, but mainly C18 fatty acid in mammalian cells. Yeast Gup1p is an ER-resident protein that is essential for the synthesis of the C26:0-containing GPI anchor [131]. *GUP1* gene was isolated based on the lipid analysis of GPI anchor in deletion mutants that lack genes bearing homology to known phospholipases and acyltransferases. It was originally found as a gene related to the glycerol uptake [132]. Gup1p is a multiple-membrane-spanning protein harboring a motif that is a characteristic of membrane-bound O-acyltransferases (MBOAT). The Gup1Δ cells show CFW sensitivity, indicating that they have a cell wall defect. In Gup1Δ cells, the incorporation of [^{3}H]inositol to GPI-APs is normal, whereas phospholipid moiety of most of the GPI-APs is lyso-PI and does not contain C26 fatty acid, nor IPCs. Gup1 function is dependent on the putative active site H447 in the MBOAT motif. The transport of GPI-APs from the ER to the Golgi is delayed and Gas1p is released from the plasma membrane into the medium in Gup1Δ cells. Although several Gup1p homologs containing MBOAT motif also exist in mammals, whether their functions are related to GPI remodeling is still unknown.

Mammalian PGAP2 is required for fatty acid remodeling, specifically for transfer of C18:0 chain to the lyso-GPI, but the precise function is still unknown. It was identified as the gene responsible for the mutant cell lines, in which GPI biosynthesis is normal, whereas the surface expression of GPI-APs is greatly decreased [133]. PGAP2 is a membrane-bound protein expressed mainly in the Golgi and weakly in the ER. The GPI lipid moiety in PGAP2-deficient mutant CHO cells is mainly lyso-GPI. The lyso-GPI-APs seem to be sensitive to unknown phospholipase D and secreted out of the cells. PGAP2 was also reported as FRAG1 that makes a fusion protein with the fibroblast growth factor receptor 2, having a constitutively elevated tyrosine kinase activity in a rat osteosarcoma cell line due to chromosomal rearrangement [134].

4. Ceramide Remodeling of GPI Anchor in Yeast

In yeast, the lipids of many GPI-APs are further changed from diacylglycerol type to ceramide type (Figure 1.3). Biochemical studies revealed that this reaction occurs in both the ER and the Golgi, but species of remodeled ceramides are different [43, 44]. In the ER, the diacylglycerol of GPI lipid moiety is replaced by ceramide consisting of phytosphingosine with C26:0 fatty acid, whereas ceramide-containing phytosphingosine and hydroxy-C26:0 fatty acid is used for the substrate at the Golgi. The yeast homolog of PGAP2, *CWH43*, has been reported to be required for cell wall integrity [135]. Cwh43p shows approximately 24% amino acid identity with human PGAP2 in its N-terminal portion. However, Cwh43p has an additional C-terminal portion of ~700 amino acid residues. Therefore, *CWH43* is expected to have other functions in addition to PGAP2 function. Recently, it has been reported that *CWH43* is required for GPI lipid remodeling to ceramides [122, 136]. In Cwh43Δ cells, GPI-containing ceramide was not detected at all. Instead, accumulation of GPI-APs bearing diacylglycerol-type GPI with very long acyl chain was found. The C-terminal domain of Cwh43p is especially important for GPI lipid remodeling to ceramides, and the N-terminal domain seems to enhance the reaction. The C-terminal domain of Cwh43p has a DNase I-like motif, which is also found in Isc1p, Inp51p, Inp52p, Inp53p, and Inp54p. Isc1p is an inositol phosphoceramide phospholipase C and Inp51/52/53/54 proteins are involved in dephosphorylation of phosphoinositides. Therefore, it is possible that this motif is related to the recognition of inositol phosphate, implying that DNase I-like motif in the C-terminal domain of Cwh43p is important for the recognition of inositol phosphate in GPI anchor. The C-terminal region of Cwh43p is also conserved in mammals. It was reported that mammalian homolog of the C-terminal region of Cwh43p complemented CFW sensitivity of cwh43Δ cells and had a partial ability to restore lipid remodeling to ceramides in yeast cells. Although there is no report about the presence of ceramide-type GPI in mammalian cells, the homolog of the C-terminal region of Cwh43p might be required for a similar lipid remodeling in mammalian cells. Gpi7p and Mcd4p, which transfer EtNP to mannoses in GPI intermediates during the biosynthesis in the ER, are reported to be indirectly required for the remodeling of diacylglycerol GPI to ceramides [80, 137]. It might be that ceramide remodelase is required for the recognition of side-chain EtNP on the GPI as a substrate.

B. LIPID REMODELING OF GPI IN PROTOZOAN PARASITES

1. *Fatty Acid Remodeling in Bloodstream Form of Trypanosoma brucei*

T. brucei is the first organism in which fatty acid remodeling of GPI anchors was found [46]. As described above, GPI anchors in the bloodstream form of *T. brucei* contain exclusively dimyristoylglycerol and their fatty acid remodeling has been well characterized. The fatty acid remodeling in trypanosomes occurs before GPI is transferred to proteins (Figure 1.2), whereas mammalian and yeast lipid remodeling occur after attachment of GPI to proteins. Inositol acylation and deacylation of *T. brucei* are described in Chapter 4. The bloodstream form of *T. brucei* possesses unique fatty acid remodeling [138]. The fatty acids of GPI intermediates are replaced by myristic acid (C14:0) through sequential deacylation and reacylation reactions at the sn-2 position first followed by reaction at the sn-1 position in the bloodstream form *T. brucei*. Trypanosome normally derives fatty acids and phospholipids from the host's bloodstream, whereas myristic acid for the fatty acid remodeling is synthesized by itself [139]. Fatty acid remodeling requires at least 4 steps. It has been identified that *TbGUP1*, a homolog of yeast *GUP1*, is required for the addition to myristic acid to the sn-2 position in this remodeling pathway [140]. *In vitro* assay demonstrated that TbGup1p prefers shorter length fatty acid, such as

C14:0 and C12:0, as a substrate. The genes involved in three other steps are not known. *T. brucei* also has a secondary pathway for GPI anchor myristoylation named myristate exchange, which is distinct from fatty acid remodeling [141]. Glycolipid A and GPI anchor of VSG undergoes a myristate exchange. The enzymes responsible for exchange appear to differ from those for fatty acid remodeling. The function of myristate exchange is still not clear, but it is likely to ensure that all synthesized VSG has a myristoylated GPI or to repair VSG recycled from the cell surface. Because GPI biosynthesis is essential for the bloodstream form of *T. brucei*, GPI lipid remodeling pathway is a potential target for the antitrypanosome chemotherapy. Actually, myristate analogs that become incorporated into the GPI anchors of VSG are toxic to trypanosomes [142, 143].

2. Lipid Remodeling in other Protozoan Parasites

It has been reported that Leishmania possesses myristate-specific fatty acid remodeling in GPIs. The longer chain fatty acids in the sn-2 position of 1-alkyl-2-acyl-PI are replaced by myristic acid [144]. American trypanosome, *T. cruzi*, has GPI anchors containing ceramides, whereas ceramides are not used for the first substrate of the GPI biosynthesis, suggesting that lipid remodeling also takes place in *T. cruzi* [145]. The *GUP1* homolog in *T. cruzi* partially complemented the phenotype of yeast gup1Δ cells [131].

VI. Enzymes Involved in Modification of GPI Glycan in GPI-APs

Enzymes involved in modification of mammalian GPI glycan after attachment to proteins, namely those for transfers of GalNAc to Man-1, Gal and sialic acid to GalNAc, and GlcNAc-phosphate to Man-2 (Figure 1.1A), are yet to be identified. Yeast mannosyltransferases for mannosylation to Man-4 (Figure 1.1B) are also to be identified.

Procyclins from procyclic *T. brucei* have a side branch containing polylactosamine and terminal sialic acids (Figure 1.1C). GlcNAc transferase (GnT) involved in the addition of β1–3 linked GlcNAc to the second Gal has been recently identified and termed TbGT8 [146]. TbGT8 is a typical Golgi-resident glycosyltransferase. *T. brucei* does not have biosynthetic pathway for sialic acid, but instead transfers sialic acids from host sialoglycoconjugates, such as sialoglycoproteins of erythrocytes included in blood meal, to the GPI using trans-sialidases. Trans-sialidases of *T. brucei* are themselves GPI-APs [147].

REFERENCES

1. Orlean, P., and Menon, A.K. (2007). Thematic review series: lipid post-translational modifications. GPI anchoring of protein in yeast and mammalian cells, or: how we learned to stop worrying and love glycophospholipids. *J Lipid Res* 48:993–1011.
2. Pittet, M., and Conzelmann, A. (2007). Biosynthesis and function of GPI proteins in the yeast *Saccharomyces cerevisiae*. *Biochim Biophys Acta* 1771:405–420.
3. Kinoshita, T., Fujita, M., and Maeda, Y. (2008). Biosynthesis, remodelling and functions of mammalian GPI-anchored proteins: recent progress. *J Biochem* 144:287–294.
4. Borner, G.H., Lilley, K.S., Stevens, T.J., and Dupree, P. (2003). Identification of glycosylphosphatidylinositol-anchored proteins in Arabidopsis. A proteomic and genomic analysis. *Plant Physiol* 132:568–577.
5. Ferguson, M.A., Kinoshita, T., and Hart, G.W. (2009). Glycosylphosphatidylinositol anchors. In Essentials of Glycobiology, A. Varki, R.D. Cummings, J.D. Esko, H.H. Freeze, P. Stanley, C.R. Bertozzi, G.W. Hart and M.E. Etzler, (eds.), pp. 143–161. Cold Spring Harbor Laboratory Press, Cold Spring Harbor, NY.
6. Mayor, S., and Riezman, H. (2004). Sorting GPI-anchored proteins. *Nat Rev Mol Cell Biol* 5:110–120.
7. Nozaki, M., Ohishi, K., Yamada, N., Kinoshita, T., Nagy, A., and Takeda, J. (1999). Developmental abnormalities of glycosylphosphatidylinositol-anchor-deficient embryos revealed by Cre/loxP system. *Lab. Invest* 79:293–299.
8. Almeida, A.M., Murakami, Y., Layton, D.M., Hillmen, P., Sellick, G.S., Maeda, Y., Richards, S., Patterson, S., Kotsianidis, I., Mollica, L., Crawford, D.H., Baker, A., et al. (2006). Hypomorphic promoter mutation in PIGM causes inherited glycosylphosphatidylinositol deficiency. *Nat Med* 12:846–851.
9. Takeda, J., Miyata, T., Kawagoe, K., Iida, Y., Endo, Y., Fujita, T., Takahashi, M., Kitani, T., and Kinoshita, T. (1993). Deficiency of the GPI anchor caused by a somatic mutation of the PIG-A gene in paroxysmal nocturnal hemoglobinuria. *Cell* 73:703–711.
10. Parker, C., Omine, M., Richards, S., Nishimura, J., Bessler, M., Ware, R., Hillmen, P., Luzzatto, L., Young, N., Kinoshita, T., Rosse, W., and Socie, G. (2005). Diagnosis and management of paroxysmal nocturnal hemoglobinuria. *Blood* 106:3699–3709.
11. De Groot, P.W., Hellingwerf, K.J., and Klis, F.M. (2003). Genome-wide identification of fungal GPI proteins. *Yeast* 20:781–796.
12. Kollar, R., Reinhold, B.B., Petráková, E., Teh, H.J.C., Ashwell, G., Dragonová, J., Kapteyn, J.C., Klis, F.M., and Cabib, E. (1997). Architecture of the yeast cell wall. *J Biol Chem* 272:17762–17775.
13. Lalanne, E., Honys, D., Johnson, A., Borner, G.H., Lilley, K.S., Dupree, P., Grossniklaus, U., and Twell, D. (2004). SETH1 and SETH2, two components of the glycosylphosphatidylinositol anchor biosynthetic pathway, are required for pollen germination and tube growth in Arabidopsis. *Plant Cell* 16:229–240.
14. Gillmor, C.S., Lukowitz, W., Brininstool, G., Sedbrook, J.C., Hamann, T., Poindexter, P., and Somerville, C. (2005). Glycosylphosphatidylinositol-anchored proteins are required for cell wall synthesis and morphogenesis in Arabidopsis. *Plant Cell* 17:1128–1140.
15. Ferguson, M.A. (1999). The structure, biosynthesis and functions of glycosylphosphatidylinositol anchors, and the contributions of trypanosome research. *J Cell Sci* 112:2799–2809.
16. Gowda, D.C., and Davidson, E.A. (1999). Protein glycosylation in the malaria parasite. *Parasitol Today* 15:147–152.

17. Nagamune, K., Nozaki, T., Maeda, Y., Ohishi, K., Fukuma, T., Hara, T., Schwarz, R.T., Sutterlin, C., Brun, R., Riezman, H., and Kinoshita, T. (2000). Critical roles of glycosyl-phosphatidylinositol for *Trypanosoma brucei*. *Proc Natl Acad Sci USA* 97:10336–10341.

18. Ferguson, M.A.J. (2000). Glycosylphosphatidylinositol biosynthesis validated as a drug target for African sleeping sickness. *Proc Natl Acad Sci USA* 97:10673–10675.

19. Smith, T.K., Crossman, A., Brimacombe, J.S., and Ferguson, M.A. (2004). Chemical validation of GPI biosynthesis as a drug target against African sleeping sickness. *EMBO J* 23:4701–4708.

20. Urbaniak, M.D., Yashunsky, D.V., Crossman, A., Nikolaev, A.V., and Ferguson, M.A. (2008). Probing enzymes late in the trypanosomal glycosylphosphatidylinositol biosynthetic pathway with synthetic glycosylphosphatidylinositol analogues. *ACS Chem Biol* 3:625–634.

21. Lillico, S., Field, M.C., Blundell, P., Coombs, G.H., and Mottram, J.C. (2003). Essential roles for GPI-anchored proteins in African trypanosomes revealed using mutants deficient in GPI8. *Mol Biol. Cell* 14:1182–1194.

22. Nagamune, K., Acosta-Serrano, A., Uemura, H., Brun, R., Kunz-Renggli, C., Maeda, Y., Ferguson, M.A., and Kinoshita, T. (2004). Surface sialic acids taken from the host allow trypanosome survival in tsetse fly vectors. *J Exp Med* 199:1445–1450.

23. Wang, Q., Fujioka, H., and Nussenzweig, V. (2005). Mutational analysis of the GPI-anchor addition sequence from the circumsporozoite protein of Plasmodium. *Cell Microbiol* 7:1616–1626.

24. Gerold, P., Schofield, L., Blackman, M.J., Holder, A.A., and Schwarz, R.T. (1996). Structural analysis of the glycosyl-phosphatidylinositol membrane anchor of the merozoite surface protein-1 and -2 of *Plasmodium falciparum*. *Mol Biochem Parasitol.* 75:131–143.

25. Homans, S.W., Ferguson, M.A., Dwek, R.A., Rademacher, T.W., Anand, R., and Williams, A.F. (1988). Complete structure of the glycosyl phosphatidylinositol membrane anchor of rat brain Thy-1 glycoprotein. *Nature* 333:269–272.

26. Brewis, I.A., Ferguson, M.A., Mehlert, A., Turner, A.J., and Hooper, N.M. (1995). Structures of the glycosyl-phosphatidylinositol anchors of porcine and human renal membrane dipeptidase. Comprehensive structural studies on the porcine anchor and interspecies comparison of the glycan core structures. *J Biol Chem* 270:22946–22956.

27. Rudd, P.M., Morgan, B.P., Wormald, M.R., Harvey, D.J., van den Berg, C.W., Davis, S.J., Ferguson, M.A., and Dwek, R.A. (1997). The glycosylation of the complement regulatory protein, human erythrocyte CD59. *J Biol Chem* 272:7229–7244.

28. Fankhauser, C., Homans, S.W., Thomas-Oates, J.E., McConville, M.J., Desponds, C., Conzelmann, A., and Ferguson, M.A.J. (1993). Structures of glycosylphosphatidylinositol membrane anchors from *Saccharomyces cerevisiae*. *J Biol Chem* 268:26365–26374.

29. Ferguson, M.A., Homans, S.W., Dwek, R.A., and Rademacher, T.W. (1988). Glycosyl-phosphatidylinositol moiety that anchors *Trypanosoma brucei* variant surface glycoprotein to the membrane. *Science* 239:753–759.

30. Oxley, D., and Bacic, A. (1999). Structure of the glycosylphosphatidylinositol anchor of an arabinogalactan protein from *Pyrus communis* suspension-cultured cells. *Proc Natl Acad Sci USA* 96:14246–14251.

31. Fujita, M., and Jigami, Y. (2008). Lipid remodeling of GPI-anchored proteins and its function. *Biochim Biophys Acta* 1780:410–420.

32. Deeg, M.A., Humphrey, D.R., Yang, S.H., Ferguson, T.R., Reinhold, V.N., and Rosenberry, T.L. (1992). Glycan components in the glycoinositol phospholipid anchor of human erythrocyte acetylcholinesterase. *J Biol Chem* 267:18573–18580.

33. Taguchi, R., Hamakawa, N., Harada Nishida, M., Fukui, T., Nojima, K., and Ikezawa, H. (1994). Microheterogeneity in glycosylphosphatidylinositol anchor structures of bovine liver 5'-nucleotidase. *Biochemistry* 33:1017–1022.

34. Taron, B.W., Colussi, P.A., Wiedman, J.M., Orlean, P., and Taron, C.H. (2004). Human Smp3p adds a fourth mannose to yeast and human glycosylphosphatidylinositol precursors *in vivo. J Biol Chem* 279:36083–36092.

35. Fukushima, K., Ikehara, Y., Kanai, M., Kochibe, N., Kuroki, M., and Yamashita, K. (2003). A beta-*N*-acetylglucosaminyl phosphate diester residue is attached to the glycosylphosphatidylinositol anchor of human placental alkaline phosphatase: a target of the channel-forming toxin aerolysin. *J Biol Chem* 278:36296–36303.

36. Redman, C.A., Thomas-Oates, J.E., Ogata, S., Ikehara, Y., and Ferguson, M.A. (1994). Structure of the glycosylphosphatidylinositol membrane anchor of human placental alkaline phosphatase. *Biochem J* 302(Pt 3):861–865.

37. Kerwin, J.L., Tuininga, A.R., and Ericsson, L.H. (1994). Identification of molecular species of glycerophospholipids and sphingomyelin using electrospray mass spectrometry. *J Lipid Res* 35:1102–1114.

38. Roberts, W.L., Myher, J.J., Kuksis, A., Low, M.G., and Rosenberry, T.L. (1988). Lipid analysis of the glycoinositol phospholipid membrane anchor of human erythrocyte acetylcholinesterase. Palmitoylation of inositol results in resistance to phosphatidylinositol-specific phospholipase C. *J Biol Chem* 263:18766–18775.

39. Imhof, I., Flury, I., Vionnet, C., Roubaty, C., Egger, D., and Conzelmann, A. (2004). Glycosylphosphatidylinositol (GPI) proteins of *Saccharomyces cerevisiae* contain ethanolamine phosphate groups on the alpha1,4-linked mannose of the GPI anchor. *J Biol Chem* 279:19614–19627.

40. Canivenc-Gansel, E., Imhof, I., Reggiori, F., Burda, P., Conzelmann, A., and Benachour, A. (1998). GPI anchor biosynthesis in yeast: phosphoethanolamine is attached to the α1,4-linked mannose of the complete precursor glycophospholipid. *Glycobiology* 8:761–770.

41. Sipos, G., Puoti, A., and Conzelmann, A. (1995). Biosynthesis of the side chain of yeast glycosylphosphatidylinositol anchors is operated by novel mannosyltransferases located in the endoplasmic reticulum and the Golgi apparatus. *J Biol Chem* 270:19709–19715.

42. Conzelmann, A., Puoti, A., Lester, R.L., and Desponds, C. (1992). Two different types of lipid moieties are present in glycophosphoinositol-anchored membrane proteins of *Saccharomyces cerevisiae. EMBO J* 11:457–466.

43. Reggiori, F., Canivenc-Gansel, E., and Conzelmann, A. (1997). Lipid remodeling leads to the introduction and exchange of defined ceramides on GPI proteins in the ER and Golgi of *Saccharomyces cerevisiae. EMBO J* 16:3506–3518.

44. Sipos, G., Reggiori, F., Vionnet, C., and Conzelmann, A. (1997). Alternative lipid remodelling pathways for glycosylphosphatidylinositol membrane anchors in *Saccharomyces cerevisiae. EMBO J* 16:3494–34505.

45. Mehlert, A., Richardson, J.M., and Ferguson, M.A. (1998). Structure of the glycosylphosphatidylinositol membrane anchor glycan of a class-2 variant surface glycoprotein from *Trypanosoma brucei. J Mol Biol* 277:379–392.

46. Masterson, W.J., Raper, J., Doering, T.L., Hart, G.W., and Englund, P.T. (1990). Fatty acid remodeling: a novel reaction sequence in the biosynthesis of trypanosome glycosyl phosphatidylinositol membrane anchors. *Cell* 62:73–80.

47. Treumann, A., Zitzmann, N., Hulsmeier, A., Prescott, A.R., Almond, A., Sheehan, J., and Ferguson, M.A. (1997). Structural characterisation of two forms of procyclic acidic repetitive protein expressed by procyclic forms of *Trypanosoma brucei. J Mol Biol* 269:529–547.

48. Vidugiriene, J., and Menon, A.K. (1993). Early lipid intermediates in glycosyl-phosphatidylinositol anchor assembly are synthesized in the ER and located in the cytoplasmic leaflet of the ER membrane bilayer. *J Cell Biol* 121:987–996.

49. Doerrler, W.T., Ye, J., Falck, J.R., and Lehrman, M.A. (1996). Acylation of glucosaminyl phosphatidylinositol revisited. *J Biol Chem* 271:27031–27038.

50. Kajiwara, K., Watanabe, R., Pichler, H., Ihara, K., Murakami, S., Riezman, H., and Funato, K. (2008). Yeast ARV1 is required for efficient delivery of an early GPI inter-mediate to the first mannosyltransferase during GPI assembly and controls lipid flow from the endoplasmic reticulum. *Mol Biol Cell* 19:2069–2082.

51. Vishwakarma, R.A., and Menon, A.K. (2005). Flip-flop of glycosylphosphatidylinositols (GPI's) across the ER. *Chem Commun (Camb)* 453–455:.

52. Houjou, T., Hayakawa, J., Watanabe, R., Tashima, Y., Maeda, Y., Kinoshita, T., and Taguchi, R. (2007). Changes in molecular species profiles of glycosylphosphatidylinositol-anchor precursors in early stages of biosynthesis. *J Lipid Res* 48:1599–1606.

53. Kang, J.Y., Hong, Y., Ashida, H., Shishioh, N., Murakami, Y., Morita, Y.S., Maeda, Y., and Kinoshita, T. (2005). PIG-V involved in transferring the second mannose in glyco-sylphosphatidylinositol. *J Biol Chem* 280:9489–9497.

54. Hirose, S., Prince, G.M., Sevlever, D., Ravi, L., Rosenberry, T.L., Ueda, E., and Medof, M.E. (1992). Characterization of putative glycoinositol phospholipid anchor precursors in mammalian cells. Localization of phosphoethanolamine. *J Biol Chem* 267:16968–16974.

55. Vidugiriene, J., and Menon, A.K. (1994). The GPI anchor of cell-surface proteins is synthesized on the cytoplasmic face of the endoplasmic reticulum. *J Cell Biol* 127:333–341.

56. Doering, T.L., Masterson, W.J., Englund, P.T., and Hart, G.W. (1989). Biosynthesis of the glycosyl phosphatidylinositol membrane anchor of the trypanosome variant surface gly-coprotein. Origin of the non-acetylated glucosamine. *J Biol Chem* 264:11168–11173.

57. Masterson, W.J., Doering, T.L., Hart, G.W., and Englund, P.T. (1989). A novel pathway for glycan assembly: biosynthesis of the glycosyl-phosphatidylinositol anchor of the trypanosome variant surface glycoprotein. *Cell* 56:793–800.

58. Sharma, D.K., Smith, T.K., Crossman, A., Brimacombe, J.S., and Ferguson, M.A. (1997). Substrate specificity of the N-acetylglucosaminyl-phosphatidylinositol de-*N*-acetylase of glycosylphosphatidylinositol membrane anchor biosynthesis in African trypanosomes and human cells. *Biochem J* 328:171–177.

59. Smith, T.K., Cottaz, S., Brimacombe, J.S., and Ferguson, M.A. (1996). Substrate specific-ity of the dolichol phosphate mannose:glucosaminyl phosphatidylinositol α1–4-mannosyl-transferase of the glycosylphosphatidylinositol biosynthetic pathway of African trypanosomes. *J Biol Chem* 271:6476–6482.

60. Guther, M.L., and Ferguson, M.A. (1995). The role of inositol acylation and inositol deacylation in GPI biosynthesis in *Trypanosoma brucei*. *EMBO J* 14:3080–3093.

61. Ikezawa, H. (2002). Glycosylphosphatidylinositol (GPI)-anchored proteins. *Biol Pharm Bull* 25:409–417.

62. Miyata, T., Takeda, J., Iida, Y., Yamada, N., Inoue, N., Takahashi, M., Maeda, K., Kitani, T., and Kinoshita, T. (1993). Cloning of PIG-A, a component in the early step of GPI-anchor biosynthesis. *Science* 259:1318–1320.

63. Inoue, N., Watanabe, R., Takeda, J., and Kinoshita, T. (1996). PIG-C, one of the three human genes involved in the first step of glycosylphosphatidylinositol biosynthesis is a homologue of *Saccharomyces cerevisiae* GPI2. *Biochem Biophys Res Comm* 226:193–199.

64. Kamitani, T., Chang, H.M., Rollins, C., Waneck, G.L., and Yeh, E.T.H. (1993). Correction of the class H defect in glycosylphosphatidylinositol anchor biosynthesis in Ltk- cells by a human cDNA clone. *J Biol Chem* 268:20733–20736.

65. Watanabe, R., Murakami, Y., Marmor, M.D., Inoue, N., Maeda, Y., Hino, J., Kangawa, K., Julius, M., and Kinoshita, T. (2000). Initial enzyme for glycosylphosphatidylinositol biosynthesis requires PIG-P and is regulated by DPM2. *EMBO J* 19:4402–4411.

66. Watanabe, R., Inoue, N., Westfall, B., Taron, C.H., Orlean, P., Takeda, J., and Kinoshita, T. (1998). The first step of glycosylphosphatidylinositol biosynthesis is mediated by a complex of PIG-A, PIG-H, PIG-C and GPI1. *EMBO J* 17:877–885.

67. Murakami, Y., Siripanyaphinyo, U., Hong, Y., Tashima, Y., Maeda, Y., and Kinoshita, T. (2005). The initial enzyme for glycosylphosphatidylinositol biosynthesis requires PIG-Y, a seventh component. *Mol Biol Cell* 16:5236–5246.

68. Eisenhaber, B., Maurer-Stroh, S., Novatchkova, M., Schneider, G., and Eisenhaber, F. (2003). Enzymes and auxiliary factors for GPI lipid anchor biosynthesis and post-translational transfer to proteins. *Bioessays* 25:367–385.

69. Maeda, Y., Tomita, S., Watanabe, R., Ohishi, K., and Kinoshita, T. (1998). DPM2 regulates biosynthesis of dolichol phosphate-mannose in mammalian cells: correct subcellular localization and stabilization of DPM1, and binding of dolichol phosphate. *EMBO J* 17:4920–4929.

70. Kostova, Z., Rancour, D.M., Menon, A.K., and Orlean, P. (2000). Photoaffinity labelling with P[3]-(4-azidoanilido)uridine 5'-triphosphate identifies Gpi3p as the UDP-GlcNAc-binding subunit of the enzyme that catalyses formation of GlcNAc-phosphatidylinositol, the first glycolipid intermediate in glycosylphosphatidylinositol synthesis. *Biochem J* 350:815–822.

71. Liu, J., and Mushegian, A. (2003). Three monophyletic superfamilies account for the majority of the known glycosyltransferases. *Protein Sci* 12:1418–1431.

72. Coutinho, P.M., Deleury, E., Davies, G.J., and Henrissat, B. (2003). An evolving hierarchical family classification for glycosyltransferases. *J Mol Biol* 328:307–317.

73. Watanabe, R., Kinoshita, T., Masaki, R., Yamamoto, A., Takeda, J., and Inoue, N. (1996). PIG-A and PIG-H, which participate in glycosylphosphatidylinositol anchor biosynthesis, form a protein complex in the endoplasmic reticulum. *J Biol Chem* 271:26868–26875.

74. Tiede, A., Nischan, C., Schubert, J., and Schmidt, R.E. (2000). Characterisation of the enzymatic complex for the first step in glycosylphosphatidylinositol biosynthesis. *Int J Biochem Cell Biol* 32:339–350.

75. Watanabe, R., Ohishi, K., Maeda, Y., Nakamura, N., and Kinoshita, T. (1999). Mammalian PIG-L and its yeast homologue Gpi12p are N-acetylglucosaminylphosphatidylinositol de-N-acetylases essential in glycosylphosphatidylinositol biosynthesis. *Biochem J* 339:185–192.

76. Murakami, Y., Siripanyapinyo, U., Hong, Y., Kang, J.Y., Ishihara, S., Nakakuma, H., Maeda, Y., and Kinoshita, T. (2003). PIG-W is critical for inositol acylation but not for flipping of glycosylphosphatidylinositol-anchor. *Mol Biol Cell* 14:4285–4295.

77. Maeda, Y., Watanabe, R., Harris, C.L., Hong, Y., Ohishi, K., Kinoshita, K., and Kinoshita, T. (2001). PIG-M transfers the first mannose to glycosylphosphatidylinositol on the lumenal side of the ER. *EMBO J.* 20:250–261.

78. Ashida, H., Hong, Y., Murakami, Y., Shishioh, N., Sugimoto, N., Kim, Y.U., Maeda, Y., and Kinoshita, T. (2005). Mammalian PIG-X and yeast Pbn1p are the essential components of glycosylphosphatidylinositol-mannosyltransferase I. *Mol Biol Cell* 16:1439–1448.

79. Hong, Y., Maeda, Y., Watanabe, R., Ohishi, K., Mishkind, M., Riezman, H., and Kinoshita, T. (1999). Pig-n, a mammalian homologue of yeast Mcd4p, is involved in transferring phosphoethanolamine to the first mannose of the glycosylphosphatidylinositol. *J Biol Chem* 274:35099–35106.

80. Benachour, A., Sipos, G., Flury, I., Reggiori, F., Canivenc-Gansel, E., Vionnet, C., Conzelmann, A., and Benghezal, M. (1999). Deletion of GPI7, a yeast gene required for addition of a side chain to the glycosylphosphatidylinositol (GPI) core structure, affects GPI protein transport, remodeling, and cell wall integrity. *J Biol Chem* 274:15251–15261.

81. Takahashi, M., Inoue, N., Ohishi, K., Maeda, Y., Nakamura, N., Endo, Y., Fujita, T., Takeda, J., and Kinoshita, T. (1996). PIG-B, a membrane protein of the endoplasmic reticulum with a large lumenal domain, is involved in transferring the third mannose of the GPI anchor. *EMBO J.* 15:4254–4261.

82. Hong, Y., Maeda, Y., Watanabe, R., Inoue, N., Ohishi, K., and Kinoshita, T. (2000). Requirement of PIG-F and PIG-O for transferring phosphoethanolamine to the third mannose in glycosylphosphatidylinositol. *J Biol Chem* 275:20911–20919.

83. Inoue, N., Kinoshita, T., Orii, T., and Takeda, J. (1993). Cloning of a human gene, PIG-F, a component of glycosylphosphatidylinositol anchor biosynthesis, by a novel expression cloning strategy. *J Biol Chem* 268:6882–6885.

84. Shishioh, N., Hong, Y., Ohishi, K., Ashida, H., Maeda, Y., and Kinoshita, T. (2005). GPI7 is the second partner of PIG-F and involved in modification of glycosylphosphatidylinositol. *J Biol Chem* 280:9728–9734.

85. Schonbachler, M., Horvath, A., Fassler, J., and Riezman, H. (1995). The yeast spt14 gene is homologous to the human PIG-A gene and is required for GPI anchor synthesis. *EMBO J* 14:1637–1645.

86. Vossen, J.H., Ram, A.F., and Klis, F.M. (1995). Identification of SPT14/CWH6 as the yeast homologue of hPIG-A, a gene involved in the biosynthesis of GPI anchors. *Biochim Biophys Acta* 1243:549–551.

87. Leidich, S.D., Kostova, Z., Latek, R.R., Costello, L.C., Drapp, D.A., Gray, W., Fassler, J. S., and Orlean, P. (1995). Temperature-sensitive yeast GPI anchoring mutants gpi2 and gpi3 are defective in the synthesis of N-acetylglucosaminyl phosphatidylinositol: cloning of the GPI2 gene. *J Biol Chem* 270:13029–13035.

88. Yan, B.C., Westfall, B.A., and Orlean, P. (2001). Ynl038wp (Gpi15p) is the Saccharomyces cerevisiae homologue of human Pig-Hp and participates in the first step in glycosylphosphatidylinositol assembly. *Yeast* 18:1383–1389.

89. Newman, H.A., Romeo, M.J., Lewis, S.E., Yan, B.C., Orlean, P., and Levin, D.E. (2005). Gpi19, the *Saccharomyces cerevisiae* homologue of mammalian PIG-P, is a subunit of the initial enzyme for glycosylphosphatidylinositol anchor biosynthesis. *Eukaryot Cell* 4:1801–1807.

90. Leidich, S.D., and Orlean, P. (1996). Gpi1, a *Saccharomyces cerevisiae* protein that participates in the first step in glycosylphosphatidylinositol anchor synthesis. *J Biol Chem* 271:27829–27837.

91. Sobering, A.K., Watanabe, R., Romeo, M.J., Yan, B.C., Specht, C.A., Orlean, P., Riezman, H., and Levin, D.E. (2004). Yeast Ras regulates the complex that catalyzes the first step in GPI-anchor biosynthesis at the ER. *Cell* 117:637–648.

92. Tsukahara, K., Hata, K., Nakamoto, K., Sagane, K., Watanabe, N.A., Kuromitsu, J., Kai, J., Tsuchiya, M., Ohba, F., Jigami, Y., Yoshimatsu, K., and Nagasu, T. (2003). Medicinal genetics approach towards identifying the molecular target of a novel inhibitor of fungal cell wall assembly. *Mol Microbiol* 48:1029–1042.

93. Umemura, M., Okamoto, M., Nakayama, K., Sagane, K., Tsukahara, K., Hata, K., and Jigami, Y. (2003). *GWT1* gene is required for inositol acylation of glycosylphosphatidylinositol anchors in yeast. *J Biol Chem* 278:23639–23647.
94. Fabre, A.L., Orlean, P., and Taron, C.H. (2005). *Saccharomyces cerevisiae* Ybr004c and its human homologue are required for addition of the second mannose during glycosylphosphatidylinositol precursor assembly. *FEBS J* 272:1160–1168.
95. Sato, K., Noda, Y., and Yoda, K. (2007). Pga1 is an essential component of Glycosylphosphatidylinositol-mannosyltransferase II of *Saccharomyces cerevisiae*. *Mol Biol Cell* 18:3472–3485.
96. Gaynor, E.C., Mondesert, G., Grimme, S.J., Reed, S.I., Orlean, P., and Emr, S.D. (1999). MCD4 encodes a conserved endoplasmic reticulum membrane protein essential for glycosylphosphatidylinositol anchor synthesis in yeast. *Mol Biol Cell* 10:627–648.
97. Sutterlin, C., Escribano, M.V., Gerold, P., Maeda, Y., Mazon, M.J., Kinoshita, T., Schwarz, R.T., and Riezman, H. (1998). *Saccharomyces cerevisiae* GPI10, the functional homologue of human PIG-B, is required for glycosylphosphatidylinositol-anchor synthesis. *Biochem J* 332:153–159.
98. Grimme, S.J., Westfall, B.A., Wiedman, J.M., Taron, C.H., and Orlean, P. (2001). The essential Smp3 protein is required for addition of the side-branching fourth mannose during assembly of yeast glycosylphosphatidylinositols. *J Biol Chem* 276:27731–27739.
99. Flury, I., Benachour, A., and Conzelmann, A. (2000). YLL031c belongs to a novel family of membrane proteins involved in the transfer of ethanolaminephosphate onto the core structure of glycosylphosphatidylinositol anchors in yeast. *J Biol Chem* 275:24458–24465.
100. Taron, C.H., Wiedman, J.M., Grimme, S.J., and Orlean, P. (2000). Glycosylphosphatidylinositol biosynthesis defects in Gpi11p- and Gpi13p-deficient yeast suggest a branched pathway and implicate gpi13p in phosphoethanolamine transfer to the third mannose. *Mol Biol Cell* 11:1611–1630.
101. Chang, T., Milne, K.G., Guther, M.L., Smith, T.K., and Ferguson, M.A. (2002). Cloning of *Trypanosoma brucei* and Leishmania major genes encoding the GlcNAc-phosphatidylinositol de-N-acetylase of glycosylphosphatidylinositol biosynthesis that is essential to the African sleeping sickness parasite. *J Biol Chem* 277:50176–50182.
102. Urbaniak, M.D., Crossman, A., Chang, T., Smith, T.K., van Aalten, D.M., and Ferguson, M.A. (2005). The *N*-acetyl-D-glucosaminylphosphatidylinositol De-*N*-acetylase of glycosylphosphatidylinositol biosynthesis is a zinc metalloenzyme. *J Biol Chem* 280:22831–22838.
103. Guther, M.L., Leal, S., Morrice, N.A., Cross, G.A., and Ferguson, M.A. (2001). Purification, cloning and characterization of a GPI inositol deacylase from *Trypanosoma brucei*. *EMBO J* 20:4923–4934.
104. Guther, M.L., Prescott, A.R., and Ferguson, M.A. (2003). Deletion of the GPIdeAc gene alters the location and fate of glycosylphosphatidylinositol precursors in Trypanosoma brucei. *Biochemistry* 42:14532–14540.
105. Hong, Y., Nagamune, K., Morita, Y.S., Nakatani, F., Ashida, H., Maeda, Y., and Kinoshita, T. (2006). Removal or maintenance of inositol-linked acyl chain in glycosylphosphatidylinositol is critical in trypanosome life cycle. *J Biol Chem* 281:11595–11602.
106. Yu, J., Nagarajan, S., Knez, J.J., Udenfriend, S., Chen, R., and Medof, M.E. (1997). The affected gene underlying the class K glycosylphosphatidylinositol (GPI) surface protein defect codes for the GPI transamidase. *Proc Natl Acad Sci USA* 94:12580–12585.
107. Hiroi, Y., Komuro, I., Chen, R., Hosoda, T., Mizuno, T., Kudoh, S., Georgescu, S.P., Medof, M.E., and Yazaki, Y. (1998). Molecular cloning of human homolog of yeast

GAA1 which is required for attachment of glycosylphosphatidylinositols to proteins. *FEBS Lett* 421:252–258.

108. Ohishi, K., Inoue, N., Maeda, Y., Takeda, J., Riezman, H., and Kinoshita, T. (2000). Gaa1p and gpi8p are components of a glycosylphosphatidylinositol (GPI) transamidase that mediates attachment of GPI to proteins. *Mol Biol Cell* 11:1523–1533.

109. Ohishi, K., Inoue, N., and Kinoshita, T. (2001). PIG-S and PIG-T, essential for GPI anchor attachment to proteins, form a complex with GAA1 and GPI8. *EMBO J* 20:4088–4098.

110. Hong, Y., Ohishi, K., Kang, J.Y., Tanaka, S., Inoue, N., Nishimura, J., Maeda, Y., and Kinoshita, T. (2003). Human PIG-U and yeast Cdc91p are the fifth subunit of GPI transamidase that attaches GPI-anchors to proteins. *Mol Biol Cell* 14:1780–1789.

111. Ohishi, K., Nagamune, K., Maeda, Y., and Kinoshita, T. (2003). Two Subunits of Glycosylphosphatidylinositol Transamidase, GPI8 and PIG-T, Form a Functionally Important Intermolecular Disulfide Bridge. *J Biol Chem* 278:13959–13967.

112. Meyer, U., Benghezal, M., Imhof, I., and Conzelmann, A. (2000). Active site determination of Gpi8p, a caspase-related enzyme required for glycosylphosphatidylinositol anchor addition to proteins. *Biochemistry* 39:3461–3471.

113. Mottram, J.C., Helms, M.J., Coombs, G.H., and Sajid, M. (2003). Clan CD cysteine peptidases of parasitic protozoa. *Trends Parasitol* 19:182–187.

114. Kang, X., Szallies, A., Rawer, M., Echner, H., and Duszenko, M. (2002). GPI anchor transamidase of *Trypanosoma brucei: in vitro* assay of the recombinant protein and VSG anchor exchange. *J Cell Sci* 115:2529–2539.

115. Hamburger, D., Egerton, M., and Riezman, H. (1995). Yeast Gaa1p is required for attachment of a completed GPI anchor onto proteins. *J Cell Biol* 129:629–639.

116. Benghezal, M., Benachour, A., Rusconi, S., Aebi, M., and Conzelmann, A. (1996). Yeast Gpi8p is essential for GPI anchor attachment onto proteins. *EMBO J* 15:6575–6583.

117. Fraering, P., Imhof, I., Meyer, U., Strub, J.M., van Dorsselaer, A., Vionnet, C., and Conzelmann, A. (2001). The GPI transamidase complex of *Saccharomyces cerevisiae* contains Gaa1p, Gpi8p, and Gpi16p. *Mol Biol Cell* 12:3295–3306.

118. Grimme, S.J., Gao, X.D., Martin, P.S., Tu, K., Tcheperegine, S.E., Corrado, K., Farewell, A.E., Orlean, P., and Bi, E. (2004). Deficiencies in the endoplasmic reticulum (ER)-membrane protein Gab1p perturb transfer of glycosylphosphatidylinositol to proteins and cause perinuclear ER-associated actin bar formation. *Mol Biol Cell* 15:2758–2770.

119. Nagamune, K., Ohishi, K., Ashida, H., Hong, Y., Hino, J., Kangawa, K., Inoue, N., Maeda, Y., and Kinoshita, T. (2003). GPI transamidase of *Trypanosoma brucei* has two previously uncharacterized (trypanosomatid transamidase 1 and 2) and three common subunits. *Proc Natl Acad Sci USA* 100:10682–10687.

120. Tanaka, S., Maeda, Y., Tashima, Y., and Kinoshita, T. (2004). Inositol deacylation of glycosylphosphatidylinositol-anchored proteins is mediated by mammalian PGAP1 and yeast Bst1p. *J Biol Chem* 279:14256–14263.

121. Maeda, Y., Tashima, Y., Houjou, T., Fujita, M., Yoko-o, T., Jigami, Y., Taguchi, R., and Kinoshita, T. (2007). Fatty acid remodeling of GPI-anchored proteins is required for their raft association. *Mol Biol Cell* 18:1497–1506.

122. Ghugtyal, V., Vionnet, C., Roubaty, C., and Conzelmann, A. (2007). *CWH43* is required for the introduction of ceramides into GPI anchors in *Saccharomyces cerevisiae. Mol Microbiol* 65:1493–1502.

123. Fujita, M., Umemura, M., Yoko-o, T., and Jigami, Y. (2006). *PER1* is required for GPI-phospholipase A2 activity and involved in lipid remodeling of GPI-anchored proteins. *Mol Biol Cell* 17:5253–5264.

124. Schuldiner, M., Collins, S.R., Thompson, N.J., Denic, V., Bhamidipati, A., Punna, T., Ihmels, J., Andrews, B., Boone, C., Greenblatt, J.F., Weissman, J.S., and Krogan, N.J. (2005). Exploration of the function and organization of the yeast early secretory pathway through an epistatic miniarray profile. *Cell* 123:507–519.

125. Ng, D.T., Spear, E.D., and Walter, P. (2000). The unfolded protein response regulates multiple aspects of secretory and membrane protein biogenesis and endoplasmic reticulum quality control. *J Cell Biol* 150:77–88.

126. Paidhungat, M., and Garrett, S. (1998). Cdc1 and the vacuole coordinately regulate Mn2+ homeostasis in the yeast *Saccharomyces cerevisiae*. *Genetics* 148:1787–1798.

127. Brown, D.A., and Rose, J.K. (1992). Sorting of GPI-anchored proteins to glycolipid-enriched membrane subdomains during transport to the apical cell surface. *Cell* 68:533–544.

128. Bagnat, M., Keranen, S., Shevchenko, A., Shevchenko, A., and Simons, K. (2000). Lipid rafts function in biosynthetic delivery of proteins to the cell surface in yeast. *Proc Natl Acad Sci USA* 97:3254–3259.

129. Watanabe, R., Funato, K., Venkataraman, K., Futerman, A.H., and Riezman, H. (2002). Sphingolipids are required for the stable membrane association of glycosylphosphatidylinositol-anchored proteins in yeast. *J Biol Chem* 277:49538–49544.

130. Yasuda, S., Kitagawa, H., Ueno, M., Ishitani, H., Fukasawa, M., Nishijima, M., Kobayashi, S., and Hanada, K. (2001). A novel inhibitor of ceramide trafficking from the endoplasmic reticulum to the site of sphingomyelin synthesis. *J Biol Chem* 276:43994–44002.

131. Bosson, R., Jaquenoud, M., and Conzelmann, A. (2006). *GUP1* of *Saccharomyces cerevisiae* encodes an O-acyltransferase involved in remodeling of the GPI anchor. *Mol Biol Cell* 17:2636–2645.

132. Holst, B., Lunde, C., Lages, F., Oliveira, R., Lucas, C., and Kielland-Brandt, M.C. (2000). *GUP1* and its close homologue *GUP2*, encoding multimembrane-spanning proteins involved in active glycerol uptake in *Saccharomyces cerevisiae*. *Mol Microbiol* 37:108–124.

133. Tashima, Y., Taguchi, R., Murata, C., Ashida, H., Kinoshita, T., and Maeda, Y. (2006). *PGAP2* is essential for correct processing and stable expression of GPI-anchored proteins. *Mol Biol Cell* 17:1410–1420.

134. Lorenzi, M.V., Horii, Y., Yamanaka, R., Sakaguchi, K., and Miki, T. (1996). *FRAG1*, a gene that potently activates fibroblast growth factor receptor by C-terminal fusion through chromosomal rearrangement. *Proc Natl Acad Sci USA* 93:8956–8961.

135. Martin-Yken, H., Dagkessamanskaia, A., De Groot, P., Ram, A., Klis, F., and Francois, J. (2001). *Saccharomyces cerevisiae YCRO17c/CWH43* encodes a putative sensor/transporter protein upstream of the BCK2 branch of the PKC1-dependent cell wall integrity pathway. *Yeast* 18:827–840.

136. Umemura, M., Fujita, M., Yoko, O.T., Fukamizu, A., and Jigami, Y. (2007). *Saccharomyces cerevisiae CWH43* is involved in the remodeling of the lipid moiety of GPI anchors to ceramides. *Mol Biol Cell* 18:4304–4316.

137. Zhu, Y., Vionnet, C., and Conzelmann, A. (2006). Ethanolaminephosphate side chain added to glycosylphosphatidylinositol (GPI) anchor by mcd4p is required for ceramide remodeling and forward transport of GPI proteins from endoplasmic reticulum to Golgi. *J Biol Chem* 281:19830–19839.

138. Ferguson, M.A., Brimacombe, J.S., Brown, J.R., Crossman, A., Dix, A., Field, R.A., Guther, M.L., Milne, K.G., Sharma, D.K., and Smith, T.K. (1999). The GPI biosynthetic pathway as a therapeutic target for African sleeping sickness. *Biochim Biophys Acta* 1455:327–340.

139. Morita, Y.S., Paul, K.S., and Englund, P.T. (2000). Specialized fatty acid synthesis in African trypanosomes: myristate for GPI anchors. *Science* 288:140–143.

140. Jaquenoud, M., Pagac, M., Signorell, A., Benghezal, M., Jelk, J., Butikofer, P., and Conzelmann, A. (2008). The Gup1 homologue of *Trypanosoma brucei* is a GPI glycosyl-phosphatidylinositol remodelase. *Mol Microbiol* 67:202–212.

141. Buxbaum, L.U., Raper, J., Opperdoes, F.R., and Englund, P.T. (1994). Myristate exchange. A second glycosyl phosphatidylinositol myristoylation reaction in African trypanosomes. *J Biol Chem* 269:30212–30220.

142. Doering, T.L., Raper, J., Buxbaum, L.U., Adams, S.P., Gordon, J.I., Hart, G.W., and Englund, P.T. (1991). An analog of myristic acid with selective toxicity for African trypanosomes. *Science* 252:1851–1854.

143. Doering, T.L., Lu, T., Werbovetz, K.A., Gokel, G.W., Hart, G.W., Gordon, J.I., and Englund, P.T. (1994). Toxicity of myristic acid analogs toward African trypanosomes. *Proc Natl Acad Sci USA* 91:9735–9739.

144. Ralton, J.E., and McConville, M.J. (1998). Delineation of three pathways of glycosylpho-sphatidylinositol biosynthesis in *Leishmania mexicana*. Precursors from different pathways are assembled on distinct pools of phosphatidylinositol and undergo fatty acid remodeling. *J Biol Chem* 273:4245–4257.

145. Bertello, L.E., Alves, M.J., Colli, W., and de Lederkremer, R.M. (2004). Inositolpho-sphoceramide is not a substrate for the first steps in the biosynthesis of glycoinositolpho-spholipids in *Trypanosoma cruzi*. *Mol Biochem Parasitol* 133:71–80.

146. Izquierdo, L., Nakanishi, M., Mehlert, A., Machray, G., Barton, G.J., and Ferguson, M.A. (2009). Identification of a glycosylphosphatidylinositol anchor-modifying beta1–3 N-acetylglucosaminyl transferase in *Trypanosoma brucei*. *Mol Microbiol* 71:478–491.

147. Montagna, G., Cremona, M.L., Paris, G., Amaya, M.F., Buschiazzo, A., Alzari, P.M., and Frasch, A.C. (2002). The trans-sialidase from the African trypanosome *Trypanosoma brucei*. *Eur J Biochem* 269:2941–2950.

148. Varki, A., and Sharon, N. (2009). Historical background and overview. In Essentials of Glycobiology, A. Varki, R.D. Cummings, J.D. Esko, H.H. Freeze, P. Stanley, C.R. Bertozzi, G.W. Hart and M.E. Etzler, (eds.), pp. 1–22. Cold Spring Harbor Laboratory Press, Cold Spring Harbor, NY.

2

The N-Acetylglucosamine-PI Transfer Reaction, the GlcNAc-PI Transferase Complex, and Its Regulation

DAVID E. LEVIN • RONALD J. STAMPER

Department of Biochemistry and Molecular Biology
The Johns Hopkins Bloomberg School of Public Health
Baltimore, MD 21205, USA

I. Abstract

Glycosylphosphatidylinositol (GPI) anchoring of cell surface proteins is the most metabolically expensive posttranslational lipid modification known, involving proteins that act through at least 11 sequential steps in the endoplasmic reticulum (ER). This chapter will focus on the first step in the biosynthesis of GPI anchors, the glycosylation of phosphatidylinositol (PI) to generate *N*-acetylglucosamine (GlcNAc)-PI. As the first committed step of the pathway, one might expect it to be subject to stringent regulation. Therefore, an important aspect of this chapter will be the growing evidence for various mechanisms that are employed to regulate GPI-anchor production at the earliest stage.

THE ENZYMES, Vol. XXVI   31   ISSN NO: 1874-6047
DOI: 10.1016/S1874-6047(09)26002-1

II.   The *N*-Acetylglucosamine-PI Transfer Reaction

The first committed step in GPI biosynthesis, which is catalyzed by the UDP-GlcNAc:PI α1–6 GlcNAc transferase (GPI-GlcNAc transferase), involves the transfer of GlcNAc from the sugar nucleotide, UDP-GlcNAc, to the inositol ring of a phosphatidylinositol (PI) molecule that resides in the endoplasmic reticulum (ER) membrane [1]. This enzyme retains the configuration at the anomeric center between the donor and acceptor glycosyl groups (Figure 2.1). The product of the GPI-GlcNAc transferase reaction, GlcNAc-PI, is accessible *in vitro* to digestion by phospholipase C in liposomes, indicating that it resides on the cytoplasmic leaflet of the ER membrane bilayer [2].

The GPI-GlcNAc transferase recognizes the fatty acyl chains on PI and the mammalian enzyme has a preference for PI acceptors that contain mainly stearic and arachidonic acids [3–5], the profile represented in free PI in animal cells. However, the acyl chains of GPI anchors are subject to extensive remodeling at later steps in biosynthesis [5, 6].

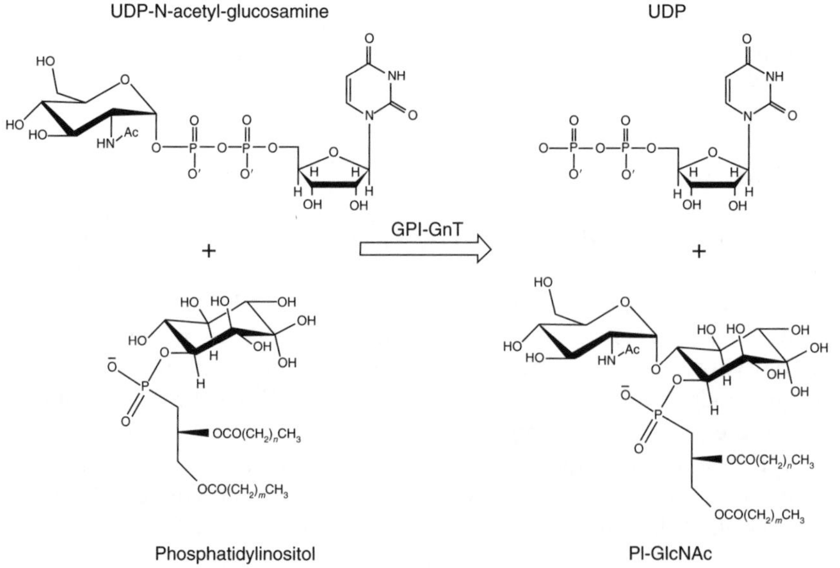

FIG. 2.1. The *N*-acetylglucosamine-PI transfer reaction. *N*-Acetylglucosamine is transferred from UDP-*N*-acetylglucosamine (UDP-GlcNAc) to PI in a retaining glycosyltransferase reaction catalyzed by the UDP-GlcNAc:PI α1–6 GlcNAc transferase (abbreviated GPI-GnT).

III.   Biological Importance of the GPI-GlcNAc Transferase

In humans, blocking of GPI biosynthesis at the first step by somatic mutation of the PIG-A gene results in paroxysmal nocturnal hemoglobinuria (PNH), an acquired hemolytic disease of hematopoietic stem cells, which results from a deficiency of GPI-anchor proteins [7–9]. Of greatest importance concerned with the pathophysiology of PNH is the absence of complement regulatory proteins, CD55 and CD59 on the surface of hematopoietic cell, which explains the hemolytic phenotype. The PNH patients often display the same PIG-A loss-of-function mutations in all hematopoietic cell types, revealing the monoclonal nature of the disease and the involvement of a multipotent hematopoietic stem cell.

The GPI-protein defects found in all PNH patients examined to date result from mutations in PIG-A, despite the fact that at least 26 genes are required for the production of GPI proteins. This observation reflects the hemizygous state of PIG-A on the X chromosome in males and functional hemizygosity in females [8]. All other genes known to be involved in GPI biosynthesis are located on autosomes [10]. Thus, a single loss-of-function mutation in PIG-A is sufficient to block GPI biosynthesis, whereas mutation of both autosomal alleles of any of the other genes would be required to cause a comparable deficiency. A perplexing question is how a PIG-A mutant clone expands through the population of blood cells. It has been suggested that immunological attack of GPI-protein proficient hematopoietic stem cells is important in selection and expansion of a PIG-A mutant clone [11], but this remains speculative.

The importance of the GPI-GlcNAc transferase to animal development was demonstrated by the finding that knockout mice in PIG-A are embryonic lethal [12]. The PIG-A mutant male embryos ceased the development beyond the ninth day of gestation. The PIG-A-disrupted females, which were mosaic for PIG-A function due to random X-inactivation, continued to develop to near full term, but displayed multiple developmental defects. Even deletion of PIG-A specifically in epidermal tissue of male mice caused defects in skin development that were lethal within the first 3 postpartum days [13].

IV.   The GPI-GlcNAc Transferase Complex

The GPI-GlcNAc transferase is best characterized from mammalian cells and from the baker's yeast, *Saccharomyces cerevisiae*. Thus, the discussion in this chapter will be restricted largely to our understanding of the enzymes from these sources. The GPI-GlcNAc transferase complex is unusually elaborate considering the relative simplicity of the reaction it catalyzes and the substrates with which it interacts. The six GPI-GlcNAc transferase complex proteins common to mammals and *S. cerevisiae* are PIG-A/Gpi3, PIG-C/Gpi2, PIG-H/Gpi15, PIG-P/Gpi19, PIG-Q(hGpi1)/Gpi1, and PIG-Y/Eri1, respectively [14]. The mammalian complex possesses an additional subunit, DPM2, which appears to play a positive regulatory role [15]. The yeast genome does not encode an ortholog of DPM2. However, the small GTPase Ras2 has emerged as a negative regulator of the yeast GPI-GnT [16]. No evidence exists for Ras regulation of the mammalian complex. Both DPM2 and Ras2 are discussed in Section IV. Aside from the catalytic subunit, PIG-A/Gpi3, little is known about the roles of other complex components. The membrane topology of some of the subunits has been established, but in other cases the topology has been modeled. A summary of this information is depicted in Figure 2.2.

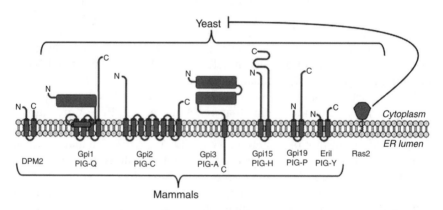

FIG. 2.2. Subunit composition and topology of the mammalian and yeast GlcNAc-PI synthetic complexes. The bulk of the protein complex is oriented on the cytoplasmic face of the ER. Six core subunits (PIG-Q/Gpi1, PIG-C/Gpi2, PIG-A/Gpi3, PIG-H/Gpi15, PIG-P/Gpi19, and PIG-Y/Eri1) are common to mammals and yeast. Membrane topology has been confirmed experimentally for PIG-A, PIG-H, PIG-Y, and Gpi19, and modeled for Gpi1 and Gpi2 [24, 26]. Mammalian GPI-GlcNAc transferase possesses a seventh subunit, DPM2, whose absence decreases enzyme activity approximately threefold, which may reflect regulation. Yeast GPI-GlcNAc transferase is negatively regulated by the small GTP-binding protein, Ras, which is tethered by lipid moieties to the cytoplasmic face of the ER. Enzyme activity is regulated over a 200-fold range depending on the state of yeast Ras2. Dark cylinders indicate transmembrane segments. Globular domains are also shown.

A. PIG-A/Gpi3

PIG-A/Gpi3 is now recognized as the catalytic subunit of the complex, a member of the glycosyltransferase family 4 [17, 18]. Early evidence that PIG-A/Gpi3 is the catalytic subunit came from the observation that it binds a photoactivatable UDP-GlcNAc analog [19]. The human and yeast PIG-A/Gpi3 genes were isolated by complementation of mutants deficient in GlcNAc-PI synthesis [7, 20], and yeast *GPI3* was also identified as an ortholog of PIG-A [21]. However, the yeast *GPI3* gene was first discovered through a genetic screen whose connection with GPI biosynthesis remains mysterious, but it was through this study the essential nature of *GPI3* (initially named *SPT14*) was revealed [22].

The yeast Gpi3 and human PIG-A orthologs share the greatest degree of amino acid sequence identity among all the GPI-GlcNAc transferase subunits (47%), perhaps reflecting its identity as the catalytic core of the complex. PIG-A/Gpi3 is a single-pass transmembrane protein with the majority of the polypeptide, the N-terminal 420 amino acids, residing on the cytoplasmic face of the ER [23, 24], consistent with the conclusion that this is the surface on which the first steps of GPI biosynthesis take place [14]. The cytoplasmic region appears, by structural similarity to other glycosyltransferases of the same class, to consist of two distinct domains with a cleft between them for binding UDP [25]. It has been suggested from structure-based comparisons that the C-terminal-most domain and between the two domains positions the UDP-GlcNAc, whereas the N-terminal-most domain binds to PI [26]. The short C-terminal domain in the ER lumen is important in targeting or retention of PIG-A to the rough ER [23].

PIG-A/Gpi3 possesses a signature motif, EX_7E, which is highly conserved among the members of several subfamilies for retaining glycosyltransferases [27]. Although the two glutamate residues have been proposed to serve as an acid–base catalyst and a nucleophile, respectively [28], the relative importance of these residues on enzyme function varies among family members [18, 29, 30]. Mutational analysis of the conserved residues in yeast Gpi3 (E289 and E297) revealed the importance of both residues on enzyme function, but only the second is essential for viability [18]. A neighboring cysteine residue that is conserved among PIG-A/Gpi3 orthologs was suggested to be important for function, perhaps for UDP-GlcNAc binding. This proposal was based on the observation that the GPI-GlcNAc transferase activity could be irreversibly inhibited by agents that alkylate thiol groups and that this inhibition was prevented in the presence of uridine nucleotides [31]. However, mutation of this Cys residue to Ala in yeast Gpi3 (C301A) had only a modest effect on GPI-GlcNAc transferase function, failing to support the earlier interpretation [18].

B. PIG-C/Gpi2

The yeast *GPI2* was isolated by complementation of a GlcNAc-PI deficient mutant [20], and its human ortholog PIG-C was identified by amino
acid sequence identity with *GPI2* [32]. The deletion of yeast *GPI2* is
lethal [20]. Both the human and yeast PIG-C/Gpi2 proteins are highly
hydrophobic and are proposed to have eight transmembrane domains
with both termini oriented in the cytoplasm [24, 26]. However, this is yet
to be determined experimentally. Because PIG-C/Gpi2 is the most hydrophobic subunit of the GPI-GlcNAc transferase complex, it has been suggested that its primary function is as a scaffold to anchor the complex to the
ER membrane [26]. However, from the results of a large-scale screen of
yeast membrane proteins, Gpi2 appears to be involved in multiple physical
interactions with other pathway enzymes [33], suggesting the possibility
that it promotes supercomplex associations, perhaps for regulatory
purposes or for substrate channeling (see Section IV.D).

C. PIG-H/Gpi15

The human PIG-H was isolated through complementation of a GlcNAc-
PI synthesis mutant [34], and its essential yeast ortholog Gpi15 was identified through amino acid sequence identity with PIG-H [35]. Although
PIG-H and Gpi15 share the lowest degree of amino acid sequence identity
(16%) among all the yeast and human orthologous pairs of GPI-GlcNAc
transferase subunits, both proteins appear to form small, hairpin-like transmembrane structures in the ER membrane. In the case of PIG-H, both
termini were shown to be cytoplasmic [23]. The N-terminus is predicted
to be unstructured, but the C-terminal 100 residues are predicted to form an
α-helical globular domain [26]. Coprecipitation experiments suggest that
PIG-H has the tightest and most direct association with PIG-A [23].

D. PIG-P/Gpi19

The human PIG-P was isolated by protein sequence determination of
peptides coprecipitating with the GPI-GlcNAc transferase complex [15].
Its essential yeast ortholog Gpi19 was identified through amino acid sequence identity with PIG-P (20%) [36]. Similar to PIG-H/Gpi15, Gpi19 (and
presumably PIG-P) forms a small hairpin-like transmembrane structure with
both termini oriented cytoplasmically in the ER membrane, but with only a
few N-terminal residues extended into the cytoplasm [36]. The N-terminus
of Gpi19 is dispensable, because an N-terminal truncation mutant of the
yeast gene that initiates translation within the C-terminal transmembrane
domain conditionally complements the lethality of a *gpi19* null mutant [36].

E. PIG-Q/Gᴘɪ1

The Yeast *GPI1* was isolated through complementation of a GlcNAc-PI synthesis mutant [37]. Its human ortholog, PIG-Q/hGPI1, was isolated on the basis of amino acid sequence identity with the yeast gene (22%) [4]. Gpi1 was predicted by Tiede et al. [24] to have six transmembrane domains with both termini oriented in the cytoplasm and by Eisenhaber et al. [26] to have four transmembrane domains with an amphipathic segment near the N-terminus that is suggested to be parallel to the membrane surface. Clearly, the true architecture of PIG-Q/Gpi1 awaits experimental determination.

Deletion of the *Schizosaccharomyces pombe GPI1* ortholog is lethal [38]. By contrast, deletion of *S. cerevisiae GPI1* results in a temperature-sensitive growth defect, indicating that this mutant is not completely defective in the GPI-GlcNAc transferase activity [37]. Consistent with this interpretation, disruption of mouse *GPI1* resulted in greatly decreased, but still detectable levels of GPI-anchored proteins [39]. For this reason, and because the GPI1 disruption causes a partial reduction in the steady-state levels of PIG-C and PIG-H, it has been suggested that the principal role of this component is to stabilize the complex [39]. This notion was supported by the additional finding that GPI1 is required to tether PIG-C to the PIG-A/PIG-H complex.

F. PIG-Y/Eʀɪ1

The yeast Eri1 was identified through the cell-wall defect displayed by a null mutant in the *ERI1* gene [40], which was subsequently characterized as a deficiency in GPI-GlcNAc transferase activity [16]. Its human ortholog, PIG-Y, was isolated through protein sequence analysis of a 6-kDa polypeptide that coprecipitated with the GPI-GlcNAc transferase [41]. The human PIG-Y is expressed from a transcript that encodes a second protein, namely PreY, upstream from PIG-Y. PIG-Y is translated by leaky scanning of the PreY ribosome-binding site. Although both the proteins are expressed together, PreY does not appear to be important for GPI biosynthesis [41].

Both the yeast Eri1 and human PIG-Y proteins are very small (68 and 71 amino acid residues, respectively), share 22% amino acid sequence identity and are predicted to have two transmembrane domains. Similar to PIG-H/Gpi15 and PIG-P/Gpi19, these polypeptides are predicted to form hairpin-like structures in the ER membrane, leaving very little sequence extending from the membrane. PIG-Y was shown experimentally to have both termini oriented in the cytoplasm [41]. A deletion mutant in the yeast *ERI1* is temperature sensitive for growth [40], similar to the phenotype of a null mutant in *GPI1*, suggesting that the mutant retains residual GPI-GlcNAc transferase activity. This is consistent with the finding that PIG-Y is not required for the other components of the mammalian GPI-GlcNAc transferase complex to assemble [41].

G. COMPLEX MODELING

Efforts to model the complex based on a data from coprecipitation experiments have yielded a few insights. Mammalian PIG-A, the catalytic subunit, appears to make its tightest association with PIG-H [4, 23, 39], raising the possibility that PIG-H plays an important regulatory role. PIG-Q/Gpi1 appears to tether PIG-C to the PIG-A:PIG-H complex [4, 39]. PIG-Q/Gpi1 seems to play a stabilizing role in the complex, because a yeast *gpi1Δ* mutant is viable, but temperature sensitive [37], and because deletion of GPI1 in mouse embryonal carcinoma F9 cells caused a strong, but not complete defect in GPI-protein production [39]. PIG-P appears to make its strongest associations with PIG-A and GPI1 [15]. The DPM2 subunit (see Section IV.A) may interact with the GPI-GlcNAc transferase complex primarily through PIG-A, PIG-C, and GPI1 [15]. PIG-Y, like PIG-H, appears to associate directly with the PIG-A catalytic subunit [41]. Additionally, its presence is not necessary for the remainder of the complex to assemble. Yeast Gpi2 self-associates in a two-hybrid setting [33], suggesting that it may exist as a dimer within the GPI-GlcNAc transferase complex. These associations are depicted in Figure 2.3. In yeast, GTP-bound Ras2 has been identified as a negative regulatory subunit of the GPI-GlcNAc transferase [16] and is discussed in greater detail in Section IV.B. Preliminary investigations suggest that Ras2 binds to the complex through a subunit other than Eri1 or Gpi1, because deletion of either of these nonessential genes does not impair Ras2 association with the remaining complex (M.J. Romeo and D.E. Levin, unpublished results). However, it seems likely, based on the observation that GTP-Ras associates with the GPI-GlcNAc transferase through its effector loop, that the association involves a single site on a single subunit of the complex.

V.   Regulation of GPI-GlcNAc Transferase Activity

A. REGULATION OF THE MAMMALIAN ENZYME BY DPM2

The mammalian GPI-GlcNAc transferase complex possesses a seventh protein, DPM2, which stimulates GPI-GlcNAc transferase activity, but its presence is not essential for enzyme activity [15]. Although DPM2 was recognized previously as one of the three subunits of Dol-P-Man synthase [42, 43], an enzyme that produces a mannosyl donor for use further down

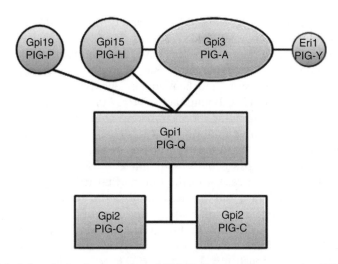

FIG. 2.3. Inferred subunit associations of GPI-GlcNAc transferase complex. This model is based on relative robustness of coprecipitation of mammalian components and yeast Gpi2 self-association by two-hybrid analysis. Lines connecting subunits indicate presumed direct interactions. Proteins with more than two transmembrane domains are indicated by squares, whereas double- and single-membrane pass proteins are indicated as circles and ovals, respectively.

the GPI biosynthetic pathway, the other two subunits were not found in association with the GPI-GlcNAc transferase, suggesting that DPM2 serves a dual function. The absence of DPM2 results in a threefold reduction in GPI-GlcNAc transferase activity, which is restored by transfection with DPM2 [15]. The mechanism by which DPM2 stimulates GPI-GlcNAc transferase activity appears to be through a combination of increased specific activity and increased level of other GPI-GlcNAc transferase subunits. Based on these results, it was suggested that DPM2 is a regulatory subunit for the GPI-GlcNAc transferase and that this enzyme may be coregulated with the Dol-P-Man synthase. However, it is not clear if this represents genuine regulation, or simply reflects the nonobligatory nature of DPM2 relative to other GPI-GlcNAc transferase subunits. No DPM2 ortholog exists in *S. cerevisiae*.

B. REGULATION OF THE YEAST GPI-GLCNAC TRANSFERASE BY RAS

The small G-protein Ras is a negative regulator of the yeast GPI-GlcNAc transferase. This was revealed through phenotypes associated with a null mutant in *ERI1* that overlap with those of hyperactive Ras mutants. The yeast Eri1 protein (*ER*-associated *R*as *I*nhibitor), recognized now as a subunit of the GPI-GlcNAc transferase, was mistakenly viewed initially as a negative regulator of Ras based on two observations. First, an *eri1* null mutant displays hyperactive Ras phenotypes, including filamentous growth, agar invasion, and heat-shock sensitivity [40]. Second, Eri1 associates *in vivo* with GTP-bound (active) Ras2 in a manner dependent on an intact Ras effector-binding loop, consistent with the interpretation that Eri1 acts as a competitive inhibitor of Ras2. However, these observations were equally consistent with Eri1 (or an Eri1-containing complex) serving as a Ras2 effector whose activity is inhibited by Ras2. This was shown, subsequently, to be the case with the finding that Eri1 is a subunit of the GPI-GlcNAc transferase and that Ras2 inhibits the activity of this enzyme over a 200-fold range depending upon the activation state of Ras2 [16]. Membranes from a *ras2Δ* mutant displayed approximately 10-fold greater GPI-GlcNAc transferase activity than wild type, and membranes from a mutant expressing constitutively active (GTP-bound) Ras2 were nearly devoid of activity.

An *eri1Δ* mutant is viable, but is deficient in GPI-GlcNAc transferase activity and like a *gpi1Δ* mutant, displays a temperature-sensitive growth defect [40]. The deficiency in GPI-GlcNAc transferase activity of an *eri1Δ* mutant was proposed to mimic phenotypically the inhibitory effect of Ras2 on the enzyme, a suggestion that was supported by the additional finding that a *gpi1Δ* mutant similarly displays hyperactive Ras phenotypes [16].

Yeast Ras1 and Ras2 control entry of the cell cycle from the quiescent state in response to nutrient signals [44]. Additionally, Ras2 drives a morphological transition on solid medium in response to nutrient limitation from ovoid, budding cells to a multicellular form consisting of filaments of elongated cells. Such filaments, known as pseudohyphae, can extend long distances from the parent colony and are capable of invasive penetration of the agar medium [45–47]. This behavior is thought to be a mechanism by which cells forage for nutrients under nutrient-limiting conditions. Until recently, all Ras functions in yeast were thought to be mediated at the plasma membrane by adenylyl cyclase, its only known effector until the discovery of the GPI-GlcNAc transferase as a Ras2 effector. The finding that Ras regulates the GPI-GlcNAc transferase not only revealed that the ER also serves as a signaling platform for Ras, but it suggested that inhibition of GPI-GlcNAc transferase activity is an important aspect of Ras signaling that acts in parallel with stimulation of adenylyl cyclase to drive filamentous/invasive growth. As noted above, Eri1 is not the GPI-GlcNAc transferase subunit with which Ras2 makes direct contact.

The functional significance of Ras2 regulation of the yeast GPI-GlcNAc transferase is not entirely clear. It is likely related to the hyperactive Ras phenotypes associated with GPI-GlcNAc transferase mutants. The Ras2-driven shift from ovoid yeast cells to elongated, filamentous, pseudohyphal cells probably requires significant changes to cell-wall organization, which may include a decrease in GPI-protein production. However, this has not been confirmed experimentally. The mammalian H-Ras, K-Ras, and R-Ras proteins failed to coprecipitate with the mammalian GPI-GlcNAc transferase, or to regulate the activity of this enzyme [41], suggesting that the GPI-GlcNAc transferase regulatory function of yeast Ras is not conserved in metazoans.

C. Feedback Regulation of UDP-GlcNAc Concentration

The deficiency in GPI biosynthesis that blocks the growth of yeast mutants in *ERI1* and *GPI1* at nonpermissive temperatures can be suppressed with exogenous glucosamine or by overexpression of *GFA1* [16], which catalyzes the first committed and rate-limiting step in the production of UDP-GlcNAc (production of glucosamine-6-phosphate) [48]. These conditions work by increasing the intracellular pool of UDP-GlcNAc, which drives GlcNAc-PI production by mass action through a weakened GPI-GlcNAc transferase complex. More interesting is the finding that a null mutant in *ERI1* responds to its biosynthetic deficiency by increasing *GFA1* expression naturally, suggesting a potential mechanism of normal regulation in response to a need for increased GPI-protein production. It appears that the system senses the deficiency in pathway output, rather than a deficiency in the first biosynthetic step, because a conditional mutant in the *GAA1* gene, which encodes a subunit of the GPI transamidase, also increases *GFA1* expression (R.J. Stamper and D.E. Levin, unpublished results). If wild-type cells increase *GFA1* expression under conditions that require greater pathway output, this would suggest that UDP-GlcNAc concentration and hence, GPI-GlcNAc transferase activity, is normally rate limiting for GPI production.

D. Communication with Later Biosynthetic Steps

Physical association of the Gpi2 GPI-GlcNAc transferase subunit with enzymes that function later in the GPI biosynthetic pathway [33] suggests physical linkage among these biosynthetic enzymes, and perhaps feedback signaling among various steps. Gpi2 associates with the Gwt1 GlcN-PI acyltransferase (Step 3), the Gpi7 ethanolamine transferase (Step 10), and the Gpi8 and Gpi16 subunits of the GPI transamidase complex (Step 11), which catalyzes transfer of the mature anchor to the recipient protein. Gwt1 also associates with Gpi8 [33], suggesting at least a three-way association among the GPI-GlcNAc transferase, the GlcN-PI acyltransferase, and the GPI transamidase. The regulatory consequences, if any, of these associations remains unexplored. It is also possible that a higher order organization of the enzymes in the GPI synthetic pathway may allow substrate channeling for increased biosynthetic efficiency. Such higher order organization has been suggested for the dolichol pathway of protein N-glycosylation [49]. If this is the correct interpretation, because the first and final steps have been linked to each other, it is expected that additional physical interactions interconnecting all of the enzymes for GPI biosynthesis should ultimately be revealed. However, these speculations should be tempered by the fact that the associations are based on results of a large-scale membrane protein interaction screen [33], and the interactions detected in that screen have not been verified in reported coprecipitation experiments.

E. POTENTIAL REGULATORY SIGNIFICANCE OF EXPRESSED,
     PROCESSED PSEUDOGENES

The genes that encode components of the human GPI-GlcNAc transferase complex are atypical in that they are highly represented by processed pseudogenes. These are generated by genomic integration of cDNAs made from mature mRNAs [50]. Hallmarks of a processed pseudogene include the absence of intronic sequences, and polyA tracts at their 3′-ends, reflecting maturation of the template mRNA.

As noted above, human PIG-A resides on the X chromosome, but a processed pseudogene of PIG-A, PIGAP1, resides on chromosome 12 [51]. Interestingly, mRNAs have been isolated for PIGAP1, raising the possibility that this pseudogene may have a regulatory function. Similarly, a processed pseudogene of PIG-C, called PIGCP1, resides on chromosome 11 [52]. As in the case of PIGAP1, there is evidence that PIGCP1 is transcribed. In this case, expressed sequence tags (ESTs) have been isolated that correspond to the pseudogene. A processed pseudogene of PIG-H is also found on chromosome 15. Here again, there is EST-based evidence that the pseudogene is expressed. Five processed PIG-P pseudogenes lay scattered throughout the human genome. Human GPI1 resides on chromosome 16, but a partial pseudogene resides on chromosome 1. However, there is no evidence that this sequence is expressed. Finally, and perhaps most intriguingly, a processed pseudogene of PreY/PIG-Y resides on chromosome 2. The identification of ESTs from this pseudogene suggests that mRNA is produced, but both initiation codons have been mutated, ensuring that they are not translated.

Although processed pseudogenes are common in the human genome, only 2–3% of these are expressed [53]. Moreover, the vast majority of processed pseudogenes arise from retrotransposition of RNAs that encode soluble cytoplasmic proteins because the reverse transcription machinery is cytoplasmic [50]. This finding is consistent with the observation that few processed pseudogenes are derived from other components of the GPI biosynthetic pathway. Notably, among 16 genes within the GPI biosynthetic pathway that do not encode subunits of the GPI-GlcNAc transferase, only PIG-F [54] and GAA-1 appear to be represented by expressed pseudogenes. Together, these facts suggest the possibility of an RNA-mediated regulatory mechanism for the first step in GPI biosynthesis. The human Makorin1-P1 pseudogene is an example of an expressed, processed pseudogene that regulates the stability of its functional homolog [53].

VI. Concluding Remarks

Recent progress in the genetics and biochemistry of the GPI-GlcNAc transferase has provided a physical description of this elaborate enzyme complex, which is conserved from yeast to mammals. However, it remains unclear why a simple glycosyl transfer reaction would require such an elaborate, multisubunit enzyme complex. Indeed, the function of most of the GPI-GlcNAc transferase subunits is an unresolved question. One possible explanation is that this enzyme represents a critical point for regulation of GPI biosynthesis. Individual subunits may thus provide sites for different sources of regulatory input, as is likely the case for regulation of the yeast enzyme by Ras. Another potential reason for the complexity of the enzyme is that it may be at the core of a superstructure that brings together some or all of the other biosynthetic enzymes in the GPI pathway for biosynthetic efficiency. Practical considerations render these various hypotheses difficult to address. However, approaches that examine the consequences of mutational ablation of specific interactions are likely to be informative. Ultimately, much work remains to be directed at uncovering the complexities of this important enzyme.

ACKNOWLEDGMENT

The authors thank Peter Orlean for helpful comments on the manuscript.

REFERENCES

1. Doering, T.L., Masterson, W.J., Englund, P.T., and Hart, G.W. (1989). Biosynthesis of the glycosyl phosphatidylinositol membrane anchor of the trypanosome variant surface glycoprotein. Origin of the non-acetylated glucosamine. *J Biol Chem* 264:11168–11173.
2. Vidugiriene, J., and Menon, A.K. (1993). Early lipid intermediates in glycosyl-phosphatidylinositol anchor assembly are synthesized in the ER and located in the cytoplasmic leaflet of the ER membrane bilayer. *J Cell Biol* 121:987–996.
3. Stevens, V.L., and Raetz, C.R. (1991). Defective glycosyl phosphatidylinositol biosynthesis in extracts of three Thy-1 negative lymphoma cell mutants. *J Biol Chem* 266:10039–10042.
4. Watanabe, R., Inoue, N., Westfall, B., Taron, C., Orlean, P., Takeda, J., and Kinoshita, T. (1998). The first step in glycosylphosphatidylinositol biosynthesis is mediated by a complex of PIG-A, PIG-H, PIG-C and GPI1. *EMBO J* 17:877–885.
5. Houjou, T., Hayakawa, J., Watanabe, R., Tashima, Y., Maeda, Y., Kinashita, T., and Taguchi, R. (2007). Changes in molecular species profiles of glycosylphosphatidylinositol-anchor precursors in early stages of biosynthesis. *Lipid Res* 4:1599–1606.
6. Kinoshita, T., Fujita, M., and Maeda, Y. (2008). Biosynthesis. Remodeling and function of mammalian GPI-achored proteins: recent progress. *Biochemistry* 144:287–294.
7. Miyata, T., Takeda, J., Iida, Y., Yamada, N., Inoue, N., Takahashi, M., Maeda, K., Kitani, T., and Kinoshita, T. (1993). The cloning of PIG-A, a component in the early step of GPI-anchor biosynthesis. *Science* 259:1318–1320.
8. Takeda, J., Miyata, T., Kawagoe, K., Iida, Y., Endo, Y., Fujita, T., Takahashi, M., Kitani, T., and Kinoshita, T. (1993). Deficiency of the GPI anchor caused by a somatic mutation of the PIG-A gene in paroxysmal nocturnal hemoglobinuria. *Cell* 73:703–711.
9. Inoue, N., Murakami, Y., and Kinoshita, T. (2003). Molecular genetics of paroxysmal nocturnal hemoglobinuria. *Int J Hematol* 77:107–112.
10. Brodsky, R.A., and Hu, R. (2006). PIG-A mutations in paroxysmal nocturnal hemoglobinuria and in normal hematopoiesis. *Leuk Lymphoma* 47:1215–1221.
11. Wanachiwanawin, W., Siripanyaphino, U., Piyawattanasakul, N., and Kinoshita, T. (2006). A cohort study of the nature of paroxysmal nocturnal hemoglobinuria clones and PIG-A mutations in patients with aplastic anemia. *Eur J Haematol* 76:502–509.
12. Nozaki, M., Ohishi, K., Yamada, N., Kinoshita, T., Nagy, A., and Takeda, J. (1999). Developmental abnormalities of glycosylphosphatidylinositol-anchor-deficient embryos revealed by Cre/loxp system. *Lab Invest* 79:293–299.
13. Tarutani, M., Itami, S., Okabe, M., Ikawa, M., Yoshikawa, K., and Kinoshita, T. (1997). Tissue-specific knockout of the mouse Pig-A gene reveals important roles for GPI-anchored proteins in skin development. *Proc Natl Acad Sci USA* 94:7400–7405.
14. Orlean, P., and Menon, A.K. (2007). GPI anchoring of protein in yeast and mammalian cells, or: how we learned to stop worrying and love glycophospholipids. *J Lipid Res* 48:993–1011.
15. Watanabe, R., Murakami, Y., Marmor, M.D., Inuoe, N., Maeda, Y., Hino, J., Kangawa, K., Julius, M., and Kinoshita, T. (2000). Initial enzyme for glycosylphosphatidylinositol biosynthesis requires PIG-P and is regulated by DPM2. *EMBO J* 19:4402–4411.
16. Sobering, A.K., Watanabe, R., Romeo, M.J., Yan, B.C., Specht, C.A., Orlean, P., Riezman, H., and Levin, D.E. (2004). Yeast Ras regulates the complex that catalyzes the first step in GPI-anchor biosynthesis as the ER. *Cell* 117:637–648.
17. Coutinho, P.M., Deleury, E., Davies, G.J., and Henrissat, B. (2003). An evolving hierarchical family classification for glycosyltransferases. *J Mol Biol* 238:307–317.

18. Kostova, Z., Yan, B.C., Vainauskas, S., Schwartz, R., Menon, A.K., and Orlean, P. (2003). Comparative importance *in vivo* of conserved glutamates in the EX_7E motif retaining glycosyltransferase Gpi3p, the UDP-GlcNAc-binding subunit of the first enzyme in glycosylphosphatidylinositol assembly. *Eur J Biochem* 270:4507–4514.

19. Kostova, Z., Rancour, D., Menon, A.K., and Orlean, P. (2000). Photoaffinity labeling with P3-(4-azidoanilido) uridine 5′-triphosphate identifies Gpi3p as the UDPGlcNAc-binding subunit of the enzyme that catalyzes formation of *N*-acetylglucosaminyl phosphatidylinositol, the first glycolipid intermediate in glycosyl phosphatidylinositol synthesis. *Biochem J* 350:815–822.

20. Leidich, S.D., Kostova, Z., Latek, R.R., Costello, L.C., Drapp, D.A., Gray, W., Fassler, J.S., and Orlean, P. (1995). Temperature-sensitive yeast GPI anchoring mutants *gpi2* and *gpi3* are defective in the synthesis of *N*-acetylglucosaminyl phosphatidylinositol. *J Biol Chem* 270:13029–13035.

21. Schönbächler, M., Horvath, A., Fassler, J., and Riezman, H. (1995). The yeast *SPT14* gene is homologous to the human PIG-A gene and is required for GPI anchor synthesis. *EMBO J* 14:1637–1645.

22. Fassler, J.S., Gray, W., Lee, J.P., Yu, G., and Gingerich, G. (1991). The *Saccharomyces cerevisiae SPT14* gene is essential for normal expression of the yeast transposon, Ty, as well as for expression of the *HIS4* gene and several genes in the mating pathway. *Mol Gen Genet* 230:310–320.

23. Watanabe, R., Kinoshita, T., Masaki, R., Yamamoto, A., Takeda, J., and Inuoe, N. (1996). PIG-A and PIG-H, which participate in glycosylphosphatidylinositol anchor biosynthesis, form a protein complex in the endoplasmic reticulum. *J Biol Chem* 271:26868–26875.

24. Tiede, A., Nichan, C., Schubert, J., and Schmidt, R.E. (2000). Characterization of the enzymatic complex for the first step in glycosylphoshatidylinositol biosynthesis. I. *J Biochem Cell Biol* 32:339–350.

25. Campbell, R.E., Mosimann, S.C., Tanner, M.E., and Strynadka, N.C. (2000). The structure of UDP-*N*-acetylglucosamine 2-epimerase reveals homology to phosphoglycosyl transferases. *Biochemistry* 39:14993–15001.

26. Eisenhaber, B., Maurer-Stroh, S., Novatchkova, M., Schneider, G., and Eisenhaber, F. (2003). Enzymes and auxiliary factors for GPI lipid anchor biosynthesis and post-translational transfer to proteins. *BioEssays* 25:367–385.

27. Campbell, J.A., Davies, G.J., Bulone, V., and Henrissat, B. (1997). A classification of nucleotide-diphospho-sugar glycosyltransferases based on amino acid sequence similarities. *Biochem J* 326:929–939.

28. Kaptinov, D., and Yu, R.K. (1999). Conserved domains of glycosyltransfersases. *Glycobiology* 9:961–978.

29. Abdian, P.L., Lellouch, A.C., Gautier, C., Ielpi, L., and Geremia, R.A. (2000). Identification of essential amino acids in the bacterial α-mannosyltransferase acea. *J Biol Chem* 275:40568–40575.

30. Cid, E., Gomis, R.R., Geremia, R.A., Guinovart, J.J., and Ferrer, J.C. (2000). Identification of two essential glutamic acid residues in glycogen synthase. *J Biol Chem* 275:33614–33621.

31. Milne, K.G., Ferguson, M.A., and Masterson, W.J. (1992). Inhibition of the GlcNAc transferase of the glycosylphosphatidylinositol anchor biosynthesis in African trypanosomes. *Eur J Biochem* 208:309–314.

32. Inuoe, N., Watanabe, R., Takeda, J., and Kinoshita, T. (1996). PIG-C, one of the three human genes involved in the first step of glycosylphosphatidylinositol biosynthesis is a homolog of *Saccharomyces cerevisiae* GPI2. *Biochem Biophys Res Commun* 226:193–199.

33. Miller, J.P., Lo, R.S., Ben-Hur, A., Desmarais, C., Stagljar, I., Noble, W.S., and Fields, S. (2005). Large-scale identification of yeast integral membrane protein interactions. *Proc Natl Acad Sci USA* 102:12123–12128.

34. Kamitani, T., Chang, H.M., Rollins, C., Waneck, G.L., and Yeh, E.T. (1993). Correction of the class H defect in glycosylphosphatidylinositol anchor biosynthesis in Ltk-cells by a human cDNA clone. *J Biol Chem* 268:20733–20736.

35. Yan, B.C., Westfall, B.A., and Orlean, P. (2001). Ynl038wp (Gpi15p) is the *Saccharomyces cerevisiae* homologue of human Pig-Hp and participates in the first step in glycosylphosphatidylinositol assembly. *Yeast* 18:1383–1389.

36. Newman, H.A., Romeo, M.J., Lewis, S.E., Yan, B.C., Orlean, P., and Levin, D.E. (2005). Gpi19, the *Saccharomycas cerevisiae* homologue of mammalian PIG-P, is a subunit of the intial enzyme for glycosylphosphatidylinositol anchor biosynthesis. *Eukaryot Cell* 4:1801–1807.

37. Leidich, S.D., and Orlean, P. (1996). Gpi1 a *S. Cerevisiae* protein that participates in the first step in GPI anchor synthesis. *J Biol Chem* 271:27829–27837.

38. Collusi, P.A., and Orlean, P. (1997). The essential *Schizosaccharomyces pombe* gpi1 + gene complements a bakers' yeast GPI anchoring mutant and is required for efficient cell separation. *Yeast* 13:139–159.

39. Hong, Y., Ohishi, K., Watanabe, R., Endo, Y., Maeda, Y., and Kinoshita, T. (1999). GPI1 stabilizes an enzyme essential in the first step of glycosylphosphatidylinositol biosynthesis. *J Biol Chem* 274:18582–18588.

40. Sobering, A.K., Romeo, M.R., Vay, H.A., and Levin, D.E. (2003). A novel Ras inhibitor, Eri1, engages yeast Ras at the endoplasmic reticulum. *Mol Cell Biol* 23:4983–4990.

41. Murakami, Y., Siripanyaphinyo, U., Hong, Y., Tashima, Y., Maeda, Y., and Kinoshita, T. (2005). The initial enzyme for glycosylphosphatidylinositol biosynthesis requires PIG-Y, a seventh subunit. *Mol Biol Cell* 16:5236–5246.

42. Maeda, Y., Tomita, S., Watanabe, R., Ohishi, K., and Kinoshita, T. (1998). DPM2 regulates biosynthesis of dolichol phosphate-mannose in mammalian cells: correct subcellular localization and stabilization of DPM1, and binding of dolichol phosphate. *EMBO J* 17:4920–4929.

43. Maeda, Y., Tanaka, S., Hino, J., Kangawa, K., and Kinoshita, T. (2000). Human dolichol-phosphate-mannose synthase consists of three subunits, DPM1, DPM2, and DPM3. *EMBO J* 19:2475–2482.

44. Thevelein, J.M., and de Winde, J.H. (1999). Novel sensing mechanisms and targets for the camp-protein kinase A pathway in the yeast *Saccharomyces cerevisiae*. *Mol Microbiol* 33:904–918.

45. Gimeno, C.J., Ljungdahl, P.O., Styles, C.A., and Fink, J.R. (1992). Unipolar cell divisions in the yeast *S. Cerevisiae* lead to filamentous growth: regulation by starvation and RAS. *Cell* 68:1077–1090.

46. Lorenz, M.C., and Heitman, J. (1997). Yeast pseudohyphal growth is regulated by GPA2, a G protein alpha homolog. *EMBO J* 16:7008–7018.

47. Gancedo, J.M. (2001). Control of pseudohyphae formation in *Saccharomyces cerevisiae*. *FEMS Microbiol Rev* 25:107–123.

48. Orlean, P. (1997). Biogenesis of yeast wall and surface components. In *Molecular and Cellular Biology of the Yeast Saccharomyces cerevisiae*, J.R.Pringle, J.R.Broach and E.W. Jones, (eds.), pp. 229–362. Cold Spring Harbor Laboratory Press, Cold Spring Harbor, NY.

49. Gao, X.D., Nishikawa, A., and Dean, N. (2004). Physical interactions between the Alg1, Alg2, and Alg11 mannosyltransferases of the endoplasmic reticulum. *Glycobiology* 14:559–570.

50. Pavlicek, A., Gentles, A.J., Paces, J., Paces, V., and Jurka, J. (2006). Retroposition of processed pseudogenes: the impact of RNA stability and translational control. *Trends Genet* 22:69–73.

51. Bressler, M., Hillman, P., Longo, L., Luzzatto, L., and Mason, P.J. (1994). Genomic organization of the X-linked gene (PIG-A) that is mutated in paroxysmal nocturnal hemoglobinurea and of a related autosomal paseudogene mapped to 12q21. *Human Mol Genet* 3:751–757.

52. Hong, Y., Ohishi, K., Inoue, N., Endo, Y., Fujita, T., Takeda, J., and Kinoshita, T. (1997). Structures and chromosomal localizations of the glycosylphosphatidylinositol synthesis gene *PIGC* and its pseudogene *PIGCP1*. *Genomics* 44:347–349.

53. Yano, Y., Saito, R., Yoshida, N., Yoshiki, A., Wynshaw-Boris, A., Tomita, M., and Hirotsune, S. (2004). A new role for expressed pseudogenes as ncRNA: regulation of mrna stability of its homologous coding gene. *J Mol Med* 82:414–422.

54. Ohishi, K., Inuoe, N., Endo, Y., Fujita, T., Takeda, J., and Kinoshita, T. (1995). Structure and chromosomal localization of the GPI-anchor synthesis gene PIG-F and its pseudogene (PIG-F). *Genomics* 29:804–807.

3

The GlcNAc-PI de-N-acetylase: Structure, Function, and Activity

MICHAEL D. URBANIAK • MICHAEL A.J. FERGUSON

Division of Biological Chemistry and Drug Discovery
College of Life Sciences
University of Dundee
Dundee DD1 5EH
United Kingdom

I. Abstract

The *N*-acetylglucosamine phosphatidylinositol (GlcNAc-PI) de-*N*-acetylase catalyzes the removal of the *N*-acetyl group from GlcNAc-PI in the second step of GPI biosynthesis. The GlcNAc-PI de-*N*-acetylase is a 252-residue integral membrane protein containing a single N-terminal membrane spanning domain, with the majority of the protein on the cytoplasmic face of the ER. Site-directed mutagenesis studies have lead to the proposal of a zinc-dependent mechanism of action analogous to zinc peptidases. The activity of the GlcNAc-PI de-*N*-acetylase can be measured both *in vivo* and *in vitro*, and active recombinant protein has been obtained. The enzyme is a potential drug target for the treatment of African sleeping sickness, and differences in substrate recognition and channeling between mammalian and trypanosomal enzymes have been exploited to produce species-specific inhibitors.

THE ENZYMES, Vol. XXVI
49
ISSN NO: 1874-6047
DOI: 10.1016/S1874-6047(09)26003-3

II. Introduction

The removal of the N-acetyl group from N-acetylglucosamine phosphatidylinositol (GlcNAc-PI) is the second committed step of the GPI biosynthetic pathway [1], and is a prerequisite for the subsequent mannosylation. The reaction is catalyzed by the enzyme GlcNAc-PI de-N-acetylase (deNAc). Nomenclature for the enzyme differs in mammalian systems, where it is known as PIG-L, whereas in protozoan and yeast it is known as GPI12; herein, we shall use the abbreviation deNAc for simplicity. The deNAc is probably the best characterized enzyme of the GPI biosynthetic pathway, and is a potential drug target for the treatment of African sleeping sickness [2].

III. Protein Structure and Function

A. IDENTIFICATION OF THE GLCNAC-PI DE-N-ACETYLASES

The deNAc gene was discovered by Nakamura *et al.* by functional complementation of a Class L mutant CHO cell line M2S2, which is defective in GPI biosynthesis and expresses the human GPI-anchored cell surface proteins CD55 and CD59 [3]. The M2S2 cells were transfected with a rat cDNA library and analyzed with anti-CD55 and anti-CD59 antibodies to identify cDNA that restored cell surface expression. A clone containing 758 bp open reading frame coding for a 252-residue protein was identified, which the authors named rPIG-L (phosphatidylinositol glycan class L). The protein was found to contain a single predicted N-terminal transmembrane region, with the majority of the protein being present on the cytoplasmic face of the ER.

Identification of homologous proteins in organisms that posses GPI anchor biosynthesis allows alignment of deNAc protein sequences to form a consensus sequence of completely conserved residues. BLAST searching [4] with rPIGL identified 11 sequences from *Anopheles gambiae*, *Cannis familiaris*, *Mus musculus*, *Homo sapiens*, *Cryptococcus neoformans*, *Arabidopsis thaliana*, *Leishmania major*, *Trypanosoma brucei*, *Schizosaccharomyces pombe*, and *Drosophila melanogaster* that could be confidently assigned as being deNAc genes. Multiple sequence alignment [5] then allowed the invariant consensus residues to be identified (Figure 3.1), which presumably must either play a critical functional or structural role in the deNAc.

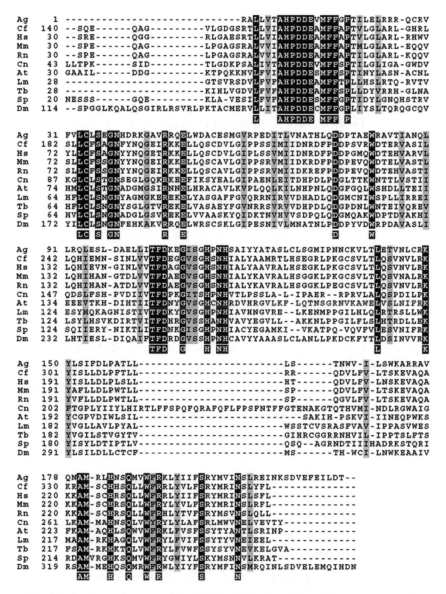

FIG. 3.1. Sequence alignment of the GlcNAc-PI de-N-acetylases. The sequences are from *Anopheles gambiae* (Ag, EAA01152), *Cannis familiaris* (Cf, XP536652), *Homo sapiens* (Hs, NP004269), *Mus musculus* (Mm, CAI24059), *Rattus norvegicus* (Rn, NP_620256) *Cryptococcus neoformans* (Cn, EAL21629), *Arabidopsis thaliana* (At, AAU15173), *Leishmania major* (Lm, AAN60998), *Trypanosoma brucei* (Tb, AAN60997), *Schizosaccharomyces pombe* (Sp, CAC21467), and *Drosophila melanogaster* (Dm, AAM50695), The completely conserved residues that form the consensus sequence are highlighted in black (and shown underneath) and similar residues are highlighted in grey. The N-terminal region of the alignment (not shown) had no significant sequence conservation. Alignment created by T-Coffee [5].

B. Structural Homology

There is currently no high-resolution structural data for any deNAc, but comparison of the deNAc primary sequence with entries in the protein data bank has identified two homologs for which structures exist, namely the *Thermus thermophilus* protein TT1542 (22% identity, 1UAN.pbd) [6] and the *Mycobacterium tuberculosis* MshB (24% identity, 1Q74.pdb and 1Q7T. pdb) [7–9]. Despite the limited homology, each protein contains unusual zinc binding motifs *H*PD*D* and H*xxH* (zinc chelating residues in italics) that appear in the deNAc consensus sequence, as well as several of the other conserved residues (Figure 3.2). The function of TT1542 is unknown, whilst MshB is a GlcNAc(α1–1)-D-*myo*-inositol de-*N*-acetylase involved in the biosynthesis of mycothiol, the major reducing agent found in actinomycetes [10].

C. The GlcNAc-PI De-*N*-acetylases are Zinc Metalloenzymes

To investigate whether the deNAc contains a tightly bound zinc ion that is essential for its catalytic activity, we used the trypanosome cell-free system and recombinant rat deNAc to perform activity assays [11]. Washing

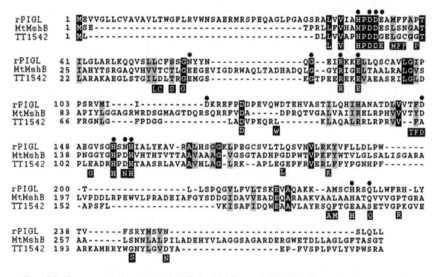

Fig. 3.2. Sequence alignment of PIG-L, MshB, and TT1534. The sequences are *Rattus norvegicus* deNAc (rPIGL), *Mycobacterium tuberculosis* MshB (MtMshB), and *Thermo thermophillus* TT1542 (TT1542); the lower consensus line gives residues totally conserved in known or putative GlcNAc-PI de-*N*-acetylases (see Figure 3.1). Conserved residues are highlighted in black (identical) or grey (similar); the dots indicate residues selected for mutagenesis. Alignment created by T-Coffee [5].

with the divalent metal chelating agent 1,10-phenanthroline to remove tightly bound metals inactivated the deNAc, and reconstitution with divalent metal cations restored activity in the order $Zn^{2+} > Cu^{2+} > Ni^{2+} > Co^{2+} > Mg^{2+}$ [11]. To investigate which residues are involved in zinc binding and catalysis, a homology model of the rat deNAc was constructed using MshB (1Q7T) as a template via the program WHAT IF [12], and used to select 14 conserved residues suitable for site-directed mutagenesis (Figure 3.2). The effect of individually mutating each residue to alanine was assessed using a semi-quantitative functional complementation assay where the M2S2 cells were transfected with GFP tagged rPIG-L mutants; the GFP signal allowed correlation of the level of protein expression with the activity of the mutant (Table 3.1). Overlaying the results of the mutagenesis on the homology model showed a clustering of important residues around the proposed zinc binding site (Figure 3.3). Each of the zinc chelating residues in the unusual zinc binding motifs *H*PD*D* and Hxx*H* (italics) were found to have a strong effect on the catalytic competence of the deNAc, whilst mutation of the proposed general base **HPDD** (bold) completely abolished activity.

TABLE 3.1

ANALYSIS OF RAT GLcNAc-PI DE-*N*-ACETYLASE ALANINE MUTANTS

Mutant	Activity	Proposed role
WT	+	
H49A	−	Zinc coordination
D51A	−*	Catalytic base
D52A	−	Zinc coordination
E53A	±	H-bonds to αNH of D51
N80A	+	
R88A	±	Substrate binding
E91A	+	
D116A	+	
D147A	+	
H154A	+	
N156A	+	
H157A	±	Zinc coordination
H226A	+	
Q229A	+	

*Activity completely abolished.
The activity of the mutants was measured using a semi-quantitative functional complementation assay where the M2S2 cells were transfected with GFP tagged rPIG-L mutants [11]. The effect of the alanine mutations is divided into three classes: − Class, residues essential for activity; ± Class, residues important for activity, + Class, residues with no effect on activity. Adapted from reference [11], © 2005 The American Society for Biochemistry and Molecular Microbiology.

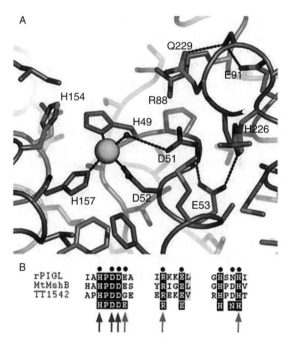

FIG. 3.3. Molecular model of the rat GlcNAc-PI de-*N*-acetylase active site. (A) The model was produced with WHAT IF [12] using MshB (PDB entry 1Q7T [9]) as the template. The active site is shown with conserved residues labelled with residue number. Zinc is represented by a sphere, hydrogen bonds are represented as dashed lines, main chain backbone atoms are represented as a C_α trace. Adapted from reference [11], © 2005 The American Society for Biochemistry and Molecular Microbiology. (B) Sections of the sequence alignment (see Fig. 2) with arrows indicating the essential (dark gray) and important residues (light gray).

D.   POSTULATED MECHANISM OF ACTION

Given the apparent similarity of the active site of the deNAc with that of MshB and zinc metalloproteases, we were able to postulate a putative catalytic mechanism consistent with our experimental observations [11]. In this mechanism (Figure 3.4), the carbonyl oxygen of acetate group of GlcNAc-PI binds directly to the zinc ion, polarizing the carbonyl bond and leaving a partial positive charge on the carbon atom. Activation of a water molecule bound to zinc by the general base D51 facilitates nucleophilic attack at the carbon atom, forming a tetrahedral intermediate stabilized by its coordination to zinc. Subsequent protonation by either D51 or H154 in a

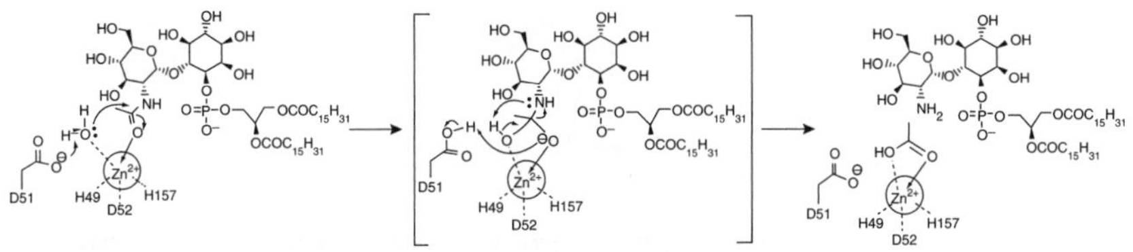

FIG. 3.4. Putative mechanism of action of the GlcNAc-PI de-*N*-acetylase. In this model, the catalytic base D51 assists the nucelophillic attack at the carbon of the carbonyl bond, forming a tetrahedral intermediate; protonation in a non-rate limiting step promotes loss of acetic acid to form GlcN-PI. Residues numbering is based on rat deNAc.

non-rate limiting step allows loss of the product GlcN-PI, with displacement of the zinc-bound acetate by an incoming water molecule completing the catalytic cycle. On the basis of this proposed mechanism of catalysis, we hypothesized that the deNAc will be inhibited by zinc chelating moieties, and have recently synthesized a hydroxymate-containing GlcNAc-PI analog to test this hypothesis [13].

IV. Recombinant Protein Expression

A. OBTAINING SOLUBLE PROTEIN

The deNAc is an integral membrane protein containing a single N-terminal membrane spanning domain, limiting the solubility of the full-length protein in the absence of detergent. Attempts to purify the native deNAc from *Trypanosoma brucei* in the presence of zwittergent 3–14 by Milne *et al.* produced active deNAc, but the enzyme could not be purified to homogeneity [14]. Wantabe *et al.* were able to express a FLAG tagged full-length rat deNAc in *E. coli*, which whilst active was soluble only when co-purified with the molecular chaperone GroEL [15].

Our initial strategy of truncating the N-terminus of the deNAc at various positions to remove the transmembrane domain yielded only insoluble protein whether expressed in *E. coli*, Sf9 insect cells, or *Pichia pastoris*. BLAST searching [4] with rat deNAc revealed a hypothetical *E. coli* protein YAIS with reasonable homology, but lacking the transmembrane domain (Figure 3.5). Whilst the *E. coli* protein is unlikely to be a GlcNAc-PI de-*N*-acetylase as *E. coli* does not produce GPI anchors, we reasoned that the proteins are likely to be structurally similar, and that the start of the

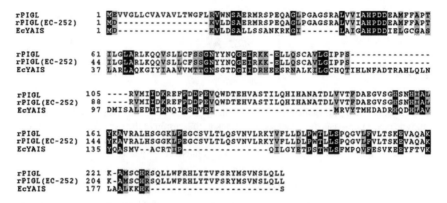

Fig. 3.5. Sequence alignment of rat PIG-L, *E. coli* YAIS, and the chimera rPIG-L(EC-252). The sequences are rat deNAc (rPIGL), *E. coli* YAIS (EcYAIS), and the chimera rPIG-L(EC-252) that consists of residues 1–6 of EcYAIS and 24–252 of rPIG-L. Conserved residues are highlighted in black (identical) or grey (similar); alignment created by T-Coffee [5].

E. coli protein may be used as a suitable replacement for the truncated transmembrane domain. We constructed a chimeric protein rPIGL(EC-252) where the first 25 residues of the rat deNAc, containing the transmembrane domain, were replaced by residues 1–6 of *E. coli* YAIS. The protein was expressed in *E. coli* with a C-terminal hexahistidine tag and purified to homogeneity by metal ion affinity chromatography in the presence of sub-CMC concentrations of the detergent *n*-octylglycopyranoside [11]. Analysis by size exclusion chromatography and analytical ultra centrifugation revealed that the chimeric protein rPIGL(EC-252) formed large, uniform aggregates (~1600 kDa); a similar sized aggregate was observed previously for full-length *T. brucei* deNAc [14]. However, despite its aggregated state, the protein was found to be highly active, demonstrating that the transmembrane domain is not directly required for catalytic activity.

B. TRUNCATIONS AND DOMAIN BOUNDARIES

The MshB and TT1542 proteins both contain second C-terminal domains that are not involved in zinc binding and that have lower levels of sequence homology to the deNAc. To determine whether the deNAc also possesses a C-terminal domain we created a truncated form of rPIGL(EC-252) with 68 residues removed from the C-terminus called rPIGL(EC-184). The truncation enhanced both the yield and the solubility of the protein, which could be purified to homogeneity in the absence of detergent. Although we were able to obtain small protein crystals of rPIGL(EC-184), the crystals did not diffract sufficiently to allow a structure to be determined, and optimization of the crystallization conditions failed to significantly improve the crystal quality. However, the ability to express, purify, and crystallize a significantly truncated deNAc suggests that the truncation is close to a domain boundary.

The purified rPIGL(EC-184) was catalytically inactive, despite all of the essential residues identified by mutagenesis being present. We reasoned that the lack of activity could either be due to the increased solubility of the protein preventing the enzyme from coming into contact with its membrane-associated GlcNAc-PI substrate or due to the loss of other as yet unidentified essential residues. In order to distinguish between these two possibilities we expressed full-length rat deNAc containing the trans-membrane domain (rPIGL(0–252)) in *E. coli*, and were able to measure significant deNAc activity in the whole cell lysates using a mass spectrometry based assay. A construct containing a catalytically inactive form of the full-length rat deNAc where the general base D51 is mutated to alanine (rPIGL(0–252)D51A) was used as a negative control to demonstrate that the observed activity was due the expression of exogenous deNAc. Having established our test system, we proceeded to express and analyze the activity of a series of constructs containing differing degrees of truncation of the C-terminus (Table 3.2). The results show that active protein is obtained only if the truncation does not remove N247, the most C-terminal conserved residue in the deNAc consensus sequence (Figure 3.1). Thus, it appears that although the N-terminal transmembrane

TABLE 3.2

THE ACTIVITY OF TRUNCATED FORMS OF RAT DENAc

Name	N-terminus	C-terminus	Active?
rPIGL(EC-252)	25*	252	Yes
rPIGL(EC-184)	25*	184	No
rPIGL(0–252)	0	252	Yes
rPIGL(0–252)D51A[†]	0	252	No
rPIGL(0–184)	0	184	No
rPIGL(0–197)	0	197	No
rPIGL(0–220)	0	220	No
rPIGL(0–234)	0	234	No
rPIGL(0–246)	0	246	No
rPIGL(0–247)	0	247	Yes

*Contains an additional six-residue at the N-terminus taken from *E. coli* YAIS.
[†]The D51A mutant is catalytically incompetent due to the loss of the general base D51. Truncated rat deNAc were expressed in *E. coli*, and the activity was measured in whole cell lysates using a mass spectrometry based assay.

domain is not required for activity, and the protein possesses a discretely folded domain from residues 25–184, the C-terminal domain is essential for its catalytic competence. However, from the present data we cannot tell whether N247 has a structural or catalytic role, and further experiments are underway to clarify this situation.

V.   Activity Assays

A.   OBTAINING SUBSTRATE

The activity of the deNAc can be measured by a variety of methods that are outlined below (Figure 3.6), and their application and limitations are discussed. The natural substrate for the deNAc, GlcNAc-PI, although not commercially available, is synthetically tractable in sufficient quantity for its use in activity assays to be feasible [16], or may be obtained by a semi-synthetic process [14]. Synthetic GlcNAc-PI usually contains palmitoyl rather than stearoyl acyl groups at positions-1 and -2 of the sn-glycerol moiety, although replacement of the diacylglycerol with the simple C_{18} alkyl moiety contained in GlcNAc-InoPC_{18} does not affect substrate recognition [17, 18]. Semi-synthetic GlcNAc-PI contains myristoyl acyl groups that are usually [^{3}H]-radiolabeled and is obtained by metabolic labeling of trypanosomes with [^{3}H]-myristic acid in the presence of phenylmethane-sulfonyl fluoride to accumulate Man$_3$GlcN-[^{3}H]PI, which can be extracted, digested with α-mannosidase and chemically N-acetylated to give GlcNAc-[^{3}H]PI.

B.   FUNCTIONAL COMPLEMENTATION

The original functional complementation assay is based on the Class L mutant CHO cell line M2S2 developed by Nakamura et $al.$, which is defective in GPI biosynthesis and expresses the GPI-anchored human cell surface proteins CD55 and CD59 [3]. Transient transfection of the M2S2 cells with an active deNAc restores GPI biosynthesis and thus cell surface expression of CD55 and CD59, which can be detected by staining with suitable antibodies and analyzed via microscopy or flow cytometry (Figure 3.6A). Fusing GFP to the deNAc enables semi-quantitative analysis of the level of activity, as the GFP signal is in proportion to the level of expression of the deNAc [11]. The strength of this in $situ$ assay is that no exogenous substrate or protein is required, but the assay is unsuitable to probe substrate specificity or determine the potency of noncell permeable inhibitors.

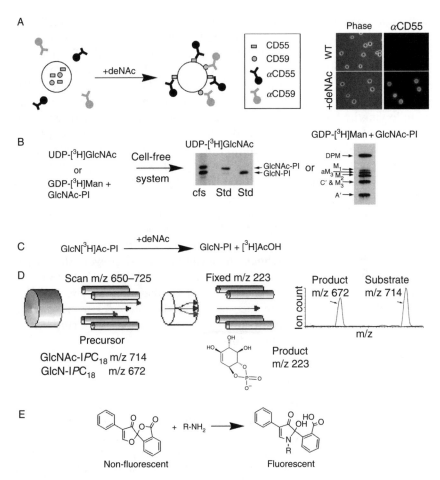

FIG. 3.6. Activity assays for the GlcNAc-PI de-N-acetylase. The assays are explained in detail in the main text. (A) Functional complementation assay; transfection of M2S2 cells with deNAc restores the cell surface expression of CD55 and CD59 as detected by immunofluorescence. (B) The cell-free system is primed with radiolabeled substrate, and the radiolabeled glycolipids extracted and analyzed by HPTLC; (C) The radiolabeled acetate assay follows the release of radiolabeled acetic acid; (D) The mass spectrometry assay uses precursor ion scanning of a common product ion to selectively monitor the analogs GlcNAc-IPC_{18} and GlcN-IPC_{18}; (E) Reaction of fluorescamine with the free amine group of GlcN-PI produces a fluorescent product.

C. THE CELL-FREE SYSTEM

Masterson *et al.* have described a method to prepare mammalian or trypanosome membranes, the so-called cell-free system, which is fully competent in GPI biosynthesis [19]. The cell-free system can be primed with radiolabeled substrates to generate a series of radiolabeled GPI-precursors that can be extracted, separated by high performance thin layer chromatography, and detected by fluorography (Figure 3.6B). The identity of the radiolabeled bands produced can be deduced via a series of chemical and enzymatic digests. Priming with UDP[^{3}H]GlcNAc generates [^{3}H] GlcNAc-PI *in situ* and allows the direct measurement of deNAc activity by following its conversion to [^{3}H]GlcN-PI. An alternative, indirect measurement of the activity of the deNAc in the cell-free system can be obtained by following the generation of radiolabeled mannosylated GPI-precursors that are formed by the processing of GlcN-PI, the product of the deNAc reaction. Priming with GDP-[^{3}H]Man generates Dol-P-[^{3}H]Man *in situ*, and combined with the addition of tunicamycin to inhibit the generation of dolichol cycle intermediates, this allows the specific labeling of GPI precursors. The indirect [^{3}H]Man deNAc assay allows the cell-free system to be primed with exogenous GlcNAc-PI or GlcNAc-PI analogs, and in the trypanosomal system *N*-ethylmaleimide can be used to prevent the formation of endogenous GlcNAc-PI by inhibition of the UDPGlcNAc:PI transferase complex, simplifying interpretation of the system. The cell-free system allows quantitation of the activity of the deNAc, can be used to examine substrate specificity albeit in an indirect manor, and is suitable for determining the potency of inhibitors.

D. RADIOLABELED ACETATE

Synthetic GlcN-PI can be radiolabeled by reaction with [^{3}H]-Ac$_2$O to produce GlcN[^{3}H]Ac-PI where the [^{3}H] label is on the acetate group [20]; [^{14}C]-Ac$_2$O has also been used to similar effect [21]. The activity of the deNAc is then followed by monitoring the release of water soluble [^{3}H] AcOH product of the deNAc reaction from the substrate GlcN[^{3}H]Ac-PI (Figure 3.6C). The radiolabeled product and substrate may be readily separated due to their differing hydrophobicity by a variety of methods such as partitioning between butanol and water or passing over a C$_8$ or C$_{18}$ solid-phase extraction cartridge. This method of measuring deNAc activity is flexible in allowing the use of cell-free system, partially purified or recombinant protein, can be used with GlcN-PI analogs and with a variety of detection methods, but does require access to synthetic GlcN-PI.

E. MASS SPECTROMETRY

An alternative label-free method of following deNAc activity is to use a triple quadrupole mass spectrometry to follow the conversion of substrate to product using a technique known as precursor ion scanning [11]. This method relies on finding a common "product" fragment present in the collision induced mass spectra of both the substrate and product. In precursor ion scanning mode, the mass spectrometer then looks only for ion "precursors" that produce the selected "product" ion, and when the GlcNAc-PI analog GlcNAc-IPC$_{18}$ is used only substrate GlcNAc-IPC$_{18}$ and product GlcN-IPC$_{18}$ are observed. Samples can be easily purified by using a C$_8$ solid-phase extraction cartridge prior to analysis to remove salt and mass spectrometry incompatible components such as detergents. The technique allows quantitative measurement of the reaction, may be used with deNAc obtained from cell-free systems, partially purified or recombinant protein, and is suitable for determining the potency of inhibitors. However, access to a triple quadrupole mass spectrometer and synthetic GlcNAc-IPC$_{18}$ is required, although the technique needs only comparable quantities of substrate to that required for the radiolabeled assays.

F. FLUORESCENCE

A final label-free technique is to detect the production of GlcN-PI using fluorescamine, an amine reactive compound that becomes a strong fluorophore upon reaction with amines but not alcohols. The method can be used with any deNAc substrate that generates a free amine and requires no separation of substrates and product. However, this technique has limited application as it requires purified recombinant deNAc and substrate in amine-free buffers, has limited sensitivity due to the background reaction with the amines of the deNAc lysine residues, and requires significantly more substrate than either the radiolabeled or mass spectrometry assays.

VI. Enzyme Substrate Specificity

The substrate specificity of the deNAc has been extensively examined using synthetic GlcNAc-PI analogs, and is summarized in Figure 3.7. The lipid portion of GlcNAc-PI appears to be important for substrate presentation but is not directly recognized; whilst the fragments GlcNAc and GlcNAc-Inositol phosphate are not recognized alone, analogs where the diacylglycerol is replaced with a C$_{18}$ alkyl chain or steroidal ring systems are substrates [17, 18, 22]. The deNAc is able to remove *N*-acetyl, propionyl or benzoyl groups readily from GlcNR-PI analogs, and even larger groups such as isobutyryl, pentanoyl, and 3-phenylthioureido are removed albeit with lower efficiency [20, 23]. However, the deNAc is unable to act upon the unnatural substrates Man$_1$GlcNAc-PI or Man$_2$GlcNAc-PI [23, 24]

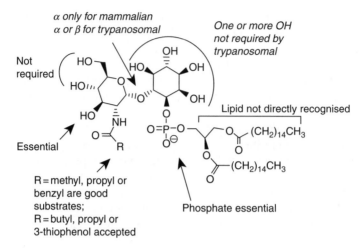

Fɪɢ. 3.7. Substrate recognition by the GlcNAc-PI de-*N*-acetylases. The natural substrate GlcNAc-PI has an α glycosidic linkage and *N*-acetate (*R* = methyl). Recognition features that differ between mammalian and trypanosomal enzymes are given in italics.

confirming that de-*N*-acetylation precedes mannosylation in GPI biosynthesis [1].

The mammalian and trypanosomal deNAc vary in their stringency of substrate recognition [20, 21, 23]; the trypanosomal system is more promiscuous in accepting various stereoisomers and both α and β glycosidic linkages. In addition, the sequence of modifications of the GlcN-PI product differs; in mammalian systems the GlcN-PI undergoes inositol acylation prior to mannosylation, whereas in trypanosomal systems the GlcN-PI is mannosylated prior to inositol acylation [25]. Indeed, the observation that GlcNAc-PI is processed more efficiently to Man_1GlcN-PI than GlcN-PI in the trypanosomal system suggests that substrate challenging occurs between the first mannose transferase and deNAc, perhaps pointing to some degree of physical association [21, 26]. As the trypanosomal deNAc is a potential drug target for the treatment of African sleeping sickness [2], the differences in substrate recognition and substrate channeling have to be exploited to produce species-specific inhibitors [20, 27].

Rᴇꜰᴇʀᴇɴᴄᴇꜱ

1. Doering, T.L., Masterson, W.J., Englund, P.T., and Hart, G.W. (1989). Biosynthesis of the glycerol phosphatidylinositol membrane anchor of the trypanosome variant surface glycoprotein. *J Biol Chem* 264:11168–11173.

2. Chang, T., Milne, K.G., Guther, M.L.S., Smith, T.K., and Ferguson, M.A.J. (2002). Cloning of *Trypanosoma brucei* and *Leishmania major* genes encoding the GlcNAc-phosphatidylinositol de-*N*-acetylase of glycosylphosphatidylinositol biosynthesis that is essential to the African sleeping sickness parasite. *J Biol Chem* 277:50176–50182.

3. Nakamura, N., Inoue, N., Wantanabe, R., Takahashi, M., Takeda, J., Stevens, V.L., and Kinoshita, T. (1997). Expression cloning of PIG-L, a candidate *N*-acetylglucosaminyl-phosphatidylinositol deacetylase. *J Biol Chem* 272:15834–15840.

4. Altschul, S.F., Gish, W., Miller, W., Myers, E.W., and Lipman, D.J. (1990). Basic local alignment search tool. *J Mol Biol* 215:403–410.

5. Notredame, C., Higgins, D., and Heringa, J. (2000). T-Coffee: A novel method for multiple sequence alignment. *J Mol Biol* 302:205–217.

6. Handa, N., Terada, T., Kamewari, Y., Hamana, H., Tame, J.R.H., Park, S.-Y., Kinoshita, K., Ota, M., Nakamura, H., Kuramitsu, S., Shirouzu, M., and Yokoyama, S. (2003). Crystal structure of the conserved protein TT1542 from *Thermus thermophilus* HB8. *Protein Sci* 12:1621–1632.

7. McCarthy, A.A., Knijff, R., Peterson, N.A., and Baker, E.N. (2003). Crystallization and preliminary X-ray analysis of *N*-acetyl-1-D-myo-inositol-2-deoxy-alpha-D-glucopyrano-side deacetylase (MshB) from *Mycobacterium tuberculosis*. *Acta Cryst D* 59:2316–2318.

8. Maynes, J.T., Garen, C., Cherney, M.M., Newton, G., Arad, D., Av-Gay, Y., Fahey, R.C., and James, M.N.G. (2003). The crystal structure of 1-D-myo-inositol 2-acetamido-2-deoxy-alpha-D-glucopyranoside deacaetylase (MshB) from *Mycobacterium tuberculosis* reveals a zinc hydrolase with a lactate dehydrogenase fold. *J Biol Chem* 278:47166–47170.

9. McCarthy, A.A., Paterson, N.A., Knijff, R., and Baker, E.N. (2004). Crystal structure of MshB from *Mycobacterium tuberculosis*, a deacetylase involved in Mycothiol biosynthesis. *J Mol Biol* 335:1131–1141.

10. Newton, G.L., Arnold, K., Price, M.S., Sherrill, C., Delcardayre, S.B., Aharonowitz, Y., Cohen, G., Davies, J., Fahey, R.C., and Davis, C. (1996). Distribution of thiols in micro-organisms: Mycothiol is a major thiol in most actinomycetes. *J Bacteriol* 178:1990–1995.

11. Urbaniak, M.D., Crossman, A., Chang, T., Smith, T.K., Aalten, D.M.F.V., and Ferguson, M.A.J. (2005). The *N*-acetyl-D-glucosaminylphosphatidylinositol de-*N*-acetyl-lase of glycosylphosphatidylinositol biosynthesis is a zinc metalloenzyme. *J Biol Chem* 280:22831–22838.

12. Vriend, G. (1990). WHAT IF: A molecular modelling and drug design program. *J Mol Graph* 8:52–56.

13. Crossman, A., Urbaniak, M.D., and Ferguson, M.A.J. (2008). Synthesis of 1-D-6-*O*-(2-*N*-hydroxyurea-2-deoxy-alpha-D-glucopyranosyl)-myo-inositol 1-(octadecyl phosphate): A potential metalloenzyme inhibitor of glycosylphosphatidylinositol biosynthesis. *Carbohydr Res* 343:1478–1481.

14. Milne, K.G., Field, R.A., Masterson, W.J., Cottaz, S., Brimacombe, J.S., and Ferguson, M. J. (1994). Partial purification and charaterization of the *N*-acetylglucosaminyl-phosphatidylinositol de-*N*-acetylase of glycosylphosphatidylinositol anchor biosynthesis in African trypanosomes. *J Biol Chem* 269:16403–16408.

15. Wantabe, R., Ohishi, K., Maeda, Y., Nakamura, N., and Kinoshita, T. (1999). Mammalian PIG-L and its yeast homologue GPI12p are *N*-acetylglucosaminylphosphatidylinositol de-*N*-acetylases essential in glycosylphosphatidylinositol biosynthesis. *Biochem J* 339:185–192.

16. Cottaz, S., Brimacombe, J.S., and Ferguson, M.A.J. (1993). Parasite glycoconjugates. Part 1. Synthesis of some early and related intermediates in the biosynthetic pathway of glycosyl-phosphatidylinositol membrane anchors. *J Chem Soc Perkin Trans* 1:2945–2951.

17. Crossman, A., Patterson, M.J., Ferguson, M.A.J., Smith, T.K., and Brimacombe, J.S. (2002). Further probing of the substrate specificities and inhibition of enzymes involved at an early stage of glycosylphosphatidylinositol (GPI) biosynthesis. *Carbohydr Res* 337:2049–2059.
18. Smith, T.K., Crossman, A., Paterson, M.J., Borissow, C.N., Brimacombe, J.S., and Ferguson, M.A.J. (2002). Specificities of enzymes of glycosylphosphatidylinositol biosynthesis in *Trypanosoma brucei* and HeLa cells. *J Biol Chem* 277:37147–37153.
19. Masterson, W.J., Doering, T.L., Hart, G.W., and Englund, P.W. (1989). A novel pathway for glycan assembly: biosynthesis of the glycosylphosphatidylinositol anchor of the trypanosome variant surface glycoprotein. *Cell* 62:73–80.
20. Smith, T.K., Crossman, A., Borrissow, C.N., Paterson, M.J., Dix, A., Brimacombe, J.S., and Ferguson, M.A.J. (2001). Specificity of GlcNAc-PI de-*N*-acetylase of GPI biosynthesis and synthesis of parasite specific suicide substrate inhibitors. *EMBO J* 20:3322–3332.
21. Sharma, D.K., Smith, T.K., Weller, C.T., Crossman, A., Brimacombe, J.S., and Ferguson, M.A.J. (1999). Differences between the trypanosomal and human GlcNAc-PI de-*N*-acetylases of glycosylphosphatidylinositol membrane anchor biosynthesis. *Glycobiology* 9:415–422.
22. Urbaniak, M.D., Crossman, A., and Ferguson, M.A.J. (2008). Probing *Trypanosoma brucei* glycosylphosphatidylinostiol biosynthesis using novel precursor-analogues. *Chem Biol Drug Des* 72:127–132.
23. Sharma, D.K., Smith, T.K., Crossman, A., Brimacombe, J.S., and Ferguson, M.A.J. (1997). Substrate specificity of the *N*-acetylglucosaminyl-phosphatidylinositol de-*N*-acetylase of glycosylphosphatidylinositol membrane anchor biosynthesis in African trypanosomes and human cells. *Biochem J* 328:171–177.
24. Urbaniak, M.D., Yashunsky, D.V., Crossman, A., Nikolaev, A.V., and Ferguson, M.A.J. (2008). Probing enzymes late in the trypanosomal glycosylphosphatidylinositol biosynthetic pathway with synthetic glycosylphosphatidylinositol analogues. *ACS Chem Biol* 3:625–634.
25. Guther, M.L.S., and Ferguson, M.A.J. (1995). The role of inositol acylation and inositol deacylation in GPI biosynthesis in *Trypanosoma brucei*. *EMBO J* 14:3080–3093.
26. Smith, T.K., Cottaz, S., Brimacombe, J.S., and Ferguson, M.A.J. (1996). Substrate specificity of the dolichol phosphate mannose: Glucosamine phosphatidylinositol alpha-1–4-mannosyltransferase of the glycosylphosphatidylinositol biosynthetic pathway of African trypanosomes. *J Biol Chem* 271:6476–6482.
27. Smith, T.K., Sharma, D.K., Crossman, A., Brimacombe, J.S., and Ferguson, M.A.J. (1999). Selective inhibitors of the glycosylphosphatidylinositol biosynthetic pathway of *Trypanosoma brucei*. *EMBO J* 18:5922–5930.

4

Inositol Acylation/Deacylation

TAKEHIKO YOKO-O • YOSHIFUMI JIGAMI

Research Center for Medical Glycoscience
National Institute of Advanced Industrial Science and Technology (AIST)
Higashi
Tsukuba 305-8566, Japan

I. Abstract

The inositol moiety of glycosylphosphatidylinositol (GPI) is often modified with a fatty acyl chain, which makes the GPI resistant to phosphatidylinositol-specific phospholipase C. The timing of inositol acylation/deacylation is different among species. In *Trypanosoma brucei*, inositol acylation occurs only after addition of the first mannose. In mammals and yeast, inositol acylation precedes addition of the first mannose, and in many cases, the fatty acyl chain is removed soon after the transfer of GPI to proteins. The inositol acyltransferases PIG-W and Gwt1p have been identified in mammals and yeast, respectively, and their homologues are found in various eukaryotes. Inositol deacylation is a prerequisite for subsequent lipid remodeling of GPI in the bloodstream form of *T. brucei*. GPIdeAc and GPIdeAc2 have been identified as the inositol deacylases of *T. brucei*, and it has been shown that the removal or maintenance of the inositol-linked acyl chain in GPI is critical in the trypanosome life cycle. The mammalian and yeast inositol deacylases are PGAP1 and Bst1p, respectively. Inositol deacylation is important not only for efficient transport of GPI-anchored proteins to the plasma membrane, but also for efficient degradation of GPI-anchored misfolded proteins. PGAP1 knockout mice show otocephaly and male infertility, indicating that inositol deacylation of GPI-anchored proteins is crucial for development and sperm function, although the molecular mechanisms have yet to be elucidated.

ISSN NO: 1874-6047
DOI: 10.1016/S1874-6047(09)26004-5

II.   Introduction: PI-PLC-Resistant GPI Molecules

The susceptibility of glycosylphosphatidylinositol (GPI) structure to bacterial (*Staphylococcus aureus* and *Bacillus thuringiensis*, etc.) phosphatidylinositol-specific phospholipase C (PI-PLC) is commonly used to detect the presence of a GPI anchor [1, 2]. Many proteins have been determined to be linked to the plasma membrane by a GPI anchor on the basis of their conversion from a hydrophobic form to a soluble form by treatment with bacterial PI-PLC. However, a number of GPI-anchored proteins are resistant to cleavage by this enzyme. Examples of this subclass include mouse and human erythrocyte acetylcholinesterase [3, 4], 5'-nucleotidase [5, 6], human CD52-II antigen [7], cell surface proteins of mouse L929 cells [8], procyclic acidic repetitive protein (PARP, also called as procyclin) of *Trypanosoma brucei* [9], and cell adhesion molecules from *Dictyostelium discoideum* [10, 11].

The best-characterized example of this subclass is the mammalian erythrocyte acetylcholinesterase. Using this protein, it has been shown for the first time that the PI-PLC resistance is due to acylation of inositol hydroxyl group(s) [12–16]. Analysis of decay-accelerating factor (DAF) and CD59 on several leukocyte types has similarly shown that the residual 5–10% of molecules that resist PI-PLC cleavage possess GPI anchors containing acylated inositol [17–19]. Another example of a GPI-anchored protein is human alkaline phosphatase (ALP). In a range of cultured human cell lines, the sensitivity of ALP to phospholipases was observed to be variable in magnitude (~20–90%), and the PI-PLC-resistance was the result of acylation of the inositol ring in the GPI anchor [20, 21]. Inositol-acylated GPIs are nevertheless sensitive to serum phosphatidylinositol-specific phospholipase D [12, 22, 23].

Not only GPI-anchored proteins, but also some GPI-anchored lipids are reported to be resistant to PI-PLC. In T cell hybridomas, an ethanolamine-containing GPI precursor has been characterized as PI-PLC-resistant [22], whilst a species with the properties of a glucosamine-(acyl) phosphatidylinositol (GlcN-(acyl)PI) has been identified in dolichol phosphomannose synthase-deficient thymoma Thy-1 expression mutant [24] and yeast *dpm1* mutant [25]. These PI-PLC-resistant precursors have also been identified in Thy-1-negative lymphoma mutants [26]. In *T. brucei*, the GPI precursor glycolipid C (also designated as P3) contains a GPI moiety and is also resistant to PI-PLC [27–29]. The structural basis of the PI-PLC resistance is acylation of inositol hydroxyl group(s).

Mass spectrometric studies of human CD52 [7], *T. brucei* glycolipid C [30], and PARP [31] have revealed that inositol acylation occurs exclusively on the 2-hydroxyl group. Inositol acylation was reported mainly as palmitoylation (C16:0) of the inositol ring in yeast and mammalian cells [2, 12, 13, 32–34], although there is some heterogeneity in the length, with both palmitoyl and myristoyl (C14:0) groups added [35]. In *T. brucei*, biosynthetic studies using labeled fatty acids showed that palmitic acid can be linked to the inositol hydroxyl group(s) in glycolipid C [29]. Other groups have shown that the acyl chain substitution on the inositol ring is considerably variable [30, 31].

III. Variations in Inositol Acylation/Deacylation Among Organisms

A. YEAST AND MAMMALS

The order of synthetic steps for the GPI core structure has been studied in yeast and mammals. Analysis of various mutants revealed that glucosamine-phosphatidylinositol (GlcN-PI) is modified by addition of an acyl group, which makes the lipid resistant to PI-PLC [22–25, 32]. All mannose-containing GPI precursors characterized in *S. cerevisiae* and mammalian cells are inositol-acylated. Therefore, it appears to be GlcN-PI that is acylated *in vivo*. The addition of an acyl chain to the inositol ring of GlcN-PI has been demonstrated in a cell-free system using yeast membranes [32]. In these organisms, the inositol acylation occurs at GlcN-PI to generate GlcN-(acyl)PI, and this strictly precedes the first mannosylation that generates Man-GlcN-(acyl)PI [23, 32, 36–43] (Figure 4.1A). Once added, the inositol-linked acyl chain remains attached until the complete GPI precursor is transferred to the protein in the endoplasmic reticulum (ER). In many mammalian cells and in yeast, the acyl chain is removed from the inositol residue immediately after the transfer of the complete GPI precursor to protein in the ER [44, 45] (Figure 4.1A). For example, DAF became 56% deacylated in the ER within 5 min [45]. However, in some mammalian cells, the acyl chain remains attached. For example, as described previously, the inositol remains acylated in human erythrocytes [12, 13, 15]. Therefore, inositol deacylation is cell- and protein-specific.

Although acylation in mammalian cells is mainly palmitoylation [12, 13, 32, 33], there is some heterogeneity in the length of the fatty acid that can be added. In addition to palmitoylation, myristoylation has also been reported [35]. It has been shown that yeast membranes can utilize acyl-CoAs containing different fatty acid lengths (C14–C20) as exogenous substrates for inositol acylation [32, 39, 46, 47]. These reports suggest that various species of mannosylated GlcN-(acyl)PI bearing different acyl chain lengths can be generated in mammals and yeast.

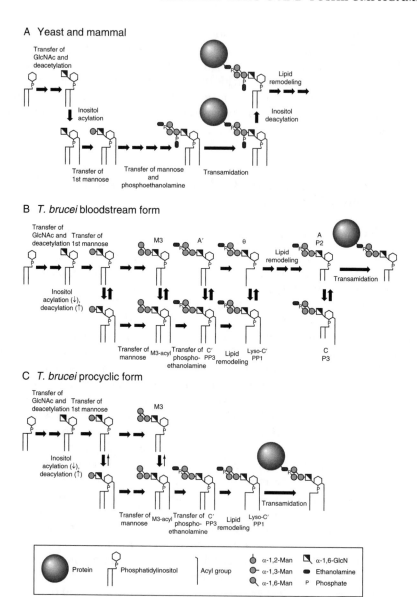

FIG. 4.1. Differences in the sequence of steps involving inositol acylation and deacylation in yeast and mammals (A), bloodstream form (B), and procyclic form (C) of *T. brucei*. For simplicity, only the conserved backbone structures in the glycan portion are depicted in this figure.

The enzymatic activity responsible for inositol acylation has been extensively investigated. In rodents, two types of acyl-transferring activities have been reported to act on GlcN-PI. Endogenous GlcN-PI in mouse and hamster microsomes can be acylated by an activity that requires free CoA and uses an unidentified endogenous acyl donor [38], suggesting that acyl-CoA is not the only donor of the acyl group on inositol. Palmitoyl-CoA-dependent activity has also been reported using a water-soluble dioctanoyl analog of GlcN-PI (GlcN-PI(C8)) in hamster microsomes [39]. Thus, both palmitoyl-CoA-independent and -dependent acyl-transferring activity can be detected in rodent microsomes with GlcN-PI(C8), demonstrating the coexistence of these activities in a single membrane preparation [48]. In yeast, an acyl-CoA-dependent activity has been reported with endogenous GlcN-PI [32].

B. TRYPANOSOMA BRUCEI

In the African trypanosome *T. brucei*, the structures of the GPI anchors in the bloodstream and procyclic forms are different [49, 50]. Variant surface glycoproteins (VSGs), the major coat protein in the bloodstream form of *T. brucei*, exhibit >99% PI-PLC sensitivity [2]. In contrast, cells of the procyclic form (insect stage) of *T. brucei* express an abundant stage-specific GPI-anchored glycoprotein, PARP. This anchor is insensitive to the action of PI-PLC, suggesting that it contains an acylated inositol [9]. Therefore, one prominent difference between the GPI structures of PARP and VSG is that the GPI inositol moiety of PARP is modified by an acyl chain, whereas the GPI inositol in the VSG is not [31, 51–53].

There are fundamental differences between the trypanosomal and mammalian pathways with respect to the timing of inositol acylation and deacylation. The structure of the GPI precursors has been extensively characterized [28, 29, 54]. *T. brucei* inositol acylation only occurs after the addition of the first mannose [39, 40, 53, 55–57] (Figure 4.1B and C). In the bloodstream form of *T. brucei*, inositol deacylation of GPI precursors occurs throughout the trypanosomal pathway (Figure 4.1B), whereas in mammalian cells, it does not occur until the complete GPI precursor is attached to the protein [45, 53, 58]. Therefore, in the bloodstream form of *T. brucei*, the mannosylated precursors can be found in both acylated and non-acylated inositol forms, and precursors from Man-GlcN-(acyl)PI onwards are in dynamic equilibrium between their inositol-acylated and non-acylated forms through the action of inositol acyltransferase(s) and inositol deacylase(s) [53, 59]. Inositol acylation is required for addition of the bridging phosphoethanolamine to the third mannose [40, 53, 56].

There are differences in the GPI biosynthetic pathway between the bloodstream and procyclic forms of *T. brucei*. In the bloodstream form, inositol deacylation appears to be a prerequisite for lipid remodeling of GPI precursors [53]. The phosphatidylinositol (PI) component of glycolipid A′ exclusively contains stearate (C18:0) at the *sn*-1 position and a complex mixture of fatty acids (including C18:0, 18:1, 18:2, 20:4, and 22:6) at *sn*-2 [60]. Glycolipid A′ is converted into glycolipid A (P2) by a highly specific lipid remodeling process, in which myristate replaces the fatty acids originally present on the glycosylated PI [61, 62] (Figure 4.1B). The inositol of glycolipid A can be acylated again to form glycolipid C (P3), and inositol acylation/deacylation maintains a dynamic equilibrium between the GPI precursor, glycolipid A and glycolipid C (Figure 4.1B) [63]. In contrast, in the procyclic form, the inositol remains acylated down to Lyso-C′ (PP1), the final precursor that is attached to proteins (Figure 4.1C). In the procyclic form, inositol deacylation of GPI precursors may not occur extensively, due to low activity of inositol deacylases [64, 65].

Enzymatic characteristics of *T. brucei* acyltransferases and deacylases differ from that of mammals. Although mammalian GPIs that have been examined in detail are mainly modified with palmitate, studies of acylated GPIs in trypanosomes demonstrate a mixture of fatty acids on inositol [30, 31]. In intact *T. brucei* [66] and with *T. brucei* membranes [56] it has been suggested that the acyl donor is not acyl-CoA. Inositol acylation is inhibited by phenylmethylsulfonyl fluoride (PMSF) in *T. brucei* but not in HeLa cells [63]. Deacylation reactions are inhibited by diisopropylfluorophosphate (DFP) [53]. Selective inhibitors of the GPI biosynthetic pathway of *T. brucei* have been reported [67]. The GPI substrate analogs, GlcN-(2-*O*-octyl)PI and GlcNAc-(2-*O*-octyl)PI, inhibit inositol acylation of $Man_{1-3}GlcN$-PI and, consequently, the addition of the ethanolamine phosphate bridge in the *T. brucei* cell-free system, suggesting that these substrate analogs may serve as the first generation of *in vitro* parasite GPI pathway-specific inhibitors [67].

C. OTHER ORGANISMS

1. *Cryptococcus neoformans*

C. neoformans, an opportunistic fungus responsible for a life-threatening infection in immunocompromised patients, is able to synthesize GPI structures. In *C. neoformans*, inositol acylation occurs on GlcN-PI as in *S. cerevisiae*, and the inositol acylation of GlcN-PI in *C. neoformans* membrane involves direct transfer from acyl-CoA [46]. *S. cerevisiae* membranes

utilize acyl-CoA containing various lengths of fatty acid *in vitro*, whereas
C. neoformans membranes have strict substrate specificity requirements for
acyl-CoA fatty acid length [46, 47].

2. Trypanosoma cruzi

The protozoan parasite *T. cruzi* has a complex life cycle, and at all stages
the majority of proteins and glycoconjugates bear a GPI feature also shared
by many pathogenic parasites. *T. brucei* glycolipid C-like GPI precursors
can also be synthesized by a cell-free system derived from *T. cruzi* [68],
indicating that this organism also has the ability to acylate inositol.

3. Plasmodium falciparum

P. falciparum is a virulent human parasite that causes malaria. These
protozoa have been shown to synthesize GPI membrane anchors, which form
their major glycoconjugates at the intraerythrocytic stage of their life cycle.
The biosynthesis of GPIs in *P. falciparum* involves a hydrophobically modified inositol [69]. Gerold and colleagues [70] showed that in *P. falciparum*,
protein-bound GPI anchors contain myristate and palmitate on the inositol
ring, which is consistent with a detailed structural analysis of mature GPI
anchors showing 10% myristate and 90% palmitate acyl chains linked to
inositol at the 2-position [71]. In *P. falciparum* as well as in yeast and mammalian cells, inositol acylation precedes the mannosylation of GPIs, since
mannosylated GPIs were not detected in the absence of acyl-CoA or CoA
[72, 73]. Interestingly, an analysis of GPI precursors formed when parasites
were incubated with glucosamine (GlcN) indicated that GlcN interferes
with the inositol acylation of GlcN-PI to form GlcN-(acyl)PI [74]. Because
the fatty acid acylation of inositol is an obligatory step for the addition of the
first mannosyl residue during the biosynthesis of GPIs, this observation
may offer a target for the development of novel antimalarial drugs [74].

4. Toxoplasma gondii

T. gondii is one of the most widespread parasites. It causes toxoplasmosis, a disease that affects humans and a wide variety of mammals. As with
many other parasitic protozoa, GPI-anchored proteins such as SAG1 dom-
inate the plasma membrane of *T. gondii* tachyzoites, a stage that is associated with the acute phase of infection. A GPI precursor with an acyl
modification on the inositol has been identified, indicating that inositol
acylation also occurs in *T. gondii* [75]. This inositol acylation is acyl-CoA
dependent and takes place before mannosylation, but unique to this class of
inositol acyltransferases, its activity is inhibited by PMSF [76]. The inositol

deacylation of fully mannosylated GPI precursors allows for further processing, that is, addition of a GalNAc side chain to the first mannose, and is
inhibited by both PMSF and DFP [76].

5. Leishmania

GPI glycolipids are major cell surface components of *Leishmania* parasites. However, no inositol-acylated GPI precursors have been observed in
L. major [77] or *L. mexicana* [78], whereas they are abundant in the
trypanosome system. This suggests that inositol acylation is not an essential
biochemical reaction in all organisms.

IV.   Enzymes Involved in Inositol Acylation/Deacylation and Their Functions

A.   INOSITOL ACYLTRANSFERASE

1. Mammalian PIG-W

Analysis of GPI precursors from two mutant CHO cell lines that are
partially defective in their surface expression of GPI-anchored proteins
revealed that they are defective in inositol acylation [79]. Expression of
cDNA for the rat *PIG-W* gene, which was isolated by expression cloning,
restored the surface expression of GPI-anchored proteins on these mutant
cells [79]. *PIG-W* encodes a 504-amino acid multispanning membrane
protein expressed in the ER. PIG-W homologues are found in various
eukaryotes. The alignment of amino acid sequences of PIG-W and their
homologues is shown in Figure 4.2. The N terminus of PIG-W is luminally
oriented and the C terminus is cytoplasmically oriented [79]. Analysis with
the TMpred program [80] suggested the presence of 13 transmembrane
domains in PIG-W. The conserved regions of the PIG-W homologues are
all oriented to the luminal side of the ER membrane [79]. When epitope-
tagged PIG-W was affinity purified from transfected human cells,
palmitoyl-CoA-dependent inositol acyltransferase activity was present,
indicating that PIG-W itself is an inositol acyltransferase [79]. Mutant
cells defective in *PIG-W* produce only very low levels of GPI-anchored
proteins, indicating that inositol acylation is important for the attachment
of GPI to proteins [79].

It was recently reported that the PI species profile was greatly changed
when the precursor was GlcN-(acyl)PI. The PI species profile became very
similar to that of GPI-anchored proteins before lipid remodeling [35].
These GPIs had unsaturated alkyl (or alkenyl)/acyl chains as the major PI

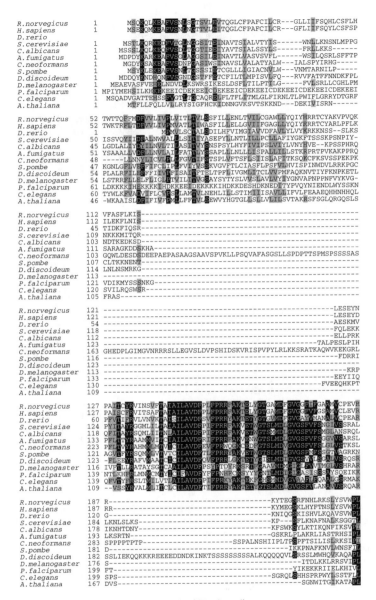

FIG. 4.2. (Continued)

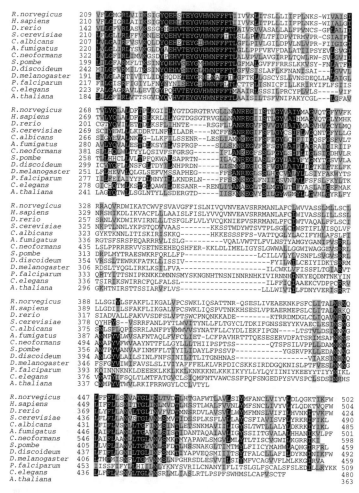

FIG. 4.2. Multiple alignment of inositol acyltransferases from various organisms. The amino acid sequences of acyltransferases were aligned by the CLUSTAL W program [108]. Identical residues are shown as white letters on black boxes, and conservative changes are shaded. The aligned amino acid sequences are as follows: *Rattus norvegicus* (PIG-W; accession number NP_919443), *Homo sapiens* (BAC04413), *Danio rerio* (NP_001108044), *Saccharomyces cerevisiae* (Gwt1p; P47026), *Candida albicans* (XP_712842), *Aspergillus fumigatus* (BAC66175), *Cryptococcus neoformans* (Q873N0), *Schizosaccharomyces pombe* (NP_594671), *Dictyostelium discoideum* (XP_637914), *Drosophila melanogaster* (NP_608679), *Plasmodium falciparum* (CAH17528), *Caenorhabditis elegans* (NP_001022407), and *Arabidopsis thaliana* (NP_193525).

species. Therefore, a specific feature of the PI moieties of mature GPI-anchored proteins, such as alkyl/acyl species being the predominant species relative to diacyl chains, is established at the stage of GlcN-(acyl)PI [35].

PIG-W may preferentially modify GlcN-PIs with alkyl/acyl species, result-ing in GlcN-(acyl)PIs with alkyl/acyl species being the predominant GPI precursors. Other possibilities, such as remodeling of the acyl chain to an alkyl chain at *sn*-1 or substitution of diacyl for alkyl/acyl diglycerol, cannot be ruled out.

2. *S. cerevisiae Gwt1p*

Gwt1p has been identified as the inositol transferase in yeast. In *S. cerevisiae,* the compound, 1-[4-butylbenzyl]isoquinoline (BIQ) inhibits cell-wall localization of GPI-anchored mannoproteins. A novel gene *GWT1* was identified as a multicopy suppressor for the action of BIQ on the growth of *S. cerevisiae* [81]. The *GWT1* gene product (Gwt1p) is the direct target of BIQ, and null mutant cells of *GWT1* (*gwt1* Δ) grow extremely slowly and show a defect in cell-wall assembly [81]. The *GWT1* gene encodes a protein of 490 amino acids that is predicted to have multiple membrane-spanning regions [81]. Umemura *et al.* [47] showed that the *GWT1* gene is involved in inositol acylation and that it is identical to the gene reported to be a homologue of *PIG-W,* which can complement the defect in mammalian *PIG-W* mutant cells [79]. *gwt1* mutant cells have normal transport of invertase and carboxypeptidase Y, but are delayed in the transport of the GPI-anchored protein, Gas1p. These cells are also defective in the maturation of Gas1p from the ER to the Golgi [47]. GPI-anchored proteins are greatly decreased in *gwt1* mutant cells, indicating that acylation is critical for the attachment of GPI to proteins [47]. A *GWT1* homologue in *C. neoformans* was also cloned and it has been confirmed that the specificity of acyl-CoA in inositol acylation, as reported in studies of endogenous membranes [46], is due to the properties of Gwt1p itself [47].

Characterization of the *GWT1* gene revealed the importance of GPI-anchored proteins for the transport of microdomain-associated membrane proteins, such as Tat2p and Fur4p. A *gwt1* mutant allele, *gwt1–10* is sensi-tive to both high and low temperatures [82]. The *gwt1–10* cells show impaired acyltransferase activity and attachment of GPI to proteins even at the permissive temperature. *TAT2*, which encodes a high affinity trypto-phan permease, was identified as a multicopy suppressor of cold sensitivity in *gwt1–10* cells. The *gwt1–10* cells are also defective in the import of tryptophan, causing low temperature sensitivity. Tat2p is not transported to the plasma membrane but is retained in the ER in *gwt1–10* cells grown under tryptophan-poor conditions. Under such conditions, Tat2p in *gwt1–10* cells was not associated with microdomains, which are required for the recruitment of Tat2p to the plasma membrane. A similar result was obtained for Fur4p, a uracil permease localized to the microdomains of the plasma membrane. These results indicate that GPI-anchored proteins are required for the recruitment of Tat2p and Fur4p to the plasma membrane via microdomains, suggesting that some membrane proteins are redistributed in the cell in response to environmental and nutritional conditions due to an association with microdomains that is dependent on GPI-anchored proteins [82].

3. Topology of Inositol Acyltransferase

The two initial reactions that generate GlcN-PI occur on the cytoplasmic side of the ER membrane [83–85], whereas PIG-M, which transfers the first mannose to GlcN-(acyl)PI, functions on the luminal side of the ER [86]. Therefore, flipping occurs before the first mannosylation. It is an important issue to clarify whether the inositol is acylated on the luminal or cytoplasmic side of the ER. Since the comparison of locations of predicted transmembrane domains and amino acid sequence of PIG-W homologues of various organisms showed that the location of conserved regions face the luminal side of the ER, it was proposed that inositol acylation occurs in the ER lumen [79]. If this prediction is correct, GlcN-PI could flip into the ER lumen and inositol acylation would not be required for this flip reaction to occur. Acyl-CoA, a substrate for the inositol acyltransferase, is synthesized in the cytosol or on the cytoplasmic side of organelle membranes [87]; the presence of acyl-CoA on the luminal side of the ER has not been demonstrated. If inositol acylation occurs on the luminal side of the ER, there must be a mechanism for translocation of acyl-CoA across the ER membrane.

Recently, yeast Arv1p was reported to be required for the delivery of GlcN-(acyl)PI to the first mannosyltransferase of GPI synthesis in the ER lumen [88]. The structure of Arv1p is consistent with it being either a GPI flippase or an accessory protein that facilitates the flipping of GPI [88]. If so, inositol acylation must occur on the cytoplasmic side of the ER, because GlcN-(acyl)PI accumulates in the *arv1*Δ mutant, and these cells have a defect in synthesis of Man-GlcN-(acyl)PI. Further analysis of Arv1p, that is, analysis of GPI flipping activity, will be a key in determining whether inositol acylation occurs on the cytoplasmic or luminal side of the ER.

4. Inositol Acyltransferase in T. brucei

As described previously, the characterization of inositol acylation in *T. brucei* has been extensively reported [39, 40, 53, 55, 56, 63]. Nevertheless, the inositol acyltransferase itself has not yet been identified in this organism. No homologue of *PIG-W* or *GWT1* has been found in the

T. brucei genome, which is consistent with the observation that inositol acylation in *T. brucei* occurs on Man-GlcN-(acyl)PI but not on GlcN-(acyl) PI, the substrate for PIG-W and Gwt1p.

B. INOSITOL DEACYLASE

1. Mammalian PGAP1

PGAP1 has been identified as the mammalian inositol deacylase using a GPI inositol-deacylase-deficient CHO cell line that was established by taking advantage of resistance to PI-PLC [89]. *PGAP1* encodes an ER-associated, 922-amino acid membrane protein bearing a lipase consensus motif. Substitution of a conserved putative catalytic serine with alanine resulted in a complete loss of function, indicating that PGAP1 is the GPI inositol deacylase [89]. The mutant cells showed a delay in the maturation of GPI-anchored proteins in the Golgi and an accumulation of GPI-anchored proteins in the ER. Thus, GPI inositol deacylation is important for efficient transport of GPI-anchored proteins from the ER to the Golgi [89]. This defect in PGAP1 function inhibits subsequent lipid remodeling that is mediated by PGAP3 and PGAP2, indicating that PGAP1-mediated deacylation from inositol of GPI-anchored proteins in the ER is required for lipid remodeling [90, 91].

Recently, PGAP1 knockout mice were established [92]. Most PGAP1 knockout mice showed otocephaly, a developmental defect, and died right after birth [92]. Mice that survived the critical 24 h after birth showed growth retardation, even though weaning and other dietary behavior appeared normal. Male knockout mice showed severely reduced fertility despite normal ejaculation and normal spermatozoa quantity and motility [92]. Mammalian fertilization consists of multiple successive events, including sperm ascent from the uterus into the oviductal tube, sperm adhesion to the zona pellucida, formation of a glycoprotein extracellular matrix surrounding the eggs, and membrane fusion of sperm and eggs. To examine which step in fertilization was disrupted in PGAP1 knockout mice, a sperm migration assay was performed [92]. Spermatozoa from PGAP1 knockout mice were able to ascend the uterus, but were unable to migrate into the oviductal tube. *In vitro*, PGAP1-deficient spermatozoa showed weak attachment to the zona pellucida [92]. The inability of spermatozoa to perform these two functions explains the infertility of PGAP1 knockout male mice. Therefore, inositol deacylation of GPI-anchored proteins seems to be crucial for development and sperm function, although the molecular mechanisms have yet to be elucidated.

2. S. cerevisiae Bst1p

In addition to mammalian PGAP1, the yeast orthologue Bst1p was also identified as a GPI inositol deacylase [89]. In a screen for suppressors of mutations in the essential COPII gene *SEC13*, *BST1* was originally identified as a gene that negatively regulates COPII vesicle formation [93]. The *BST1* gene encodes an integral membrane protein that resides predominantly in the ER [93]. *BST1* was also identified as a gene required for the transport of misfolded proteins to the Golgi [94]. A mutant allele of the *BST1* gene called *per17–1* prevents the ER-to-Golgi transport of misfolded proteins while preserving the transport of most normal proteins [94]. Maturation of Gas1p, a GPI-anchored β-1,3-glucanosyltransferase, which normally exits the ER in a class of vesicles distinct from those used by non-GPI proteins [95–97], is reduced as well [94]. These findings suggest a role for Bst1p in sorting GPI-anchored cargo proteins and in ER quality control [93, 94]. Therefore, the inositol deacylation of GPI is important for the efficient transport of GPI-anchored proteins from the ER to the Golgi [89, 94]. *BST1* partially complements PGAP1-deficient cells, indicating that the function of inositol deacylation is conserved among yeast and mammals [89]. Removal of the fatty acid from inositol by Bst1p seems to be a prerequisite for subsequent GPI lipid remodeling in yeast [98].

GPI inositol deacylation also plays important roles in quality control and in ER-associated degradation of GPI-anchored proteins [99, 100]. To elucidate the process by which misfolded GPI-anchored proteins are degraded, a model misfolded GPI-anchored protein was constructed using Gas1p. A mutant of Gas1p, designated Gas1*p (G291R), was modified with a GPI anchor but retained in the ER and degraded via the proteasome pathway. To determine whether *BST1* is involved in the degradation of GPI-anchored proteins, the effect of the deletion of *BST1* in *gas1**-expressing cells was investigated. In *bst1*-deleted cells (*bst1* △), Gas1*p showed a notable stabilization as compared to wild-type cells. The stabilization of Gas1*p in *bst1* △ cells was due to the loss of GPI inositol deacylation activity. The degradation rate of SHg*, a misfolded Gas1p lacking a GPI signal, was the same in wild-type and *bst1* △ cells, indicating that Bst1p affects the exit of only GPI-anchored proteins from the ER. These results suggest that the delay in degradation of Gas1*p in *bst1* △ cells is because of the persistence of the inositol-acylated form of GPI-anchored Gas1*p caused by the defect in Bst1p function, and that the inositol deacylation of the misfolded GPI-anchored protein is required for its efficient degradation in the ER. Gas1*p remains associated with the molecular chaperone BiP/Kar2p, further supporting the notion that Gas1*p is misfolded and that Kar2p is involved in the folding of Gas1*p. Moreover, Gas1*p associates with Bst1p, suggesting that Bst1p associates with misfolded GPI-anchored proteins and that deacylation activity is required for their degradation. The Bst1p function appears to occur at the junction of protein folding in the ER and ER exit of the GPI-anchored proteins. One attractive hypothesis is that Bst1p acts as a gatekeeper that allows GPI-anchored proteins to exit the ER after ascertaining their folding status [99].

Mechanisms for quality control and degradation of GPI-anchored proteins are important in diseases mediated by improper folding of proteins, including prion diseases and transmissible spongiform encephalopathies, both of which are caused by a conformational modification of the GPI-anchored prion protein [101]. In addition, impaired intracellular transport of GPI-anchored mutant tissue-non-specific ALP due to its aggregation causes lethal hypophosphatasia [102, 103]. A mutant tissue-non-specific ALP was found to form a disulfide-bonded high-molecular-mass aggregate and was rapidly degraded within the cell, although the mutant protein was modified by GPI [103]. Further investigation into the mechanisms of quality control and inositol deacylation of GPI-anchored proteins may provide strategies for designing interventions to overcome these diseases.

3.   T. brucei GPIdeAc and GPIdeAc2

In *T. brucei*, two GPI inositol deacylases, GPIdeAc and GPIdeAc2, have been identified. GPIdeAc was purified by affinity labeling with [^{3}H]DFP and the corresponding gene was cloned [64]. *GPIdeAc* encodes a protein with significant sequence and hydropathy similarity to mammalian acyloxyacyl hydrolase, an enzyme that removes fatty acids from bacterial lipopolysaccharide [64] (Figure 4.3).

GPIdeAc seems to be extensively produced only in the bloodstream form and not in the procyclic form, since GPIdeAc that is affinity labeled with [^{3}H]DFP is found only in the bloodstream form [64]. Affinity-purified HA-tagged GPIdeAc was shown to have inositol deacylase activity [64]. *GPIdeAc*-knockout trypanosomes are viable *in vitro* and in animals and synthesize galactosylated forms of the mature GPI precursor glycolipid A at an accelerated rate [64, 104]. These free GPIs accumulate at the cell surface as metabolic end products, indicating that deletion of the *GPIdeAc* gene alters the location and fate of GPI precursors [104]. Null mutants of GPIdeAc are still capable of expressing normal mature GPI anchors on their VSGs. Some GPI inositol deacylase activity remains in the mutants, suggesting that there was an additional unidentified enzyme whose sequence is not closely related to GPIdeAc [64, 104]. In agreement with this idea, a second *T. brucei* GPI inositol deacylase, GPIdeAc2, was identified based on sequence homology to the mammalian GPI inositol

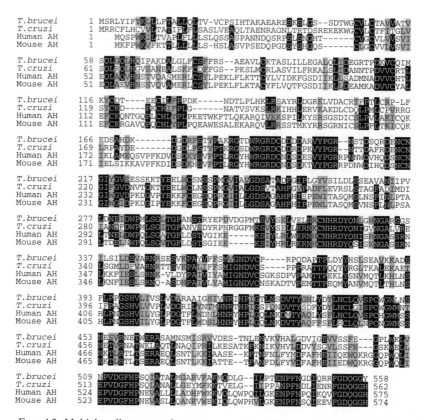

FIG. 4.3. Multiple alignment of trypanosome inositol deacylases (GPIdeAc) and mammalian acyloxyacyl hydrolases. The amino acid sequences of *T. brucei* GPIdeAc (accession number CAC50079), a putative inositol deacylase of *T. cruzi* (XP_809682), and acyloxyacyl hydrolases of human (EAL23977) and mouse (EDL32676) were aligned by CLUSTAL W.

deacylase, PGAP1 [65]. The amino acid sequence alignment of Bst1p, PGAP1, and GPIdeAc2 is shown in Figure 4.4. Consistent with the acylation/deacylation profile, the level of *GPIdeAc2* mRNA is sixfold higher in the bloodstream form than in the procyclic form [65]. It is notable that the difference in inositol deacylase activity between the bloodstream and procyclic forms was predicted in 1992 [52].

Knockdown of *GPIdeAc2* in the bloodstream form causes accumulation of inositol-acylated GPI, decreased VSG expression on the cell surface, and slower growth, indicating that inositol deacylation is essential for the growth of the bloodstream form [65]. Overexpression of *GPIdeAc2* in the

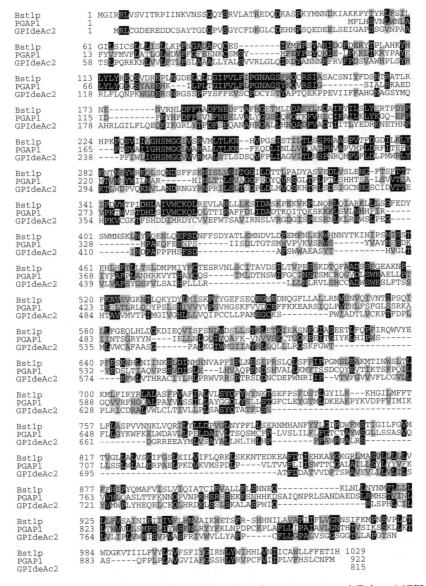

FIG. 4.4. Multiple alignment of inositol deacylases from yeast, rat, and *T. brucei* (GPI-deAc2). The amino acid sequences of *S. cerevisiae* Bst1p (accession number NP_116628), rat PGAP1 (BAD08353), and *T. brucei* GPIdeAc2 (BAE71372) were aligned by CLUSTAL W.

procyclic form causes an accumulation of GPI biosynthetic precursors lacking inositol-linked acyl chains and decreased cell surface PARP, which is released into the culture medium, indicating that overexpression of *GPIdeAc2* is deleterious to the surface coat of the procyclic form [65]. Therefore, regulation of *GPIdeAc2* expression appears to be critical in the life cycle of *T. brucei*.

V.   Perspective and Concluding Remarks

In this chapter, we reviewed the research history and recent progress in inositol acylation and deacylation. An important issue is elucidating the physiological role of inositol acylation/deacylation. As described previously, the timing of inositol acylation is quite strict; the inositol must be acylated prior to the addition of the first mannose residue in mammals and yeast, and prior to the addition of phosphoethanolamine to the third mannose residue in *T. brucei*. It is conceivable that the strictness of the timing of inositol acylation may assure the production of GPI anchor precursors with correct structure. The fatty acyl chain is removed by inositol deacylase. Inositol acylation/deacylation activity is reminiscent of the system of glucosylation/deglucosylation of N-linked oligosaccharides, which is involved in the quality control of glycoproteins. Indeed, Bst1p, the inositol deacylase of *S. cerevisiae*, is closely related to the degradation of misfolded GPI-anchored proteins [99]. In mammals, yeast, and *T. brucei*, inositol deacylation seems to be essential for subsequent lipid remodeling. In mammalian cells, a defect in PGAP1 inhibits lipid remodeling mediated by PGAP3 and PGAP2 [90]. Removal of fatty acid chains from inositol by Bst1p seems to be a prerequisite for subsequent GPI lipid remodeling in yeast [98]. In the bloodstream form of *T. brucei*, inositol deacylation is required for lipid remodeling [53]. Nevertheless, inositol deacylation is not always essential for maturation of GPI-anchored proteins, such as acetyl-cholinesterase in mammalian cells and PARP in *T. brucei* procyclic form. In contrast, it has been reported that an extra acyl chain in GPI anchors causes severe deleterious effects in development and sperm function [92]. These reports indicate that the importance of inositol deacylation depends on individual GPI-anchored proteins. Since inositol deacylation is required for lipid remodeling of GPI-anchored proteins, which is important for the association of GPI-anchored proteins with microdomains [90, 105], inositol deacylation may participate in the microdomain association of individual GPI-anchored proteins. In *T. brucei*, VSG can be detergent-extracted with 1% Triton X-100 at 4 °C [106] but not with 0.35% Triton X-100 at 4 °C [107]. Thus, it is unclear whether VSG is indeed associated with microdomains, since the lipid composition of *T. brucei* microdomains may be somewhat different from that in mammalian and yeast cells, which are usually resistant to 1% Triton X-100 at 4 °C. Continued investigation of the microdomains in *T. brucei* will help elucidate the relationship between inositol deacylation and the microdomain association of GPI-anchored proteins. It is noteworthy that the replacement of unsaturated fatty acids at the *sn*-2 position of the PI moiety with saturated fatty acids by lipid remodeling is conserved among mammals, yeast, and trypanosomes, and that this replacement is important for microdomain association. Therefore, the function of inositol deacylation, which triggers the microdomain association of GPI-anchored proteins, seems to be conserved among all eukaryotes. Future studies will be necessary to thoroughly understand the physiological role of inositol acylation/deacylation.

ACKNOWLEDGMENTS

We are grateful to Kisaburo Nagamune (University of Tsukuba), Morihisa Fujita (Osaka University), and Mariko Umemura (Tokyo University of Pharmacy and Life Sciences) for critical reading of the manuscript and helpful discussions.

REFERENCES

1. Low, M.G., and Saltiel, A.R. (1988). Structural and functional roles of glycosyl-phosphatidylinositol in membranes. *Science* 239:268–275.
2. Ferguson, M.A., and Williams, A.F. (1988). Cell-surface anchoring of proteins via glycosyl-phosphatidylinositol structures. *Annu Rev Biochem* 57:285–320.
3. Low, M.G., and Finean, J.B. (1977). Non-lytic release of acetylcholinesterase from erythrocytes by a phosphatidylinositol-specific phospholipase C. *FEBS Lett* 82:143–146.
4. Futerman, A.H., Low, M.G., Michaelson, D.M., and Silman, I. (1985). Solubilization of membrane-bound acetylcholinesterase by a phosphatidylinositol-specific phospholipase C. *J Neurochem* 45:1487–1494.
5. Low, M.G., and Finean, J.B. (1978). Specific release of plasma membrane enzymes by a phosphatidylinositol-specific phospholipase C. *Biochim Biophys Acta* 508:565–570.
6. Shukla, S.D., Coleman, R., Finean, J.B., and Michell, R.H. (1980). Selective release of plasma-membrane enzymes from rat hepatocytes by a phosphatidylinositol-specific phospholipase C. *Biochem J* 187:277–280.
7. Treumann, A., Lifely, M.R., Schneider, P., and Ferguson, M.A. (1995). Primary structure of CD52. *J Biol Chem* 270:6088–6099.
8. Singh, N., Singleton, D., and Tartakoff, A.M. (1991). Anchoring and degradation of glycolipid-anchored membrane proteins by L929 versus by LM-TK⁻ mouse fibroblasts: implications for anchor biosynthesis. *Mol Cell Biol* 11:2362–2374.

9. Clayton, C.E., and Mowatt, M.R. (1989). The procyclic acidic repetitive proteins of *Trypanosoma brucei*. Purification and post-translational modification. *J Biol Chem* 264:15088–15093.

10. Sadeghi, H., da Silva, A.M., and Klein, C. (1988). Evidence that a glycolipid tail anchors antigen 117 to the plasma membrane of *Dictyostelium discoideum* cells. *Proc Natl Acad Sci USA* 85:5512–5515.

11. Stadler, J., Keenan, T.W., Bauer, G., and Gerisch, G. (1989). The contact site A glycoprotein of *Dictyostelium discoideum* carries a phospholipid anchor of a novel type. *EMBO J* 8:371–377.

12. Roberts, W.L., Myher, J.J., Kuksis, A., Low, M.G., and Rosenberry, T.L. (1988). Lipid analysis of the glycoinositol phospholipid membrane anchor of human erythrocyte acetylcholinesterase. Palmitoylation of inositol results in resistance to phosphatidylinositol-specific phospholipase C. *J Biol Chem* 263:18766–18775.

13. Roberts, W.L., Santikarn, S., Reinhold, V.N., and Rosenberry, T.L. (1988). Structural characterization of the glycoinositol phospholipid membrane anchor of human erythrocyte acetylcholinesterase by fast atom bombardment mass spectrometry. *J Biol Chem* 263:18776–18784.

14. Toutant, J.P., Richards, M.K., Krall, J.A., and Rosenberry, T.L. (1990). Molecular forms of acetylcholinesterase in two sublines of human erythroleukemia K562 cells. Sensitivity or resistance to phosphatidylinositol-specific phospholipase C and biosynthesis. *Eur J Biochem* 187:31–38.

15. Deeg, M.A., Humphrey, D.R., Yang, S.H., Ferguson, T.R., Reinhold, V.N., and Rosenberry, T.L. (1992). Glycan components in the glycoinositol phospholipid anchor of human erythrocyte acetylcholinesterase. Novel fragments produced by trifluoroacetic acid. *J Biol Chem* 267:18573–18580.

16. Richier, P., Arpagaus, M., and Toutant, J.P. (1992). Glycolipid-anchored acetylcholinesterases from rabbit lymphocytes and erythrocytes differ in their sensitivity to phosphatidylinositol-specific phospholipase C. *Biochim Biophys Acta* 1112:83–88.

17. Walter, E.I., Roberts, W.L., Rosenberry, T.L., Ratnoff, W.D., and Medof, M.E. (1990). Structural basis for variations in the sensitivity of human decay accelerating factor to phosphatidylinositol-specific phospholipase C cleavage. *J Immunol* 144:1030–1036.

18. Ratnoff, W.D., Knez, J.J., Prince, G.M., Okada, H., Lachmann, P.J., and Medof, M.E. (1992). Structural properties of the glycoplasmanylinositol anchor phospholipid of the complement membrane attack complex inhibitor CD59. *Clin Exp Immunol* 87:415–421.

19. Rudd, P.M., Morgan, B.P., Wormald, M.R., Harvey, D.J., van den Berg, C.W., Davis, S.J., Ferguson, M.A., and Dwek, R.A. (1997). The glycosylation of the complement regulatory protein, human erythrocyte CD59. *J Biol Chem* 272:7229–7244.

20. Wong, Y.W., and Low, M.G. (1992). Phospholipase resistance of the glycosyl-phosphatidylinositol membrane anchor on human alkaline phosphatase. *Clin Chem* 38:2517–2525.

21. Wong, Y.W., and Low, M.G. (1994). Biosynthesis of glycosylphosphatidylinositol-anchored human placental alkaline phosphatase: evidence for a phospholipase C-sensitive precursor and its post-attachment conversion into a phospholipase C-resistant form. *Biochem J* 301:205–209.

22. DeGasperi, R., Thomas, L.J., Sugiyama, E., Chang, H.M., Beck, P.J., Orlean, P., Albright, C., Waneck, G., Sambrook, J.F., Warren, C.D., and Yeh, E.T.H. (1990). Correction of a defect in mammalian GPI anchor biosynthesis by a transfected yeast gene. *Science* 250:988–991.

23. Urakaze, M., Kamitani, T., DeGasperi, R., Sugiyama, E., Chang, H.M., Warren, C.D., and Yeh, E.T. (1992). Identification of a missing link in glycosylphosphatidylinositol anchor biosynthesis in mammalian cells. *J Biol Chem* 267:6459–6462.

24. Sugiyama, E., DeGasperi, R., Urakaze, M., Chang, H.M., Thomas, L.J., Hyman, R., Warren, C.D., and Yeh, E.T. (1991). Identification of defects in glycosylphosphatidylinositol anchor biosynthesis in the Thy-1 expression mutants. *J Biol Chem* 266:12119–12122.

25. Orlean, P. (1990). Dolichol phosphate mannose synthase is required *in vivo* for glycosyl phosphatidylinositol membrane anchoring, *O* mannosylation, and *N* glycosylation of protein in *Saccharomyces cerevisiae*. *Mol Cell Biol* 10:5796–5805.

26. Lemansky, P., Gupta, D.K., Meyale, S., Tucker, G., and Tartakoff, A.M. (1991). Atypical mannolipids characterize Thy-1-negative lymphoma mutants. *Mol Cell Biol* 11:3879–3885.

27. Krakow, J.L., Doering, T.L., Masterson, W.J., Hart, G.W., and Englund, P.T. (1989). A glycolipid from *Trypanosoma brucei* related to the variant surface glycoprotein membrane anchor. *Mol Biochem Parasitol* 36:263–270.

28. Mayor, S., Menon, A.K., Cross, G.A., Ferguson, M.A., Dwek, R.A., and Rademacher, T.W. (1990). Glycolipid precursors for the membrane anchor of *Trypanosoma brucei* variant surface glycoproteins. I. Can structure of the phosphatidylinositol-specific phospholipase C sensitive and resistant glycolipids. *J Biol Chem* 265:6164–6173.

29. Mayor, S., Menon, A.K., and Cross, G.A. (1990). Glycolipid precursors for the membrane anchor of *Trypanosoma brucei* variant surface glycoproteins. II. Lipid structures of phosphatidylinositol-specific phospholipase C sensitive and resistant glycolipids. *J Biol Chem* 265:6174–6181.

30. Güther, M.L., Treumann, A., and Ferguson, M.A. (1996). Molecular species analysis and quantification of the glycosylphosphatidylinositol intermediate glycolipid C from *Trypanosoma brucei*. *Mol Biochem Parasitol* 77:137–145.

31. Treumann, A., Zitzmann, N., Hülsmeier, A., Prescott, A.R., Almond, A., Sheehan, J., and Ferguson, M.A. (1997). Structural characterisation of two forms of procyclic acidic repetitive protein expressed by procyclic forms of *Trypanosoma brucei*. *J Mol Biol* 269:529–547.

32. Costello, L.C., and Orlean, P. (1992). Inositol acylation of a potential glycosyl phosphoinositol anchor precursor from yeast requires acyl coenzyme A. *J Biol Chem* 267:8599–8603.

33. Sevlever, D., Humphrey, D.R., and Rosenberry, T.L. (1995). Compositional analysis of glucosaminyl(acyl)phosphatidylinositol accumulated in hela S3 cells. *Eur J Biochem* 233:384–394.

34. McConville, M.J., and Ferguson, M.A. (1993). The structure, biosynthesis and function of glycosylated phosphatidylinositols in the parasitic protozoa and higher eukaryotes. *Biochem J* 294:305–324.

35. Houjou, T., Hayakawa, J., Watanabe, R., Tashima, Y., Maeda, Y., Kinoshita, T., and Taguchi, R. (2007). Changes in molecular species profiles of glycosylphosphatidylinositol anchor precursors in early stages of biosynthesis. *J Lipid Res* 48:1599–1606.

36. Hirose, S., Prince, G.M., Sevlever, D., Ravi, L., Rosenberry, T.L., Ueda, E., and Medof, M.E. (1992). Characterization of putative glycoinositol phospholipid anchor precursors in mammalian cells. Localization of phosphoethanolamine. *J Biol Chem* 267:16968–16974.

37. Puoti, A., and Conzelmann, A. (1993). Characterization of abnormal free glycophosphatidylinositols accumulating in mutant lymphoma cells of classes B, E, F, and H. *J Biol Chem* 268:7215–7224.

38. Stevens, V.L., and Zhang, H. (1994). Coenzyme A dependence of glycosylphosphatidylinositol biosynthesis in a mammalian cell-free system. *J Biol Chem* 269:31397–31403.

39. Doerrler, W.T., Ye, J., Falck, J.R., and Lehrman, M.A. (1996). Acylation of glucosaminyl phosphatidylinositol revisited. Palmitoyl-coa dependent palmitoylation of the inositol

residue of a synthetic dioctanoyl glucosaminyl phosphatidylinositol by hamster membranes permits efficient mannosylation of the glucosamine residue. *J Biol Chem* 271:27031–27038.

40. Smith, T.K., Sharma, D.K., Crossman, A., Dix, A., Brimacombe, J.S., and Ferguson, M.A. (1997). Parasite and mammalian GPI biosynthetic pathways can be distinguished using synthetic substrate analogues. *EMBO J* 16:6667–6675.

41. Kinoshita, T., and Inoue, N. (2000). Dissecting and manipulating the pathway for glycosylphos-phatidylinositol-anchor biosynthesis. *Curr Opin Chem Biol* 4:632–638.

42. Pittet, M., and Conzelmann, A. (2007). Biosynthesis and function of GPI proteins in the yeast *Saccharomyces cerevisiae*. *Biochim Biophys Acta* 1771:405–420.

43. Orlean, P., and Menon, A.K. (2007). Thematic review series: lipid posttranslational modifications. GPI anchoring of protein in yeast and mammalian cells, or: how we learned to stop worrying and love glycophospholipids. *J Lipid Res* 48:993–1011.

44. Sipos, G., Puoti, A., and Conzelmann, A. (1994). Glycosylphosphatidylinositol membrane anchors in *Saccharomyces cerevisiae*: absence of ceramides from complete precursor glycolipids. *EMBO J* 13:2789–2796.

45. Chen, R., Walter, E.I., Parker, G., Lapurga, J.P., Millan, J.L., Ikehara, Y., Udenfriend, S., and Medof, M.E. (1998). Mammalian glycophosphatidylinositol anchor transfer to proteins and posttransfer deacylation. *Proc Natl Acad Sci USA* 95:9512–9517.

46. Franzot, S.P., and Doering, T.L. (1999). Inositol acylation of glycosylphosphatidylinositols in the pathogenic fungus *Cryptococcus neoformans* and the model yeast *Saccharomyces cerevisiae*. *Biochem J* 340:25–32.

47. Umemura, M., Okamoto, M., Nakayama, K., Sagane, K., Tsukahara, K., Hata, K., and Jigami, Y. (2003). GWT1 gene is required for inositol acylation of glycosylphosphatidylinositol anchors in yeast. *J Biol Chem* 278:23639–23647.

48. Doerrler, W.T., and Lehrman, M.A. (2000). A water-soluble analogue of glucosaminylphosphatidylinositol distinguishes two activities that palmitoylate inositol on GPI anchors. *Biochem Biophys Res Commun* 267:296–299.

49. Englund, P.T. (1993). The structure and biosynthesis of glycosyl phosphatidylinositol protein anchors. *Annu Rev Biochem* 62:121–138.

50. Mehlert, A., Zitzmann, N., Richardson, J.M., Treumann, A., and Ferguson, M.A. (1998). The glycosylation of the variant surface glycoproteins and procyclic acidic repetitive proteins of *Trypanosoma brucei*. *Mol Biochem Parasitol* 91:145–152.

51. Field, M.C., Menon, A.K., and Cross, G.A. (1991). Developmental variation of glycosylphosphatidylinositol membrane anchors in *Trypanosoma brucei*. Identification of a candidate biosynthetic precursor of the glycosylphosphatidylinositol anchor of the major procyclic stage surface glycoprotein. *J Biol Chem* 266:8392–8400.

52. Field, M.C., Menon, A.K., and Cross, G.A. (1992). Developmental variation of glycosylphosphatidylinositol membrane anchors in *Trypanosoma brucei*. *In vitro* biosynthesis of intermediates in the construction of the GPI anchor of the major procyclic surface glycoprotein. *J Biol Chem* 267:5324–5329.

53. Güther, M.L., and Ferguson, M.A. (1995). The role of inositol acylation and inositol deacylation in GPI biosynthesis in *Trypanosoma brucei*. *EMBO J* 14:3080–3093.

54. Menon, A.K., Mayor, S., Ferguson, M.A., Duszenko, M., and Cross, G.A. (1988). Candidate glycophospholipid precursor for the glycosylphosphatidylinositol membrane anchor of *Trypanosoma brucei* variant surface glycoproteins. *J Biol Chem* 263:1970–1977.

55. Masterson, W.J., Doering, T.L., Hart, G.W., and Englund, P.T. (1989). A novel pathway for glycan assembly: biosynthesis of the glycosyl-phosphatidylinositol anchor of the trypanosome variant surface glycoprotein. *Cell* 56:793–800.

56. Menon, A.K., Schwarz, R.T., Mayor, S., and Cross, G.A. (1990). Cell-free synthesis of glycosyl-phosphatidylinositol precursors for the glycolipid membrane anchor of *Trypanosoma brucei* variant surface glycoproteins. Structural characterization of putative biosynthetic intermediates. *J Biol Chem* 265:9033–9042.

57. Smith, T.K., Cottaz, S., Brimacombe, J.S., and Ferguson, M.A. (1996). Substrate specificity of the dolichol phosphate mannose: glucosaminyl phosphatidylinositol 1–4-mannosyltransferase of the glycosylphosphatidylinositol biosynthetic pathway of African trypanosomes. *J Biol Chem* 271:6476–6482.

58. Ferguson, M.A. (1999). The structure, biosynthesis and functions of glycosylphosphatidylinositol anchors, and the contributions of trypanosome research. *J Cell Sci* 112:2799–2809.

59. Ferguson, M.A., Brimacombe, J.S., Brown, J.R., Crossman, A., Dix, A., Field, R.A., Güther, M.L., Milne, K.G., Sharma, D.K., and Smith, T.K. (1999). The GPI biosynthetic pathway as a therapeutic target for African sleeping sickness. *Biochim Biophys Acta* 1455:327–340.

60. Doering, T.L., Pessin, M.S., Hart, G.W., Raben, D.M., and Englund, P.T. (1994). The fatty acids in unremodelled trypanosome glycosyl-phosphatidylinositols. *Biochem J* 299:741–746.

61. Masterson, W.J., Raper, J., Doering, T.L., Hart, G.W., and Englund, P.T. (1990). Fatty acid remodeling: a novel reaction sequence in the biosynthesis of trypanosome glycosyl phosphatidylinositol membrane anchors. *Cell* 62:73–80.

62. Doering, T.L., Masterson, W.J., Hart, G.W., and Englund, P.T. (1990). Biosynthesis of glycosyl phosphatidylinositol membrane anchors. *J Biol Chem* 265:611–614.

63. Güther, M.L., Masterson, W.J., and Ferguson, M.A. (1994). The effects of phenylmethylsulfonyl fluoride on inositol-acylation and fatty acid remodeling in African trypanosomes. *J Biol Chem* 269:18694–18701.

64. Güther, M.L., Leal, S., Morrice, N.A., Cross, G.A., and Ferguson, M.A. (2001). Purification, cloning and characterization of a GPI inositol deacylase from *Trypanosoma brucei*. *EMBO J* 20:4923–4934.

65. Hong, Y., Nagamune, K., Morita, Y.S., Nakatani, F., Ashida, H., Maeda, Y., and Kinoshita, T. (2006). Removal or maintenance of inositol-linked acyl chain in glycosylphosphatidylinositol is critical in trypanosome life cycle. *J Biol Chem* 281:11595–11602.

66. Field, M.C., Menon, A.K., and Cross, G.A. (1991). A glycosylphosphatidylinositol protein anchor from procyclic stage *Trypanosoma brucei*: lipid structure and biosynthesis. *EMBO J* 10:2731–2739.

67. Smith, T.K., Sharma, D.K., Crossman, A., Brimacombe, J.S., and Ferguson, M.A. (1999). Selective inhibitors of the glycosylphosphatidylinositol biosynthetic pathway of *Trypanosoma brucei*. *EMBO J* 18:5922–5930.

68. Heise, N., Raper, J., Buxbaum, L.U., Peranovich, T.M., and de Almeida, M.L. (1996). Identification of complete precursors for the glycosylphosphatidylinositol protein anchors of *Trypanosoma cruzi*. *J Biol Chem* 271:16877–16887.

69. Gerold, P., Dieckmann-Schuppert, A., and Schwarz, R.T. (1994). Glycosylphosphatidylinositols synthesized by asexual erythrocytic stages of the malarial parasite, *Plasmodium falciparum*. Candidates for plasmodial glycosylphosphatidylinositol membrane anchor precursors and pathogenicity factors. *J Biol Chem* 269:2597–2606.

70. Gerold, P., Schofield, L., Blackman, M.J., Holder, A.A., and Schwarz, R.T. (1996). Structural analysis of the glycosyl-phosphatidylinositol membrane anchor of the merozoite surface proteins-1 and -2 of *Plasmodium falciparum*. *Mol Biochem Parasitol* 75:131–143.

71. Naik, R.S., Branch, O.H., Woods, A.S., Vijaykumar, M., Perkins, D.J., Nahlen, B.L., Lal, A. A., Cotter, R.J., Costello, C.E., Ockenhouse, C.F., Davidson, E.A., and Gowda, D.C. (2000). Glycosylphosphatidylinositol anchors of *Plasmodium falciparum*: molecular characterization and naturally elicited antibody response that may provide immunity to malaria pathogenesis. *J Exp Med* 192:1563–1576.
72. Gerold, P., Jung, N., Azzouz, N., Freiberg, N., Kobe, S., and Schwarz, R.T. (1999). Biosynthesis of glycosylphosphatidylinositols of *Plasmodium falciparum* in a cell-free incubation system: inositol acylation is needed for mannosylation of glycosylphosphatidylinositols. *Biochem J* 344:731–738.
73. Smith, T.K., Gerold, P., Crossman, A., Paterson, M.J., Borissow, C.N., Brimacombe, J.S., Ferguson, M.A., and Schwarz, R.T. (2002). Substrate specificity of the *Plasmodium falciparum* glycosylphosphatidylinositol biosynthetic pathway and inhibition by species-specific suicide substrates. *Biochemistry* 41:12395–12406.
74. Naik, R.S., Krishnegowda, G., and Gowda, D.C. (2003). Glucosamine inhibits inositol acylation of the glycosylphosphatidylinositol anchors in intraerythrocytic *Plasmodium falciparum*. *J Biol Chem* 278:2036–2042.
75. Kimmel, J., Smith, T.K., Azzouz, N., Gerold, P., Seeber, F., Lingelbach, K., Dubremetz, J.F., and Schwarz, R.T. (2006). Membrane topology and transient acylation of *Toxoplasma gondii* glycosylphosphatidylinositols. *Eukaryot Cell* 5:1420–1429.
76. Smith, T.K., Kimmel, J., Azzouz, N., Shams-Eldin, H., and Schwarz, R.T. (2007). The role of inositol acylation and inositol deacylation in the *Toxoplasma gondii* glycosylphosphatidylinositol biosynthetic pathway. *J Biol Chem* 282:32032–32042.
77. Smith, T.K., Milne, F.C., Sharma, D.K., Crossman, A., Brimacombe, J.S., and Ferguson, M.A. (1997). Early steps in glycosylphosphatidylinositol biosynthesis in *Leishmania major*. *Biochem J* 326:393–400.
78. Ralton, J.E., and McConville, M.J. (1998). Delineation of three pathways of glycosylphosphatidylinositol biosynthesis in *Leishmania mexicana*. Precursors from different pathways are assembled on distinct pools of phosphatidylinositol and undergo fatty acid remodeling. *J Biol Chem* 273:4245–4257.
79. Murakami, Y., Siripanyapinyo, U., Hong, Y., Kang, J.Y., Ishihara, S., Nakakuma, H., Maeda, Y., and Kinoshita, T. (2003). PIG-W is critical for inositol acylation but not for flipping of glycosylphosphatidylinositol-anchor. *Mol Biol Cell* 14:4285–4295.
80. Smith, R.F., Wiese, B.A., Wojzynski, M.K., Davison, D.B., and Worley, K.C. (1996). BCM Search Launcher—an integrated interface to molecular biology data base search and analysis services available on the World Wide Web. *Genome Res* 6:454–462.
81. Tsukahara, K., Hata, K., Nakamoto, K., Sagane, K., Watanabe, N.A., Kuromitsu, J., Kai, J., Tsuchiya, M., Ohba, F., Jigami, Y., Yoshimatsu, K., and Nagasu, T. (2003). Medicinal genetics approach towards identifying the molecular target of a novel inhibitor of fungal cell wall assembly. *Mol Microbiol* 48:1029–1042.
82. Okamoto, M., Yoko-o, T., Umemura, M., Nakayama, K., and Jigami, Y. (2006). Glycosylphosphatidylinositol-anchored proteins are required for the transport of detergent-resistant microdomain-associated membrane proteins Tat2p and Fur4p. *J Biol Chem* 281:4013–4023.
83. Vidugiriene, J., and Menon, A.K. (1993). Early lipid intermediates in glycosylphosphatidylinositol anchor assembly are synthesized in the ER and located in the cytoplasmic leaflet of the ER membrane bilayer. *J Cell Biol* 121:987–996.
84. Watanabe, R., Kinoshita, T., Masaki, R., Yamamoto, A., Takeda, J., and Inoue, N. (1996). PIG-A and PIG-H, which participate in glycosylphosphatidylinositol anchor biosynthesis, form a protein complex in the endoplasmic reticulum. *J Biol Chem* 271:26868–26875.

85. Nakamura, N., Inoue, N., Watanabe, R., Takahashi, M., Takeda, J., Stevens, L., and Kinoshita, T. (1997). Expression cloning of PIG-L, a candidate *N*-acetylglucosaminyl-phosphatidylinositol deacetylase. *J Biol Chem* 272:15834–15840.

86. Maeda, Y., Watanabe, R., Harris, C.L., Hong, Y., Ohishi, K., Kinoshita, K., and Kinoshita, T. (2001). PIG-M transfers the first mannose to glycosylphosphatidylinositol on the lumenal side of the ER. *EMBO J* 20:250–261.

87. Black, P.N., and DiRusso, C.C. (2007). Yeast acyl-coa synthetases at the crossroads of fatty acid metabolism and regulation. *Biochim Biophys Acta* 1771:286–298.

88. Kajiwara, K., Watanabe, R., Pichler, H., Ihara, K., Murakami, S., Riezman, H., and Funato, K. (2008). Yeast *ARV1* is required for efficient delivery of an early GPI interme-diate to the first mannosyltransferase during GPI assembly and controls lipid flow from the endoplasmic reticulum. *Mol Biol Cell* 19:2069–2082.

89. Tanaka, S., Maeda, Y., Tashima, Y., and Kinoshita, T. (2004). Inositol deacylation of glycosylphosphatidylinositol-anchored proteins is mediated by mammalian PGAP1 and yeast Bst1p. *J Biol Chem* 279:14256–14263.

90. Maeda, Y., Tashima, Y., Houjou, T., Fujita, M., Yoko-o, T., Jigami, Y., Taguchi, R., and Kinoshita, T. (2007). Fatty acid remodeling of GPI-anchored proteins is required for their raft association. *Mol Biol Cell* 18:1497–1506.

91. Kinoshita, T., Fujita, M., and Maeda, Y. (2008). Biosynthesis, remodelling and functions of mammalian GPI-anchored proteins: recent progress. *J Biochem* 144:287–294.

92. Ueda, Y., Yamaguchi, R., Ikawa, M., Okabe, M., Morii, E., Maeda, Y., and Kinoshita, T. (2007). PGAP1 knock-out mice show otocephaly and male infertility. *J Biol Chem* 282:30373–30380.

93. Elrod-Erickson, M.J., and Kaiser, C.A. (1996). Genes that control the fidelity of endo-plasmic reticulum to Golgi transport identified as suppressors of vesicle budding mutations. *Mol Biol Cell* 7:1043–1058.

94. Vashist, S., Kim, W., Belden, W.J., Spear, E.D., Barlowe, C., and Ng, D.T. (2001). Distinct retrieval and retention mechanisms are required for the quality control of endoplasmic reticulum protein folding. *J Cell Biol* 155:355–368.

95. Muñiz, M., Morsomme, P., and Riezman, H. (2001). Protein sorting upon exit from the endoplasmic reticulum. *Cell* 104:313–320.

96. Morsomme, P., Prescianotto-Baschong, C., and Riezman, H. (2003). The ER v-snares are required for GPI-anchored protein sorting from other secretory proteins upon exit from the ER. *J Cell Biol* 162:403–412.

97. Watanabe, R., and Riezman, H. (2004). Differential ER exit in yeast and mammalian cells. *Curr Opin Cell Biol* 16:350–355.

98. Ghugtyal, V., Vionnet, C., Roubaty, C., and Conzelmann, A. (2007). *CWH43* is required for the introduction of ceramides into GPI anchors in *Saccharomyces cerevisiae*. *Mol Microbiol* 65:1493–1502.

99. Fujita, M., Yoko-o, T., and Jigami, Y. (2006). Inositol deacylation by Bst1p is required for the quality control of glycosylphosphatidylinositol-anchored proteins. *Mol Biol Cell* 17:834–850.

100. Fujita, M., and Jigami, Y. (2008). Lipid remodeling of GPI-anchored proteins and its function. *Biochim Biophys Acta* 1780:410–420.

101. Prusiner, S.B. (1998). Prions. *Proc Natl Acad Sci USA* 95:13363–13383.

102. Shibata, H., Fukushi, M., Igarashi, A., Misumi, Y., Ikehara, Y., Ohashi, Y., and Oda, K. (1998). Defective intracellular transport of tissue-nonspecific alkaline phosphatase with an Ala[162] Thr mutation associated with lethal hypophosphatasia. *J Biochem* 123:968–977.

103. Fukushi, M., Amizuka, N., Hoshi, K., Ozawa, H., Kumagai, H., Omura, S., Misumi, Y., Ikehara, Y., and Oda, K. (1998). Intracellular retention and degradation of tissue-nonspecific alkaline phosphatase with a Gly^{317} Asp substitution associated with lethal hypophosphatasia. *Biochem Biophys Res Commun* 246:613–618.
104. Güther, M.L., Prescott, A.R., and Ferguson, M.A. (2003). Deletion of the *gpideac* gene alters the location and fate of glycosylphosphatidylinositol precursors in *Trypanosoma brucei. Biochemistry* 42:14532–14540.
105. Fujita, M., Umemura, M., Yoko-o, T., and Jigami, Y. (2006). *PER1* is required for GPI-phospholipase A$_2$ activity and involved in lipid remodeling of GPI-anchored proteins. *Mol Biol Cell* 17:5253–5264.
106. Benting, J., Rietveld, A., Ansorge, I., and Simons, K. (1999). Acyl and alkyl chain length of GPI-anchors is critical for raft association *in vitro. FEBS Lett* 462:47–50.
107. Denny, P.W., Field, M.C., and Smith, D.F. (2001). GPI-anchored proteins and glycoconjugates segregate into lipid rafts in Kinetoplastida. *FEBS Lett* 491:148–153.
108. Thompson, J.D., Higgins, D.G., and Gibson, T.J. (1994). CLUSTAL W: improving the sensitivity of progressive multiple sequence alignment through sequence weighting, position-specific gap penalties and weight matrix choice. *Nucleic Acids Res* 22:4673–4680.

5

Mannosylation

YUSUKE MAEDA • YASU S. MORITA

Department of Immunoregulation
Research Institute for Microbial Diseases and
Department of Immunoglycobiology
WPI Immunology Frontier Research Center
Osaka University
Suita, Osaka 565-0871, Japan

I. Abstract

The backbone structure of GPI is EtNP-6Manα1–2Manα1–6Manα1–4GlcNα1–6*myo*-inositol-phospholipid (where EtNP is ethanolamine phosphate, Man is mannose, and GlcN is glucosamine), which is highly conserved among eukaryotes such as mammals, plants, yeast, and protozoan parasites. In general, the addition of three mannoses (Man-1, Man-2, and Man-3 in the order of their attachment to the GPI intermediates) is indispensable for the biosynthesis of GPI and is essential for the attachment of mature GPI to proteins. In yeast and some mammalian GPI-anchored proteins (GPI-APs), the fourth Man (Man-4) is linked via an α1–2 linkage to Man-3. This fourth mannose is essential in yeast. Each mannose is transferred by GPI-Man-TI, -TII, -TIII, and -TIV, respectively. In addition, in mycobacteria polyprenol-phosphate-mannose-dependent mannosyltransferases which are homologous to PIG-M (GPI-Man-TI) are involved in the biosynthesis of glycolipids structurally similar to GPIs known as phosphatidylinositol mannosides (PIMs) and lipomannan/lipoarabinomannan (LM/LAM). This chapter focuses on the mannosylation of GPI and introduces mycobacterial polyprenol-phosphate-mannose-dependent mannosyltransferases to illustrate the evolutionary conservation of mannosyltransferases that are involved in the biosynthesis of "glycosylated" phosphatidylinositols.

II.   Overview of Biosynthetic Pathway for GPI Mannosylation

Three or four mannoses are sequentially transferred to GPI intermediates from dolichol-phosphate-mannose (Dol-P-Man) in the luminal side of the endoplasmic reticulum (ER). Man-1 and Man-2 are sequentially transferred to GlcN-(acyl)PI (PI: *myo*-inositol-phospholipid or phosphatidylinositol) to generate Man-Man-GlcN-(acyl)PI. Man-1 is then modified with an EtNP side branch by mammalian PIG-N/yeast Mcd4p, GPI-EtNP-TIs [1, 2]. Although defects in GPI-Man-TII result in accumulation of (EtNP) Man-GlcN-(acyl)PI, it seems likely that the actual order of biosynthesis is Man-Man-GlcN-(acyl)PI followed by Man-(EtNP)Man-GlcN-(acyl)PI in mammalian cells (see Section IV.B.). The third mannose is added to Man-(EtNP)Man-GlcN-(acyl)PI by GPI-Man-TIII. Because Man-(EtNP)Man-GlcN-(acyl)PI is accumulated in mutants of mammalian PIG-B, a GPI-Man-TIII, and its yeast homolog GPI10 [3, 4], and because Man-Man-GlcN-(acyl)PI is accumulated in both PIG-N and yeast Mcd4p mutant cells [1, 2], Man-(EtNP)Man-GlcN-(acyl)PI is a substrate for GPI-Man-TIII. The addition of Man-4 is catalyzed by SMP3/PIG-Z. In yeast, this step is essential for the subsequent attachment of EtNP to Man-3, but it is not required in mammalian cells. In SMP3 mutant yeast, the so-called "bridging" EtNP that connects GPI with proteins via an amide bond is not added to Man-3, which results in the lack of mature GPI. In transamidase-deficient mammalian cells, (EtNP)Man-Man-(EtNP)Man-GlcN-(acyl)PI is predominantly accumulated and only a small portion of the accumulated GPIs have Man-4.

III.   Dol-P-Man as a Substrate for GPI-Man-Ts

Dol-P-Man is considered to be the sole mannose donor in the lumen of the ER because of the absence of a transporter of GDP-mannose, a common donor in the cytosolic face of the ER. Dol-P-Man is utilized in mannosylation of *N*-glycans, protein *O*- and *C*-mannosylation as well as GPI [5, 6]. Dolichol-phosphate-glucose (Dol-P-Glc), another dolichol-phosphate-sugar, is a glucose donor for the biosynthesis of *N*-glycans [5]. Defective Dol-P-Man biosynthesis/usage causes a lack of mature GPI, which in turn causes degradation or abnormal processing/secretion of precursor proteins that would normally be modified with GPI and expressed on the cell surface [7, 8]. Dol-P-Man is synthesized from GDP-Man and Dol-P on the cytosolic side of the ER by DPM1, a Dol-P-Man synthase (GDP-Man:Dol-P mannosyltransferase (EC 2.4.1.83)). DPM1s can be divided into two classes based on their protein structures [9, 10]. The first class has a hydrophobic region at the carboxyl-terminus and includes *Saccharomyces cerevisiae*, *Ustilago maydis*, *Trypanosoma brucei*, and *Leishmania mexicana* DPM1s. The second class lacks a hydrophobic stretch and includes *Schizosaccharomyces pombe*, *Caenorhabditis briggsiae*, *Trichoderma reesei*,and mammalian (human and mouse) DPM1s. The second class appears to be soluble. DPM1s that belong to the first class with a hydrophobic tail seem to be sufficient for Dol-P-Man synthase activity by themselves, because the DPM1 of *S. cerevisiae* complemented human mutant cells and lethal DPM1-null *S. pombe* mutant cells, which belong to the second class [10] and because the DPM1s of *S. cerevisiae*, *U. maydis*, and *T. brucei* were all functional when expressed in *Escherichia coli* [10–13]. In contrast, neither human nor *S. pombe* DPM1 restored the viability of *S. cerevisiae dpm1* mutant cells, and recombinant human DPM1 expressed in *E. coli* did not show DPM synthase activity [10]. Moreover, two mammalian mutant cells that were defective in Dol-P-Man synthase activity were classified into different complementation groups by somatic cell hybridization analysis [14]. These findings indicated that the second class of DPM1s require additional components to exhibit their Dol-P-Man synthase activity, such as components that tether DPM1 to the membrane for efficient interaction with Dol-P, a substrate buried in the ER membrane. In fact, two components, DPM2 and DPM3, have been identified as factors in mammalian Dol-P-Man synthase [15–17]. Thus, the mammalian DPM synthase complex is composed of three components, which seems to be sufficient to confer activity. DPM1, DPM2, and DPM3 are conserved in *S. pombe* and probably in *Aspergillus*, *Neosartorya*, *Dictyostelium*, *Dario*, *Xenopus*, and *Arabidopsis* according to the genome database. In these organisms, the DPM1s lack the hydrophobic carboxy-terminal domain. The synthesized Dol-P-Man on the cytosolic side of the ER is subsequently translocated to the luminal side, presumably by a putative flippase machinery [18–20], and utilized as a donor for mannosylation in four protein glycosylation pathways occurring in the lumen of the ER [5]. MPDU1/SL15 was identified as a gene responsible for Lec35 mutant cells, which exhibit defective Dol-P-Man utilization, and was thought to be a candidate for the flippase, but biochemical analyses using mannosylphosphorylcitronellol suggested the existence of a flippase other than MPDU1/SL15 [18].

IV. GPI-Mannosyltransferases (GPI-MAN-Ts)

A. GPI-MAN-TI

GPI-Man-TI, mammalian PIG-M/yeast GPI14p, catalyzes the transfer of Man-1 from Dol-P-Man to GlcN of GlcN-(acyl)PI via α1,4 linkage, which results in Manα1–4GlcNAc-(acyl)PI. They belong to glycosyltransferase family 50 (GT50) in the CAZy classification and share significant homology in terms of primary and secondary structure with ALG3 (Dol-P-Man:Man$_5$GlcNAc$_2$-PP-Dol α1–3 mannosyltransferase), which is classified into the GT58 family and utilizes Dol-P-Man as a substrate for the N-glycosylation pathway (Figure 5.2). PIG-M was identified as a gene responsible for GPI-AP defects in human Burkitt lymphoma mutant (Ramos) cells [21]. The PIG-M protein is predicted to have around 10 transmembrane domains (TMDs) [21–23] and has a large loop portion in the luminal side of the ER between the first and second TMDs, where the catalytic sites are believed to be located [23, 24]. The DXD sequence is conserved in the loop in PIG-Ms from various organisms (Figure 5.2) and is considered to be functionally important, because mutagenesis of either one of the two Ds (aspartic acid) abolished or impaired its ability to restore the defective phenotype of PIG-M mutant cells [21]. It remains to be determined whether the DXD motif serves a ligand-binding function.

The unique characteristic of GPI-Man-TI is that this enzyme is composed of two proteins. Many glycosyltransferases that are involved in other pathways such as N- and O-glycans are thought to be able to function by themselves. As a unique characteristic of GPI biosynthesis pathway, many enzymes catalyzing the addition of GlcNAc, mannose and EtNP, make complexes that are composed of more than two proteins.

The second component, PIG-X, was identified as a gene responsible for GPI-APs defects in Chinese hamster ovary (CHO) mutant cells and encodes an ER-resident type I transmembrane protein of 252 amino acids (rat PIG-X) with a large luminal portion [25]. PIG-X utilizes an atypical start codon CTG instead of ATG. Putative homologs are found in *Arabidopsis thaliana*, *Anopheles gambiae*, *Caenorhabditis elegans*, and *S. pombe*, although the similarities are not high (<20% identity within the overlapping region). Of these, *S. cerevisiae* Pbn1p, which has been reported to be involved in autoprocessing of proproteinase B in the ER, was confirmed to be a functional homolog of PIG-X [25]. Although either yeast Pbn1p or Gpi14p alone did not restore function in mammalian PIG-X- or PIG-M-deficient mutant cells, co-expression of Pbn1p and Gpi14p in these mutant cells did restore the surface expression of GPI-APs. This indicates that the combination of PIG-M and PIG-X in mammals and of Gpi14p and Pbn1p in yeast are sufficient for GPI-Man-TI activity, although the components in each combination are not interchangeable, most likely due to impaired ability to form a complex. This speculation is supported by the fact that PIG-X associates with and stabilizes PIG-M [25]. Nevertheless, analysis of Pbn1p-deficient yeast strains showed that Pbn1p is involved in the processing of diverse proteins other than GPI-APs and the attachment of GPI was not apparently affected in Gas1p. The precise functions of Pbn1p remain to be fully elucidated [26].

B. GPI-MAN-TII

GPI-Man-TII, mammalian PIG-V/yeast Gpi18p, catalyzes the addition of Man-2 from Dol-P-Man to Man-1 of Man-GlcN-(acyl)PI and (EtNP)Man-GlcN-(acyl)PI via α1,6 linkage, resulting in the formation of Manα1–6Man-GlcN-(acyl)PI and Manα1–6(EtNP)Man-GlcN-(acyl)PI, respectively. PIG-V was identified as a gene responsible for GPI-APs defects in CHO cells [27], and yeast Gpi18p was identified based on a bioinformatics-based strategy [28]. PIG-V-defective mammalian cells and gpi18-defective yeast accumulate (EtNP)Man-GlcN-(acyl)PI, whereas defective Mcd4p causes accumulation of Man-Man-GlcN-(acyl)PI [2], indicating that GPI-Man-TII may use Man-GlcN-(acyl)PI, irrespective of the addition of EtNP, as the substrate. However, mammalian PIG-V mutant cells predominantly accumulated Man-GlcN-(acyl)PI rather than (EtNP)Man-GlcN-(acyl)PI, and transient expression of PIG-V in GPI-transamidase (PIG-U)-deficient cells decreased the amount of Man-GlcN-(acyl)PI and increased the equivalent amount of Man-Man-GlcN-(acyl)PI [27]. Therefore, Man-GlcN-(acyl)PI could be first mannosylated to Man-Man-GlcN-(acyl)PI followed by the addition of EtNP, which results in Man-(EtNP)Man-GlcN-(acyl)PI in mammalian cells.

GPI-Man-TIIs belong to GT76 in the CAZy classification and do not show obvious resemblance to members of the established glycosyltransferase family in the CAZy database, including GPI-Man-TI, -TIII, and -TIV (Figure 5.3). Consistent with this, yeast Gpi18p could not complement lethal null mutations of the *GPI14*, *GPI10*, or *SMP3* genes encoding GPI-Man-TI, -TIII, and -TIV, respectively [28]. GPI-Man-TIIs are reported to have around eight TMD. Several amino acids facing the luminal side of the ER are well conserved among PIG-V orthologs from various species, and were found to be functionally important by site-directed mutagenesis ([27], see below). Further analysis is required to elucidate the function of these regions. Expression of human PIG-V restored the viability of the yeast *gpi18* deletion mutant, indicating that the PIG-V/Gpi18p is most likely to be the enzyme itself. Although human PIG-V restored viability, the yeast *gpi18* deletion mutant transformed with PIG-V grew more slowly than wild-type yeast, suggesting that PIG-V/Gpi18 requires additional cofactors, similar to the PIG-M/PIG-X complex. Supporting this idea, a new factor,

Pga1p, was recently identified as an essential component of GPI-Man-TII in yeast [29]. The yeast *PGA1* encodes a type I transmembrane protein of 198 amino acids with the N-terminal signal peptide. *GPI18* functioned as a high-copy suppressor of the temperature sensitivity of *pga1^{ts}*, and *gpi18^{ts}* and *pga1^{ts}* mutants accumulated the same GPI synthetic intermediate at restrictive temperatures [29]. Moreover, both Gpi18p and Pga1p were in the ER and were coprecipitated [29]. Thus, Pga1p is an essential component of yeast GPI-Man-TII. However, no structural homolog of Pga1 has been identified in the mammalian gene/protein database, and it remains to be solved whether mammalian cells have a functional homolog of Pga1p or whether PIG-V alone is sufficient.

C. GPI-Man-TIII

GPI-Man-TIII, such as mammalian PIG-B/yeast GPI10p, catalyzes the addition of Man-3 from Dol-P-Man to Man-2 of Manα1–6(EtNP)Man-GlcN-(acyl)PI via an α1,2 linkage, which results in Man_2-(EtNP)Man-GlcN-(acyl)PI. Therefore, Man-(EtNP)Man-GlcN-(acyl)PI accumulates in PIG-B mutant cells. PIG-B was identified as a gene responsible for GPI-APs defects in murine T-lymphoma mutant (S1A) cells [3]. PIG-B was predicted to have around eight to ten TMDs based on *in silico* analysis [22, 23], while results of a proteinase K protection assay suggested that PIG-B might be a single-pass transmembrane protein whose large C-terminal portion is localized to the luminal side of the ER [3]. PIG-B and GPI10p share significant homology in primary and secondary structures with GPI-Man-TIV (PIG-Z/Smp3p) and ALG9 (Dol-P-Man:Man_8-$GlcNAc_2$-PP-Dol and $Man_6GlcNAc_2$-PP-Dol α1–2 mannosyltransferase) (Figure 5.2 and Figure 5.3). These three groups of mannosyltransferases are all classified into GT22 in the CAZy database, and utilize Dol-P-Man as the substrate and catalyze the α1–2 linkage (Figure 5.1). In addition to these, ALG12, a Dol-P-Man:$Man_7GlcNAc_2$-PP-Dol α1–6 mannosyltransferase that also belongs to GT22, PIG-M/Gpi14p and Alg3p are structurally related and classified into the GT-C superfamily (see below). GT-C superfamily has a long loop portion in the first luminal side of the ER after the first TMD, where catalytic sites are most likely to be located [24]. The DE or EE sequence is considered to be involved in the active site of PIG-B, ALG9 and ALG12, instead of the DXD motif that is present in PIG-M (Figure 5.2).

D. GPI-Man-TIV

GPI-Man-TIV, mammalian PIG-Z/yeast Smp3, catalyzes the addition of Man-4 from Dol-P-Man to Man-3 via an α1,2 linkage [30, 31]. Smp3p was identified as a Man-4 transferase based on its similarity with PIG-B, which

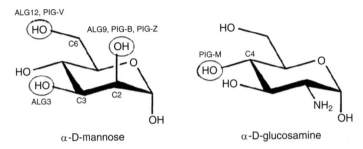

FIG. 5.1 Structure of the acceptor sugars for mannosyltransferases using Dol-P-Man as the substrate in the ER. Ovals represent the OH sites that are catalyzed by α2-mannosyltransferases (ALG9, PIG-B, and PIG-Z), α3-mannosyltransferase (ALG3), α4-mannosyltransferase (PIG-M), and α6-mannosyltransferases (ALG12 and PIG-V).

was the first enzyme identified as GPI-Man-T [30]. Being quite different from other GPI-Man-Ts, PIG-Z is considered non-essential for the synthesis of mature GPI and GPI-APs in mammalian cells, whereas Smp3 is essential in yeast, because the addition of Man-4 is required for the subsequent addition of EtNP to the Man-3 that links GPI to proteins [30, 32]. Smp3p shares homology with PIG-B in terms of peptide sequence and secondary structure, and also belongs to GT22, like PIG-B. The addition of Man-4 has scarcely been detected in many mammalian cell lines, and GPI-transamidase-deficient cells mainly accumulate trimannosyl GPIs, indicating that mammalian PIG-Z is not essential, at least in cell lines [33, 34], although PIG-Z complemented *smp3* mutant yeast and resulted in *in vivo* mannosylation of trimannosyl GPIs, indicating that PIG-Z is indeed a functional homolog of Smp3p [31]. Expression analysis of mRNA in human tissues indicated that PIG-Z is expressed in most tissues with the highest levels in brain and colon [31]. Thus, the Man-4 modification may play a specific role in protein function. Because Man-4-containing GPI-APs are normally formed in yeast and *Plasmodium falciparum*, the difference in the requirement of Man-4 between these and mammalian cells may represent a target candidate for antifungal and antimalarial drugs.

V.   Structural Consideration of Glycosyltransferases that use Dol-P-monosaccharides as Donor Substrates

A new classification of glycosyltransferases was recently proposed, in addition to the CAZy classification, and was based on iterative searches of sequence databases, motif extraction, and structural comparison analysis of completely sequenced genomes [24]. More than 75% of the recognized glycosyltransferases fell into three monophyletic superfamilies of proteins namely (1) a nucleoside-diphosphosugar transferase (GT-A) superfamily, which is characterized by a DxD sequence signature; (2) a GT-B super-family, which is characterized by a GPGTF (glycogen phosphorylase/glyco-syl transferase) motif; and (3) a GT-C superfamily of integral membrane glycosyltransferases with a modified DxD sequence in the first extracellular loop. The GT-C links at least 10 CaZy families (GT22, 39, 50, 57, 58, 59, 66, 83, 85, and 87) and possibly GT76 in which GPI-Man-II is classified. They are embedded in the ER or plasma membranes with predicted multiple transmembrane regions and consistent with the use of lipid-linked sugars as donor substrates [23]. ALG3, ALG9, ALG12, GPI-Man-I, GPI-Man-III, GPI-Man-IV, and presumably GPI-Man-II, which utilize Dol-P-Man as the substrate, belong to the GT-C superfamily. One of the common features of GT-C is the presence of DxD, ExD, DxE, DD, or DE residues in the first extracellular loop, which is considered to be part of the catalytic site, although there is no evidence for a common evolutionary origin of the putative DxD sequence in GT-C and the DxD motif in GT-A, and the mechanistic basis of catalysis in the GT-C superfamily is not known [24].

The peptide sequences in the first extracellular loop of the mannosyl-transferases are aligned (Figure 5.2) and its phylogram is shown in Figure 5.3. The phylogram derived from the loop sequences may not be identical to one derived from whole peptide sequences, although the loop sequences are thought to be important for their catalytic activities. ALG9, PIG-B (GPI-Man-III), and PIG-Z (GPI-Man-IV) belong to α2-mannosyl-transferases, ALG3 and PIG-M (GPI-Man-I) are α3- and α4-mannosyl-transferases, respectively, and ALG12 and PIG-V (GPI-Man-II) belong to α6-mannosyltransferases (Figure 5.1). The NCBI database accession numbers of each protein referred to here are also shown in Table 5.1. PIG-M is evolutionarily close to ALG3 irrespective of the differences between the α3- and α4-mannosyltransferases, which may be due to similar spatial arrangement of the reactive –OH group in acceptor sugars (Figure 5.1) [23]. The DxD and ExD sequences are located at a similar position in the loops of both proteins and the adjacent sequences are also well conserved (Figure 5.2, A (I)). It is not known whether the sequences bind a divalent cation. The conversion of one of either aspartic acid in human PIG-M to alanine by site-directed mutagenesis abolished its activity, although mutation of the former aspartic acid was less effective [21]. Panel B shows the alignment of loop sequences of the PIG-B and PIG-Z α2-mannosyltransferases together with ALG12, an α6-mannosyltransfer-ase. They have two well conserved, short sequences (II) and (III) at similar positions in the loops. The sequence of (II) is DE or EE, which meets the

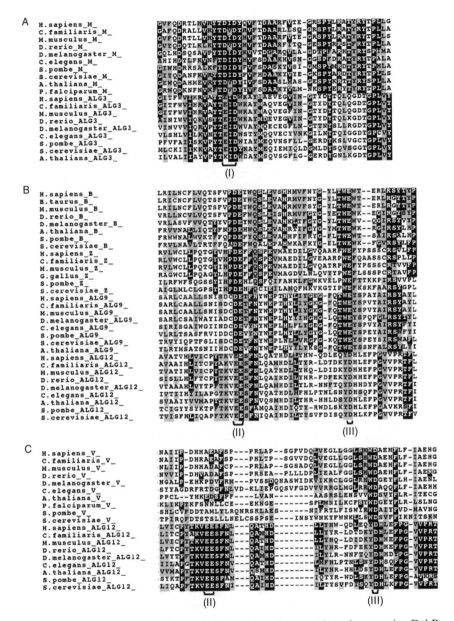

FIG. 5.2 Alignment of the first extracellular loop of mannosyltransferases using Dol-P-monosaccharides as the donor substrate. Proteins described in Table 5.1 were aligned using the ClustalW program (http://www.ch.embnet.org/software/ClustalW.html) with Gonnet (A and C) and Pam (B) matrices.

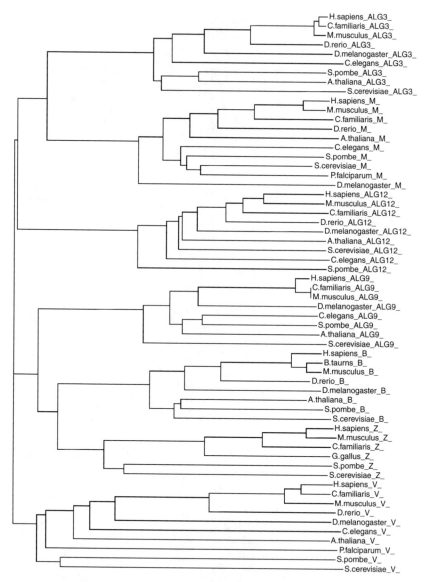

F1G. 5.3 Phylogram tree of the first extracellular loop of mannosyltransferases using Dol-P-monosaccharides as the donor substrate. Proteins described in Table 5.1 were aligned using ClustalW (http://www.ebi.ac.uk/Tools/clustalw2/index.html) with a Pam matrix.

TABLE 5.1

Protein Accession Numbers

	ALG2	ALG9	ALG12	PIG-B	PIG-M	PIG-V	PIG-Z
Homo sapiens	NP_005778.1	NP_079016.2	NP_077010.1	NP_0004846.4	NP_660150.1	NP_060207.2	NP_079429.2
Canis familiaris	XP_516912.2	XP_862216.1	XP_528217.2		XP_852677.1	XP_525246.1	XP_545157.2
Bos taurus				XP_001254700.1			
Mus musculus	NP_666051.1	NP_598742.1	NP_662452.1	NP_061277.1	NP_080510.1	NP_848812.2	NP_766410.1
Gallus gallus							XP_426699.2
Danio rerio	NP_001018522.1		NP_001092219.1	NP_956770.1	NP_956684.1	XP_001922691.1	
Drosophila melanogaster	NP_522829.2	NP_651252.1	NP_649929.1	NP_995991.1	NP_611600.1	NP_725620.1	
Caenorhabditis elegans	NP_496950.2	NP_496282.2	NP_505071.1		NP_496426.1	NP_491782.1	
Oryza sativa							
Schizosaccharomyces pombe	NP_592852.1	NP_594684.1	NP_595429.2	NP_587992.1	NP_596260.1	NP_592878.1	NP_592072.1
Saccharomyces cerevisiae	NP_009471.1	NP_014180.1	NP_014427.1	NP_011272.1	NP_012547.2	NP_009558.1	NP_014792.1
Arabidopsis thaliana	NP_182297.1	NP_172124.2	NP_001077448.1	NP_568205.1	NP_680199.1	NP_172652.2	
Plasmodium falciparum					XP_001250517.1	XP_001250858.1	

These accession numbers were obtained using the NCBI HomoloGene program.

GT-C criteria, and (III) is (W/Y)(D/E). The sequence equivalent to (III) is not present in PIG-M and ALG3. Interestingly, PIG-V and ALG12, both of which are α6-mannosyltransferases, have a short sequence (III), which is followed by xxx(F/Y), but the sequence equivalent to (II) is not present in PIG-V. Moreover, the conversion of one of either tryptophan or aspartic acid in the short sequence (III) of PIG-V to alanine by site-directed mutagenesis completely abolished the activity [27]. This evidence indicates that sequence (III) is directly involved in catalytic activity in PIG-V and possibly ALG12 as well as ALG9, PIG-B, and PIG-Z. Alternatively, PIG-V may be evolutionarily separated from the other glycosyltransferases, as deduced by the phylogram tree (Figure 5.3), and have created a new active site.

VI.   Substrate Specificities and Inhibitors for Mannosylations

T. brucei is the causative agent of human sleeping sickness. There are significant differences in the biosynthethic pathways during mannosylation of GPI intermediates between mammalian cells and *T. brucei*. Inositol acylation occurs before mannosylation and is not obligatory for mannosylation, and EtNP on the first mannose is required for the addition of the third mannose in mammalian cells [1, 35]. In contrast, inositol acylation in *T. brucei* occurs only after the addition of the first mannose and is essential for the addition of EtNP to the third mannose [36]. These differences indicate that understanding the basis for substrate specificities for mannosylations might facilitate the development of specific inhibitors for the GPI pathway in parasites. Several papers have reported the structural requirements of GPI intermediates that affect mannosylation using chemically synthesized GPI analogs, especially in the biosynthetic pathway of *T. brucei* [37–40]. GlcNAcα1–6-L-*myo*-inositol-1-phosphate-diacylglycerol, a diastereoisomer of a natural substrate that contains D-*myo*-inositol, and L-*myo*-inositol-1-phosphate both inhibited the first mannosylation system presumably by binding to GPI-MTI in *T. brucei* but not in a human (HeLa) cell-free system [37]. A series of synthetic Man-Man-GlcN-PI analogs containing systematic modifications of the mannose residues were analyzed in a trypanosomal GPI biosynthetic pathway to reveal which portions of the natural substrate are important for the GPI biosynthesis and to identify the inhibitors of GPI biosynthesis [40]. First, the study revealed that the presence of a hydroxyl group at position 4 of the first mannose and a free amine on the glucosamine residue are required for the inositol acylation, explaining the molecular basis for the fact that mannosylation is required prior to inositol acylation [36]. This order differs from that in *P. falciparum* as well as mammalian cells. In *P. falciparum*, modification of the inositol ring with myristate is required for the mannosylation of GPI, because, in the absence of acyl-CoA or CoA, mannosylated GPIs were not detected in the cell-free GPI biosynthetic system [41]. Second, hydroxyl groups at positions 2 and 3 but not at 4 and 6 of the second mannose were recognized by GPI-Man-TIII. Similarly, positions 2 and 3 of the first mannose and a free amine on the glucosamine residue were not recognized by GPI-Man-TIII. Third, substitution of a hydroxyl group at position 2 of the second mannose with an amine inhibited GPI-Man-TIII activity, which is consistent with previous reports that used mannosamine. Mannosamine has been reported to inhibit the biosynthesis of GPI in *T. brucei* and mammalian cells [42, 43]. This inhibitor results in the accumulation of ManN-Man-GlcN-PI or ManN-Man-GlcN-(acyl)PI and inhibited the addition of the third mannose by GPI-Man-TIII, an α1,2-mannosyltransferase, due to the loss of the –OH group at position 2 of the second mannose [42, 43].

Another substrate for GPI-Man-TI is Dol-P-Man, formally known as Dol-P-β-Man. GPI-Man-TI exhibited strict stereospecificity for the β-mannosyl-phosphoryl linkage, when compared with Dol-P-α-Man. Dol-P-β-Mans containing 11 or 19 isoprenyl units were equally effective substrates for GPI-Man-TI, indicating that the length of lipid portion was less important [44]. The presence of a saturated α-isoprene unit in the dolichyl moiety is required for optimal GPI-Man-TI activity because polyprenol-phospho-mannose (PPM) was only 50% as effective as Dol-P-Man as a mannosyl donor [44]. Compatible with the ability of GPI-Man-TI to utilize of PPM, GlcN-(acyl)PI was not accumulated in Lec9 CHO mutant cells that synthesize very little Dol-P-Man but accumulate Poly-P-Man due to an apparent lack of polyprenol reductase activity [44].

VII. GPI Mannosylation-Related Diseases

Congenital disorders of glycosylation (CDG) are a growing group of multi-systemic disorders caused by defective glycosylation and have severe clinical implications in infancy and early childhood. Dol-P-Man is required for the biosynthesis of GPI as well as *N*-glycans, protein *O*- and *C*-mannosylation [5, 6]. Therefore, the defect in this system causes CDG. At present, five genes involved in Dol-P-Man synthesis or utilization for GPI biosynthesis were reported to be responsible for CDG; PMM2 (phosphomannomutase2), MPI (phosphomannose isomerase), DPM1, MPDU1/SL15, and DK1 (dolichol kinase1). Defects in these enzymes cause CDG-1a, -Ib,

-1e, -If, and, -Im, respectively [45–51]. Clinical features in seven CDG-1e patients were complicated but included encephalopathy, ataxia, seizures, dysmorphic features, and developmental delay. The patients produced a dwarfed dolichol-linked precursor oligosaccharide, which is required for N-glycoprotein biosynthesis, with five mannose residues, instead of the normal precursor that has nine mannose residues [47, 48].

In addition to paroxysmal nocturnal hemoglobinuria, a well known acquired GPI-deficient disease caused by the defect of PIG-A [52], the defective expression of PIG-M was recently reported as the cause of a newly identified inherited GPI deficiency [53] (see Chapter 16). The defective expression of PIG-M was due to weakened promoter activity induced by a point mutation in a transcription factor Sp1 binding site. The main clinical manifestations were venous thrombosis and seizures [53]. PIG-M transcription and the surface GPI expression *in vitro* as well as *in vivo* were significantly restored by butyrate, a histone deacetylase inhibitor, through enhanced histone acetylation in a Sp1-dependent manner, which dramatically improved the intractable seizures [54].

VIII. Polyprenol-Phosphate-Mannose (PPM)-Dependent Mannosyltransferases in Mycobacteria

In bacteria, phosphatidylinositol is an uncommon component of plasma membrane. Therefore, glycosylated PIs are rarely found in bacteria. Mycobacteria are an exception because phosphatidylinositol is an essential component of the plasma membrane [55] and can be modified by multiple numbers of mannoses. Dimannosyl and hexamannosyl PIs are termed phosphatidylinositol mannosides (PIMs) [56, 57]. Furthermore, mycobacterial polysaccharides, lipomannan and lipoarabinomannan (LM/LAM), are anchored to the plasma membrane by a lipid that is structurally identical to dimannosyl PI (Figure 5.4) [58]. Therefore, although the structures of PIMs and LAM are distinct from eukaryotic GPIs, PIMs can be considered to be functionally equivalent to eukaryotic free GPIs, while LM/LAM are equivalent to a GPI-anchored polysaccharide. GPI-anchored proteins are the most common form of GPIs found in eukaryotes, but mycobacterial PIMs do not appear to anchor proteins.

PIMs and LAM are currently a subject of intense research because mycobacteria cause important diseases such as tuberculosis and leprosy, and these glycosylated PIs seem to play important roles in pathogen survival in the intracellular compartments of host macrophages. In particular, LAM from pathogenic species exerts a number of immuno-suppressive activities including inhibition of the production of pro-inflammatory

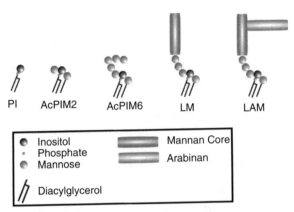

F<small>IG</small>. 5.4 Structure of PI (phosphatidylinositol), AcPIM2, AcPIM6 (AcPIM: acyl phosphatidylinositol mannosides), LM (lipomannan), and LAM (lipoarabinomannan).

cytokines, inhibition of apoptosis, and block of phagosome maturation. Phagosome maturation is a process of endocytic vesicle transport to deliver phagocytosed materials to the lysosome for digestion. *Mycobacterium tuberculosis* blocks this process and survives in the phagosomes of alveolar macrophages. How LAM can have so many different biological activities is an interesting question, and each aspect is currently being dissected at molecular levels. A number of review articles have described the biological activities of PIMs and LM/LAM for interested readers [59, 60–62].

PIMs and LM/LAM can be modified by up to four fatty acids. In addition to the two fatty acids in the PI moiety, 6-OH of mannose that is attached to the 2-OH of the *myo*-inositol ring is modified by a third fatty acid [63, 64]. Furthermore, 3-OH of inositol can be acylated [62]. PIMs and LAM are structurally distinct from their eukaryotic counterparts in that mannoses are directly attached to inositol without the bridging glucosamine. Furthermore, the linkages of mannoses are not identical. Therefore, one may assume that the evolutions of PIM/LAM and eukaryotic GPIs are independent events. Nevertheless, some mycobacterial PIM/LAM mannosyltransferases were found to be homologous to eukaryotic PIG-M. So far, six PIG-M homologs have been identified in *M. tuberculosis* genome. These homologs are found widely in various *Mycobacterium* species including *Mycobacterium leprae*, the causative agent of leprosy, and *Mycobacterium smegmatis*, a non-pathogenic experimentally tractable species.

We will describe the biosynthetic pathway of PIMs and LM/LAM briefly, and then focus the discussion on the functions of PPM-dependent mannosyltransferases that are homologous to eukaryotic GPI mannosyltransferase PIG-M. For more complete coverage of the biosynthetic pathway and glycosyltransferases involved in the pathway, please refer to recent reviews [60, 65, 66]. The current understanding of the biosynthetic pathway is shown in Figure 5.5.

AcPIM4 is thought to be the intermediate at which the AcPIM6 biosynthesis and LM/LAM biosynthesis diverge. The structural features of PIMs and LM/LAM support this branch point; the terminal two mannoses attached to AcPIM6 are α1,2-linked, while LM/LAM continues to elongate the α1,6-linked mannose backbone. In other words, if the AcPIM4 intermediate is catalyzed by an α1,2 mannosyltransferase (i.e., PimE, see below for detail), the pathway is committed to AcPIM6 synthesis. On the other hand, if AcPIM4 is catalyzed by an α1,6 mannosyltransferase, the pathway will lead to LM/LAM synthesis. Studies manipulating the PIM/LAM biosynthetic genes also support the conclusion that AcPIM4 is the branch point (see below).

The first two mannose transfers are GDP-mannose-dependent [67, 68]. A GDP-mannose-dependent mannosyltransferase, PimA, mediates the transfer of the first mannose to the 2-OH position of *myo*-inositol, producing an intermediate PIM1 [69]. PIM1 can be modified by an acyltransferase, which transfers a fatty acid to the 6-OH of the mannose residue, producing AcPIM1. MSMEG_2934 mediates this reaction in *M. smegmatis* [70], and

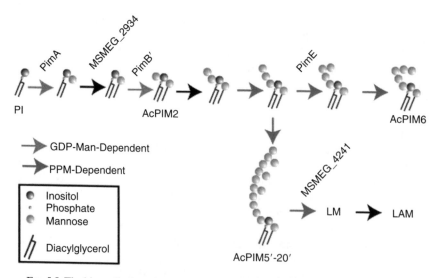

Fɪɢ. 5.5 The biosynthetic pathway of PIMs and LM/LAM. PimA and PimB′ catalyze GDP-mannose dependent mannosylation. AcPIM4 is the intermediate where AcPIM6 and LM/LAM biosynthesis diverge. PimE catalyzes α1,2-linked mannosylation, whereas LM/LAM synthesis diverges from AcPIM4 by α1,6-linked mannosylation.

its homologs are found in all mycobacteria. Interestingly, the deletion mutant did not completely abolish the acyltransferase activity, suggesting that there may be a redundant enzyme involved in this step. The second mannose is transferred by a GDP-mannose-dependent mannosyltransferase in mycobacteria, and PimB has been suggested to be responsible for this reaction [71]. However, in corynebacteria, PimB (MgtA) is involved in the biosynthesis of unrelated glycolipids [72], and PimB' was identified as the second mannosyltransferase of PIM biosynthesis [73]. While PimB is a pseudo gene in *M. leprae*, the PimB' homologs are widely conserved in mycobacteria including *M. leprae*. Formal proof is needed to verify that PimB' mediates the second mannose transfer in mycobacteria. The second mannose transfer and acylation seems to also take place in a reverse order, and substrate specificities of these enzymes remain to be determined. A third GDP-mannose-dependent mannosyltransferase, PimC, has been identified in some clinical strains of *M. tuberculosis* [74], but homologs do not appear to exist in many species of mycobacteria, including *M. smegmatis*.

The fourth mannose transfer and onwards are dependent on PPM rather than GDP-mannose [75, 76]. PPM in mycobacteria is a functional equivalent of Dol-P-Man in eukaryotes, and their structures are similar. Therefore, these reactions are similar to eukaryotic GPI mannosyltransferases in terms of substrate specificity. As mentioned above, a number of genes homologous to eukaryotic GPI mannosyltransferase PIG-M have been characterized. We will describe each of these PIG-M homologs below.

IX.   Mycobacteria Genes Homologous to Mammalian PIG-M

A. PimE

The gene encoding PimE is found widely in mycobacteria species. When the *pimE* gene was deleted from *M. smegmatis* genome, the mutant accumulated AcPIM4, suggesting that it mediates the fifth mannose transfer [77]. Human PIG-M carries a DXD amino acid motif, which is critically important for the enzyme activity, within the loop between the first and second transmembrane domains. One of these aspartic acid residues is conserved in mycobacterial PimE homologs and, when this residue was mutated to alanine, catalytic activity was abolished. These data suggest that PimE is the fifth mannosyltransferase of PIM biosynthesis.

Introduction of the *M. tuberculosis Rv1159* gene restored the *M. smegmatis* mutant phenotype, suggesting that Rv1159 is a functional homolog. Importantly, deletion of PimE did not alter biosynthesis of LM/LAM, indicating that the PIM biosynthetic pathway beyond the fifth mannose transfer is separate from the LM/LAM biosynthetic pathway. This observation supports the notion that AcPIM4 is the branch point intermediate. Further support for this conclusion came from studies on mutants of a lipoprotein LpqW. Although the molecular mechanisms are unknown, disruption of the LpqW gene resulted in highly compromised LM/LAM biosynthesis. The mutant grows slowly under certain culture conditions [78], and spontaneous mutation of PimE rescues the growth phenotype [79]. These secondary mutants accumulate AcPIM4 and apparently divert it to LM/LAM synthesis, resulting in normalized LM/LAM synthesis. The crystal structure of LpqW suggests that LpqW may bind to AcPIM4 and channel this intermediate to LM/LAM biosynthesis [80].

B. MSMEG_3120

MSMEG_3120 is a PIG-M homolog found in *M. smegmatis* and is conserved in mycobacteria and corynebacteria. When this gene is deleted in *M. smegmatis*, no phenotype was observed in PIM/LM/LAM biosynthesis. However, when its ortholog was deleted in *Corynebacterium glutamicum*, LM and LAM were no longer synthesized [81]. These findings indicate that this mannosyltransferase mediates an early part of LM/LAM elongation, at least in corynebacteria. It is not clear why the deletion of MSMEG_3120 does not result in obvious phenotypes in *M. smegmatis*, but one possibility is that MSMEG_3120 is a redundant gene.

C. MSMEG_4241

MSMEG_4241 is another PIG-M homolog, which has been characterized in *M. smegmatis* and *C. glutamicum*. When MSMEG_4241 is deleted in *M. smegmatis*, mutant LM bearing only 5–20 mannose residues accumulated in contrast to normal LM, which carries 21–34 mannose residues [82]. This suggests that MSMEG_4241 mediates the elongation of an LM intermediate, which carries ~20 mannose residues. While this mutation grossly changed the biosynthetic profile of LM/LAM, it did not affect PIM biosynthesis. A homologous gene, MptA, was deleted in *C. glutamicum* and the mutant showed a similar LM/LAM-deficient phenotype [83], suggesting that the general scheme of the LM/LAM biosynthetic pathway is conserved between mycobacteria and corynebacteria.

D. MSMEG_4247

The α1,6 mannose backbone of LM/LAM is further elaborated by α1,2 mono-mannosyl side chains. This modification is mediated by MSMEG_4247 in *M. smegmatis* [84]. The mutant accumulates a mutant LAM, which lacks α1,2 mono-mannosyl side chains. For unknown reasons, LM does not accumulate efficiently in this mutant, suggesting that the α1,2 side chain is necessary for the stable accumulation of LM.

There are several steps for which the enzymes responsible remain to be identified. One PIG-M homolog was recently reported to be involved in the transfer of arabinose [85], but it is unclear why this enzyme can still maintain the primary amino acid sequence homologous to eukaryotic PIG-M. In pathogenic mycobacteria, terminal arabinose residues of arabinan are modified by mannose caps. A mannosyltransferase involved in the synthesis of mannose capping is Rv1635c [86]. A homolog of this gene is found in all pathogenic mycobacteria, but is absent in non-pathogenic species. While this enzyme is likely to be PPM-dependent, its homology to PIG-M is not significant, compared with that of the other mannosyltransferases discussed above. We speculate that Rv1635c does not recognize the structural features of the lipid anchor part. On the other hand, PIG-M homologs may possess a structural feature, which is involved in the recognition of PI lipid moiety, and this structural requirement may be a key reason why the homology between eukaryotic PIG-M and mycobacterial PIG-M homologs were maintained (or independently invented) through evolution.

REFERENCES

1. Hong, Y., Maeda, Y., Watanabe, R., Ohishi, K., Mishkind, M., Riezman, H., and Kinoshita, T. (1999). PIG-N, a mammalian homologue of yeast Mcd4p, is involved in transferring phosphoethanolamine to the first mannose of the glycosylphosphatidylinositol. *J Biol Chem* 274:35099–35106.
2. Gaynor, E.C., Mondesert, G., Grimme, S.J., Reed, S.I., Orlean, P., and Emr, S.D. (1999). MCD4 encodes a conserved endoplasmic reticulum membrane protein essential for glycosylphosphatidylinositol anchor synthesis in yeast. *Mol Biol Cell* 10:627–648.
3. Takahashi, M., Inoue, N., Ohishi, K., Maeda, Y., Nakamura, N., Endo, Y., Fujita, T., Takeda, J., and Kinoshita, T. (1996). PIG-B, a membrane protein of the endoplasmic reticulum with a large lumenal domain, is involved in transferring the third mannose of the GPI anchor. *EMBO J* 15:4254–4261.
4. Sutterlin, C., Escribano, M.V., Gerold, P., Maeda, Y., Mazon, M.J., Kinoshita, T., Schwarz, R.T., and Riezman, H. (1998). *Saccharomyces cerevisiae* GPI10, the functional homologue of human PIG-B, is required for glycosylphosphatidylinositol-anchor synthesis. *Biochem J* 332:153–159.

5. Lehle, L., Strahl, S., and Tanner, W. (2006). Protein glycosylation, conserved from yeast to man: a model organism helps elucidate congenital human diseases. *Angew Chem Int Ed Engl* 45:6802–6818.

6. Maeda, Y., and Kinoshita, T. (2008). Dolichol-phosphate mannose synthase: structure, function and regulation. *Biochim Biophys Acta* 1780:861–868.

7. Delahunty, M.D., Stafford, F.J., Yuan, L.C., Shaz, D., and Bonifacino, J.S. (1993). Uncleaved signals for glycosylphosphatidylinositol anchoring cause retention of precursor proteins in the endoplasmic reticulum. *J Biol Chem* 268:12017–12027.

8. Nagamune, K., Acosta-Serrano, A., Uemura, H., Brun, R., Kunz-Renggli, C., Maeda, Y., Ferguson, M.A., and Kinoshita, T. (2004). Surface sialic acids taken from the host allow trypanosome survival in tsetse fly vectors. *J Exp Med* 199:1445–1450.

9. Tomita, S., Inoue, N., Maeda, Y., Ohishi, K., Takeda, J., and Kinoshita, T. (1998). A homologue of *Saccharomyces cerevisiae* Dpm1p is not sufficient for synthesis of dolichol-phosphate-mannose in mammalian cells. *J Biol Chem* 273:9249–9254.

10. Colussi, P.A., Taron, C.H., Mack, J.C., and Orlean, P. (1997). Human and *Saccharomyces cerevisiae* dolichol phosphate mannose synthases represent two classes of the enzyme, but both function in *Schizosaccharomyces pombe*. *Proc Natl Acad Sci USA* 94:7873–7878.

11. Mazhari-Tabrizi, R., Eckert, V., Blank, M., Mueller, R., Mumberg, D., Funk, M., and Schwarz, R.T. (1996). Cloning and functional expression of glycosyltransferases from parasitic protozoans by heterologous complementation in yeast: the dolichol phosphate mannose synthase from *Trypanosoma brucei brucei*. *Biochem J* 316:853–858.

12. Zimmerman, J.W., Specht, C.A., Cazares, B.X., and Robbins, P.W. (1996). The isolation of a Dol-P-Man synthase from *Ustilago maydis* that functions in *Saccharomyces cerevisiae*. *Yeast* 12:765–771.

13. Schutzbach, J.S., Zimmerman, J.W., and Forsee, W.T. (1993). The purification and characterization of recombinant yeast dolichyl-phosphate-mannose synthase. Site-directed mutagenesis of the putative dolichol recognition sequence. *J Biol Chem* 268:24190–24196.

14. Singh, N., and Tartakoff, A.M. (1991). Two different mutants blocked in synthesis of dolichol-phosphoryl-mannose do not add glycophospholipid anchors to membrane proteins: quantitative correction of the phenotype of a CHO cell mutant with tunicamycin. *Mol Cell Biol* 11:391–400.

15. Maeda, Y., Tomita, S., Watanabe, R., Ohishi, K., and Kinoshita, T. (1998). DPM2 regulates biosynthesis of dolichol phosphate-mannose in mammalian cells: correct subcellular localization and stabilization of DPM1, and binding of dolichol phosphate. *EMBO J* 17:4920–4929.

16. Maeda, Y., Tanaka, S., Hino, J., Kangawa, K., and Kinoshita, T. (2000). Human dolichol-phosphate-mannose synthase consists of three subunits, DPM1, DPM2 and DPM3. *EMBO J* 19:2475–2482.

17. Ashida, H., Maeda, Y., and Kinoshita, T. (2006). DPM1, the catalytic subunit of dolichol-phosphate mannose synthase, is tethered to and stabilized on the endoplasmic reticulum membrane by DPM3. *J Biol Chem* 281:896–904.

18. Anand, M., Rush, J.S., Ray, S., Doucey, M.A., Weik, J., Ware, F.E., Hofsteenge, J., Waechter, C.J., and Lehrman, M.A. (2001). Requirement of the Lec35 gene for all known classes of monosaccharide-P-dolichol-dependent glycosyltransferase reactions in mammals. *Mol Biol Cell* 12:487–501.

19. Rush, J.S., and Waechter, C.J. (1995). Transmembrane movement of a water-soluble analogue of mannosylphosphoryldolichol is mediated by an endoplasmic reticulum protein. *J Cell Biol* 130:529–536.

20. Ware, F.E., and Lehrman, M.A. (1996). Expression cloning of a novel suppressor of the Lec15 and Lec35 glycosylation mutations of chinese hamster ovary cells. *J Biol Chem* 271:13935–13938.

21. Maeda, Y., Watanabe, R., Harris, C.L., Hong, Y., Ohishi, K., Kinoshita, K., and Kinoshita, T. (2001). PIG-M transfers the first mannose to glycosylphosphatidylinositol on the lumenal side of the ER. *EMBO J* 20:250–261.

22. Pittet, M., and Conzelmann, A. (2007). Biosynthesis and function of GPI proteins in the yeast *Saccharomyces cerevisiae*. *Biochim Biophys Acta* 1771:405–420.

23. Oriol, R., Martinez-Duncker, I., Chantret, I., Mollicone, R., and Codogno, P. (2002). Common origin and evolution of glycosyltransferases using Dol-P-monosaccharides as donor substrate. *Mol Biol Evol* 19:1451–1463.

24. Liu, J., and Mushegian, A. (2003). Three monophyletic superfamilies account for the majority of the known glycosyltransferases. *Protein Sci* 12:1418–1431.

25. Ashida, H., Hong, Y., Murakami, Y., Shishioh, N., Sugimoto, N., Kim, Y.U., Maeda, Y., and Kinoshita, T. (2005). Mammalian PIG-X and yeast Pbn1p are the essential components of glycosylphosphatidylinositol-mannosyltransferase I. *Mol Biol Cell* 16:1439–1448.

26. Naik, R.R., and Jones, E.W. (1998). The PBN1 gene of *Saccharomyces cerevisiae*: an essential gene that is required for the post-translational processing of the protease B precursor. *Genetics* 149:1277–1292.

27. Kang, J.Y., Hong, Y., Ashida, H., Shishioh, N., Murakami, Y., Morita, Y.S., Maeda, Y., and Kinoshita, T. (2005). PIG-V involved in transferring the second mannose in glycosylphosphatidylinositol. *J Biol Chem* 280:9489–9497.

28. Fabre, A.L., Orlean, P., and Taron, C.H. (2005). *Saccharomyces cerevisiae* Ybr004c and its human homologue are required for addition of the second mannose during glycosylphosphatidylinositol precursor assembly. *FEBS J* 272:1160–1168.

29. Sato, K., Noda, Y., and Yoda, K. (2007). Pga1 is an essential component of glycosylphosphatidylinositol-mannosyltransferase II of *Saccharomyces cerevisiae*. *Mol Biol Cell* 18:3472–3485.

30. Grimme, S.J., Westfall, B.A., Wiedman, J.M., Taron, C.H., and Orlean, P. (2001). The essential Smp3 protein is required for addition of the side-branching fourth mannose during assembly of yeast glycosylphosphatidylinositols. *J Biol Chem* 276:27731–27739.

31. Taron, B.W., Colussi, P.A., Grimme, J.M., Orlean, P., and Taron, C.H. (2004). Human Smp3p adds a fourth mannose to yeast and human glycosylphosphatidylinositol precursors *in vivo*. *J Biol Chem* 279:36083–36092.

32. Grimme, S.J., Colussi, P.A., Taron, C.H., and Orlean, P. (2004). Deficiencies in the essential Smp3 mannosyltransferase block glycosylphosphatidylinositol assembly and lead to defects in growth and cell wall biogenesis in *Candida albicans*. *Microbiology* 150:3115–3128.

33. Homans, S.W., Ferguson, M.A., Dwek, R.A., Rademacher, T.W., Anand, R., and Williams, A.F. (1988). Complete structure of the glycosyl phosphatidylinositol membrane anchor of rat brain Thy-1 glycoprotein. *Nature* 333:269–272.

34. Hong, Y., Maeda, Y., Watanabe, R., Inoue, N., Ohishi, K., and Kinoshita, T. (2000). Requirement of PIG-F and PIG-O for transferring phosphoethanolamine to the third mannose in glycosylphosphatidylinositol. *J Biol Chem* 275:20911–20919.

35. Murakami, Y., Siripanyapinyo, U., Hong, Y., Kang, J.Y., Ishihara, S., Nakakuma, H., Maeda, Y., and Kinoshita, T. (2003). PIG-W is critical for inositol acylation but not for flipping of glycosylphosphatidylinositol-anchor. *Mol Biol Cell* 14:4285–4295.

36. Guther, M.L., and Ferguson, M.A. (1995). The role of inositol acylation and inositol deacylation in GPI biosynthesis in *Trypanosoma brucei*. *EMBO J* 14:3080–3093.

37. Smith, T.K., Paterson, M.J., Crossman, A., Brimacombe, J.S., and Ferguson, M.A. (2000). Parasite-specific inhibition of the glycosylphosphatidylinositol biosynthetic pathway by stereoisomeric substrate analogues. *Biochemistry* 39:11801–11807.

38. Smith, T.K., Sharma, D.K., Crossman, A., Dix, A., Brimacombe, J.S., and Ferguson, M.A. (1997). Parasite and mammalian GPI biosynthetic pathways can be distinguished using synthetic substrate analogues. *EMBO J* 16:6667–6675.

39. Smith, T.K., Sharma, D.K., Crossman, A., Brimacombe, J.S., and Ferguson, M.A. (1999). Selective inhibitors of the glycosylphosphatidylinositol biosynthetic pathway of *Trypanosoma brucei*. *EMBO J* 18:5922–5930.

40. Urbaniak, M.D., Yashunsky, D.V., Crossman, A., Nikolaev, A.V., and Ferguson, M.A. (2008). Probing enzymes late in the trypanosomal glycosylphosphatidylinositol biosynthetic pathway with synthetic glycosylphosphatidylinositol analogues. *ACS Chem Biol* 3:625–634.

41. Gerold, P., Jung, N., Azzouz, N., Freiberg, N., Kobe, S., and Schwarz, R.T. (1999). Biosynthesis of glycosylphosphatidylinositols of *Plasmodium falciparum* in a cell-free incubation system: inositol acylation is needed for mannosylation of glycosylphosphatidylinositols. *Biochem J* 344:731–738.

42. Ralton, J.E., Milne, K.G., Guether, M.L.S., Field, R.A., and Ferguson, M.A.J. (1993). The mechanism of inhibition of glycosylphosphatidylinositol anchor biosynthesis in *Trypanosoma brucei* by mannosamine. *J Biol Chem* 268:24183–24189.

43. Sevlever, D., and Rosenberry, T.L. (1993). Mannosamine inhibits the synthesis of putative glycoinositol phospholipid anchor precursors in mammalian cells without incorporating into an accumulated intermediate. *J Biol Chem* 268:10938–10945.

44. DeLuca, A.W., Rush, J.S., Lehrman, M.A., and Waechter, C.J. (1994). Mannolipid donor specificity of glycosylphosphatidylinositol mannosyltransferase-I (GPIMT-I) determined with an assay system utilizing mutant CHO-K1 cells. *Glycobiol* 4:909–916.

45. Van Schaftingen, E., and Jaeken, J. (1995). Phosphomannomutase deficiency is a cause of carbohydrate-deficient glycoprotein syndrome type I. *FEBS Lett* 377:318–320.

46. Niehues, R., Hasilik, M., Alton, G., Korner, C., Schiebe-Sukumar, M., Koch, H.G., Zimmer, K.P., Wu, R., Harms, E., Reiter, K., von Figura, K., Freeze, H.H., *et al.* (1998). Carbohydrate-deficient glycoprotein syndrome type Ib. Phosphomannose isomerase deficiency and mannose therapy. *J Clin Invest* 101:1414–1420.

47. Kim, S., Westphal, V., Srikrishna, G., Mehta, D.P., Peterson, S., Filiano, J., Karnes, P.S., Patterson, M.C., and Freeze, H.H. (2000). Dolichol phosphate mannose synthase (DPM1) mutations define congenital disorder of glycosylation Ie (CDG-Ie). *J Clin Invest* 105:191–198.

48. Imbach, T., Schenk, B., Schollen, E., Burda, P., Stutz, A., Grunewald, S., Bailie, N.M., King, M.D., Jaeken, J., Matthijs, G., Berger, E.G., Aebi, M., and Hennet, T. (2000). Deficiency of dolichol-phosphate-mannose synthase-1 causes congenital disorder of glycosylation type Ie. *J Clin Invest* 105:233–239.

49. Kranz, C., Denecke, J., Lehrman, M.A., Ray, S., Kienz, P., Kreissel, G., Sagi, D., Peter-Katalinic, J., Freeze, H.H., Schmid, T., Jackowski-Dohrmann, S., Harms, E., and Marquardt, T. (2001). A mutation in the human MPDU1 gene causes congenital disorder of glycosylation type If (CDG-If). *J Clin Invest* 108:1613–1619.

50. Schenk, B., Imbach, T., Frank, C.G., Grubenmann, C.E., Raymond, G.V., Hurvitz, H., Korn-Lubetzki, I., Revel-Vik, S., Raas-Rotschild, A., Luder, A.S., Jaeken, J., Berger, E. G., *et al.* (2001). MPDU1 mutations underlie a novel human congenital disorder of glycosylation, designated type If. *J Clin Invest* 108:1687–1695.

51. Kranz, C., Jungeblut, C., Denecke, J., Erlekotte, A., Sohlbach, C., Debus, V., Kehl, H.G., Harms, E., Reith, A., Reichel, S., Grobe, H., Hammersen, G., *et al.* (2007). A defect in

dolichol phosphate biosynthesis causes a new inherited disorder with death in early infancy. *Am J Hum Genet* 80:433–440.

52. Takeda, J., Miyata, T., Kawagoe, K., Iida, Y., Endo, Y., Fujita, T., Takahashi, M., Kitani, T., and Kinoshita, T. (1993). Deficiency of the GPI anchor caused by a somatic mutation of the PIG-A gene in paroxysmal nocturnal hemoglobinuria. *Cell* 73:703–711.

53. Almeida, A.M., Murakami, Y., Layton, D.M., Hillmen, P., Sellick, G.S., Maeda, Y., Richards, S., Patterson, S., Kotsianidis, I., Mollica, L., Crawford, D.H., Baker, A., *et al.* (2006). Hypomorphic promoter mutation in PIGM causes inherited glycosylphosphatidylinositol deficiency. *Nat Med* 12:846–851.

54. Almeida, A.M., Murakami, Y., Baker, A., Maeda, Y., Roberts, I.A., Kinoshita, T., Layton, D.M., and Karadimitris, A. (2007). Targeted therapy for inherited GPI deficiency. *N Engl J Med* 356:1641–1647.

55. Jackson, M., Crick, D.C., and Brennan, P.J. (2000). Phosphatidylinositol is an essential phospholipid of mycobacteria. *J Biol Chem* 275:30092–30099.

56. Ballou, C.E., Vilkas, E., and Lederer, E. (1963). Structural studies on the *myo*-inositol phospholipids of *Mycobacterium tuberculosis* (var. bovis, strain BCG). *J Biol Chem* 238:69–76.

57. Lee, Y.C., and Ballou, C.E. (1965). Complete structures of the glycophospholipids of mycobacteria. *Biochemistry* 4:1395–1404.

58. Chatterjee, D., Hunter, S.W., McNeil, M., and Brennan, P.J. (1992). Lipoarabinomannan. Multiglycosylated form of the mycobacterial mannosylphosphatidylinositols. *J Biol Chem* 267:6228–6233.

59. Quesniaux, V., Fremond, C., Jacobs, M., Parida, S., Nicolle, D., Yeremeev, V., Bihl, F., Erard, F., Botha, T., Drennan, M., Soler, M.N., Le Bert, M., *et al.* (2004). Toll-like receptor pathways in the immune responses to mycobacteria. *Microbes Infect* 6:946–959.

60. Brennan, P.J. (2003). Structure, function, and biogenesis of the cell wall of *Mycobacterium tuberculosis*. *Tuberculosis (Edinb)* 83:91–97.

61. Briken, V., Porcelli, S.A., Besra, G.S., and Kremer, L. (2004). Mycobacterial lipoarabinomannan and related lipoglycans: from biogenesis to modulation of the immune response. *Mol Microbiol* 53:391–403.

62. Nigou, J., Gilleron, M., and Puzo, G. (1999). Lipoarabinomannans: characterization of the multiacylated forms of the phosphatidyl-*myo*-inositol anchor by NMR spectroscopy. *Biochem J* 337:453–460.

63. Khoo, K.H., Dell, A., Morris, H.R., Brennan, P.J., and Chatterjee, D. (1995). Structural definition of acylated phosphatidylinositol mannosides from *Mycobacterium tuberculosis*: definition of a common anchor for lipomannan and lipoarabinomannan. *Glycobiology* 5:117–127.

64. Nigou, J., Gilleron, M., Cahuzac, B., Bounery, J.D., Herold, M., Thurnher, M., and Puzo, G. (1997). The phosphatidyl-*myo*-inositol anchor of the lipoarabinomannans from *Mycobacterium bovis* bacillus Calmette Guerin. Heterogeneity, structure, and role in the regulation of cytokine secretion. *J Biol Chem* 272:23094–23103.

65. Berg, S., Kaur, D., Jackson, M., and Brennan, P.J. (2007). The glycosyltransferases of *Mycobacterium tuberculosis*—roles in the synthesis of arabinogalactan, lipoarabinomannan, and other glycoconjugates. *Glycobiology* 17:35–56R.

66. Nigou, J., Gilleron, M., and Puzo, G. (2003). Lipoarabinomannans: from structure to biosynthesis. *Biochimie* 85:153–166.

67. Brennan, P., and Ballou, C.E. (1967). Biosynthesis of mannophosphoinositides by *Mycobacterium phlei*. The family of dimannophosphoinositides. *J Biol Chem* 242:3046–3056.

68. Brennan, P., and Ballou, C.E. (1968). Biosynthesis of mannophosphoinositides by *Mycobacterium phlei*. Enzymatic acylation of the dimannophosphoinositides. *J Biol Chem* 243:2975–2984.

69. Kordulakova, J., Gilleron, M., Mikusova, K., Puzo, G., Brennan, P.J., Gicquel, B., and Jackson, M. (2002). Definition of the first mannosylation step in phosphatidylinositol mannoside synthesis. PimA is essential for growth of mycobacteria. *J Biol Chem* 277:31335–31344.

70. Kordulakova, J., Gilleron, M., Puzo, G., Brennan, P.J., Gicquel, B., Mikusova, K., and Jackson, M. (2003). Identification of the required acyltransferase step in the biosynthesis of the phosphatidylinositol mannosides of *Mycobacterium* species. *J Biol Chem* 278:36285–36295.

71. Schaeffer, M.L., Khoo, K.H., Besra, G.S., Chatterjee, D., Brennan, P.J., Belisle, J.T., and Inamine, J.M. (1999). The pimB gene of *Mycobacterium tuberculosis* encodes a mannosyltransferase involved in lipoarabinomannan biosynthesis. *J Biol Chem* 274:31625–31631.

72. Tatituri, R.V., Illarionov, P.A., Dover, L.G., Nigou, J., Gilleron, M., Hitchen, P., Krumbach, K., Morris, H.R., Spencer, N., Dell, A., Eggeling, L., and Besra, G.S. (2007). Inactivation of *Corynebacterium glutamicum* NCgl0452 and the role of MgtA in the biosynthesis of a novel mannosylated glycolipid involved in lipomannan biosynthesis. *J Biol Chem* 282:4561–4572.

73. Lea-Smith, D.J., Martin, K.L., Pyke, J.S., Tull, D., McConville, M.J., Coppel, R.L., and Crellin, P.K. (2008). Analysis of a new mannosyltransferase required for the synthesis of phosphatidylinositol mannosides and lipoarbinomannan reveals two lipomannan pools in corynebacterineae. *J Biol Chem* 283:6773–6782.

74. Kremer, L., Gurcha, S.S., Bifani, P., Hitchen, P.G., Baulard, A., Morris, H.R., Dell, A., Brennan, P.J., and Besra, G.S. (2002). Characterization of a putative alpha-mannosyltransferase involved in phosphatidylinositol trimannoside biosynthesis in *Mycobacterium tuberculosis*. *Biochem J* 363:437–447.

75. Besra, G.S., Morehouse, C.B., Rittner, C.M., Waechter, C.J., and Brennan, P.J. (1997). Biosynthesis of mycobacterial lipoarabinomannan. *J Biol Chem* 272:18460–18466.

76. Morita, Y.S., Patterson, J.H., Billman-Jacobe, H., and McConville, M.J. (2004). Biosynthesis of mycobacterial phosphatidylinositol mannosides. *Biochem J* 378:589–597.

77. Morita, Y.S., Sena, C.B., Waller, R.F., Kurokawa, K., Sernee, M.F., Nakatani, F., Haites, R.E., Billman-Jacobe, H., McConville, M.J., Maeda, Y., and Kinoshita, T. (2006). PimE is a polyprenol-phosphate-mannose-dependent mannosyltransferase that transfers the fifth mannose of phosphatidylinositol mannoside in mycobacteria. *J Biol Chem* 281:25143–25155.

78. Kovacevic, S., Anderson, D., Morita, Y.S., Patterson, J., Haites, R., McMillan, B.N., Coppel, R., McConville, M.J., and Billman-Jacobe, H. (2006). Identification of a novel protein with a role in lipoarabinomannan biosynthesis in mycobacteria. *J Biol Chem* 281:9011–9017.

79. Crellin, P.K., Kovacevic, S., Martin, K.L., Brammananth, R., Morita, Y.S., Billman-Jacobe, H., McConville, M.J., and Coppel, R.L. (2008). Mutations in pimE restore lipoarabinomannan synthesis and growth in a *Mycobacterium smegmatis* lpqW mutant. *J Bacteriol* 190:3690–3699.

80. Marland, Z., Beddoe, T., Zaker-Tabrizi, L., Lucet, I.S., Brammananth, R., Whisstock, J.C., Wilce, M.C., Coppel, R.L., Crellin, P.K., and Rossjohn, J. (2006). Hijacking of a substrate-binding protein scaffold for use in mycobacterial cell wall biosynthesis. *J Mol Biol* 359:983–997.

81. Mishra, A.K., Alderwick, L.J., Rittmann, D., Wang, C., Bhatt, A., Jacobs, W.R., Jr., Takayama, K., Eggeling, L., and Besra, G.S. (2008). Identification of a novel alpha (1→6) mannopyranosyltransferase MptB from *Corynebacterium glutamicum* by deletion of a conserved gene, NCgl1505, affords a lipomannan- and lipoarabinomannan-deficient mutant. *Mol Microbiol* 68:1595–1613.

82. Kaur, D., McNeil, M.R., Khoo, K.H., Chatterjee, D., Crick, D.C., Jackson, M., and Brennan, P.J. (2007). New insights into the biosynthesis of mycobacterial lipomannan arising from deletion of a conserved gene. *J Biol Chem* 282:27133–27140.

83. Mishra, A.K., Alderwick, L.J., Rittmann, D., Tatituri, R.V., Nigou, J., Gilleron, M., Eggeling, L., and Besra, G.S. (2007). Identification of an alpha(1→6) mannopyranosyl-transferase (MptA), involved in *Corynebacterium glutamicum* lipomanann biosynthesis, and identification of its orthologue in *Mycobacterium tuberculosis*. *Mol Microbiol* 65:1503–1517.

84. Kaur, D., Berg, S., Dinadayala, P., Gicquel, B., Chatterjee, D., McNeil, M.R., Vissa, V.D., Crick, D.C., Jackson, M., and Brennan, P.J. (2006). Biosynthesis of mycobacterial lipoar-abinomannan: role of a branching mannosyltransferase. *Proc Natl Acad Sci USA* 103:13664–13669.

85. Birch, H.L., Alderwick, L.J., Bhatt, A., Rittmann, D., Krumbach, K., Singh, A., Bai, Y., Lowary, T.L., Eggeling, L., and Besra, G.S. (2008). Biosynthesis of mycobacterial arabi-nogalactan: identification of a novel alpha(1→3) arabinofuranosyltransferase. *Mol Microbiol* 69:1191–1206.

86. Dinadayala, P., Kaur, D., Berg, S., Amin, A.G., Vissa, V.D., Chatterjee, D., Brennan, P.J., and Crick, D.C. (2006). Genetic basis for the synthesis of the immunomodulatory man-nose caps of lipoarabinomannan in *Mycobacterium tuberculosis*. *J Biol Chem* 281:20027–20035.

6

Phosphoethanolamine Addition to Glycosylphosphatidylinositols

PETER ORLEAN

Department of Microbiology
University of Illinois at Urbana-Champaign
Urbana, IL 61801, USA

I. Abstract

A phosphoethanolamine (Etn-P) moiety can be attached to up to three of the mannoses (Man) of human and yeast GPI precursors during their assembly in the membrane of the endoplasmic reticulum. All the three Etn-Ps originate from phosphatidylethanolamine. Mammals and yeast have a family of three related proteins, of which orthologous mammalian PIG-N and yeast Mcd4 are likely to catalyze transfer of Etn-P to the α1,4-linked Man, PIG-G and yeast Gpi7 are responsible for Etn-P transfer to the α1,6-linked Man, and PIG-O/Gpi13—predominantly in mammals and exclusively in yeast—are involved in Etn-P transfer to the third, α1,2-linked Man. The PIG-G/Gpi7 and PIG-O/Gpi13 proteins are both partnered by the small, hydrophobic PIG-F/Gpi11 protein. The three GPI-Etn-P transferases (Etn-P-T) are large proteins with an N-terminal, lumenally oriented catalytic domain that resembles members of the alkaline phosphatase superfamily, and with multiple transmembrane domains towards their C-termini. The GPI anchors made by *Trypanosoma brucei* and *Plasmodium falciparum* bear a single Etn-P on the α1,2-linked Man, and these parasites' genomes encode just one GPI-Etn-P-T. The Etn-P on the α1,2-linked Man becomes amide linked to protein, but the molecular function of the Etn-Ps on the other two Man is unclear, although the severe growth defects of yeast mutants deficient in Mcd4 and Gpi7 indicate that these side branches are of great importance. This chapter focuses on the properties and proposed functions of GPI-Etn-P-T in mammals and yeast, and reviews the phenotypes of yeast GPI-Etn-P-T mutants.

II. Sites of Etn-P Modification on Protein-Bound and Free GPIs

The protein-bound GPIs from yeast and mammalian cells can be modified with up to three phosphodiester-linked ethanolamines (Etn-Ps) [1–4]. The first, α1,4-linked mannose (Man-1) receives Etn-P on its 2′-OH, whereas the second, α1,6-Man (Man-2), and third, α1,2-Man (Man-3), are modified on their 6′-OH [1, 2] (Figure 6.1). In trypanosomes, only the third Man receives Etn-P [9]. The Etn-P on Man-3 is indispensable because it

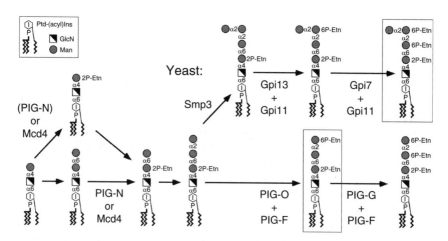

Fig. 6.1. Etn-P addition steps in GPI-precursor biosynthesis in mammalian and yeast cells, and the proteins involved. The assembly intermediates shown are those in the likely major assembly route for GPIs from unmodified Man₁-GPI onwards. The possibility that PIG-N and Mcd4 can also act on a Man₁-GPI is illustrated. In yeast, Smp3-dependent addition of a fourth Man obligatorily precedes Gpi13/Gpi11-dependent addition of Etn-P to Man-3. Mammalian PIG-O does not require a fourth Man on its acceptor, but Man₄-GPIs can be detected in radiolabeling experiments [5, 6], and a subset of the GPIs on mammalian proteins have four Man [1]. Yeast Gpi7 likely normally acts after Gpi13, however, aberrant Man₃- and Man₄-GPIs bearing a single Etn-P on Man-2 (not depicted) have been detected [7, 8]. The boxed GPI structures are the likely preferred substrates for the mammalian and yeast GPI transamidases.

provides the nucleophile for the transamidase reaction in which the GPI is attached to its target protein, but the roles of the Etn-Ps on Man-1 and Man-2, which are not invariably present on protein-bound GPIs, are unclear. There is no evidence that mammalian or yeast GPIs can become linked to protein through the Etn-Ps on Man-1 or Man-2 [10].

Analyses of the headgroups of the free GPIs that can be detected in *in vivo* radiolabeling experiments with mutant mammalian or yeast cells indicate that a range of Etn-P-modified Man_2–Man_4-GPIs can be made. These include a Man_2-GPI with Etn-P on Man-1; Man_3- and Man_4-GPIs with a single Etn-P either on Man-1, Man-2, or Man-3 [5, 7, 8, 11, 12]; Man_3- and Man_4-GPIs with two Etn-Ps, which could be either on Man-1 and Man-3, Man-1 and Man-2, or Man-2 and Man-3 [5, 11, 13–15]; and Man_3- and Man_4-GPIs with Etn-Ps on Man-1, Man-2, and Man-3 [5, 6, 11, 14–16]. Though these many alternative structures complicate schemes for the GPI-assembly pathway, it is not yet clear whether all these Etn-P-modified, free GPIs are physiologically relevant.

III.   Phosphoethanolamine Donor

Addition of Etn-P moieties occurs during GPI-precursor biosynthesis in the endoplasmic reticulum (ER). The results of *in vitro* and *in vivo* radiolabeling experiments with the mammalian, trypanosomal, and yeast systems indicate that the Etn-P moieties are transferred from phosphatidyl-ethanolamine (Ptd-Etn), rather than cytidine diphosphate (CDP)-Etn. Thus, Etn-P bearing free GPIs are made by washed microsomal membranes upon incubation with radiolabeled sugar nucleotides but in the absence of any added CDP-Etn [10, 11, 16, 17], suggesting CDP-Etn is not the direct donor, and leaving Ptd-Etn as a likely alternative. *In vivo* radiolabeling experiments using yeast mutants unable to make CDP-Etn and CDP-Cho from exogenous Etn, but which still make Ptd-Etn by decarboxylation of phosphatidylserine (Ptd-Ser), fail to incorporate [3H]Etn into protein-bound GPIs or into a Man_2-GPI with Etn-P on Man-1 [18, 19], whereas the radiolabel supplied as [3H]Ser is incorporated into the latter lipid [19] after the formation and decarboxylation of Ptd-[3H]Ser. These *in vivo* labeling experiments, and similar [3H]Ser labeling experiments in trypano-somes [20], indicate that the Etn-Ps on Man-1 and Man-3 most likely originate from Ptd-Etn. This notion is supported by the finding that triple mutants between a conditional allele of the α1,2-Man-specific Etn-P-T gene (*gpi13*) and null alleles of yeast's two Ptd-Ser decarboxylase genes (*psd1Δ* and *psd2Δ*) are nonviable [21]. Moreover, [3H]inositol labeling experiments with a *psd1Δ psd2Δ* double null mutants show that this strain accumulates small amounts of GPI with the same chromatographic mobility as the Man$_4$-GPI with one Etn-P that accumulates in Gpi13-depleted cells, suggesting that GPI-Etn-P-T-III preferentially uses Ptd-Etn generated, which in turn derived from Ptd-Ser decarboxylation (J. Wiedman and P. Orlean, unpublished results). Interestingly, *gpi13 psd2Δ* mutants grow more slowly than *gpi13 psd1Δ* strains, suggesting that the primary source of Ptd-Etn used by Gpi13 is Psd2, which is localized in Golgi and vacuole. This raises the possibility that Gpi13p acts in a section of the ER juxtaposed with the Golgi or vacuolar membranes [21]. In mammalian cells, the GPI-assembly reactions from GlcNAc-PI de-*N*-acetylation to formation of Etn-P-modified Man$_1$-GPI are enriched in a mitochondria-associated ER compartment [22], suggesting that PIG-N might function near a source of mitochondrially generated Ptd-Etn. Because all Etn-P additions are likely to occur in the ER lumen, they are dependent on flipping of Ptd-Etn from the cytoplasmic face of the ER to the lumenal face.

The direct transfer of Etn-P from Ptd-Etn to a GPI has yet to be demonstrated *in vitro*.

IV.   Proteins Involved in Etn-P Addition

A.   GENERAL FEATURES AND PROPERTIES OF GPI-ETN-P-TRANSFERASES

Four proteins have, so far, been directly implicated in Etn-P addition during GPI biosynthesis in mammals and yeast. The presumed catalytic proteins are the related mammalian PIG-N, PIG-G, and PIG-O proteins and their yeast orthologs Mcd4, Gpi7, and Gpi13, which are referred to as GPI-Etn-P-T-I, II, and III, and which add Etn-P to Man-1, Man-2, and Man-3, respectively. GPI-Etn-P-T are 830–1100 amino acid proteins predicted to have 10–14 membrane-spanning domains and a large lumenal loop of some 400 amino acids between transmembrane sequences 1 and 2 at their N-terminus (Figure 6.2). The lumenal loop contains sequences characteristic of members of the alkaline phosphatase superfamily [5, 14, 23, 24], consistent with the proteins being involved in breakage and formation of a phosphodiester linkage. This domain is necessary for protein function because the G227E mutation in the *mcd4–174* allele that confers temperature-sensitive growth and a conditional block in the GPI-assembly pathway [24, 25] lies in one of the two metal—likely Zn^{2+}—binding sites in alkaline phosphatase family proteins [23]. Moreover, Etn-P addition *in vitro* is zinc dependent [26, 27], and the temperature sensitivity of a yeast *gpi13* allele is suppressed by Zn^{2+} (as well as Mn^{2+}) [21].

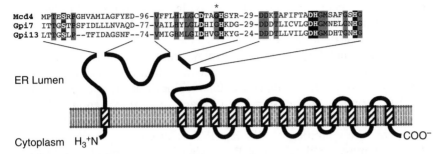

FIG. 6.2. Transmembrane organization of a representative GPI-Etn-P-T. Shown are the predicted transmembrane topology of yeast Mcd4 and the approximate positions of the amino acid sequence motifs common to alkaline phosphatase superfamily enzymes [23] in the amino terminal lumenal domain. Amino acid residues in white on a black background are involved in metal binding, the putative active site serine is in white, underlined, and on a dark grey background, and amino acids identical in Mcd4, Gpi7, and Gpi13 are on a light gray background. The numbers of amino acid residues separating the three motifs in each of the yeast GPI-Etn-P-T are indicated. Glycine 227, which is mutated to glutamate in the temperature-sensitive *mcd4-174* mutant, is underlined and indicated with an asterisk (adapted from Ref. [24], Figure 1A).

Among the three members of the GPI-Etn-P-T family, mammalian PIG-O and PIG-G resemble one another comparatively more closely than either resembles PIG-N, and yeast Gpi13 and Gpi7 likewise are slightly more similar to one another than they are to Mcd4 [28]. These similarities correlate with the facts that PIG-O/Gpi13 and PIG-G/Gpi7 are involved in Etn-P transfer to the 6'-OH of Man and also interact with PIG-F/Gpi11, whereas PIG-N/Mcd4 transfer to the 2'-OH and may not associate with PIG-F/Gpi11. The proteomes of the parasites *Trypanosoma brucei* and *Plasmodium*, which add Etn-P only onto Man-3, encode a single GPI-Etn-P-T family member, which most closely resembles PIG-O/Gpi13 [29]. Genomes of Gram-negative bacteria encode proteins that transfer Etn-P from Ptd-Etn to 3-deoxy-D-manno-octulosonic acid moieties in lipopolysaccharide [30]. However, the relationship of these proteins to the eukaryal GPI-Etn-P-T seems remote because hydropathy plots of their amino acid sequences show four or five potential transmembrane segments in the N-terminal halves of these proteins, and the alkaline phosphatase-related domain is located at the C-terminus.

In addition to three alkaline phosphatase domain-containing proteins, small, hydrophobic PIG-F/Gpi11 acts in concert with GPI-Etn-P-T-II and III, but its relative importance to these two Etn-P-T may vary between the yeast and mammalian cells. The three alkaline phosphatase domain-containing proteins are all localized in the ER [5, 12, 15, 24, 28], although yeast Gpi7 has been reported to be in the plasma membrane [14]. None of the GPI-Etn-P-T has been purified, and detailed enzymological studies remain to be carried out on them.

The Zn^{2+} chelator 1,10-phenanthroline inhibits, at least partially, all three GPI-Etn-P-transfer reactions in mammalian cells and probably in yeast. The 1,10-phenanthroline treatment blocks the formation of GPIs with Etn-P on Man-2 and Man-3 *in vivo* and *in vitro* in mammalian cells, and such cells also show an accumulation of an unmodified Man_1-GPI, consistent with inhibition of GPI-Etn-P-T-I as well, though the formation of a trimannosyl GPI with Etn-P on Man-1 is not completely blocked [26, 27]. In the presence of 1,10-phenanthroline, yeast membranes show diminished formation of a Man_4-GPI with Etn-P on its first three Man, and an accumulation of unmodified Man_2-GPI [26].

Interestingly, *T. brucei*'s sole GPI-Etn-P-T shows differential sensitivity to two inhibitors. In contrast to mammalian and yeast cells, 1,10-phenanthroline had no effect on the formation of Man_3-GPI bearing the sole Etn-P on Man-3 in *T. brucei* lysates [26]. The reverse was true for phenyl methyl sulfonyl fluoride. This serine protease inhibitor prevented the addition of Etn-P to Man-3 in *T. brucei* [31], but had no impact on the formation of GPIs with Etn-Ps on Man-1 or Man-3 in HeLa cell lysates [32]. The differential sensitivities of parasite and human GPI-Etn-P-T to two types of compound suggests that selective inhibitors of parasite GPI-Etn-P-T could, in principle, be developed.

Features of mammalian and yeast GPI-Etn-P-T, and phenotypes of yeast cells deficient in them, are described further.

B.   GPI-ETN-P-T-I

The PIG-N/Mcd4 proteins are involved in addition of Etn-P to the 2′-OH of Man-1 in mammalian cells and yeast, respectively, based on the absence of Etn-P on Man-1 of the GPIs that accumulate in *pig-n* and *mcd4* mutants. Thus, PIG-N-deficient mammalian cells synthesize GPI precursors with three Man (and traces of Man_4-GPIs) lacking Etn-P on Man-1 and fail to transfer Etn-P to Man-1 *in vitro* [5], whereas a conditionally lethal yeast *mcd4* allele accumulates a Man_2-GPI lacking Etn-P [25], and *in vitro* synthesis of GPIs by *mcd4* microsomes does not proceed beyond Man_2-GPI [3]. PIG-N deficiency reduces, though does not abolish, surface expression of GPI proteins [5], indicating that GPI structures optimal for transfer to protein are not made. The effect of an Mcd4 deficiency is more severe in yeast because temperature-sensitive yeast *mcd4* mutants are blocked in GPI anchoring [24], although slow growth on osmotically supported medium has been reported for an *mcd4* deletion mutation in one strain background [33].

The preferred Etn-P acceptor for PIG-N and Mcd4 *in vivo* seems to be Man_2-GPI, because mammalian GPI-mannosyltransferase-II (Man-T-II) (PIG-V) mutants accumulate predominantly an unmodified Man_1-GPI [34], and yeast *mcd4* mutants, in which blockage of the GPI pathway is much more complete than in *pig-n* mutants, accumulate a Man_2-GPI. However, because mammalian, and particularly yeast mutants in GPI-Man-T-II also accumulate Man_1-GPI with Etn-P on its Man [34, 35], both PIG-N and Mcd4 (if they are indeed responsible for transfer of this Etn-P) can use a Man_1-GPI as acceptor as well. It is not known whether PIG-N/Mcd4 can act on Man_3- or Man_4-GPIs that are either unmodified, or which have already received Etn-Ps on Man-2 or Man-3.

Phosphoethanolamine transfer to Man-1 is inhibited by the terpenoid lactone YW3548, a compound identified in a screen for inhibitors of GPI-dependent processing of the yeast glycoprotein, Gas1. Yeast cells treated with YW3548 accumulate unmodified Man_2-GPI, consistent with the notion that the compound blocks Mcd4-dependent addition of Etn-P to Man-1 [36, 37]. YW3548 treatment also inhibits Etn-P addition to Man-1 of mammalian GPI precursors [5], but not to Man-2 or Man-3, indicating that this compound is specific for PIG-N/Mcd4. Consistent with this, YW3548 does not inhibit growth of *T. brucei* or *Plasmodium* parasites, whose GPIs lack Etn-P on Man-1 [36].

Why might the Etn-P moiety on Man-1 of GPIs be important? Although mammalian *pig-n* mutants are not blocked in any of the mannosylation steps, the finding that yeast Mcd4-deficient cells accumulate a Man_2-GPI lacking Etn-P on Man-1 suggests that the Etn-P on Man-1 is required for GPI-Man-T-III to add Man-3. However, this requirement is not an absolute one, for *mcd4* disruptants can be partially rescued by high-level expression of yeast GPI-Man-T-III (Gpi10p) [25]. In human cells, the Etn-P moiety on Man-1 is important for the interactions between GPIs and the GPI transamidase complex. GPIs bearing Etn-P on Man-1 were efficiently pulled down in experiments using antibodies to GPI transamidase subunits, in contrast to GPIs lacking Etn-P on Man-1 [54], and surprisingly, the Etn-P on Man-3 did not contribute to GPI recognition. The Etn-P on Man-1 of yeast GPIs may have functions in addition to enhancing the efficiency of Man-3 addition by Gpi10. Thus, *MCD4*-disrupted cells expressing mammalian or trypanosomal GPI-Man-T-III, proteins that efficiently mannosylate Man_2-GPIs lacking Etn-P, grow, albeit slowly [25, 38]. *mcd4* disruptants harboring the *T. brucei* GPI-Man-T-III gene transfer GPIs to protein less efficiently, show retarded export of GPI-anchored proteins from the ER, do not carry out remodeling of the GPIs lipid moiety to ceramide, and have a defect in selection of axial sites of budding [38]. The walls of viable *mcd4* disruptants may lack their outer mannan protein layer [39], though this could be attributable to a general perturbation in GPI anchoring. How the Etn-P on Man-1 influences these diverse processes at the molecular level in yeast is unknown.

The Mcd4 protein itself may have additional functions in yeast. Thus, the *mcd4*-P301L allele, but not the G227E variant, has a defect in Ptd-Ser transport to the Golgi and vacuole for subsequent decarboxylation, but does not show an obvious defect in GPI anchoring, suggesting Mcd4p may have an additional role in transport-dependent metabolism of Ptd-Ser [40]. In addition, high-level expression of Mcd4p results in extracellular release of ATP from yeast cells, and a membrane fraction from Mcd4p-overexpressing cells containing the Golgi vesicles was enriched in Mcd4 and exhibited elevated ATP uptake. These findings led to the suggestion that Mcd4 normally might mediate symport of ATP and Ptd-Etn into the lumen of the ER [41]. Lumenally accumulated ATP is presumably released from overexpressing cells following vesicle movement along the secretory pathway and vesicle fusion with the plasma membrane. Overexpression of the yeast *GPI7* and *GPI13* genes, as well as human PIG-N cDNA, also stimulated ATP release from cells [41], suggesting that all three GPI-Etn-P-T may have this transport activity.

C. GPI-ETN-P-T-II

The orthologous PIG-G and Gpi7 proteins are involved in Etn-P addition to Man-2 in mammals and yeast, respectively. Mutant human cells depleted of PIG-G accumulate a Man_3-GPI bearing Etn-P on Man-1 and Man-3, but not on Man-2 [28], and analogously, yeast *gpi7* disruptants, which are viable though temperature sensitive, accumulate a Man_4-GPI with Etn-P on Man-1 and Man-3, but lacking one on Man-2 [14].

PIG-G is bound and stabilized by PIG-F, a 219 amino acid protein [42] predicted to have five or six transmembrane domains, but which shows no obvious resemblance to known enzymes [28]. PIG-F also interacts with PIG-O (see below). PIG-F's yeast ortholog, essential Gpi11, was proposed to contribute to Gpi7 function because the profile of Man_4-GPI precursors that accumulates in *gpi7* disruptants resembles that seen in Gpi11-depleted yeast, as well as in *gpi11* disruptants complemented by PIG-F, which have a temperature-sensitive phenotype [7]. A physical interaction between Gpi7 and Gpi11 has not been demonstrated.

From the fact that loss of PIG-G/Gpi7 function leads to accumulation of Man_3- or Man_4-GPIs with Etn-Ps on Man-1 and Man-3, it seems likely that these GPIs are the normal acceptors and that GPI-Etn-P-T-II adds Etn-P to

Man-2 after GPI-Etn-P-T-III has added the bridging Etn-P to Man-3. However, Man_3- and Man_4-GPI isoforms with a single Etn-P on Man-2 accumulate in yeast *gpi11* and GPI-Man-T-IV mutants [7, 8], suggesting that yeast Gpi7 could have transferred Etn-P to unmodified mannosylated GPIs. The acceptor requirements of GPI-Etn-P-T-II have not been explored *in vitro*.

Yeast *gpi7* null mutants exhibit various defects, suggesting that the Etn-P moiety on Man-2 may have a range of functions. First, because a Man_4-GPI with Etn-P on Man-1 and Man-3 accumulates in yeast *gpi7Δ* mutants, and because a double mutant between a *gpi7* null and a *gpi8* transamidase mutant has a synthetic growth defect [14], an Etn-P on Man-2 may be necessary for optimal transfer of GPIs to protein. The possible preference of the yeast GPI transamidase for GPIs with Etn-P on Man-2 contrasts with the situation in mammalian cells, in which Man_3-GPIs with Etn-Ps on Man-1 and Man-3 have been suggested to be the preferred transamidase substrate [28] (Figure 6.1). Second, yeast *gpi7* disruptants are impaired in ER to Golgi transport of GPI proteins and remodeling of the diacylglycerol-based GPIs initially transferred to protein to ceramides is much reduced [14]. Third, *gpi7Δ* cells exhibit cell-wall defects [43], and proteins predicted normally to become crosslinked to the cell wall in a GPI-dependent manner are released into the growth medium by *gpi7Δ* cells [44], indicating the inefficient transfer of proteins from their GPI-anchor to an acceptor polysaccharide in the wall. Finally, *S. cerevisiae gpi7* null mutants exhibit a cell separation defect that has been attributed to mistargeting of Egt2p, a daughter cell-specific, predicted GPI protein that normally is involved in degradation of the division septum [45]. Although mechanistic details are unknown, an interpretation of the diverse phenotypes conferred by Gpi7 deficiency is that the Etn-P moiety on Man-2 of GPIs is recognized by GPI transamidase, by components of the intracellular transport machinery, by GPI-lipid remodeling enzymes, and by proteins involved in the proposed transglycosylation reaction(s) in which the GPI glycan of a membrane-anchored GPI protein becomes crosslinked to an acceptor glycan in the cell wall [46, 47]. It is possible that Gpi7, like Mcd4, has additional functions [40, 41], and that their loss contributes to the phenotypes of *gpi7* mutants. The importance of PIG-G and the Etn-P moiety on Man-2 of GPIs in multicellular organisms remains to be explored.

D. GPI-ETN-P-T-III

The PIG-O and Gpi13 proteins are required for addition of Etn-P to the 6′-OH of the α1,2-linked Man-3 in mammalian and yeast cells, respectively [7, 12, 15]. This Etn-P moiety subsequently serves as nucleophile in the transamidase reaction in which the GPI precursor is transferred to protein. The major GPI species accumulated by mammalian cells lacking PIG-O is a Man$_3$-GPI lacking Etn-P on Man-3, but bearing Etn-P on Man-1 [15], and analogously, yeast strains depleted of the essential Gpi13 protein accumulate a Man$_4$-GPI with a single Etn-P on Man-1 [7, 12]. *GPI13* is an essential gene, and neither Gpi7, which is also involved in Etn-P transfer to the 6′-OH of a mannosyl residue, nor Mcd4, can substitute for Gpi13 [12]. Mammalian *pig-o* mutants, however, make minor amounts of a free GPI that bears Etn-P on Man-3 and express low levels of GPI proteins on their surface [15], indicating the occurrence of PIG-O-independent Etn-P transfer to Man-3. Whether PIG-G is involved is unknown.

PIG-O acts together with PIG-F. The structure of the major GPI that accumulates in *pig-f* mutants—a Man$_3$-GPI with Etn-P on Man-1—indicates they are blocked in Etn-P addition to Man-3 [11, 48–50]. Pull-down experiments show that PIG-O and PIG-F interact physically, and PIG-F is important for stability of PIG-O, as it is for PIG-G [15, 28]. However, PIG-O and PIG-G do not interact with PIG-F at the same time; rather, it appears they compete for PIG-F. It has, therefore, been proposed that PIG-F has a regulatory role in GPI anchoring, whereby, for example, levels of PIG-F and PIG-G impact PIG-O levels, and in turn, the formation of the optimal transamidase substrate, a Man$_3$-GPI with Etn-Ps on Man-1 and Man-3 [28] (Figure 6.1).

pig-f mutant cell lines accumulate GPIs with one, two, or three Man [11, 49–51], and it is not yet clear whether the accumulation of Man$_1$- and Man$_2$-GPIs can be accounted for by deficiencies in GPI-Etn-P-II and III, or whether a PIG-F defect has an impact earlier in the pathway, for example, on PIG-N activity. In contrast to PIG-G and PIG-O, however, PIG-F does not coprecipitate with PIG-N [15].

The apparent extent of involvement of *S. cerevisiae*'s PIG-F homolog, Gpi11, in Etn-P addition to Man-3 in yeast varies depending on the type of mutant the test is carried out in. Gpi11 is a 219 amino acid protein whose sequence is 26% identical and 49% similar to human PIG-F [7]. Yeast *gpi11* null mutants are nonviable, but are rescued by expression of human PIG-F, although the complemented *gpi11* null mutants are temperature-sensitive. At nonpermissive temperature, yeast *gpi11* disruptants harboring PIG-F accumulated at least four lipids, the two most polar of which are a Man$_4$-GPI bearing two Etn-Ps, of which one is on Man-3 and the other on Man-1 or Man-2, and a Man$_4$-GPI isoform with its single Etn-P on Man-2 [7]. GPIs with the same chromatographic mobility as the latter two lipids also accumulated in *gpi11* disruptants in which expression of native *GPI11* was repressed, although it is not certain whether these have the same distribution of Etn-Ps. These findings suggested that Gpi11 is not obligatory for addition of the bridging Etn-P to Man-3, although scenarios were considered in which heterologously expressed PIG-F promoted Etn-P addition to Man-3 [7]. Interestingly, the recently isolated temperature-sensitive *gpi11-1* allele accumulates predominantly a species with the mobility of a Man$_4$-GPI with one Etn-P moiety (K. Willis and P. Orlean, unpublished results), implicating Gpi11 in the Gpi13-dependent step in yeast.

With the recognition that PIG-F physically interacts with PIG-O or PIG-G [15, 28], the lipid accumulation phenotypes of various types of yeast *gpi11* mutant can in part be rationalized in terms of the differential abilities of PIG-F, wild-type Gpi11, and mutant Gpi11-1 to stabilize Gpi13 and Gpi7. However, the close resemblance of the lipid accumulation phenotypes of Gpi11-depleted yeast cells and *gpi7Δ* mutants [7] cannot yet be reconciled with a disruption of a Gpi11/Gpi13 interaction. One possibility is that the Man$_4$-GPI with two Etn-Ps that accumulates in Gpi11-depleted cells is an isoform that bears its Etn-Ps on Man-1 and Man-2 [7].

Gpi11 may have roles in addition to adding Etn-P to Man-2 and Man-3. Thus, [^{3}H]Man labeling experiments with *gpi11Δ* strains complemented by PIG-F accumulate two further, less polar mannolipids [7]. The structures of these additional glycolipids are unknown, but if they are GPIs with one or two Man, this would implicate Gpi11 in Etn-P addition to Man-1 as well.

Because mammalian cells transfer Man$_3$-GPIs to protein [1] and make trimannosyl GPI precursors bearing Etn-P on Man-3 [5, 11, 13, 15, 16], PIG-O can use a Man$_3$-GPI as acceptor. In contrast, yeast Gpi13 requires a Man$_4$-GPI. Thus, radiolabeling experiments with a yeast double mutant defective in both GPI-Man-T-IV (*smp3*) and Gpi13 showed that addition of a side-branching fourth Man to the 2′-OH of Man-3 obligatorily precedes addition of Etn-P to the 6′-OH of Man-3 [8]. This may hold in the human pathogenic fungus *Candida albicans*, in which addition of a fourth Man is also essential [52], and the difference in the acceptor specificities of human and fungal GPI-Etn-P-T-IIIs could therefore be exploited in antifungal drug therapy. PIG-O, but not PIG-N, appears to prefer an acyl chain on the inositol moiety, because GPIs with Etn-P on Man-3 were not detected in the inositol acyltransferase defective *pig-w* mutants [53].

V. Concluding Remarks

The identification of the proteins involved in Etn-P addition to GPIs will allow intriguing questions about this critical modification to be explored. Solubilization and purification of GPI-Etn-P-T will allow their Ptd-Etn-dependent transferase activity to be verified and their acceptor requirements to be probed using panels of synthetic GPIs. The possibility that

GPI-Etn-P-T carry out symport of ATP and Ptd-Etn into the ER lumen might be explored by reconstituting these proteins in liposomes, if necessary together with PIG-F or Gpi11. It is not known whether any of the four proteins involved in Etn-P addition to GPIs interact with other components of the GPI biosynthetic machinery. However, because the PIG-N/Mcd4 step occurs between two mannosylation steps, there may be a loose, higher order organization of the membrane-bound enzymes involved in the three successive additions to the GPI to permit substrate channeling.

The biological importance of side-branching Etn-Ps in multicellular organisms is as yet unexplored. No human diseases associated with a defect in Etn-P modification of Man-1 or Man-2 of GPIs have yet been described; however, PIG-N- or PIG-G-deficient model organisms may prove to exhibit developmental defects. Determination of the Etn-P distribution on individual protein-bound GPIs may reveal whether there is protein- or tissue-specific variation in the number and position of Etn-Ps, which in turn might impact protein function and localization.

Determination of why side-branching Etn-Ps are important for the reactions in fungi in which the GPI-lipid moiety is remodeled or the GPI glycan becomes crosslinked to the cell wall must await development of assays for these reactions, and in the latter case, identification of the protein(s) responsible. Of particular interest is whether Etn-P moieties on Man-1 and Man-2 of protein-bound GPIs participate directly in the crosslinking reaction in fungi. Clearly, much remains to be studied about GPI-Etn-P-T themselves and the functions of side-branching Etn-Ps.

REFERENCES

1. Homans, S.W., Ferguson, M.A., Dwek, R.A., Rademacher, T.W., Anand, R., and Williams, A.F. (1988). Complete structure of the glycosyl phosphatidylinositol membrane anchor of rat brain Thy-1 glycoprotein. *Nature* 333:269–272.
2. Roberts, W.L., Santikarn, S., Reinhold, V.N., and Rosenberry, T.L. (1988). Structural characterization of the glycoinositol phospholipid membrane anchor of human erythrocyte acetylcholinesterase by fast atom bombardment mass spectrometry. *J Biol Chem* 263:18776–18784.
3. Imhof, I., Flury, I., Vionnet, C., Roubaty, C., Egger, D., and Conzelmann, A. (2004). Glycosylphosphatidylinositol (GPI) proteins of *Saccharomyces cerevisiae* contain ethanolamine phosphate groups on the alpha1,4-linked mannose of the GPI anchor. *J Biol Chem* 279:19614–19627.
4. Deeg, M.A., Humphrey, D.R., Yang, S.H., Ferguson, T.R., Reinhold, V.N., and Rosenberry, T.L. (1992). Glycan components in the glycoinositol phospholipid anchor of human erythrocyte acetylcholinesterase. Novel fragments produced by trifluoroacetic acid. *J Biol Chem* 267:18573–18580.
5. Hong, Y., Maeda, Y., Watanabe, R., Ohishi, K., Mishkind, M., Riezman, H., and Kinoshita, T. (1999). Pig-N, a mammalian homologue of yeast Mcd4p, is involved in

transferring phosphoethanolamine to the first mannose of the glycosylphosphatidylinositol. *J Biol Chem* 274:35099–35106.

6. Taron, B.W., Colussi, P.A., Wiedman, J.M., Orlean, P., and Taron, C.H. (2004). Human Smp3p adds a fourth mannose to yeast and human glycosylphosphatidylinositol precursors *in vivo*. *J Biol Chem* 279:36083–36092.

7. Taron, C.H., Wiedman, J.M., Grimme, S.J., and Orlean, P. (2000). Glycosylphosphatidylinositol biosynthesis defects in Gpi11p- and Gpi13p-deficient yeast suggest a branched pathway and implicate Gpi13p in phosphoethanolamine transfer to the third mannose. *Mol Biol Cell* 11:1611–1630.

8. Grimme, S.J., Westfall, B.A., Wiedman, J.M., Taron, C.H., and Orlean, P. (2001). The essential Smp3 protein is required for addition of the side-branching fourth mannose during assembly of yeast glycosylphosphatidylinositols. *J Biol Chem* 276:27731–27739.

9. Ferguson, M.A.J., Homans, S.W., Dwek, R.A., and Rademacher, T.W. (1988). Glycosylphosphatidylinositol moiety that anchors *Trypanosoma bruceii* variant surface glycosprotein to the membrane. *Science* 239:753–759.

10. Canivenc-Gansel, E., Imhof, I., Reggiori, F., Burda, P., Conzelmann, A., and Benachour, A. (1998). GPI anchor biosynthesis in yeast: phosphoethanolamine is attached to the α1,4-linked mannose of the complete precursor glycophospholipid. *Glycobiology* 8:761–770.

11. Hirose, S., Prince, G.M., Sevlever, D., Ravi, L., Rosenberry, T.L., Ueda, E., and Medof, M.E. (1992). Characterization of putative glycoinositol phospholipid anchor precursors in mammalian cells. Localization of phosphoethanolamine. *J Biol Chem* 267:16968–16974.

12. Flury, I., Benachour, A., and Conzelmann, A. (2000). YLL031c belongs to a novel family of membrane proteins involved in the transfer of ethanolaminephosphate onto the core structure of glycosylphosphatidylinositol anchors in yeast. *J Biol Chem* 275:24458–24465.

13. Mohney, R.P., Knez, J.J., Ravi, L., Sevlever, D., Rosenberry, T.L., Hirose, S., and Medof, M.E. (1994). Glycoinositol phospholipid anchor-defective K562 mutants with biochemical lesions distinct from those in Thy-1-murine lymphoma mutants. *J Biol Chem* 269:6536–6542.

14. Benachour, A., Sipos, G., Flury, I., Reggiori, F., Canivenc-Gansel, E., Vionnet, C., Conzelmann, A., and Benghezal, M. (1999). Deletion of *GPI7*, a yeast gene required for addition of a side chain to the glycosylphosphatidylinositol (GPI) core structure, affects GPI protein transport, remodeling, and cell wall integrity. *J Biol Chem* 274:15251–15261.

15. Hong, Y., Maeda, Y., Watanabe, R., Inoue, N., Ohishi, K., and Kinoshita, T. (2000). Requirement of PIG-F and PIG-O for transferring phosphoethanolamine to the third mannose in glycosylphosphatidylinositol. *J Biol Chem* 275:20911–20919.

16. Ueda, E., Sevlever, D., Prince, G.M., Rosenberry, T.L., Hirose, S., and Medof, M.E. (1993). A candidate mammalian glycoinositol phospholipid precursor containing three phosphoethanolamines. *J Biol Chem* 268:9998–10002.

17. Masterson, W.J., Doering, T.L., Hart, G.W., and Englund, P.T. (1989). A novel pathway for glycan assembly: biosynthesis of the glycosylphosphatidylinositol anchor of the trypanosome variant surface glycoprotein. *Cell* 56:793–800.

18. Menon, A.K., and Stevens, V.L. (1992). Phosphatidylethanolamine is the donor of the ethanolamine residue linking a glycosylphosphatidylinositol anchor to protein. *J Biol Chem* 267:15277–15280.

19. Imhof, I., Canivenc-Gansel, E., Meyer, U., and Conzelmann, A. (2000). Phosphatidylethanolamine is the donor of the phosphorylethanolamine linked to the α1,4-linked mannose of yeast GPI structures. *Glycobiology* 10:1271–1275.

20. Menon, A.K., Eppinger, M., Mayor, S., and Schwarz, R.T. (1993). Phosphatidylethanol-amine is the donor of the terminal phosphoethanolamine group in trypanosome glyco-sylphosphatidylinositols. *EMBO J* 12:1907–1914.

21. Toh-e, A., and Oguchi, T. (2002). Genetic characterization of genes encoding enzymes catalyzing addition of phospho-ethanolamine to the glycosylphosphatidylinositol anchor in *Saccharomyces cerevisiae*. *Genes Genet Syst* 77:309–322.

22. Vidugiriene, J., and Menon, A.K. (1993). Early lipid intermediates in glycosyl-phosphatidylinositol anchor assembly are synthesized in the ER and located in the cytoplasmic leaflet of the ER membrane bilayer. *J Cell Biol* 121:987–996.

23. Galperin, M.Y., and Jedrzejas, M.J. (2001). Conserved core structure and active site residues in alkaline phosphatase superfamily enzymes. *Proteins* 45:318–324.

24. Gaynor, E.C., Mondesert, G., Grimme, S.J., Reed, S.I., Orlean, P., and Emr, S.D. (1999). *MCD4* encodes a conserved endoplasmic reticulum membrane protein essential for glycosylphosphatidylinositol anchor synthesis in yeast. *Mol Biol Cell* 10:627–648.

25. Wiedman, J.M., Fabre, A.-L., Taron, B.W., Taron, C.H., and Orlean, P. (2007). *In vivo* characterization of the GPI assembly defect in yeast *mcd4–174* mutants and bypass of the Mcd4p-dependent step in *mcd4* null mutants. *FEMS Yeast Res* 7:78–83.

26. Sevlever, D., Mann, K.J., and Medof, M.E. (2001). Differential effect of 1,10-phenanthro-line on mammalian, yeast, and parasite glycosylphosphatidylinositol anchor synthesis. *Biochem Biophys Res Commun* 288:1112–1118.

27. Mann, K.J., and Sevlever, D. (2001). 1,10-Phenanthroline inhibits glycosylphosphatidyli-nositol anchoring by preventing phosphoethanolamine addition to glycosylphosphatidy-linositol anchor precursors. *Biochemistry* 40:1205–1213.

28. Shishioh, N., Hong, Y., Ohishi, K., Ashida, H., Maeda, Y., and Kinoshita, T. (2005). GPI7 is the second partner of PIG-F and involved in modification of glycosylphosphatidylino-sitol. *J Biol Chem* 280:9728–9734.

29. Delorenzi, M., Sexton, A., Shams-Elsin, H., Schwarz, R.T., Speed, T., and Schofield, L. (2002). Genes for glycosylphosphatidylinositol toxin biosynthesis in *Plasmodium falciparum*. *Infect Immun* 70:4510–4522.

30. Reynolds, C.M., Kalb, S.R., Cotter, R.J., and Raetz, C.R. (2005). A phosphoethanola-mine transferase specific for the outer 3-deoxy-D-manno-octulosonic acid residue of *Escherichia coli* lipopolysaccharide. Identification of the eptB gene and Ca^{2+} hypersensi-tivity of an eptB deletion mutant. *J Biol Chem* 280:21202–21211.

31. Masterson, W.J., and Ferguson, M.A. (1991). Phenylmethanesulphonyl fluoride inhibits GPI anchor biosynthesis in the African trypanosome. *EMBO J* 10:2041–2045.

32. Güther, M.L., Masterson, W.J., and Ferguson, M.A. (1994). The effects of phenylmethyl-sulfonyl fluoride on inositol-acylation and fatty acid remodeling in African trypanosomes. *J Biol Chem* 269:18694–18701.

33. Maneesri, J., Azuma, M., Sakai, Y., Igarashi, K., Matsumoto, T., Fukuda, H., Kondo, A., and Ooshima, H. (2005). Deletion of *MCD4* involved in glycosylphosphatidylinositol (GPI) anchor synthesis leads to an increase in β-1,6-glucan level and a decrease in GPI-anchored protein and mannan levels in the cell wall of *Saccharomyces cerevisiae*. *J Biosci Bioeng* 99:354–360.

34. Kang, J.Y., Hong, Y., Ashida, H., Shishioh, N., Murakami, Y., Morita, Y.S., Maeda, Y., and Kinoshita, T. (2005). PIG-V involved in transferring the second mannose in glyco-sylphosphatidylinositol. *J Biol Chem* 280:9489–9497.

35. Fabre, A.-L., Orlean, P., and Taron, C.H. (2005). *Saccharomyces cerevisiae* Ybr004c and its human homologue are required for addition of the second mannose during glycosyl-phosphatidylinositol precursor assembly. *FEBS J* 272:1160–1168.

36. Sutterlin, C., Horvath, A., Gerold, P., Schwarz, R.T., Wang, Y., Dreyfuss, M., and Riezman, H. (1997). Identification of a species-specific inhibitor of glycosylphosphatidylinositol synthesis. *EMBO J* 16:6374–6383.

37. Sütterlin, C., Escribano, M.V., Gerold, P., Maeda, Y., Mazon, M.J., Kinoshita, T., Schwarz, R.T., and Riezman, H. (1998). *Saccharomyces cerevisiae GPI10*, the functional homologue of human PIG-B, is required for glycosylphosphatidylinositol-anchor synthesis. *Biochem J* 332:153–159.

38. Zhu, Y., Vionnet, C., and Conzelmann, A. (2006). Ethanolaminephosphate side chain added to GPI anchor by Mcd4p is required for ceramide remodeling and forward transport of GPI proteins from ER to Golgi. *J Biol Chem* 281:19830–19839.

39. Sakai, Y., Azuma, M., Takada, Y., Umeyama, T., Kaneko, A., Fujita, T., Igarashi, K., and Ooshima, H. (2007). *Saccharomyces cerevisiae* mutant displaying beta-glucans on cell surface. *J Biosci Bioeng* 103:161–166.

40. Storey, M.K., Wu, W.I., and Voelker, D.R. (2001). A genetic screen for ethanolamine auxotrophs in *Saccharomyces cerevisiae* identifies a novel mutation on Mcd4p, a protein implicated in glycosylphosphatidylinositol anchor synthesis. *Biochim Biophys Acta* 1532:234–247.

41. Zhong, X., Malhotra, R., and Guidotti, G. (2003). ATP uptake in the Golgi and extracellular release require Mcd4 protein and the vacuolar H^+-ATPase. *J Biol Chem* 278:33436–33444.

42. Inoue, N., Kinoshita, T., Orii, T., and Takeda, J. (1993). Cloning of a human gene, PIG-F, a component of glycosyl-phosphatidylinositol anchor biosynthesis, by a novel expression cloning strategy. *J Biol Chem* 268:6882–6885.

43. Tohe, A., and Oguchi, T. (1999). Las21 participates in extracellular/cell surface phenomena in *Saccharomyces cerevisiae*. *Genes Genet Syst* 74:241–256.

44. Richard, M., De Groot, P., Courtin, O., Poulain, D., Klis, F., and Gaillardin, C. (2002). *GPI7* affects cell-wall protein anchorage in *Saccharomyces cerevisiae* and *Candida albicans*. *Microbiology* 148:2125–2133.

45. Fujita, M., Yoko-o, T., Okamoto, M., and Jigami, Y. (2004). *GPI7* involved in glycosylphosphatidylinositol biosynthesis is essential for yeast cell separation. *J Biol Chem* 279:51869–51879.

46. Lu, C.F., Montijn, R.C., Brown, J.L., Klis, F., Kurjan, J., Bussey, H., and Lipke, P.N. (1995). Glycosyl phosphatidylinositol-dependent cross-linking of α-agglutinin and β1,6-glucan in the *Saccharomyces cerevisiae* cell wall. *J Cell Biol* 128:333–340.

47. Kapteyn, J.C., Montijn, R.C., Vink, E., de la Cruz, J., Llobell, A., Douwes, J.E., Shimoi, H., Lipke, P.N., and Klis, F.M. (1996). Retention of *Saccharomyces cerevisiae* cell wall proteins through a phosphodiester-linked β-1,3-/β-1,6-glucan heteropolymer. *Glycobiology* 6:337–345.

48. Sugiyama, E., DeGasperi, R., Urakaze, M., Chang, H.M., Thomas, L.J., Hyman, R., Warren, C.D., and Yeh, E.T. (1991). Identification of defects in glycosylphosphatidylinositol anchor biosynthesis in the Thy-1 expression mutants. *J Biol Chem* 266:12119–12122.

49. Kamitani, T., Menon, A.K., Hallaq, Y., Warren, C.D., and Yeh, E.T. (1992). Complexity of ethanolamine phosphate addition in the biosynthesis of glycosylphosphatidylinositol anchors in mammalian cells. *J Biol Chem* 267:24611–24619.

50. Puoti, A., and Conzelmann, A. (1993). Characterization of abnormal free glycophosphatidylinositols accumulating in mutant lymphoma cells of classes B, E, F, and H. *J Biol Chem* 268:7215–7224.

51. Lemansky, P., Gupta, D.K., Meyale, S., Tucker, G., and Tartakoff, A.M. (1991). Atypical mannolipids characterize Thy-1-negative lymphoma mutants. *Mol Cell Biol* 11:3879–3885.

52. Grimme, S.J., Colussi, P.A., Taron, C.H., and Orlean, P. (2004). Deficiencies in the essential Smp3 mannosyltransferase block glycosylphosphatidylinositol assembly and lead to defects in growth and cell wall biogenesis in *Candida albicans*. *Microbiology* 150:3115–3128.

53. Murakami, Y., Siripanyapinyo, U., Hong, Y., Kang, J.Y., Ishihara, S., Nakakuma, H., Maeda, Y., and Kinoshita, T. (2003). PIG-W is critical for inositol acylation but not for flipping of glycosylphosphatidylinositol-anchor. *Mol Biol Cell* 14:4285–4295.

54. Vainauskas, S., and Menon, A.K. (2006). Ethanolamine phosphate linked to the first mannose residue of glycosylphosphatidylinositol (GPI) lipids is a major feature of the GPI structure that is recognized by human GPI transamidase. *J Biol Chem* 281:38358–38364.

7

Attachment of a GPI Anchor to Protein

AITA SIGNORELL • ANANT K. MENON

Department of Biochemistry
Weill Cornell Medical College
New York, NY 10065, USA

I.   Abstract

Proteins destined to receive a GPI anchor are synthesized with two signal sequences: an N-terminal signal sequence that targets the preproprotein to the endoplasmic reticulum (ER), and a C-terminal signal sequence that directs the attachment of a preassembled GPI anchor. The attachment of a GPI anchor to the protein's C-terminus occurs via a transamidation reaction that is catalyzed by the GPI transamidase (GPIT), a membrane-bound, penta-subunit enzyme complex. The different GPIT subunits are conserved from protozoa to yeast to mammals either by sequence or membrane topology.

While GPI8 has been shown to be the catalytic subunit, the function of the other proteins within the GPIT complex is still unclear. PIG-U and GAA1 have been proposed to be involved in recognizing and binding of the GPI, while comparison of PIG-T with proteins of known 3D structure predicts that it forms a funnel-like structure, leading to speculations about a role of this subunit in gating the proprotein to the GPI anchor for transamidation.

THE ENZYMES, Vol. XXVI   133   ISSN NO: 1874-6047
DOI: 10.1016/S1874-6047(09)26007-0

Subjects for future research include testing the involvement of these subunits in recognition and binding of the GPI anchor and/or the protein substrate, as well as structural analysis of GPIT.

II. The GPI Signal Sequence

The C-terminal GPI addition signal sequence is both necessary and sufficient to direct GPI attachment once the target protein is translocated into the ER lumen [1]. As illustrated in Figure 7.1A, the sequence has certain conserved features [1, 4–12]. The site of GPI attachment is designated as ω site; amino acids N-terminal to it are referred to as ω-minus and those C-terminal to it as ω-plus. Comparison of known and predicted GPI addition sites [8, 10, 11, 13, 14], as well as mutational analysis [6, 13] suggest that the general features of a signal anchor sequence are (i) a stretch of 10 polar amino acids (ω-10 to ω-1) that form a flexible linker region; (ii) the ω amino acid, typically Ser, Asn, Asp, Gly, Ala, or Cys; (iii) the ω + 1 amino acid consisting of any amino acid except Pro and Trp [6, 7]; (iv) Ala, Gly, Ser, and less often Thr, Arg, Cys, Met, and Trp at the ω + 2 position [6, 7]; (v) a spacer region of moderately polar amino acids (ω + 3 to ω + 9 or more); and finally (vi) a stretch of hydrophobic amino acids, variable in length but capable—in some cases—of spanning the membrane at the C-terminal end of the protein.

Mutagenesis studies indicate differences between organisms in their ability to process GPI addition signal sequences. For instance, the GPI addition signal sequence of *Trypanosoma brucei* variant surface glycoproteins (VSG) is not recognized by the transamidation machinery in mammalian cells [15]. This difference in signal sequence recognition is likely due to the ω and ω + 2 amino acids, which in parasite proteins are larger than those found in mammalian proteins [15]. Interestingly, replacement of the GPI attachment signal of VSG with that of mammalian decay accelerating factor (DAF), a regulator of the cell surface complement system, produced a protein that was appropriately processed, GPI anchored, and displayed on the mammalian cell surface [15, 16]. These studies suggest differences in the active site of the enzyme involved in GPI transfer to proteins between higher eukaryotes and protozoans [15, 17]. Interestingly, when GPIs are limiting there is a preferential transfer of the available GPIs to proteins with certain GPI signal sequences [18].

In *T. brucei*, GPI signal sequences on VSG are highly conserved; however, it has been shown that this conservation is not required for efficient VSG synthesis and anchoring, as mutagenesis of the most conserved residues had neither an effect on the efficiency of GPI anchoring nor on the abundance of the protein [17].

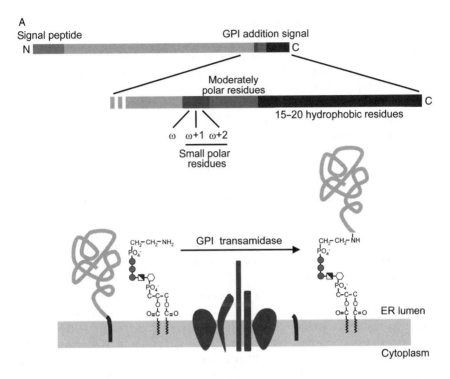

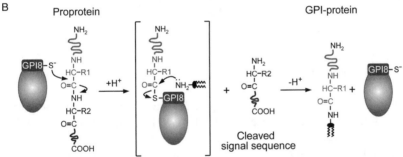

FIG. 7.1. The GPI signal sequence and attachment of a GPI anchor to proteins. (A) GPI-anchored proteins contain an N-terminal ER import signal (orange) and a C-terminal GPI addition sequence; the C-terminal signal is replaced by GPI in a reaction catalyzed by GPIT (modified from Ref. [2]). See text for details. (B) The GPI-anchoring reaction. Proproteins are recognized by GPIT on the luminal face of the ER. The C-terminal signal sequence is cleaved between ω (in red) and ω + 1 residues by the GPIT subunit GPI8, thereby activating the carbonyl group of the ω amino acid. An amide bond is formed by nucleophilic attack on the activated carbonyl by the amino group of the ethanolamine-phosphate cap of a GPI [3]. (See color plate section in the back of the book.)

III. The Transamidation Reaction

After translation and translocation to the luminal face of the ER, the proprotein's C-terminal GPI signal sequence is recognized by GPI transamidase (GPIT), a multisubunit membrane-bound enzyme complex. GPIT catalyzes the removal of the GPI signal sequence and its replacement with a GPI anchor. Displacing the proprotein's GPI signal sequence (residues $\omega + 1$ through the C-terminus) by GPIT activates the carbonyl group of the ω amino acid. The metabolic fate of the cleaved signal sequence is unknown, but it may be processed in a manner analogous to that described for cleaved ER-targeting N-terminal signals [19]. The activated carbonyl is susceptible to nucleophilic attack by the amino group of an ethanolamine-phosphate (Etn-P)-capped GPI yielding a GPI-anchored protein, and, in the process, regenerating GPIT [19–22] (Figure 7.1B). This process is ATP and GTP independent [20]. Evidence supporting carbonyl activation comes from microsome-based assays of GPI anchoring where small nucleophiles like hydrazine and hydroxylamine, were shown to be able to substitute for GPI; use of biotin hydrazide in one example, resulted in the C-terminal biotinylation of the target protein [21–23]. In such assays, nucleophilic attack by water is also seen [21], raising the hypothesis that a small percentage of protein is simply secreted from the ER without receiving a GPI anchor. Indeed, inhibition of GPI precursor biosynthesis by mannosamine in MDCK cells led to the hydrolysis of the proprotein and the secretion of a GPI-free protein into the medium [24].

IV. GPI Transamidase (GPIT)

Velocity gradient sedimentation [25, 26] and native gel electrophoretic analyses [25–28] demonstrate that GPIT is a large (>500 kDa) complex. Genetic evidence and coimmunoprecipitation studies indicate that GPIT is composed of five subunits; in mammals and yeast these are GPI8 (or PIG-K)/Gpi8p, GAA1/Gaa1p, PIG-S/Gpi17p, PIG-T/Gpi16p, and PIG-U/Gab1p, respectively (also see Table 7.1).

The subunits may be organized into two subcomplexes: one containing Gpi8p, Gpi16p, and Gaa1p and the other being composed of the two remaining subunits, Gab1p and Gpi17p [26, 27, 37]. It has been shown in yeast that the GPIT complex persists even after the depletion of GPIs and precursor proteins [27].

GPITs from *Drosophila melanogaster*, *Caenorhabditis elegans*, and *Arabidopsis thaliana* are closely related to the mammalian and yeast

TABLE 7.1

GPIT Subunits in Mammals, Yeast and *T. brucei* (Adapted from Ref. [3, 29])

Mammals (kDa)	Yeast (kDa)	*T. brucei* (kDa)
PIG-K (45.3) [30]	Gpi8p (47.4) [31]	TbGPI8 (36.7) [32]
GAA1 (67.6) [33]	Gaa1p (69.2) [34]	TbGAA1 (51.2) [28]
PIG-S (61.7) [35]	Gpi17p (60.8) [35]	–
PIG-T (65.7) [35]	Gpi16p (68.8) [35]	TbGPI16 (75.8) [28]
PIG-U (50.1) [36]	Gab1p/Cdc91p (44.7) [36]	–
–	–	TTA1 (41.9) [28]
–	–	TTA2 (45.6) [28]

enzyme. In contrast, GPITs from trypanosomatids such as *T. brucei* share three subunits with the mammalian and yeast complexes (homologs of GPI8/Gpi8p, GAA1/Gaa1p, and PIG-T/Gpi16p termed TbGPI8, TbGaa1, and TbGPI16, respectively) but have two novel subunits, trypanosomatid transamidase 1 (TTA1) and TTA2 in place of PIG-S/Gpi17p and PIG-U/Gab1p. Nevertheless, their proposed topology closely resembles the ones predicted for PIG-S/Gpi17p and PIG-U/Gab1p [28] (Figure 7.2).

All five subunits are essential for human and trypanosomatid GPIT function, as all have to be present for the nucleophilic attack on the ω residue that initiates the transamidation reaction [35, 36, 38]. While two of the subunits, GPI8/Gpi8p and PIG-U/Gab1p, share some sequence homology with proteins of known function, the functions of the other subunits of GPIT cannot be deduced from their primary sequence. These subunits resemble only their counterparts in other eukaryotes and share no sequence similarity with proteins of known function. The situation is similar for the ER-localized oligosaccharyltransferase (OST), a complex that in yeast and mammals is composed of eight subunits, many of which are essential [39]. OST, like GPIT, engages protein and lipid (dolichol-PP-oligosaccharide) in order to generate an *N*-glycosylated protein. Notwithstanding the multi-subunit complex, a single OST subunit, STT3, is able to carry out oligosaccharyl transfer alone in a bacterial system [40]. Furthermore, in the trypanosomatid *Leishmania major*, STT3 is the only OST subunit that can be identified, and leishmanial STT3 can complement stt3 deficiency in *S. cerevisiae* as a monomeric or homodimeric enzyme [41, 42], supporting the one-subunit model predicted for the *L. major* OST enzyme. Because of these observations, the question about the variety and essentiality of the other OST subunits in eukaryotes arises. The same is the case for trypanosomatid GPI8 which is catalytically active on its own, at least as a protease [32].

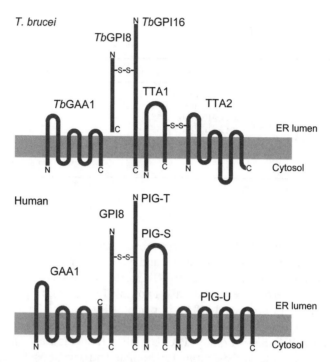

Fig. 7.2. Comparison of the five subunits of *T. brucei* and human GPIT. Trypanosomes share three subunits with humans (GPI8, GAA1, and GPI16/PIG-T), but have two novel subunits, TTA1 and TTA2 in place of PIG-S and PIG-U [28].

For both GPIT and OST, it seems likely that the noncatalytic subunits are necessary for selecting and recruiting the two very different substrates, lipid and protein, that are subsequently handled by the enzymes, or they might be involved in regulating substrate access to the catalytic site.

V. GPI8/Gpi8p

GPI8/Gpi8p is presumed to be the catalytic center of the enzyme, because (i) it shares homology (25–28% in yeast) with a family of cysteine proteases, one member of which, jack bean asparaginyl endopeptidase, displays transamidase activity *in vitro* [31, 43]; (ii) mutagenesis of a putative cysteine–histidine catalytic dyad in the human, yeast, and *Leishmania mexicana* GPI8/Gpi8p sequence inactivates GPIT activity [38, 43, 44]; and

(iii) it is in physical proximity to the proprotein substrate [45, 46]. *T. brucei* GPI8, when expressed in *Escherichia coli*, cleaves a tetrapeptide substrate [32], indicating that TbGPI8, in the absence of other GPIT subunits, is catalytically active, at least as a protease.

In mammals and yeast, GPI8/Gpi8p is a type I membrane protein with the majority of its sequence located in the ER lumen [30, 31]; in contrast, in many lower eukaryotes such as *D. melanogaster*, nematodes (*C. elegans*), and trypanosomatids (*T. brucei* and *L. major*), GPI8 is a soluble protein of the ER lumen, suggesting that the transmembrane (TM) domain is functionally dispensable [28, 47, 48]. Consistent with this, a human GPI8 construct lacking the TM domain is able to rescue GPI anchoring in GPI8-deficient K562 cells [38]. In yeast, heterologous expression of a soluble truncation mutant of Gpi8p forms a homodimer, thereby resembling caspases that share some sequence similarity with Gpi8p (Figure 7.3). However, this Gpi8p dimer is catalytically inactive [49].

Yeast Gpi8p contains three *N*-glycosylation sites [31], whereas the *T. brucei* GPI8 sequence predicts one single *N*-linked glycosylation site near the N-terminus that is not conserved in other GPI8 molecules of

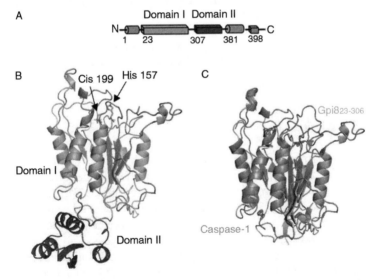

FIG. 7.3. Predicted structure for yeast Gpi8p. The protein sequence predicts an N-terminal signal sequence and a C-terminal TM domain (gray tubes, panel (A)), and two soluble domains (domain I in turquoise and domain II in purple). Panel (B) depicts the predicted structure of yeast Gpi8p, showing the two domains in the same color as in the top panel. The putative catalytic dyad, consisting of Cys199 and His157, is highlighted by arrows. (C) Overlay of a model of the caspase-like domain of yeast Gpi8p (turquoise) with the resolved crystal structure of Caspase-1 from *Spodoptera frugiperda* (PDB: 1M72, orange) (adapted from Ref. [49]). (See color plate section in the back of the book.)

other species [32]. Moreover, GPI8 /Gpi8p is disulfide-linked to PIG-T/ Gpi16p [50]. Although the association between these two GPIT subunits is not essential, it is required for full transamidase activity *in vivo* [50, 51]. Furthermore, Gpi8p is stabilized by this incorporation into the transami- dase complex: depletion of yeast Gpi16p leads to a concomitant loss of Gpi8p protein, and *vice versa* [27].

In procyclic *T. brucei*, deletion of the gene encoding GPI8 resulted in parasites deficient in GPI-anchored proteins and accumulation of mature GPIs and their display on the cell surface [48]. Accordingly, in the absence of EP and GPEET procyclins, the major protein substrate for procyclic *T. brucei* GPIT, free GPIs are upregulated on the surface of the parasite [52]. It is believed that the cell surface expression of GPIs provides the parasites with a protease-resistant glycocalyx that enables them to survive in the hostile environment of the lumen of the tsetse fly midgut. Similarly, in the $\Delta GPI8$ mutant of *L. mexicana* GPI-anchored proteins are no longer detectable on the cell surface while expression of nonprotein linked GPIs like lipophosphoglycans (LPGs) and glycosylinositol phospholipids (GIPLs) is not affected [53]. These null mutants display normal growth in *in vitro* culture as well as within macrophages and can establish an infection in mice. In contrast, in the bloodstream form of *T. brucei*, GPI8 depletion by RNA interference leads to a block in cytokinesis and to cell death, suggesting that GPI-anchored proteins are essential in the bloodstream form trypanosome and that GPI8 is important for proper cell cycle progres- sion [48].

VI.   GAA1/Gaa1p

GAA1/Gaa1p, named for its role in GPI anchor attachment, was the first component of the GPIT complex to be identified by genetic approaches [34] and is one of the two most hydrophobic subunits of GPIT. The gene is predicted to encode a ~68 kDa protein, with an uncleaved, cytoplasmically oriented N-terminal signal sequence, and a large, hydrophilic, luminal domain, followed by several TM domains (six in yeast and trypanosomes and seven in human cells) [25, 28, 35]. Two putative *N*-linked glycosylation sites were identified in the yeast Gaa1p sequence [54]. Strict conservation of two stretches of polar residues located at the interface between ER membrane and lumen (one at the C-terminal side of the first TM region and the other at the N-terminal side of the second TM span) suggests a possible involvement in binding of the free GPI lipid [55]. There are also two other conserved motifs ($RXPRX_3TE$ and NGX_2PNXDX_2N) within the globular, luminal part of the protein that might play a role in this context.

Furthermore, site-specific photocrosslinking experiments indicate that mammalian GAA1 is physically close to proproteins that are bound to GPIT [46], and coimmunoprecipitation analyses show that GAA1 is required for mammalian GPI8 to recognize its protein substrate [56].

GAA1/Gaa1p coimmunoprecipitates with GPI8/Gpi8p in mammals and yeast, respectively [35, 56]. The luminal loop of the protein, located between the first and second TM segments, is required for interaction with GPI8, PIG-S, and PIG-T [25]. The cytoplasmic N-terminus of GAA1, however, is not required for the formation of a functional GPIT complex but may act as a membrane-sorting determinant directing GAA1 and associated GPIT subunits to an ER membrane domain [25].

GPIT complexes containing C-terminally truncated GAA1 possess a full complement of subunits and are able to interact with the proprotein substrate but fail to coimmunoprecipitate GPI [25]. In this context, a conserved proline in the last TM segment of the C-terminus is of particular interest. Prolines with a GXXP or GXP motif located in the middle of a TM α-helix have been shown to behave as a molecular hinge capable of dynamically kinking and swiveling the helix. Mutation of this highly conserved Pro609 in the mammalian GAA1 sequence also resulted in the inability of the resulting GPIT complex to coimmunoprecipitate GPI [57]. These results indicate that GAA1 plays a key role in GPI recognition by GPIT.

The immunoprecipitation methodology applied to identify a role for GAA1 in GPI binding was also used to identify the molecular features of GPI that are recognized by human GPIT. Surprisingly, the Etn-P cap linked to the third mannose (Man-3) of the GPI core structure, that is present in all species and required for GPI attachment to protein, was shown not to be a prerequisite for GPI recognition, but instead, the Etn-P residue linked to Man-1 was critical [58]. Thus, in an *in vitro* assay, human GPIT was able to pull down the minimal GPI (Etn-P)2Manα1–4GlcN-acylPI [58] and its more elaborate variants. However, Etn-P-capped GPIs lacking the Etn-P modification on Man-1 were poorly recognized. This is a point of interest for future work, because trypanosomatids do not modify GPIs with Etn-P on Man-1 and, as described above, trypanosomatid GPIT complexes are compositionally distinct from the mammalian and yeast family of GPITs, thereby representing an attractive potential antitrypanosomatid drug target.

VII.   PIG-T/Gpi16p

Coimmunoprecipitation studies suggest that PIG-T/Gpi16p may be critical for formation and stability of the GPIT complex [35]. PIG-T/Gpi16p is predicted to encode a type I membrane protein with two *N*-glycosylation

Fig. 7.4. N-terminal 430 amino acids of prolyl oligopeptidase from porcine brain (PDB: 1H2W, [60]). The protein is shown along the axis of the β-propeller to visualize the pore structure. The N-terminus is directed to the left, the C-terminus is at the top of the structure.

sites, an N-terminal ER-targeting signal sequence, a large luminal segment with a high sequence complexity and probable globular structure, and a single C-terminal TM domain followed by a short, polar cytoplasmic tail. According to secondary structure predictions, the luminal region consists N-terminally of two β–strands and a preferentially α-helical part followed almost exclusively by two β-strands at the N-terminal region [12]. Compatibility tests with known three-dimensional (3D) structures generate a single reasonable hit, that with the N-terminal chain of the 3D structure of the prolyl oligopeptidase from porcine muscle, essentially a seven-bladed β-propeller structure [59]. It has been proposed that this β-propeller forms a funnel that gates the access of proteins to the active site of GPI8/Gpi8p contributing to the specificity of the GPI anchor addition [12] (Figure 7.4).

VIII.   PIG-U/Gab1p

PIG-U/Gab1p encodes a highly hydrophobic protein with a number of TM domains [36, 37]. Interestingly, its sequence contains a motif found in mammalian and yeast fatty acid elongases [36]. This motif has been found to be important for mammalian PIG-U function and it has been suggested to play a role in recognizing long chain fatty acids of the GPI lipid. When the conserved aromatic amino acids in the elongase motif were mutated to leucine, the mutant recombinant protein failed to restore GPI anchoring in mammalian PIG-U mutants [36].

PIG-U is not essential for the formation of the GPIT complex, as high-molecular weight complexes of GPI8, GAA1, GPIT, and PIG-S were purified from mammalian class U mutants. However, the PIG-U-deficient GPIT complex is nonfunctional and lacks GPI anchoring activity [36].

IX. PIG-S/Gpi17

PIG-S/Gpi17 consists of a hydrophobic region close to the N-terminus followed by a relatively polar segment, an increasingly hydrophobic C-terminal half and a final, lysine-rich part at the C-terminus [26, 35], an arrangement that predicts two putative TM regions, one close to each of the two termini. The absence of the leading methionine in the mature form of the human protein is consistent with a cytoplasmic orientation of the N-terminus. The sequence between the two TM regions is located in the ER lumen and can interact with other components of GPIT. Surprisingly, there are only very few conserved residues with polar, functional side chains among PIG-S/Gpi17p orthologs; thus, binding of ligands with many polar groups such as the GPI anchor seems highly unlikely. It is possible that PIG-S/Gpi17p plays a role in maintaining the structure of the transamidase complex or in the selectivity of species-specific substrate proteins.

X. TTA1 and TTA2

As mentioned above, GPITs from *T. brucei* share three subunits with mammalian/yeast GPIT (GPI8/Gpi8p, GAA1/ Gaa1p, and PIG-T/Gpi16), but have two novel subunits in place of PIG-S/Gpi17p and PIG-U/Gab1p [28]. Homologs of TTA1 and TTA2 are present in *Leishmania sp.* and *Trypanosoma cruzi*, but not in mammals, yeasts, flies, nematodes, plants or malaria parasites, suggesting that these components may play unique roles in GPI attachment in trypanosomatid parasites.

TTA1 has two potential *N*-glycosylation sites; the size of TTA1 decreased after treatment with peptide *N*-glycanase F indicating that one or both sites were *N*-glycosylated, and that the central hydrophilic part faces the luminal side of the ER. In addition, the N-terminal methionine was removed, which is consistent with its cytoplasmic orientation. TTA2, on the other hand, has six predicted TM regions, and TTA1 and TTA2 are linked via disulfide bonds. Remarkably, hydrophobicity profiles and membrane orientations of TTA1 and TTA2 are similar to those of PIG-S and PIG-U (compare Figure 7.2).

XI. Final Remarks

In addition to their involvement in GPI anchoring, components of GPIT have yet other, unknown functions. Depletion of Gab1p or Gpi8p in yeast, but not of Gaa1p, Gpi16p, or Gpi17p, results in the accumulation of bar-like structures of actin that are closely associated with the perinuclear ER and that are decorated with the actin-binding protein cofilin [37]. Such actin bars were not seen in other mutants with a defect in GPI assembly; therefore, they might not reflect a consequence of a defect in GPI anchoring. Based on these findings, it was speculated that Gab1p and Gpi8p, among other proteins residing in the ER membrane, are involved in a functional interaction between the ER and the actin cytoskeleton, although it is not clear whether this interaction is direct or indirect.

Furthermore, several subunits of GPIT are involved in the development of different cancers. PIG-U is overexpressed in bladder cancer cell lines, and is associated with malignant transformations *in vitro* and *in vivo* [61]. PIG-T, GAA1, and PIG-U are overexpressed and have increased gene copy numbers in human breast cancer cell lines and primary tumors [62]. By screening expression patterns, overexpression of GPIT subunits was also seen in lymphoma, nonsmall cell lung carcinoma, and various other cancers [63]. Whether the components of the GPIT complex function independently or as a functional group of oncogenes remains to be elucidated.

XII. Future Work

Roughly 1% of all eukaryotic proteins are modified with a GPI anchor, among them are cell surface receptors (e.g., folate receptor, CD14), cell adhesion molecules (e.g., NCAMs, carcinoembryonic antigen variants, fasciclin I), cell surface hydrolases (e.g., 5′-nucleotidase, aceltylcholinesterase), complement regulatory proteins (e.g., DAF), and protozoal coat antigens (e.g., *T. brucei* VSG). All these proteins require GPI in order to be expressed at the cell surface, and to function. It appears that most, if not all, of the proteins involved in the transfer of GPI to protein have been identified. However, there is no high-resolution structural information available for any of the GPIT subunits, and membrane-topological information relies mostly on prediction tools. Also, further research has to be undertaken to understand the mechanism of GPI transfer to protein. While GPI8 has been shown to be the catalytic subunit, the function of other proteins within the GPIT complex is still unclear. Different roles have been proposed but remain to be demonstrated. Subjects for future research include the identification of the subunits involved in recognition and binding of the GPI anchor and of the protein substrate. In this context, the specificity of the recognition and binding of GPI structures by GPIT will have to be resolved in order to determine possible differences between organisms, as there is variation not only in the GPIT complex but also in GPI structures. This aspect is particularly interesting as GPI anchoring has been identified as a potential drug target against the causative agent of African sleeping sickness, *T. brucei* [64]. The blood stream form trypanosomes depend on GPI-anchored proteins; therefore, GPI anchoring represents a potential target for antitrypanosomal drugs. In light of this goal, the two GPIT subunits that are unique to trypanosomes are of special interest, as well as the fact that differences exist in the recognition of GPI addition signals between humans and trypanosomes.

ACKNOWLEDGEMENTS

This chapter was developed after several reviews [3, 12, 29, 54, 65]. Supported by Swiss National Science Foundation Grant PBBEP3–123662 to AS and National Institutes of Health Grant GM55427 to AKM.

REFERENCES

1. Caras, I.W., Weddell, G.N., Davitz, M.A., Nussenzweig, V., and Martin, D.W., Jr. (1987). Signal for attachment of a phospholipid membrane anchor in decay accelerating factor. *Science* 238:1280–1283.
2. Ferguson, M.A.J., Kinoshita, T., and Hart, G.W. (2009). Glycosylphosphatidylinositol anchors. In Essentials of Glycobiology, A. Varki, R.D. Cummings, J.D. Esko, H.H. Freeze, P. Stanley, C.R. Bertozzi, G.W. Hart and M.E. Etzler, (eds.). Cold Spring Harbor Laboratory Press, Cold Spring Harbor, NY.
3. Orlean, P., and Menon, A.K. (2007). Thematic review series: lipid posttranslational modifications. GPI anchoring of protein in yeast and mammalian cells, or: how we learned to stop worrying and love glycophospholipids. *J Lipid Res* 48:993–1011.
4. Micanovic, R., Gerber, L.D., Berger, J., Kodukula, K., and Udenfriend, S. (1990). Selectivity of the cleavage/attachment site of phosphatidylinositol-glycan-anchored membrane proteins determined by site-specific mutagenesis at Asp-484 of placental alkaline phosphatase. *Proc Natl Acad Sci USA* 87:157–161.
5. Moran, P., and Caras, I.W. (1991). A nonfunctional sequence converted to a signal for glycophosphatidylinositol membrane anchor attachment. *J Cell Biol* 115:329–336.

6. Gerber, L.D., Kodukula, K., and Udenfriend, S. (1992). Phosphatidylinositol glycan (PI-G) anchored membrane proteins. Amino acid requirements adjacent to the site of cleavage and PI-G attachment in the COOH-terminal signal peptide. *J Biol Chem* 267:12168–12173.
7. Nuoffer, C., Horvath, A., and Riezman, H. (1993). Analysis of the sequence requirements for glycosylphosphatidylinositol anchoring of *Saccharomyces cerevisiae* Gas1 protein. *J Biol Chem* 268:10558–10563.
8. Caro, L.H., Tettelin, H., Vossen, J.H., Ram, A.F., van den Ende, H., and Klis, F.M. (1997). *In silicio* identification of glycosyl-phosphatidylinositol-anchored plasma-membrane and cell wall proteins of *Saccharomyces cerevisiae*. *Yeast* 13:1477–1489.
9. Eisenhaber, B., Bork, P., and Eisenhaber, F. (1998). Sequence properties of GPI-anchored proteins near the omega-site: constraints for the polypeptide binding site of the putative transamidase. *Protein Eng* 11:1155–1161.
10. Hamada, K., Fukuchi, S., Arisawa, M., Baba, M., and Kitada, K. (1998). Screening for glycosylphosphatidylinositol (GPI)-dependent cell wall proteins in *Saccharomyces cerevisiae*. *Mol Gen Genet* 258:53–59.
11. De Groot, P.W., Hellingwerf, K.J., and Klis, F.M. (2003). Genome-wide identification of fungal GPI proteins. *Yeast* 20:781–796.
12. Eisenhaber, B., Maurer-Stroh, S., Novatchkova, M., Schneider, G., and Eisenhaber, F. (2003). Enzymes and auxiliary factors for GPI lipid anchor biosynthesis and post-translational transfer to proteins. *Bioessays* 25:367–385.
13. Furukawa, Y., Tsukamoto, K., and Ikezawa, H. (1997). Mutational analysis of the C-terminal signal peptide of bovine liver 5′-nucleotidase for GPI anchoring: a study on the significance of the hydrophilic spacer region. *Biochim Biophys Acta* 1328:185–196.
14. Eisenhaber, B., Schneider, G., Wildpaner, M., and Eisenhaber, F. (2004). A sensitive predictor for potential GPI lipid modification sites in fungal protein sequences and its application to genome-wide studies for *Aspergillus nidulans*, *Candida albicans*, *Neurospora crassa*, *Saccharomyces cerevisiae* and *Schizosaccharomyces pombe*. *J Mol Biol* 337:243–253.
15. Moran, P., and Caras, I.W. (1994). Requirements for glycosylphosphatidylinositol attachment are similar but not identical in mammalian cells and parasitic protozoa. *J Cell Biol* 125:333–343.
16. Caras, I.W., and Moran, P. (1994). The requirements for GPI-attachment are similar but not identical in mammalian cells and parasitic protozoa. *Braz J Med Biol Res* 27:185–188.
17. Böhme, U., and Cross, G.A. (2002). Mutational analysis of the variant surface glycoprotein GPI-anchor signal sequence in *Trypanosoma brucei*. *J Cell Sci* 115:805–816.
18. Thomas, L.J., Urakaze, M., DeGasperi, R., Kamitani, T., Sugiyama, E., Chang, H.M., Warren, C.D., and Yeh, E.T. (1992). Differential expression of glycosylphosphatidylinositol-anchored proteins in a murine T cell hybridoma mutant producing limiting amounts of the glycolipid core. Implications for paroxysmal nocturnal hemoglobinuria. *J Clin Invest* 89:1172–1177.
19. Weihofen, A., Lemberg, M.K., Ploegh, H.L., Bogyo, M., and Martoglio, B. (2000). Release of signal peptide fragments into the cytosol requires cleavage in the transmembrane region by a protease activity that is specifically blocked by a novel cysteine protease inhibitor. *J Biol Chem* 275:30951–30956.
20. Mayor, S., Menon, A.K., and Cross, G.A. (1991). Transfer of glycosyl-phosphatidylinositol membrane anchors to polypeptide acceptors in a cell-free system. *J Cell Biol* 114:61–71.
21. Maxwell, S.E., Ramalingam, S., Gerber, L.D., Brink, L., and Udenfriend, S. (1995). An active carbonyl formed during glycosylphosphatidylinositol addition to a protein is evidence of catalysis by a transamidase. *J Biol Chem* 270:19576–19582.

22. Sharma, D.K., Vidugiriene, J., Bangs, J.D., and Menon, A.K. (1999). A cell-free assay for glycosylphosphatidylinositol anchoring in African trypanosomes. Demonstration of a transamidation reaction mechanism. *J Biol Chem* 274:16479–16486.

23. Chen, R., Udenfriend, S., Prince, G.M., Maxwell, S.E., Ramalingam, S., Gerber, L.D., Knez, J., and Medof, M.E. (1996). A defect in glycosylphosphatidylinositol (GPI) transamidase activity in mutant K cells is responsible for their inability to display GPI surface proteins. *Proc Natl Acad Sci USA* 93:2280–2284.

24. Lisanti, M.P., Field, M.C., Caras, I.W., Menon, A.K., and Rodriguez-Boulan, E. (1991). Mannosamine, a novel inhibitor of glycosylphosphatidylinositol incorporation into proteins. *EMBO J* 10:1969–1977.

25. Vainauskas, S., Maeda, Y., Kurniawan, H., Kinoshita, T., and Menon, A.K. (2002). Structural requirements for the recruitment of Gaa1 into a functional glycosylphosphatidylinositol transamidase complex. *J Biol Chem* 277:30535–30542.

26. Zhu, Y., Fraering, P., Vionnet, C., and Conzelmann, A. (2005). Gpi17p does not stably interact with other subunits of glycosylphosphatidylinositol transamidase in *Saccharomyces cerevisiae*. *Biochim Biophys Acta* 1735:79–88.

27. Fraering, P., Imhof, I., Meyer, U., Strub, J.M., van Dorsselaer, A., Vionnet, C., and Conzelmann, A. (2001). The GPI transamidase complex of *Saccharomyces cerevisiae* contains Gaa1p, Gpi8p, and Gpi16p. *Mol Biol Cell* 12:3295–3306.

28. Nagamune, K., Ohishi, K., Ashida, H., Hong, Y., Hino, J., Kangawa, K., Inoue, N., Maeda, Y., and Kinoshita, T. (2003). GPI transamidase of *Trypanosoma brucei* has two previously uncharacterized (trypanosomatid transamidase 1 and 2) and three common subunits. *Proc Natl Acad Sci USA* 100:10682–10687.

29. Pittet, M., and Conzelmann, A. (2007). Biosynthesis and function of GPI proteins in the yeast *Saccharomyces cerevisiae*. *Biochim Biophys Acta* 1771:405–420.

30. Yu, J., Nagarajan, S., Knez, J.J., Udenfriend, S., Chen, R., and Medof, M.E. (1997). The affected gene underlying the class K glycosylphosphatidylinositol (GPI) surface protein defect codes for the GPI transamidase. *Proc Natl Acad Sci USA* 94:12580–12585.

31. Benghezal, M., Benachour, A., Rusconi, S., Aebi, M., and Conzelmann, A. (1996). Yeast Gpi8p is essential for GPI anchor attachment onto proteins. *EMBO J* 15:6575–6583.

32. Kang, X., Szallies, A., Rawer, M., Echner, H., and Duszenko, M. (2002). GPI anchor transamidase of *Trypanosoma brucei*: *in vitro* assay of the recombinant protein and VSG anchor exchange. *J Cell Sci* 115:2529–2539.

33. Hiroi, Y., Komuro, I., Chen, R., Hosoda, T., Mizuno, T., Kudoh, S., Georgescu, S.P., Medof, M.E., and Yazaki, Y. (1998). Molecular cloning of human homolog of yeast GAA1 which is required for attachment of glycosylphosphatidylinositols to proteins. *FEBS Lett* 421:252–258.

34. Hamburger, D., Egerton, M., and Riezman, H. (1995). Yeast Gaa1p is required for attachment of a completed GPI anchor onto proteins. *J Cell Biol* 129:629–639.

35. Ohishi, K., Inoue, N., and Kinoshita, T. (2001). PIG-S and PIG-T, essential for GPI anchor attachment to proteins, form a complex with GAA1 and GPI8. *EMBO J* 20:4088–4098.

36. Hong, Y., Ohishi, K., Kang, J.Y., Tanaka, S., Inoue, N., Nishimura, J., Maeda, Y., and Kinoshita, T. (2003). Human PIG-U and yeast Cdc91p are the fifth subunit of GPI transamidase that attaches GPI-anchors to proteins. *Mol Biol Cell* 14:1780–1789.

37. Grimme, S.J., Gao, X.D., Martin, P.S., Tu, K., Tcheperegine, S.E., Corrado, K., Farewell, A.E., Orlean, P., and Bi, E. (2004). Deficiencies in the endoplasmic reticulum (ER)-membrane protein Gab1p perturb transfer of glycosylphosphatidylinositol to proteins and cause perinuclear ER-associated actin bar formation. *Mol Biol Cell* 15:2758–2770.

148   AITA SIGNORELL AND ANANT K. MENON

38. Ohishi, K., Inoue, N., Maeda, Y., Takeda, J., Riezman, H., and Kinoshita, T. (2000). Gaa1p and gpi8p are components of a glycosylphosphatidylinositol (GPI) transamidase that mediates attachment of GPI to proteins. *Mol Biol Cell* 11:1523–1533.
39. Kelleher, D.J., and Gilmore, R. (2006). An evolving view of the eukaryotic oligosaccharyltransferase. *Glycobiology* 16:47R–62R.
40. Wacker, M., Linton, D., Hitchen, P.G., Nita-Lazar, M., Haslam, S.M., North, S.J., Panico, M., Morris, H.R., Dell, A., Wren, B.W., and Aebi, M. (2002). N-linked glycosylation in *Campylobacter jejuni* and its functional transfer into *E. coli*. *Science* 298:1790–1793.
41. Nasab, F.P., Schulz, B.L., Gamarro, F., Parodi, A.J., and Aebi, M. (2008). All in one: *Leishmania major* STT3 proteins substitute for the whole oligosaccharyltransferase complex in *Saccharomyces cerevisiae*. *Mol Biol Cell* 19:3758–3768.
42. Hese, K., Otto, C., Routier, F.H., and Lehle, L. (2009). The yeast oligosaccharyltransferase complex can be replaced by STT3 from *Leishmania major*. *Glycobiology* 19:160–171.
43. Meyer, U., Benghezal, M., Imhof, I., and Conzelmann, A. (2000). Active site determination of Gpi8p, a caspase-related enzyme required for glycosylphosphatidylinositol anchor addition to proteins. *Biochemistry* 39:3461–3471.
44. Ellis, M., Sharma, D.K., Hilley, J.D., Coombs, G.H., and Mottram, J.C. (2002). Processing and trafficking of *Leishmania mexicana* GP63. Analysis using GP18 mutants deficient in glycosylphosphatidylinositol protein anchoring. *J Biol Chem* 277:27968–27974.
45. Spurway, T.D., Dalley, J.A., High, S., and Bulleid, N.J. (2001). Early events in glycosylphosphatidylinositol anchor addition. Substrate proteins associate with the transamidase subunit gpi8p. *J Biol Chem* 276:15975–15982.
46. Vidugiriene, J., Vainauskas, S., Johnson, A.E., and Menon, A.K. (2001). Endoplasmic reticulum proteins involved in glycosylphosphatidylinositol-anchor attachment: photo-crosslinking studies in a cell-free system. *Eur J Biochem* 268:2290–2300.
47. Sharma, D.K., Hilley, J.D., Bangs, J.D., Coombs, G.H., Mottram, J.C., and Menon, A.K. (2000). Soluble GPI8 restores glycosylphosphatidylinositol anchoring in a trypanosome cell-free system depleted of lumenal endoplasmic reticulum proteins. *Biochem J* 351 (3):717–722.
48. Lillico, S., Field, M.C., Blundell, P., Coombs, G.H., and Mottram, J.C. (2003). Essential roles for GPI-anchored proteins in African trypanosomes revealed using mutants deficient in GPI8. *Mol Biol Cell* 14:1182–1194.
49. Meitzler, J.L., Gray, J.J., and Hendrickson, T.L. (2007). Truncation of the caspase-related subunit (Gpi8p) of *Saccharomyces cerevisiae* GPI transamidase: dimerization revealed. *Arch Biochem Biophys* 462:83–93.
50. Ohishi, K., Nagamune, K., Maeda, Y., and Kinoshita, T. (2003). Two subunits of glycosylphosphatidylinositol transamidase, GPI8 and PIG-T, form a functionally important intermolecular disulfide bridge. *J Biol Chem* 278:13959–13967.
51. Hong, Y., Nagamune, K., Ohishi, K., Morita, Y.S., Ashida, H., Maeda, Y., and Kinoshita, T. (2006). TbGPI16 is an essential component of GPI transamidase in *Trypanosoma brucei*. *FEBS Lett* 580:603–606.
52. Vassella, E., Bütikofer, P., Engstler, M., Jelk, J., and Roditi, I. (2003). Procyclin null mutants of *Trypanosoma brucei* express free glycosylphosphatidylinositols on their surface. *Mol Biol Cell* 14:1308–1318.
53. Hilley, J.D., Zawadzki, J.L., McConville, M.J., Coombs, G.H., and Mottram, J.C. (2000). *Leishmania mexicana* mutants lacking glycosylphosphatidylinositol (GPI):protein transamidase provide insights into the biosynthesis and functions of GPI-anchored proteins. *Mol Biol Cell* 11:1183–1195.

54. Zacks, M.A., and Garg, N. (2006). Recent developments in the molecular, biochemical and functional characterization of GPI8 and the GPI-anchoring mechanism [review]. *Mol Membr Biol* 23:209–225.

55. Eisenhaber, B., Bork, P., and Eisenhaber, F. (2001). Post-translational GPI lipid anchor modification of proteins in kingdoms of life: analysis of protein sequence data from complete genomes. *Protein Eng* 14:17–25.

56. Chen, R., Anderson, V., Hiroi, Y., and Medof, M.E. (2003). Proprotein interaction with the GPI transamidase. *J Cell Biochem* 88:1025–1037.

57. Vainauskas, S., and Menon, A.K. (2004). A conserved proline in the last transmembrane segment of Gaa1 is required for glycosylphosphatidylinositol (GPI) recognition by GPI transamidase. *J Biol Chem* 279:6540–6545.

58. Vainauskas, S., and Menon, A.K. (2006). Ethanolamine phosphate linked to the first mannose residue of glycosylphosphatidylinositol (GPI) lipids is a major feature of the GPI structure that is recognized by human GPI transamidase. *J Biol Chem* 281:38358–38364.

59. Fülöp, V., Böcskei, Z., and Polgár, L. (1998). Prolyl oligopeptidase: an unusual beta-propeller domain regulates proteolysis. *Cell* 94:161–170.

60. Szeltner, Z., Rea, D., Renner, V., Fülöp, V., and Polgár, L. (2002). Electrostatic effects and binding determinants in the catalysis of prolyl oligopeptidase. Site specific mutagenesis at the oxyanion binding site. *J Biol Chem* 277:42613–42622.

61. Guo, Z., Linn, J.F., Wu, G., Anzick, S.L., Eisenberger, C.F., Halachmi, S., Cohen, Y., Fomenkov, A., Hoque, M.O., Okami, K., Steiner, G., Engles, J.M., et al. (2004). CDC91L1 (PIG-U) is a newly discovered oncogene in human bladder cancer. *Nat Med* 10:374–381.

62. Wu, G., Guo, Z., Chatterjee, A., Huang, X., Rubin, E., Wu, F., Mambo, E., Chang, X., Osada, M., Sook Kim, M., Moon, C., Califano, J.A., et al. (2006). Overexpression of glycosylphosphatidylinositol (GPI) transamidase subunits phosphatidylinositol glycan class T and/or GPI anchor attachment 1 induces tumorigenesis and contributes to invasion in human breast cancer. *Cancer Res* 66:9829–9836.

63. Nagpal, J.K., Dasgupta, S., Jadallah, S., Chae, Y.K., Ratovitski, E.A., Toubaji, A., Netto, G.J., Eagle, T., Nissan, A., Sidransky, D., and Trink, B. (2008). Profiling the expression pattern of GPI transamidase complex subunits in human cancer. *Mod Pathol* 21:979–991.

64. Ferguson, M.A. (2000). Glycosylphosphatidylinositol biosynthesis validated as a drug target for African sleeping sickness. *Proc Natl Acad Sci USA* 97:10673–10675.

65. Udenfriend, S., and Kodukula, K. (1995). How glycosylphosphatidylinositol-anchored membrane proteins are made. *Annu Rev Biochem* 64:563–591.

8

Split Topology of GPI Biosynthesis

ANANT K. MENON

Department of Biochemistry
Weill Cornell Medical College
1300 York Avenue
New York
NY 10065, USA

I. Abstract

The first two reactions of GPI biosynthesis occur on the cytoplasmic face of the endoplasmic reticulum (ER), while subsequent reactions (inositol acylation, mannosylation, phosphoethanolamine addition, and the attachment of a GPI anchor to proteins) occur in the ER lumen. This implies that GlcN-PI crosses (flips) from the cytoplasmic face of the ER to the lumenal leaflet in the course of GPI assembly. In this chapter, we survey the evidence for the split topology of the GPI biosynthetic pathway, and discuss the nascent subject of lipid flipping across the ER membrane. There are several review articles that cover this topic [1–3].

II. Synthesis and De-*N*-Acetylation of GlcNAc-PI Occur on the Cytoplasmic Face of the ER

GlcNAc-PI is synthesized from UDP-GlcNAc and phosphatidylinositol (PI) by GlcNAc-PI synthase, a multi-subunit, membrane-bound enzyme. Although UDP-GlcNAc is synthesized in the cytoplasm, it can be transported into the ER lumen [4]. Likewise, PI is synthesized on the cytoplasmic face of the ER but can be translocated to the lumenal leaflet by an unidentified glycerophospholipid transporter (flippase) located in the ER membrane [3, 5]. Despite the availability of both substrates on both sides of the ER membrane, it is clear that GlcNAc-PI is synthesized on the cytopasmic face of the ER. This is because PIG-A/Gpi3, the catalytic component of GlcNAc-PI synthase, is a monotopic membrane protein with the bulk of its sequence—including the active site motif EX7E [6, 7]—located in the cytoplasm [8].

The de-N-acetylation of GlcNAc-PI is catalyzed by PIG-L/Gpi12, a membrane protein that has a transmembrane domain close to its N-terminus and the majority of its sequence in the cytoplasm [9]. Thus, the catalytic sites of enzymes required for the first two steps of GPI synthesis are oriented towards the cytoplasm; consequently, GlcN-PI is generated on the cytoplasmic face of the ER.

When sealed ER vesicles were incubated with UDP-[^{3}H]GlcNAc to generate [^{3}H]GlcNAc-PI and [^{3}H]GlcN-PI *in situ*, then treated with a bacterial phosphatidylinositol-specific phospholipase C, both lipids were quantitatively hydrolyzed [10, 11]; this indicates that both lipids can be found on the cytoplasmic face of the ER, consistent with their site of synthesis.

III. Inositol Acylation Probably Occurs in the ER Lumen

Inositol acylation, the third reaction of GPI biosynthesis, likely occurs in the lumen of the ER but this point remains to be conclusively established. The acylating enzyme, PIG-W/Gwt1 [12], is a polytopic membrane protein with 13 predicted transmembrane spans. Regions of the protein that are conserved amongst the various PIG-W homologs are on the same side of the membrane as the N-terminus, which was experimentally established to be located in the ER lumen [12]. This suggests that inositol acylation occurs in the ER lumen.

Fatty acids for the PIG-W-catalyzed acyl transfer reaction are contributed by fatty acyl CoAs. These molecules are synthesized in the cytoplasm and must be transported across the ER membrane. Although an early report concluded that the ER presents a barrier to the transport of fatty acyl CoAs [13], it is known that a number of secretory proteins are acylated in the ER lumen by members of the acyl CoA-dependent membrane-bound O-acyltransferase (MBOAT) family of proteins [14] and that fatty acid remodeling of the GPI anchor in the ER lumen requires the acyl CoA-utilizing MBOAT family members Per1 and Gup1. While these points suggest that fatty acyl-CoAs must somehow be available in the ER lumen, the mechanism by which they are transported across the ER membrane remains to be identified.

IV.   Beyond Inositol Acylation: Later Reactions of GPI Assembly (Mannosylation, Phosphoethanolamine Addition, and GPI Transfer to Protein) Occur in the ER Lumen

GPI mannose residues are contributed by dolichol-phosphate mannose. This lipid is synthesized on the cytoplasmic face of the ER from GDP-mannose and dolichol phosphate, and then translocated across the ER membrane by an ATP-independent process that requires a specific transporter or flippase [15]. The dolichol-phosphate mannose flippase has yet to be identified, although its activity has been biochemically reconstituted in proteoliposomes generated from detergent-solubilized ER membrane proteins [16]. By analogy with the dolicholphosphate mannose-dependent mannosylation reactions in the dolichol pathway of protein N-glycosylation [17–19] as well as in O-mannosylation of yeast glycoproteins [20], GPI mannosylation reactions are expected to occur in the ER lumen. A functionally important "DXD" motif in PIG-M/Gpi14, the first GPI mannosyltransferase, was shown to be oriented towards the ER lumen consistent with the predicted lumenal orientation of mannosylation [29].

Mannosylated GPIs are modified by one or more phosphoethanolamine residues. The phosphoethanolamine transferases are polytopic membrane proteins with a large lumenal segment that contains a domain that is critical for function [21]. The lumenal location of this domain suggests that phosphoethanolamine transfer reactions occur in the ER lumen. The transferases use phosphatidylethanolamine (PE) as the phosphoethanolamine donor. This lipid is synthesized by the Kennedy pathway on the cytoplasmic face of the ER, or delivered to the cytoplasmic face of the ER after being produced in mitochondria via phosphatidylserine decarboxylation. Like PI, PE can be translocated across the ER membrane by the ER glycerophospholipid flippase [3], ensuring a lumenal supply of PE for the GPI phosphoethanolamine transferases.

Mannosylated GPIs with a phosphoethanolamine cap (linked to the third mannose residue) are substrates for the GPI anchoring reaction that occurs in the ER lumen. The proteins that receive a GPI anchor are synthesized on membrane-bound ribosomes and translocated into the ER lumen via the ER protein translocon. In the lumen, their C-terminal GPI-directing signal sequence is recognized by GPI transamidase, a multisubunit, membrane-bound enzyme complex with a lumenally-oriented active site contained within its Gpi8 subunit [22]. GPI transamidase cleaves the GPI signal sequence from the target protein and replaces it with a GPI anchor to generate a GPI-anchored protein in the lumenal leaflet of the ER.

Although it is likely that mannosylation and phosphoethanolamine addition occur on the lumenal face of the ER, mannosylated intermediates and mature GPIs can be detected on the cytoplasmic face of sealed ER vesicles by membrane topological probes such as the mannose-binding lectin Concanavalin A [11, 23]. One explanation for this result is that the lipids are flipped from their lumenal site of synthesis back to the cytoplasmic side where they are trapped upon binding to Concanavalin A.

V.   GlcN-PI Flips Across the ER Membrane During GPI Biosynthesis

The evidence compiled in the previous sections indicates that GlcN-PI is synthesized on the cytoplasmic face of the ER and must be flipped across the ER membrane into the lumenal leaflet in order to be elaborated into a mature GPI (Figure 8.1). The topological split in GPI biosynthesis resembles that seen in the assembly of $Glc_3Man_9GlcNAc_2$-PP-dolichol, the oligosaccharide donor for protein N-glycosylation [18, 24]. The first seven steps in the synthesis of this complex glycolipid occur on the cytoplasmic face of the ER, whereas the final seven steps occur in the ER lumen. This requires the heptasaccharide lipid intermediate $Man_5GlcNAc_2$-PP-dolichol to be flipped from the cytoplasmic face to the lumenal leaflet of the ER. The cytoplasmically oriented reactions of the dolichol pathway use nucleotide sugars (UDP-GlcNAc and GDP-Man) as sugar donors, whereas the lumenal reactions require dolichol-phosphate mannose and dolichol-phosphate glucose; these glycolipids are synthesized on the cytoplasmic face of the ER from dolichol phosphate and the corresponding nucleotide sugar, then flipped into the ER lumen. Thus, as in the GPI pathway, three lipid-flipping events are required in order to assemble the mature glycolipid; the flipping of dolichol-phosphate mannose is common to both pathways.

Flipping of GlcN-PI and other phospholipids is energetically expensive since the polar headgroup of the lipid must be transported through the hydrophobic interior of the bilayer. For the common phospholipid phosphatidylcholine (PC), the energy barrier to flipping is ~20–50 kcal/mol [3], equivalent to the energy of hydrolysis of ~3–7 molecules of ATP. A similar barrier likely exists for flipping of GlcN-PI. For flipping to occur at a physiologically relevant rate, it must be catalyzed. The ER has dedicated

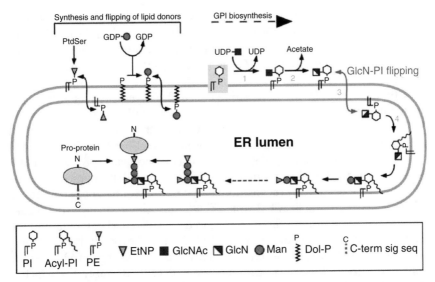

FIG. 8.1. The figure (adapted from reference [2]) depicts key steps of GPI assembly in mammalian cells. Assembly starts with the synthesis and de-*N*-acetylation of GlcNAc-PI on the cytoplasmic face of the ER to generate GlcN-PI. GlcN-PI flips to the lumenal leaflet of the ER membrane where it is inositol acylated, mannosylated and modified by phosphoethanolamine residues to yield a mature GPI anchor precursor capable of being attached to proteins bearing a C-terminal GPI signal sequence. Dolichol-phosphate mannose and phosphatidylethanolamine (PE) contribute the mannose and phosphoethanolamine residues of the GPI structure; these lipids are synthesized on the cytoplasmic face of the ER (top left of figure) and flipped into the lumenal leaflet where the corresponding mannosyltransferases and phosphoethanolamine transferases function. Flipping of GlcN-PI, dolichol-phosphate mannose and PE are mediated by specific flippases that have yet to be identified although their activities have been biochemically reconstituted.

lipid transporters or flippases that specifically transport glycerophospholipids (such as PC, PE, and PI) as well as dolicholphosphate mannose, dolichol-phosphate glucose, and dolichol-linked oligosaccharides [25]. Although none of these ER flippases have been identified, they have been functionally reconstituted and preliminary characterizations have been done. For example, it is likely that the ER flippases appear to function as facilitators of transverse diffusion since they promote bi-directional movement of lipids in an ATP-independent fashion. Also, it is clear that the flippases are specific for a particular class of lipid; for example, the Man5GlcNAc2-PP-dolichol flippase does not transport glycerophospholipids while the glycerophospholipid flippase does not transport dolichol-linked oligosaccharides [25, 26]. However, while the glycerophospholipid flippase appears relatively unspecific since it is able to translocate all the common glycerophospholipids, including stereoisomers (and possibly GPIs (see below)), the Man5GlcNAc2-PP-dolichol flippase is able to discriminate between structural isomers of Man5GlcNAc2-PP-dolichol and translocates higher order structures (such as Man7GlcNAc2-PP-dolichol) more slowly [26].

A flippase capable of translocating GlcN-PI has yet to be identified; it is likely that the flippase is capable of flipping most GPI structures since even lumenally synthesized GPIs can be reacted with membrane topological probes (see above). The ER membrane protein Arv1 has been implicated in some aspect of the conversion of GlcN-PI to a mannosylated product, but it is unlikely to be the GlcN-PI flippase [27]. ATP-independent flipping of GlcN-PI has been biochemically reconstituted in proteoliposomes generated from a detergent extract of rat liver ER vesicles [28]. Curiously, both the GPI flippase and the ER phospholipid flippase appeared to be similarly abundant in the detergent extract used for reconstitution, raising the possibility that they could be one and the same protein [28]. This result would explain why, despite extensive efforts, flipping is the only step of GPI assembly for which a mammalian cell mutant has not been identified; this could be because the GPI flippase also operates in a process—such as flipping of bulk phospholipids—that is essential for mammalian cell viability in culture.

VI. Conclusion

A variety of lipids must be translocated across the ER membrane as part of complex lipid biosynthetic pathways or simply to propagate the ER membrane bilayer [3]. These lipids include "membrane-building" glycerophospholipids (such as phosphatidylcholine), dolichol-phosphate mannose, dolichol-phosphate glucose, Man5GlcNAc2-PP-dolichol, and GPIs. With the possible exception of a wider substrate specificity that may enable the glycerophospholipid flippase to translocate GPIs, it is likely that a dedicated transporter/flippase exists for each of these lipid classes; future work must be directed at the identification of these transport proteins and an understanding of the transport mechanism.

Acknowledgments

This work was supported by NIH grant GM55427. Sections of the text were adapted from Orlean, P., and Menon, A.K. (2007). Thematic review series: Lipid posttranslational modifications. GPI anchoring of protein in yeast and mammalian cells, or: How we learned to stop worrying and love glycophospholipids. *J Lipid Res* 48:993–1011.

REFERENCES

1. McConville, M.J., and Menon, A.K. (2000). Recent developments in the cell biology and biochemistry of glycosylphosphatidylinositol lipids (review). *Mol Membr Biol* 17:1–16.
2. Orlean, P., and Menon, A.K. (2007). Thematic review series: Lipid posttranslational modifications. GPI anchoring of protein in yeast and mammalian cells, or: How we learned to stop worrying and love glycophospholipids. *J Lipid Res* 48:993–1011.
3. Pomorski, T., and Menon, A.K. (2006). Lipid flippases and their biological functions. *Cell Mol Life Sci* 63:2908–2921.
4. Abeijon, C., and Hirschberg, C.B. (1992). Topography of glycosylation reactions in the endoplasmic reticulum. *Trends Biochem Sci* 17:32–36.
5. Vishwakarma, R.A., Vehring, S., Mehta, A., Sinha, A., Pomorski, T., Herrmann, A., and Menon, A.K. (2005). New fluorescent probes reveal that flippase-mediated flip-flop of phosphatidylinositol across the endoplasmic reticulum membrane does not depend on the stereochemistry of the lipid. *Org Biomol Chem* 3:1275–1283.
6. Kostova, Z., Rancour, D.M., Menon, A.K., and Orlean, P. (2000). Photoaffinity labelling with P3-(4-azidoanilido)uridine 5'-triphosphate identifies gpi3p as the UDP-GlcNAc-binding subunit of the enzyme that catalyses formation of GlcNAc-phosphatidylinositol, the first glycolipid intermediate in glycosylphosphatidylinositol synthesis. *Biochem J* 350 (Pt. 3):815–822.
7. Kostova, Z., Yan, B.C., Vainauskas, S., Schwartz, R., Menon, A.K., and Orlean, P. (2003). Comparative importance *in vivo* of conserved glutamate residues in the EX7E motif retaining glycosyltransferase Gpi3p, the UDPGlcNAc-binding subunit of the first enzyme in glycosylphosphatidylinositol assembly. *Eur J Biochem* 270:4507–4514.
8. Watanabe, R., Kinoshita, T., Masaki, R., Yamamoto, A., Takeda, J., and Inoue, N. (1996). PIG-A and PIG-H, which participate in glycosylphosphatidylinositol anchor biosynthesis, form a protein complex in the endoplasmic reticulum. *J Biol Chem* 271:26868–26875.
9. Pottekat, A., and Menon, A.K. (2004). Subcellular localization and targeting of *N*-acetylglucosaminyl phosphatidylinositol de-*N*-acetylase, the second enzyme in the glycosylphosphatidylinositol biosynthetic pathway. *J Biol Chem* 279:15743–15751.
10. Vidugiriene, J., and Menon, A.K. (1993). Early lipid intermediates in glycosylphosphatidylinositol anchor assembly are synthesized in the ER and located in the cytoplasmic leaflet of the ER membrane bilayer. *J Cell Biol* 121:987–996.
11. Vidugiriene, J., and Menon, A.K. (1994). The GPI anchor of cell-surface proteins is synthesized on the cytoplasmic face of the endoplasmic reticulum. *J Cell Biol* 127:333–341.
12. Murakami, Y., Siripanyapinyo, U., Hong, Y., Kang, J.Y., Ishihara, S., Nakakuma, H., Maeda, Y., and Kinoshita, T. (2003). PIG-W is critical for inositol acylation but not for flipping of glycosylphosphatidylinositolanchor. *Mol Biol Cell* 14:4285–4295.
13. Polokoff, M.A., and Bell, R.M. (1978). Limited palmitoyl-CoA penetration into microsomal vesicles as evidenced by a highly latent ethanol acyltransferase activity. *J Biol Chem* 253:7173–7178.
14. Hofmann, K. (2000). A superfamily of membrane-bound *O*-acyltransferases with implications for wnt signaling. *Trends Biochem Sci* 25:111–112.
15. Rush, J.S., and Waechter, C.J. (1995). Transmembrane movement of a water-soluble analogue of mannosylphosphoryldolichol is mediated by an endoplasmic reticulum protein. *J Cell Biol* 130:529–536.
16. Rush, J.S., and Waechter, C.J. (2004). Functional reconstitution into proteoliposomes and partial purification of a rat liver ER transport system for a water-soluble analogue of mannosylphosphoryldolichol. *Biochemistry* 43:7643–7652.

17. Anand, M., Rush, J.S., Ray, S., Doucey, M.A., Weik, J., Ware, F.E., Hofsteenge, J., Waechter, C.J., and Lehrman, M.A. (2001). Requirement of the Lec35 gene for all known classes of monosaccharide-P-dolicholdependent glycosyltransferase reactions in mammals. *Mol Biol Cell* 12:487–501.

18. Schenk, B., Fernandez, F., and Waechter, C.J. (2001). The ins(ide) and out(side) of dolichyl phosphate biosynthesis and recycling in the endoplasmic reticulum. *Glycobiology* 11:61R–70R.

19. Helenius, J., and Aebi, M. (2002). Transmembrane movement of dolichol linked carbohydrates during *N*-glycoprotein biosynthesis in the endoplasmic reticulum. *Semin Cell Dev Biol* 13:171–178.

20. Strahl-Bolsinger, S., Gentzsch, M., and Tanner, W. (1999). Protein omannosylation. *Biochim Biophys Acta* 1426:297–307.

21. Gaynor, E.C., Mondesert, G., Grimme, S.J., Reed, S.I., Orlean, P., and Emr, S.D. (1999). MCD4 encodes a conserved endoplasmic reticulum membrane protein essential for glycosylphosphatidylinositol anchor synthesis in yeast. *Mol Biol Cell* 10:627–648.

22. Benghezal, M., Benachour, A., Rusconi, S., Aebi, M., and Conzelmann, A. (1996). Yeast Gpi8p is essential for GPI anchor attachment onto proteins. *EMBO J* 15:6575–6583.

23. Mensa-Wilmot, K., LeBowitz, J.H., Chang, K.P., al-Qahtani, A., McGwire, B.S., Tucker, S., and Morris, J.C. (1994). A glycosylphosphatidylinositol (GPI)-negative phenotype produced in Leishmania major by GPI phospholipase C from *Trypanosoma brucei*: Topography of two GPI pathways. *J Cell Biol* 124:935–947.

24. Hirschberg, C.B., and Snider, M.D. (1987). Topography of glycosylation in the rough endoplasmic reticulum and Golgi apparatus. *Annu Rev Biochem* 56:63–87.

25. Sanyal, S., Frank, C.G., and Menon, A.K. (2008). Distinct flippases translocate glycerophospholipids and oligosaccharide diphosphate dolichols across the endoplasmic reticulum. *Biochemistry* 47:7937–7946.

26. Sanyal, S., and Menon, A.K. (2009). Specific transbilayer translocation of dolichol-linked oligosaccharides by an endoplasmic reticulum flippase. Proc Natl Acad Sci USA 106:767–772.

27. Kajiwara, K., Watanabe, R., Pichler, H., Ihara, K., Murakami, S., Riezman, H., and Funato, K. (2008). Yeast ARV1 is required for efficient delivery of an early GPI intermediate to the first mannosyltransferase during GPI assembly and controls lipid flow from the endoplasmic reticulum. *Mol Biol Cell* 19:2069–2082.

28. Vishwakarma, R.A., and Menon, A.K. (2005). Flip-flop of glycosylphosphatidylinositols (GPI's) across the ER. *Chem Commun (Camb)*453–455.

29. Maeda, Y., Watanabe, R., Harris, C.L., Hong, Y., Ohishi, K., Kinoshita, K., and Kinoshita, T. (2001). PIG-M transfers the first mannose to glycosylphosphatidylinositol on the lumenal side of the ER. *EMBO J* 20:250–261.

9

Gpis of Apicomplexan Protozoa

HOSAM SHAMS-ELDIN[a] • FRANÇOISE DEBIERRE-
GROCKIEGO[a] • JÜRGEN KIMMEL[a] • RALPH T. SCHWARZ[a,b]

[a]*Institut für Virologie, AG Parasitologie*
Philipps-Universität Marburg
Hans-Meerwein-Strasse 2
35043 Marburg, Germany

[b]*Unité de Glycobiologie Structurale et Fonctionnelle*
UMR 8576 CNRS
Université des Sciences et Technologies de Lille
59655 Villeneuve d'Ascq cedex, France

I. Abstract

Apicomplexan protozoa are a phylum of parasites that includes medically and agriculturally important pathogens. Some apicomplexans, such as
Eimeria, Babesia, and *Theileria*, only infect agricultural animals and have
profound indirect effects on human welfare in developing countries. Members of this group include *Toxoplasma gondii*, the pathogen causing toxoplasmosis, and *Plasmodium falciparum*, the causative agent of the most
severe form of malaria. No vaccine is currently available to protect against
malaria and parasites have evolved to be resistant to many of the drugs
used. Glycoconjugate pathways of apicomplexan protozoa might represent
a novel target for the development of new drugs.

This chapter describes the biosynthesis and structures of GPI in *T. gondii*
and *P. falciparum*, and points out the relevant genes. Two immediate GPI-
anchor precursors for *P. falciparum* with the structures ethanolamine-
phosphate-6(Manα1–2)Manα1–2Manα1–6Manα1–4GlcN-PI and ethanol-
amine-phosphate-6Manα1–2Manα1–6Man-α1–4-GlcN-PI are synthesized

ISSN NO: 1874-6047
DOI: 10.1016/S1874-6047(09)26009-4

by *P. falciparum* in the asexual erythrocytic stage. Two main structures are synthesized by *T. gondii*: ethanolamine-phosphate-6Manα1–2Manα1–6(GalNAcβ1–4)Manα1–4GlcN-PI and ethanolamine-phosphate-6Manα1–2Manα1–6(Glcα1–4GalNAcβ1–4)Manα1–4GlcN-PI. *T. gondii* GPIs bearing the unique glucose-*N*-acetyl-galactosamine side branch are immunogenic in humans and are widely distributed among *T. gondii* isolates.

GPIs from parasitic protozoa have been recognized as pathogenicity factors and their new aspects will be discussed here. Parasite GPIs are recognized by toll-like receptor 2 (TLR2) or TLR4 leading to the activation of adaptor molecules and the induction of inflammatory cytokines, which play a key role in host resistance to infection from parasites.

II.   Introduction: Apicomplexan Parasites

Apicomplexan protozoa are a phylum of parasites that includes medically and agriculturally important pathogens. They are named for their cell apex that contains a number of organelles (rhoptries, micronemes, conoid, and apical polar ring), important for their invasion and development within host cells. Rhoptries and micronemes are secretory organelles that contain products required for motility, adhesion to host cells, and invasion of host cells and establishment of the parasitophorous vacuole. The conoid is a small cone-shaped structure that is thought to play a mechanical role in the invasion of host cells. The apical polar ring, a structure unique to apicomplexans, serves as one of the two microtubule-organizing centers in these parasites. The apicomplexa have acquired during their evolution the apicoplast, a chloroplast-derived organelle with an algal origin. Apicomplexan parasites are surrounded by the pellicle, a structure consisting of the plasma membrane and the closely apposed inner membrane complex that consists of flattened vesicles (Figure 9.1).

Apicomplexans undergo haploid and diploid stages during their life cycle. In many cases, their life cycle is distributed between two hosts, an intermediate host and a definitive host. Rapidly multiplying haploid stages of the parasites cause the acute and deleterious symptoms of infection. In addition to asexual mitotic proliferation, these parasites differentiate into gamete forms that produce zygotes after fusion. Differentiation and fertilization occurs in the intestinal epithelia of the respective definitive host. *Plasmodium* gametogenesis and fertilization occur in the mosquito intestine, whereas the sexual part of the life cycle of *Toxoplasma* takes place in the cat intestine. The apicomplexan protozoal parasites replicate within the cells of their host and, thus, are obligate intracellular parasites. The merozoites (in the case of *Toxoplasma* called tachyzoites) released by

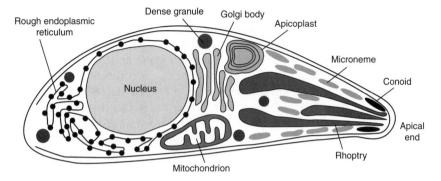

FIG. 9.1. Ultrastructure of an apicomplexan parasite (*Toxoplasma gondii* tachyzoite).

lysis of the host cell must invade new host cells in order to stay viable. Parasite replication occurs within the host cell in a so-called parasitophorous vacuole, derived from the host cell by invagination of the plasma membrane, which is modified by the parasite.

A. APICOMPLEXAN PROTOZOA CAUSE SEVERAL IMPORTANT HUMAN AND ANIMAL DISEASES

Although *Toxoplasma* infects about one-third of the world population, immunocompetent individuals show no or very mild symptoms of the disease. On the other hand, immunocompromised individuals (e.g., patients who have AIDS or who undergo organ transplantation or cancer chemotherapy) develop serious diseases such as cerebral meningitis. Another group at risk is women infected for the first time during pregnancy. The parasites can cause birth defects or abortion of the fetus. Infection of farm animals (particularly sheep) with *Toxoplasma* leads to abortion. The apicomplexan *Cryptosporidium* is also an opportunistic parasite infesting immunocompromised individuals and is a common cause of waterborne diarrhea. *Cryptosporidium*-induced diarrhea is also a widespread problem in the cattle industry. Some apicomplexans, such as *Eimeria, Babesia*, and *Theileria*, only infect agricultural animals, but these parasites have profound indirect effects on human welfare, particularly in developing countries. *Neospora caninum* was identified as a species in 1988 and is a close relative of *T. gondii* due to their structural similarities. *N. caninum* is an important cause of spontaneous abortion in infected livestock. Malaria is one of the most common infectious diseases and an enormous public health problem. Four types of the *Plasmodium* parasite can infect humans; the most serious form of the disease is caused by *P. falciparum* and to a lesser extent by *P. vivax*. Other related species (*P. ovale* and *P. malariae*) also affect humans but their geographical distribution is restricted.

No vaccine is currently available to protect against malaria and parasites have evolved to be resistant to many of the drugs employed. Therefore, glycoconjugate pathways of apicomplexan protozoa are being unraveled and might represent targets for the development of new drugs.

B. GLYCOSYLATION IN APICOMPLEXA

It has been believed that *O*-glycosylation is the major carbohydrate modification in the intraerythrocytic stage of *P. falciparum* and that the parasite has no *N*-glycosylation capacity. However, later studies have shown that *P. falciparum* has a low *N*-glycosylation capability, and *O*-glycosylation is either absent or present at an extremely low level, whereas glycosylphosphatidylinositol (GPI) anchor modification is common and is the major carbohydrate modification in parasite proteins. For more details see papers on glycoconjugates of *Trypanosoma brucei* [15], *P. falciparum* [6, 24, 46, 58], *T. gondii* [47], and on protozoal glycosylation [40].

The GPI anchor moieties are essential for parasite survival. The parasite GPI anchors can activate signaling pathways in host cells, and thereby induce the expression of inflammatory cytokines, adhesion molecules, and induced nitric oxide synthase (iNOS). This might cause erythrocyte sequestration, hypoglycemia, triglyceride lipogenesis, and immune dysregulation. Thus, the parasite GPI anchor structure and biosynthetic pathways are attractive targets for antimalarial and/or antiparasite drug development [20].

III. GPI Structures of *Plasmodium falciparum* and *Toxoplasma gondii*

A. STRUCTURE OF *PLASMODIUM* GPIs

Biochemical studies were focused mainly on the asexual, intraerythrocytic stages of *P. falciparum* because of their role in the clinical phase of the disease and also the possibility of their propagation in a cell culture system. GPIs were identified by labeling with [³H]mannose, [³H]glucosamine, and [³H]ethanolamine as well as by their sensitivity toward GPI-specific phospholipase D (GPI-PLD), phospholipase A_2, and nitrous acid. The malarial

GPIs were shown to be unaffected by treatment with phosphatidylinositol-specific phospholipase C (PI-PLC), indicating permanent inositol acylation of their core glycans. Two candidates for putative GPI-anchor precursors to malarial GPI-membrane proteins with the structures ethanolamine-phosphate-6(Manα1–2)Manα1–2Manα1–6Manα1–4GlcN-PI (Pf$_{gl}\alpha$) and ethanolamine-phosphate-6Manα1–2Manα1–6Man-α1–4-GlcN-PI (Pf$_{gl}\beta$) were identified [16] and confirmed in an independent investigation [21]. *P. falciparum* accumulates two merozoite surface proteins-1 and -2 (MSP-1 and MSP-2) during schizogony. Both proteins are anchored in membranes with GPIs. Detailed structural analysis of the core glycans showed that the GPI anchors of both proteins possess an extra α1–2 linked mannose at the evolutionary conserved trimannosyl-core-glycan. MSP-1 and MSP-2 labeled with tritiated myristic acid possess primarily radioactive myristic acid at inositol rings in both GPI anchors. Additionally, the hydrophobic fragments released from [^{3}H]myristic acid labeled GPI anchors were identified as diacylglycerol, carrying preferentially [^{3}H]palmitic acid in an ester linkage. An analysis of the acyl substituent on C-2 of the inositol of the nonlabeled GPIs of *P. falciparum* showed palmitic acid ($\sim$90%) and myristic acid ($\sim$10%). The diacylglycerol moiety released on treatment of GPI with hydrofluoric acid contained predominantly C18:0 and C18:1 (two isomers), minor proportions of C14:0, C16:0, C18:2, C20:0, and C22:0, and unidentified acids. Additional GC–MS analysis of phospholipase A$_2$-released fatty acids showed the presence of predominantly oleic acid ($\sim$85%) with minor amounts of *cis*-vaccenic ($\sim$6%) and linoleic ($\sim$9%) acids at the *sn*-2 position [38] (Figure 9.2A).

Only one of the precursors (Pf$_{gl}\alpha$) serves as an anchor for MSP-1 and MSP-2 [17]. The synthesis of two GPI membrane anchor precursors and the protein-bound GPI anchors is tightly regulated and varies markedly throughout the intraerythrocytic development of the asexual stages of *P. falciparum*. The GPI membrane anchor precursor Pf$_{gl}\beta$ is synthesized and transferred to protein predominantly in trophozoite stages about 30 h p.i. [16, 42]. Interestingly, the core glycans of nine different *P. falciparum* isolates from different geographic regions show a universal core glycan structure [3]. Thus, the use of inhibitors should be universally applicable. Similarly, studies on GPI structure of two rodent parasites (*P. chabaudi chabaudi* and *P. yoelii yoelii*) show that core glycans are also identical to *P. falciparum* GPIs [18, 29].

Recently, a study on identification and stoichiometry of GPI-anchored membrane proteins of *P. falciparum* was performed using a combination of proteomic and computational approaches. Proteomic analysis of proteins labeled with radioactive glucosamine identified GPI anchoring on 11 proteins: MSP-1, -2, -4, -5, -10, rhoptry-associated membrane antigen, apical

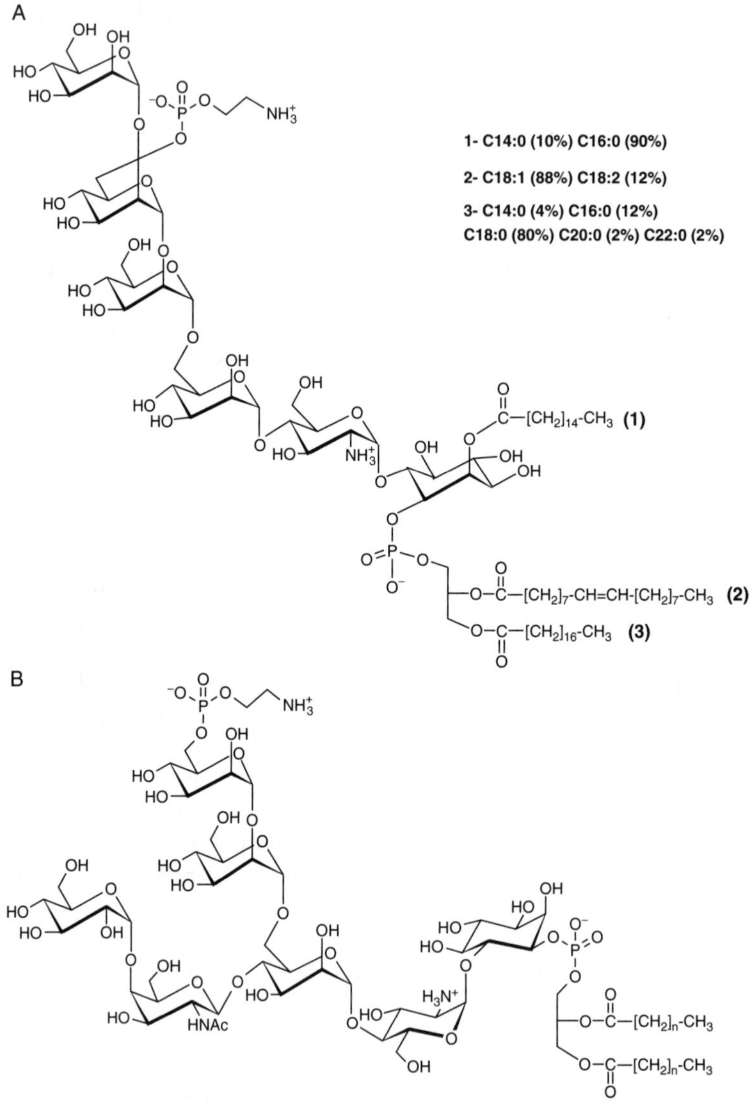

Fig. 9.2. Structures of the GPIs of *P. falciparum* (A) and *T. gondii* (B).

sushi protein, Pf92, Pf38, Pf12, and Pf34. These proteins represent approximately 94% of the GPI-anchored schizont/merozoite proteome and constitute the largest set of GPI-anchored proteins in this organism. MSP-1 and MSP-2 are present in a similar copy number, and are estimated to comprise approximately two-thirds of the total membrane-associated surface coat. This work identified 11 and predicted further 19 GPI-anchored proteins in *P. falciparum* [19].

Chemical synthesis of malaria candidate GPIs was performed by several laboratories. A strategy for fully inositol acylated and phosphorylated GPIs was elaborated by the group of Fraser Reid [35] and a highly convergent synthesis of a fully lipidated GPI anchor of *P. falciparum* was established by the laboratory of Peter Seeberger [32, 33]. A rapid access to the target GPIs in a highly efficient manner in sufficient quantities for the biological studies has been achieved. (For details see Chapter 11 of Peter Seeberger.)

B. STRUCTURES OF *TOXOPLASMA GONDII* GPIS

GPIs of *T. gondii* have been shown to be identical with the "low molecular weight antigen" identified by the group around Remington [13] and they elicit an early immunoglobulin M immune response in humans. The following structures of *T. gondii* GPIs were identified: ethanolamine-phosphate-6Manα1–2Manα1–6(GalNAcβ1–4)Manα1–4GlcN-PI and the novel structure ethanolamine-phosphate-6Manα1–2Manα1–6(Glcα1–4GalNAcβ1–4)Manα1–4GlcN-PI both with and without its terminal ethanolamine phosphate. Only *T. gondii* GPIs bearing the unique glucose-*N*-acetylgalactosamine side branch are immunogenic in humans and are widely distributed among *T. gondii* isolates [54] (Figure 9.2B).

In another study, GPI anchor peptides were isolated from [³H]glucosamine-labeled SAG1 (P30) major surface protein from *T. gondii* cultivated in Vero cells using protease digestion and phase partitioning. Neutral glycans were prepared from this material by dephosphorylation and deamination. Two glycoforms were characterized by gel filtration and high-performance ion exchange chromatography in combination with exoglycosidase treatment. Both forms were shown to have an *N*-acetylgalactosamine side-chain modification bound to the first mannose of the conserved three-mannosyl core. The second glycoform carries an additional terminal hexose linked to GalNAc [65]. Comparison of these structures with free GPI glycolipid precursors characterized in *T. gondii* suggests that core modification of the anchor takes place prior to transfer to the protein. In infected human cells, the pool of *T. gondii* GPIs having only GalNAc residue linked to the evolutionary conserved trimannosyl core glycan is competent for transfer to nascent surface GPI proteins, whereas the pool of GPIs having an additional Glc linked to GalNAc side branch accumulates at the cell surface as protein-free metabolic end products [1].

A synthesis of a fully phosphorylated toxoplasmal GPI anchor pseudo-hexasaccharide was also achieved by the group around Richard R. Schmidt, Konstanz [39].

Similarly, scanning of the thin layer chromatograms (TLC) for radioactivity revealed 7 [^{3}H]glucosamine-labeled peaks in *N. caninum*, where glycolipid peaks II-VII represent GPIs and that all but one (glycolipid peak II) have a modified inositol structure rendering them insensitive to PI–PLC treatment. Peak I, regardless of the labeling experiments, could not be identified as GPI as it was insensitive towards PI–PLC and GPI–PLD [41].

C.   BIOSYNTHESIS OF *P. FALCIPARUM* GPIs

A detailed understanding of GPI synthesis in protozoa is a prerequisite for identifying differences present in the biosynthetic pathways of parasites and host cells. A comparison of the biosynthetic pathway of GPIs has revealed differences between mammalian cells and parasitic protozoans. A cell-free incubation system prepared from asexual erythrocytic stages of *P. falciparum* is capable of synthesizing the same spectrum of GPIs as that found in metabolically labeled parasites. The formation of mannosylated GPIs in the cell-free system is shown to be inhibited by GTP and, unexpectedly, micromolar concentrations of guanosine diphosphate mannose (GDP-Man). Lower concentrations of GDP-Man affect the spectrum of GPIs synthesized. An acyl group modifies the inositol ring of GPIs of *P. falciparum*. The preferred donor of this fatty acid at the inositol ring is myristoyl-CoA. Inositol acylation has to precede the mannosylation of GPIs, because, mannosylated GPIs were not detected in the absence of acyl-CoA or CoA. Inositol acylation is a salient feature of plasmodial GPIs and thus might provide a potential target for drug therapy.

The substrate specificities of the early GPI biosynthetic enzymes of *Plasmodium* were determined using substrate analogs of D-GlcNα1–6-D-myo-inositol-1-HPO$_4$-*sn*-1,2-dipalmitoylglycerol (GlcN-PI). Similarities between the *Plasmodium* and mammalian (HeLa) enzymes were observed. These are as follows: (i) The presence and orientation of the 2′-acetamido/ amino and 3′-OH groups are essential for substrate recognition for the de-*N*-acetylase, inositol acyltransferase, and first mannosyltransferase enzymes, (ii) The 6′-OH group of the GlcN is dispensable for the de-*N*-acetylase, inositol acyltransferase, all four of the mannosyltransferases, and the ethanolamine phosphate transferase, (iii) The 4′-OH group of GlcNAc is not required for recognition, but substitution interferes with binding to the de-*N*-acetylase, the 4′-OH group of GlcN is essential for the inositol acyltransferase and first mannosyltransferase, (iv) The carbonyl group of the natural 2-O-hexadecanoyl ester of GlcN-(acyl)PI is essential for substrate recognition by the first mannosyltransferase. However, several differences were also discovered: (i) *Plasmodium*-specific inhibition of the inositol acyltransferase was detected with GlcN-[L]-PI, while GlcN-(2-O-alkyl)PI weakly inhibited the first mannosyltransferase in a competitive manner, (ii) The *Plasmodium* de-N-acetylase can act on analogs containing N-benzoyl, GalNAc, or βGlcNAc, whereas the human enzyme cannot. Using the parasite specificity of the later two analogs with the known nonspecific de-N-acetylase suicide inhibitor [49], GalNCONH$_2$-PI and GlcNCONH$_2$-β-PI were designed and found to be potent (IC$_{50}$ ~0.2 μM) *Plasmodium*-specific suicide substrate inhibitors. These inhibitors could be potential compounds for the development of antimalaria drugs [50]. For a recent survey on inhibitors of GPI synthesis see Ref. [7].

Another target could also be the dolichol phosphate mannose synthase (DPM), which is a key enzyme catalyzing the reaction between dolichol phosphate (Dol-P) and GDP-Man to generate dolichol-phosphate-mannose (Dol-P-Man) [5, 22]. Dol-P-Man is the main mannosyl donor providing not only four mannosyl residues for the synthesis of the lipid-linked precursor oligosaccharide Dol-PP-GlcNAc$_2$Man$_9$Glc$_3$ [23, 25], but also offers three (four in yeast and *Plasmodium*) mannosyl residues for the synthesis of GPIs [11, 12, 34].

Evaluation of the Dol-P-Man synthases from diverse eukaryotes by using HMMTOP Software (http://www.enzim.hu/hmmtop/) revealed that the clade of DPM could be divided into six main groups [48]. The baker's yeast group contains enzymes having a C-terminal transmembrane helix (TMH) (*Pyrococcus furiosus, Entamoeba histolytica, Bradyrhizobium japonicum, Erythrobacter* sp., and *Ustilago maydis*). The fission yeast group contains enzymes showing no TMH at all (moulds, flies, worm, *T. gondii, P. vivax*, and poultry). The mammalian group comprises enzymes with a predicted TMH near the middle of the proteins (human, mammalian, *Arabidopsis thaliana*). The *P. falciparum* group contains enzymes having a TMH near their N-terminal ends (*P. falciparum* and *Oryza sativa*). The trypanosome group contains enzymes that are similar to those present in the baker's yeast group, but contain at least two predicted transmembrane helices in their C-terminal domains (*T. brucei* and *P. y. yoelii*). The *Legionella* group contains enzymes predicted to be integral membrane protein with at least nine transmembrane helices (*Legionella pneumophila, Nitrosomonas europaea*). The fact that the *P. falciparum* and *P. vivax* belong to two different groups appears to be consistent with the idea that the *P. malariae* and *P. vivax* branches diverged ~100 million years ago, while the branch that gave rise to *P. falciparum* split even earlier [14]. Furthermore, comparisons of the available DNA sequences from several species suggest that *P. falciparum* became infective to humans by a recent lateral transfer from avian hosts [37, 59]. It was also shown that the Dol-P-Man synthases from several *Plasmodium* species fall into very different classes, whereby differences between the Dol-P-Man synthases of human and *Plasmodium* species could be exploited in the development of antimicrobial agents. It is a straightforward strategy to predict what effects a compound will have on the parasite.

Progress was made in characterizing glycosyltransferases involved in the biosynthesis of the core glycans of the GPI-anchor. For example: PIG-M that encodes the mammalial GPI-MT-I, the first mannosyltransferase. Mammalian GPI-MT-I consists of two components, PIG-M and PIG-X, which are homologous to Gpi14p and Pbn1p in *Saccharomyces cerevisiae*, respectively. PfPIG-M partially restored cell surface expression of the GPI-anchored protein CD59 in PIG-M deficient mammalian cells, and first mannose transfer activity *in vitro*; however, this was not the case for GPI14 (*S. cerevisiae*) [28]. In another report, it was shown that the association of PIG-X and PIG-M for GPI-MT-I activity is not interchangeable between mammals and the other lower eukaryotes [27]. Furthermore, the parasite PfPIG-B mannosyltransferase-III of GPI biosynthesis is novel in that its signature sequence HKEHKI is unique and only partially conserved as compared to HKEXRF signature motif of mammalian PIG-B enzymes [2].

D.   BIOSYNTHESIS OF *T. GONDII* GPIs

The biosynthesis of GPI is essential for the survival of *T. gondii* [63]. As in yeast or mammalian cells, the pathway starts with the transfer of GlcNAc derivate from UDP-GlcNAc to PI, followed by the de-*N*-acetylation of the GlcNAc and the addition of three mannose residues. The Man_3GlcN-PI structure is then modified by the GalNAc residue linked to the mannose adjacent to glucosamine. This GalNAc-containing intermediate plays two roles: first as precursor for free Glc-GalNAc-containing GPIs and second as precursor for the GalNAc-containing GPIs. In infected human cells, the mechanism leading to the selection of the GalNAc-containing GPI, which is transferred to nascent GPI proteins of the parasite, is not clearly understood, but it is supposed that it is linked to the substrate specificity of the parasite transamidase complex. This specificity is lost in non-human cells and recovered only after the transfer of the parasite into the human cells. In contrast to trypanosome GPIs, side-chain modification in *T. gondii* GPIs takes place before addition to protein in the ER. Therefore, the biosynthesis of the side-chain modifications was studied in an *in vitro* system prepared from hypotonically lysed *T. gondii* parasites. Radiolabeled glucose-containing GPI precursors were synthesized by *T. gondii* membrane preparations upon incubation with uridine diphosphate-[^{3}H]glucose. Synthesis of glucosylated glycolipids was shown to take place only in the presence of exogenous uridine diphosphateglucose and can be stimulated by unlabeled uridine diphosphate-glucose in a dose-dependent manner. In contrast to GPI mannosylation, glucosylation was shown to be insensitive to amphomycin treatment. In addition, the glucose analog 2-deoxy-D-glucose was used to trace the GPI glucosylation pathway. Detailed analysis of glycolipids synthesized *in vitro* in the presence of UDP and GDP derivatives of D-glucose and 2-deoxy-D-glucose ruled out an involvement of dolichol phosphate-glucose and demonstrates direct transfer of glucose from uridine diphosphateglucose [53]. Furthermore, using hypotonically permeabilized *T. gondii* tachyzoites, the topology of the free GPIs within the ER membrane was investigated. The morphology and permeability of parasites were checked by electron microscopy and release of a cytosolic protein. The membrane integrity of organelles (ER and rhoptries) was checked by protease protection assays. In initial experiments, GPI biosynthetic intermediates were labeled with UDP-[6-^{3}H]GlcNAc in permeabilized parasites, and the transmembrane distribution of the radiolabeled lipids was probed with PI-PLC. A new early intermediate with an acyl modification on the inositol was identified, indicating that inositol acylation also occurs in *T. gondii*. A significant portion of the early GPI intermediates (GlcN-PI and GlcNAc-PI) could be hydrolyzed following PI-PLC treatment, indicating that these glycolipids are predominantly present in the cytoplasmic leaflet of the ER. Permeabilized *T. gondii* parasites labeled with either GDP-[2-^{3}H]mannose or UDP-[6-^{3}H]glucose showed that the more mannosylated and side chain (Glc-GalNAc)-modified GPI intermediates are also preferentially localized in the cytoplasmic leaflet of the ER [30].

In addition, the serine protease inhibitors phenylmethylsulfonyl fluoride (PMSF) and diisopropyl fluoride (DFP) were found to have a profound effect on the *T. gondii* GPI biosynthetic pathway, leading to the observation and characterization of novel inositol-acylated mannosylated GPI intermediates. This inositol acylation is acyl-CoA-dependent and takes place before mannosylation, but uniquely for this class of inositol acyltransferase, it is inhibited by PMSF. The subsequent inositol deacylation of fully mannosylated GPI intermediates is inhibited by both PMSF and DFP. The use of these serine protease inhibitors allows observations as to the timing of inositol acylation and subsequent inositol deacylation of the GPI intermediates. Inositol acylation of the nonmannosylated GPI intermediate D-GlcNα1–6-D-myo-inositol-1-HPO$_4$-*sn*-lipid precedes mannosylation. Inositol deacylation of the fully mannosylated GPI intermediate allows further processing, i.e., addition of GalNAc side chain to the first mannose. Characterization of the phosphatidylinositol moieties present on both free GPIs and GPI-anchored proteins shows the presence of a diacylglycerol lipid, whose *sn*-2 position contains almost exclusively a C18:1 acyl chain. The data identify key novel inositol-acylated mannosylated intermediates, allowing the formulation of an updated *T. gondii* GPI biosynthetic pathway along with identification of the putative genes involved. Most of the genes involved in the biosythesis of *T. gondii* and *P. falciparum* were cloned and are shown in Table 9.1. Putative *T. gondii* GPI biosynthetic genes for steps 1 (PIG-A, PIG-C, and GPI1), 2 (PIG-L), 3 (PIG-W), 4 (PIG-M), 5 (PIG-M), 7 (PIG-V), 8 (PIG-B), 9 (PIG-O and PIG-F) and 11 (GPI8 and GAA-1) and for the generation of Dol-P-Man (DPM1) were identified by searching the *Toxoplasma* genome for homologs of known *P. falciparum*, human, and yeast GPI biosynthetic genes. No homologs for the other mammalian genes implicated the first step (PIG-H, PIG-P, and DPM2) could be found [51].

IV.   Immunological Functions of Protozoan GPIs

A.   ROLE IN INFLAMMATION

The GPIs exhibit a variety of functions other than the mere anchoring of membrane proteins. Many biological and immunological properties have been shown to be associated with this glycolipid modification. Acute infection with protozoan parasites results in the production of high levels of inflammatory cytokines, such as interleukins (ILs) and tumor necrosis factor (TNF)-α by macrophages, dendritic cells, and neutrophils. These cytokines initiate various effector mechanisms that are responsible for the control of parasite growth and pathology. GPI of *Plasmodium* was first shown to act as a parasite toxin by its ability to induce TNF-α and IL-1 production by macrophages [43]. GPI purified from the variant surface glycoprotein of *Trypanosoma* has similar activities in macrophage activation, and could thus also account for the high level of IL-1 and TNF-α found in trypanosomiasis [57]. In the same way, it was shown that *T. gondii* GPIs are able to induce the production of TNF-α in macrophages [8]. The GPI from *Plasmodium* increases secretion of nitric oxide (NO) and expression of intercellular adhesion molecule-1 (ICAM-1) and vascular cell adhesion molecule-1 (VCAM-1) in host cells, which are implicated in the etiology of the cerebral malaria syndrome [45, 55]. These studies have concluded that GPIs are likely to be the dominant agents responsible for inflammatory cytokine production that these parasites generate. Although endogenous inflammatory cytokines are important for resistance against *T. gondii*, an

TABLE 9.1

SUMMARY OF MOST OF THE GENES FOR THE GPI BIOSYNTHETIC PATHWAY OF *S. CEREVISIAE*, MAMMALS, *T. GONDII* AND *P. FALCIPARUM*

Step	Enzyme	Donor	Genes			
			Mammals	Yeast	*T. gondii*	*P. falciparum*
1	GPI-GlcNAc transferase	UDP-GlcNAc	PIG-A PIG-H PIG-C PIG-Q PIG-P DPM2	GPI3/SPT14/CWH6 GPI15 GPI2 GPI1 GPI19	PIG-A PIG-C (CAI84649) GPI1	PIG-A GPI1
2	de-N-acetylase		PIG-L	GPI12	PIG-L (CAI91276)	
3	Acyltransferase	Palmitoyl-CoA	PIG-W	GWT1	PIG-W (CAH18643)	PIG-W
4	Flippase					
5	GPI-α1-4-mannosyltransferase I (MT-I)	Dol-P-Man	PIG-M	GPI14	PIG-M (CAI84647)	PIG-M
6	Ethanolamine-P-transferase		PIG-N	MCD4		
7	GPI-α1-6-mannosyltransferase II (MT-II)	Dol-P-Man	PIG-V	GPI18	PIG-V (CAI91278)	
8	GPI-α1-2-mannosyltransferase III (MT-III)	Dol-P-Man	PIG-B	GPI10	PIG-B	PIG-B
9	Ethanolamine-P-transferase	PE	PIG-F PIG-O	GPI11 GPI13	PIG-F PIG-O	PIG-O

(Continued)

TABLE 9.1 (*Continued*)

Step	Enzyme	Donor	Genes			
			Mammals	Yeast	*T. gondii*	*P. falciparum*
10	Ethanolamine-P-transferase			GPI7		
11	Transamidase		GAA1 GPI8 PIG-T PIG-S PIG-U	GAA1 GPI8 GPI16 GPI17 Cdc91p	GAA1 (CAI91277) GPI8 (CAD44992) PIG-T	GAA1 GPI8
	Dol-P-Man synthase	GDP-Man	DPM1 DPM2 DPM3	DPM1	DPM1 (CAI84648)	DPM1

See also Ref. [66].

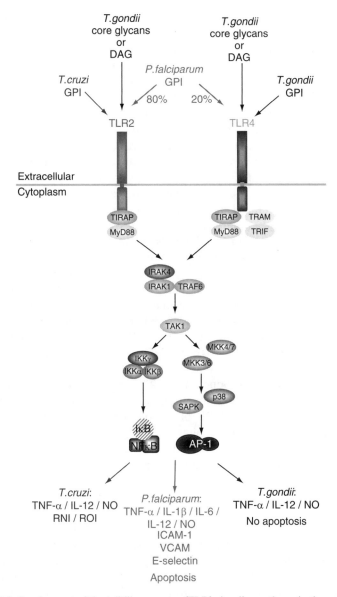

FIG. 9.3. Involvement of the toll-like receptor (TLR) signaling pathway in the activation of the immune response by parasites GPIs. Parasite GPIs are recognized by TLR2 or TLR4 leading to the activation of adaptor molecules. The myeloid differentiation factor 88 (MyD88) by inducing inflammatory cytokines plays a key role in host resistance to infection with parasites. *Trypanosoma* and *Plasmodium* GPIs trigger phosphorylation of kinases (extracellular regulated kinase-1 (ERK-1)/ERK-2, MAPK kinases (MKK), and stress-activated protein kinases (SAPK)/p38), while numerous cytoplasmic proteins are phosphorylated on tyrosine residues in response to GPIs of *Toxoplasma*. GPIs of all the three species induce the nuclear translocation of the transcription factor NF-κB responsible for the expression of inflammatory cytokine genes.

excessive production leads to the death of the host [26]. For this reason, the control of the GPI-induced inflammation may be a strategy to reduce pathogenicity and mortality due to toxoplasmosis.

In mammalian cells, the toll-like receptors (TLRs) are the first instruments of the innate immune system to recognize every known category of microorganisms that cause human diseases. Indeed, TLRs are critical for the recruitment of phagocytes to infected tissue and subsequent microbial killing. To date, 13 TLRs have been described, and for most of them, the microbial molecules that they recognize have been identified. The GPIs of *Plasmodium* and *Trypanosoma* were shown to exert their inflammatory effects by activation of a signaling pathway in host cells mainly through TLR2 [4, 31] (Figure 9.3). This pathway involves kinases like p38, and c-Jun N-terminal kinases (JNK1 and JNK2) in macrophages. The expression of host loci implicated in parasite pathogenesis (e.g., TNF-α, IL-1, NO, ICAM) in response to GPIs, depends on the phosphorylation of these kinases and on the activation of the transcription factor NF-κB [56, 64]. As the GPIs of *Plasmodium* and *Trypanosoma*, the GPIs of *Toxoplasma* induce the production of TNF-α in macrophages through NF-κB activation. However, GPIs of *Toxoplasma*, as well as the glycan moiety alone, activate CHO cells via TLR4, while the lipid moiety cleaved from the GPIs activate these cells via both TLR4 and TLR2 [9]. TLR2 and TLR4 were individually not absolutely needed to trigger TNF-α production by macrophages exposed to *T. gondii* GPIs. When macrophages were deficient for both TLR2 and TLR4, the production of TNF-α in response to GPIs was completely abrogated [9]. Phospholipases expressed at the surface of macrophages might cleave GPIs of *T. gondii* and liberated lipid moieties could activate TLR4$^{-/-}$ macrophages through TLR2. This indicates that both TLR2 and TLR4 are involved in the signaling pathway leading to production of TNF-α by macrophages exposed to *T. gondii* GPIs.

B. ROLE IN APOPTOSIS

Apoptosis is the process of programmed cell death that may occur in multicellular organisms and is caused by a series of biochemical events that lead to morphological changes (blebbing, changes to the cell membrane with loss of membrane asymmetry, cell shrinkage, nuclear fragmentation, chromatin condensation, chromosomal DNA fragmentation, etc.). *T. gondii* renders infected cells resistant to programmed cell death triggered by multiple apoptotic stimuli [36]. On the other hand, increased apoptosis of lymphocytes and granulocytes after *in vivo* infection with *T. gondii* may suppress the immune response against the parasite [60]. *T. gondii* GPIs fail to block apoptosis that was triggered in human-derived cells via extrinsic or intrinsic apoptotic pathways [10]. Furthermore, characteristics of apoptosis, e.g., caspase-3/7 activity, phosphatidylserine exposition at the cell surface, or DNA strand breaks, were not observed in the presence of *T. gondii* GPIs [10]. These results indicate that *T. gondii* GPIs are not involved in survival or in apoptosis of host cells. In contrast, the GPI extracted from *Plasmodium* was shown to induce apoptosis in spleen and liver of mice *in vivo* [62]. The plasmodial GPI induced direct apoptosis in rat cardiomyocytes by regulating the expression of pro- and antiapoptotic genes [61]. Furthermore, apoptosis was observed in the heart biopsy of a patient infected with *P. falciparum* who succumbed to cardiac impairment. Thus, the GPI of *P. falciparum* might play a role in myocardial impairment observed in patients suffering from severe malaria.

C.   POTENT TARGET FOR ANTIDISEASE VACCINE

A new antidisease concept using the host response to GPIs could protect infected hosts from the fatal development of parasitic diseases. On the basis of the sequence of *P. falciparum* GPI glycan, the nontoxic analog ethanolamine-phosphate-6(Manα1–2)Manα1–2Manα1–6Manα1–4GlcNα1–6myo-inositol-1,2-cyclic-phosphate was chemically synthesized, conjugated to carriers, and used to immunize mice [44]. In this study, it was shown that parasites continue to proliferate; however, toxic activity of GPIs was abrogated. Indeed, recipients were substantially protected against malarial acidosis, pulmonary edema, cerebral syndrome, and fatality, confirming that GPI is a highly conserved endotoxin of malarial parasite origin. Anti-GPI antibodies also neutralized proinflammatory activity of *P. falciparum in vitro*. Thus, a nontoxic GPI oligosaccharide coupled to carrier protein is immunogenic and provides significant protection against malarial pathogenesis and fatalities in a preclinical rodent model. Therefore, GPI could serve as the basis for an antidisease vaccine alleviating the most severe symptoms of malaria. In addition, animals treated with intact GPI (encompassing both the lipid and the carbohydrate moiety) of the African trypanosomes (*T. brucei*) before infection with this parasite were significantly protected against host clinical manifestations of *T. brucei*-induced pathology like anemia, acidosis, weight loss, and locomotor activity, without influencing initial parasite development [52]. In addition, GPI-based treatment resulted in reduced circulating serum levels of the inflammatory cytokines TNF-α and IL-6, increased circulating IL-10, and abrogation of infection-induced lipopolysaccharide hypersensitivity. Thus, this kind of vaccine, which prevents the most severe consequences of the disease (without inhibiting the multiplication of the parasite) is a new strategically concept. It might be the method of choice for the treatment of toxoplasmosis, which is difficult to control with classical vaccines because of immune evasion mechanisms developed by the parasite.

References

1. Azzouz, N., Shams-Eldin, H., Niehus, S., Debierre-Grockiego, F., Bieker, U., Schmidt, J., Mercier, C., Delauw, M.F., Dubremetz, J.F., Smith, T.K., and Schwarz, R.T. (2006). *Toxoplasma gondii* grown in human cells uses GalNAc-containing glycosylphosphatidyl-inositol precursors to anchor surface antigens while the immunogenic Glc-GalNAc-con-taining precursors remain free at the parasite cell surface. *Int J Biochem Cell Biol* 38:1914–1925.

2. Basagoudanavar, S.H., Feng, X., Krishnegowda, G., Muthusamy, A., and Gowda, D.C. (2007). *Plasmodium falciparum* GPI mannosyltransferase-III has novel signature sequence and is functional. *Biochem Biophys Res Commun* 364:748–754.

3. Berhe, S., Schofield, L., Schwarz, R.T., and Gerold, P. (1999). Conservation of structure among glycosylphosphatidylinositol toxins from different geographic isolates of *Plasmodium falciparum. Mol Biochem Parasitol* 103:273–278.

4. Campos, M.A., Almeida, I.C., Takeuchi, O., Akira, S., Valente, E.P., Procopio, D.O., Travassos, L.R., Smith, J.A., Golenbock, D.T., and Gazzinelli, R.T. (2001). Activation of Toll-like receptor-2 by glycosylphosphatidylinositol anchors from a protozoan parasite. *J Immunol* 167:416–423.

5. Costello, L.C., and Orlean, P. (1992). Inositol acylation of a potential glycosyl phospho-inositol anchor precursor from yeast requires acyl coenzyme A. *J Biol Chem* 267:8599–8603.

6. Davidson, E.A., and Gowda, D.C. (2001). Glycobiology of *Plasmodium falciparum. Biochimie* 83:601–604.

7. de Macedo, C.S., Shams-Eldin, H., Smith, T.K., Schwarz, R.T., and Azzouz, N. (2003). Inhibitors of glycosyl-phosphatidylinositol anchor biosynthesis. *Biochimie* 85:465–472.

8. Debierre-Grockiego, F., Azzouz, N., Schmidt, J., Dubremetz, J.F., Geyer, H., Geyer, R., Weingart, R., Schmidt, R.R., and Schwarz, R.T. (2003). Roles of glycosylphosphatidyl-inositols of *Toxoplasma gondii*. Induction of tumor necrosis factor-alpha production in macrophages. *J Biol Chem* 278:32987–32993.

9. Debierre-Grockiego, F., Campos, M.A., Azzouz, N., Schmidt, J., Bieker, U., Resende, M.G., Mansur, D.S., Weingart, R., Schmidt, R.R., Golenbock, D.T., Gazzinelli, R.T., and Schwarz, R.T. (2007). Activation of TLR2 and TLR4 by glycosylphosphatidylinositols derived from *Toxoplasma gondii. J Immunol* 179:1129–1137.

10. Debierre-Grockiego, F., Hippe, D., Schwarz, R.T., and Luder, C.G. (2007). *Toxoplasma gondii* glycosylphosphatidylinositols are not involved in *T. gondii*-induced host cell survival. *Apoptosis* 12:781–790.

11. DeGasperi, R., Thomas, L.J., Sugiyama, E., Chang, H.M., Beck, P.J., Orlean, P., Albright, C., Waneck, G., Sambrook, J.F., Warren, C.D., and Yeh, E.T.H. (1990). Correction of a defect in mammalian GPI anchor biosynthesis by a transfected yeast gene. *Science* 250:988–991.

12. Endo, M., Beatty, P.G., Vreeke, T.M., Wittwer, C.T., Singh, S.P., and Parker, C.J. (1996). Syngeneic bone marrow transplantation without conditioning in a patient with paroxys-mal nocturnal hemoglobinuria: *in vivo* evidence that the mutant stem cells have a survival advantage. *Blood* 88:742–750.

13. Erlich, H.A., Rodgers, G., Vaillancourt, P., Araujo, F.G., and Remington, J.S. (1983). Identification of an antigen-specific immunoglobulin M antibody associated with acute *Toxoplasma* infection. *Infect Immun* 41:683–690.

14. Escalante, A.A., Barrio, E., and Ayala, F.J. (1995). Evolutionary origin of human and primate malarias: evidence from the circumsporozoite protein gene. *Mol Biol Evol* 12:616–626.

15. Ferguson, M.A. (1992). Colworth Medal Lecture. Glycosyl-phosphatidylinositol membrane anchors: the tale of a tail. *Biochem Soc Trans* 20:243–256.

16. Gerold, P., Dieckmann-Schuppert, A., and Schwarz, R.T. (1994). Glycosylphosphatidyl-inositols synthesized by asexual erythrocytic stages of the malarial parasite, *Plasmodium falciparum*. Candidates for plasmodial glycosylphosphatidylinositol membrane anchor precursors and pathogenicity factors. *J Biol Chem* 269:2597–2606.

17. Gerold, P., Schofield, L., Blackman, M.J., Holder, A.A., and Schwarz, R.T. (1996). Structural analysis of the glycosyl-phosphatidylinositol membrane anchor of the merozoite surface proteins-1 and -2 of *Plasmodium falciparum*. *Mol Biochem Parasitol* 75:131–143.

18. Gerold, P., Vivas, L., Ogun, S.A., Azzouz, N., Brown, K.N., Holder, A.A., and Schwarz, R.T. (1997). Glycosylphosphatidylinositols of *Plasmodium chabaudi chabaudi*: a basis for the study of malarial glycolipid toxins in a rodent model. *Biochem J* 328(Pt 3): 905–911.

19. Gilson, P.R., Nebl, T., Vukcevic, D., Moritz, R.L., Sargeant, T., Speed, T.P., Schofield, L., and Crabb, B.S. (2006). Identification and stoichiometry of glycosylphosphatidylinositol-anchored membrane proteins of the human malaria parasite *Plasmodium falciparum*. *Mol Cell Proteomics* 5:1286–1299.

20. Gowda, D.C., and Davidson, E.A. (1999). Protein glycosylation in the malaria parasite. *Parasitol Today* 15:147–152.

21. Gowda, D.C., Gupta, P., and Davidson, E.A. (1997). Glycosylphosphatidylinositol anchors represent the major carbohydrate modification in proteins of intraerythrocytic stage *Plasmodium falciparum*. *J Biol Chem* 272:6428–6439.

22. Haselbeck, A., and Tanner, W. (1982). Dolichyl phosphate-mediated mannosyl transfer through liposomal membranes. *Proc Natl Acad Sci USA* 79:1520–1524.

23. Herscovics, A., and Orlean, P. (1993). Glycoprotein biosynthesis in yeast. *Faseb J* 7:540–550.

24. Hoessli, D.C., Davidson, E.A., Schwarz, R.T., and Nasir, R.T. (1996). Glycobiology of *Plasmodium falciparum*: An emerging area of research. *Glycoconjugate J* 13:1–3.

25. Hoflack, B., and Kornfeld, S. (1985). Lysosomal enzyme binding to mouse P388D1 macrophage membranes lacking the 215-kDa mannose 6-phosphate receptor: evidence for the existence of a second mannose 6-phosphate receptor. *Proc Natl Acad Sci* 82:4428–4432.

26. Hunter, C.A., Suzuki, Y., Subauste, C.S., and Remington, J.S. (1996). Cells and cytokines in resistance to *Toxoplasma gondii*. *Curr Top Microbiol Immunol* 219:113–125.

27. Kim, Y.U., Ashida, H., Mori, K., Maeda, Y., Hong, Y., and Kinoshita, T. (2007). Both mammalian PIG-M and PIG-X are required for growth of GPI14-disrupted yeast. *J Biochem* 142:123–129.

28. Kim, Y.U., and Hong, Y. (2007). Functional analysis of the first mannosyltransferase (PIG-M) involved in glycosylphosphatidylinositol synthesis in *Plasmodium falciparum*. *Mol Cells* 24:294–300.

29. Kimmel, J., Ogun, S.A., de Macedo, C.S., Gerold, P., Vivas, L., Holder, A.A., Schwarz, R.T., and Azzouz, N. (2003). Glycosylphosphatidyl-inositols in murine malaria: *Plasmodium* yoelii yoelii. *Biochimie* 85:473–481.

30. Kimmel, J., Smith, T.K., Azzouz, N., Gerold, P., Seeber, F., Lingelbach, K., Dubremetz, J.F., and Schwarz, R.T. (2006). Membrane topology and transient acylation of *Toxoplasma gondii* glycosylphosphatidylinositols. *Eukaryot Cell* 5:1420–1429.
31. Krishnegowda, G., Hajjar, A.M., Zhu, J., Douglass, E.J., Uematsu, S., Akira, S., Woods, A.S., and Gowda, D.C. (2005). Induction of proinflammatory responses in macrophages by the glycosylphosphatidylinositols of *Plasmodium falciparum*: cell signaling receptors, glycosylphosphatidylinositol (GPI) structural requirement, and regulation of GPI activity. *J Biol Chem* 280:8606–8616.
32. Kwon, Y.U., Soucy, R.L., Snyder, D.A., and Seeberger, P.H. (2005). Assembly of a series of malarial glycosylphosphatidylinositol anchor oligosaccharides. *Chemistry* 11:2493–2504.
33. Liu, X., Kwon, Y.U., and Seeberger, P.H. (2005). Convergent synthesis of a fully lipidated glycosylphosphatidylinositol anchor of *Plasmodium falciparum*. *J Am Chem Soc* 127:5004–5005.
34. Low, P., Dallner, G., Mayor, S., Cohen, S., Chait, B.T., and Menon, A.K. (1991). The mevalonate pathway in the bloodstream form of *Trypanosoma brucei*. Identification of dolichols containing 11 and 12 isoprene residues. *J Biol Chem* 266:19250–19257.
35. Lu, J., Jayaprakash, K.N., Schlueter, U., and Fraser-Reid, B. (2004). Synthesis of a malaria candidate glycosylphosphatidylinositol (GPI) structure: a strategy for fully inositol acylated and phosphorylated GPIs. *J Am Chem Soc* 126:7540–7547.
36. Luder, C.G., and Gross, U. (2005). Apoptosis and its modulation during infection with *Toxoplasma gondii*: molecular mechanisms and role in pathogenesis. *Curr Top Microbiol Immunol* 289:219–237.
37. McCutchan, T.F., Dame, J.B., Miller, L.H., and Barnwell, J. (1984). Evolutionary relatedness of *Plasmodium* species as determined by the structure of DNA. *Science* 225:808–811.
38. Naik, R.S., Branch, O.H., Woods, A.S., Vijaykumar, M., Perkins, D.J., Nahlen, B.L., Lal, A.A., Cotter, R.J., Costello, C.E., Ockenhouse, C.F., Davidson, E.A., and Gowda, D.C. (2000). Glycosylphosphatidylinositol anchors of *Plasmodium falciparum*: molecular characterization and naturally elicited antibody response that may provide immunity to malaria pathogenesis. *J Exp Med* 192:1563–1576.
39. Pekari, K., Tailler, D., Weingart, R., and Schmidt, R.R. (2001). Synthesis of the fully phosphorylated GPI anchor pseudohexasaccharide of *Toxoplasma gondii*. *J Org Chem* 66:7432–7442.
40. Samuelson, J., Banerjee, S., Magnelli, P., Cui, J., Kelleher, D.J., Gilmore, R., and Robbins, P.W. (2005). The diversity of dolichol-linked precursors to Asn-linked glycans likely results from secondary loss of sets of glycosyltransferases. *Proc Natl Acad Sci USA* 102:1548–1553.
41. Schares, G., Zinecker, C.F., Schmidt, J., Azzouz, N., Conraths, F.J., Gerold, P., and Schwarz, R.T. (2000). Structural analysis of free and protein-bound glycosylphosphatidylinositols of *Neospora caninum*. *Mol Biochem Parasitol* 105:155–161.
42. Schmidt, A., Schwarz, R.T., and Gerold, P. (1998). *Plasmodium falciparum*: asexual erythrocytic stages synthesize two structurally distinct free and protein-bound glycosylphosphatidylinositols in a maturation-dependent manner. *Exp Parasitol* 88:95–102.
43. Schofield, L., and Hackett, F. (1993). Signal transduction in host cells by a glycosylphosphatidylinositol toxin of malaria parasites. *J Exp Med* 177:145–153.
44. Schofield, L., Hewitt, M.C., Evans, K., Siomos, M.A., and Seeberger, P.H. (2002). Synthetic GPI as a candidate anti-toxic vaccine in a model of malaria. *Nature* 418:785–789.
45. Schofield, L., Novakovic, S., Gerold, P., Schwarz, R.T., McConville, M.J., and Tachado, S.D. (1996). Glycosylphosphatidylinositol toxin of *Plasmodium* up-regulates intercellular adhesion molecule-1, vascular cell adhesion molecule-1, and E-selectin expression in vascular

endothelial cells and increases leukocyte and parasite cytoadherence via tyrosine kinase-dependent signal transduction. *J Immunol* 156:1886–1896.

46. Schuppert, A.A., Gerold, P., and Schwarz, R.T. (1996). Glycoproteins of malaria parasites in "New Comprehensive Biochemistry," in Montreuil, J., Vliegenthart, J.F.G. and Schachter, H. (eds.), *Glycoproteins and Disease*, Vol. 30, Amsterdam: Elsevier, pp. 125–158.

47. Schwarz, R.T., Tomavo, S., Odenthal-Schnittler, M., Striepen, B., Becker, D., Eppinger, M., Zinecker, C.F., and Dubremetz, J.F. (1993). Recent advances in the glycobiology of *Toxoplasma gondii*, in Smith, J.E. (ed.), *Toxoplasmosis* Berlin, Heidelberg: Springer-Verlag, pp. 109–121.

48. Shams-Eldin, H., de Macedo, C.S., Niehus, S., Dorn, C., Kimmel, J., Azzouz, N., and Schwarz, R.T. (2008). *Plasmodium falciparum* dolichol phosphate mannose synthase represents a novel clade. *Biochem Biophys Res Commun* 370:388–393.

49. Smith, T.K., Crossman, A., Borissow, C.N., Paterson, M.J., Dix, A., Brimacombe, J.S., and Ferguson, M.A. (2001). Specificity of GlcNAc-PI de-N-acetylase of GPI biosynthesis and synthesis of parasite-specific suicide substrate inhibitors. *Embo J* 20:3322–3332.

50. Smith, T.K., Gerold, P., Crossman, A., Paterson, M.J., Borissow, C.N., Brimacombe, J.S., Ferguson, M.A., and Schwarz, R.T. (2002). Substrate specificity of the *Plasmodium falciparum* glycosylphosphatidylinositol biosynthetic pathway and inhibition by species-specific suicide substrates. *Biochemistry* 41:12395–12406.

51. Smith, T.K., Kimmel, J., Azzouz, N., Shams-Eldin, H., and Schwarz, R.T. (2007). The role of inositol acylation and inositol deacylation in the *Toxoplasma gondii* glycosylphosphatidylinositol biosynthetic pathway. *J Biol Chem* 282:32032–32042.

52. Stijlemans, B., Baral, T.N., Guilliams, M., Brys, L., Korf, J., Drennan, M., Van Den Abbeele, J., De Baetselier, P., and Magez, S. (2007). A glycosylphosphatidylinositol-based treatment alleviates trypanosomiasis-associated immunopathology. *J Immunol* 179:4003–4014.

53. Striepen, B., Dubremetz, J.F., and Schwarz, R.T. (1999). Glucosylation of glycosylphosphatidylinositol membrane anchors: identification of uridine diphosphate-glucose as the direct donor for side chain modification in *Toxoplasma gondii* using carbohydrate analogues. *Biochemistry* 38:1478–1487.

54. Striepen, B., Zinecker, C.F., Damm, J.B., Melgers, P.A., Gerwig, G.J., Koolen, M., Vliegenthart, J.F., Dubremetz, J.F., and Schwarz, R.T. (1997). Molecular structure of the "low molecular weight antigen" of Toxoplasma gondii: a glucose alpha 1–4 N-acetylgalactosamine makes free glycosyl-phosphatidylinositols highly immunogenic. *J Mol Biol* 266:797–813.

55. Tachado, S.D., Gerold, P., McConville, M.J., Baldwin, T., Quilici, D., Schwarz, R.T., and Schofield, L. (1996). Glycosylphosphatidylinositol toxin of *Plasmodium* induces nitric oxide synthase expression in macrophages and vascular endothelial cells by a protein tyrosine kinase-dependent and protein kinase C-dependent signaling pathway. *J Immunol* 156:1897–1907.

56. Tachado, S.D., Gerold, P., Schwarz, R., Novakovic, S., McConville, M., and Schofield, L. (1997). Signal transduction in macrophages by glycosylphosphatidylinositols of *Plasmodium*, *Trypanosoma*, and *Leishmania*: activation of protein tyrosine kinases and protein kinase C by inositolglycan and diacylglycerol moieties. *Proc Natl Acad Sci USA* 94:4022–4027.

57. Tachado, S.D., and Schofield, L. (1994). Glycosylphosphatidylinositol toxin of *Trypanosoma brucei* regulates IL-1 alpha and TNF-alpha expression in macrophages by protein tyrosine kinase mediated signal transduction. *Biochem Biophys Res Commun* 205:984–991.

58. von Itzstein, M., Plebanski, M., Cooke, B.M., and Coppel, R.L. (2008). Hot, sweet and sticky: the glycobiology of *Plasmodium falciparum*. *Trends Parasitol* 24:210–218.
59. Waters, A.P., Higgins, D.G., and McCutchan, T.F. (1991). *Plasmodium falciparum* appears to have arisen as a result of lateral transfer between avian and human hosts. *Proc Natl Acad Sci USA* 88:3140–3144.
60. Wei, S., Marches, F., Borvak, J., Zou, W., Channon, J., White, M., Radke, J., Cesbron-Delauw, M.F., and Curiel, T.J. (2002). *Toxoplasma gondii*-infected human myeloid dendritic cells induce T-lymphocyte dysfunction and contact-dependent apoptosis. *Infect Immun* 70:1750–1760.
61. Wennicke, K., Debierre-Grockiego, F., Wichmann, D., Brattig, N.W., Pankuweit, S., Maisch, B., Schwarz, R.T., and Ruppert, V. (2008). Glycosylphosphatidylinositol-induced cardiac myocyte death might contribute to the fatal outcome of *Plasmodium falciparum* malaria. *Apoptosis* 13:857–866.
62. Wichmann, D., Schwarz, R.T., Ruppert, V., Ehrhardt, S., Cramer, J.P., Burchard, G.D., Maisch, B., and Debierre-Grockiego, F. (2007). *Plasmodium falciparum* glycosylphosphatidylinositol induces limited apoptosis in liver and spleen mouse tissue. *Apoptosis* 12:1037–1041.
63. Wichroski, M.J., and Ward, G.E. (2003). Biosynthesis of glycosylphosphatidylinositol is essential to the survival of the protozoan parasite *Toxoplasma gondii*. *Eukaryot Cell* 2:1132–1136.
64. Zhu, J., Krishnegowda, G., and Gowda, D.C. (2005). Induction of proinflammatory responses in macrophages by the glycosylphosphatidylinositols of *Plasmodium falciparum*: the requirement of extracellular signal-regulated kinase, p38, c-Jun N-terminal kinase and NF-kappaB pathways for the expression of proinflammatory cytokines and nitric oxide. *J Biol Chem* 280:8617–8627.
65. Zinecker, C.F., Striepen, B., Geyer, H., Geyer, R., Dubremetz, J.F., and Schwarz, R.T. (2001). Two glycoforms are present in the GPI-membrane anchor of the surface antigen 1 (P30) of Toxoplasma gondii. *Mol Biochem Parasitol* 116:127–135.
66. Pittet, M., and Conzelmann, A. (2007). Biosynthesis and function of GPI proteins in the yeast *Saccharomyces cerevisiae*. *Biochim Biophys Acta* 1771:405–420.

10

Chemical Synthesis of Glycosylphosphatidylinositol (GPI) Anchors

RAM VISHWAKARMA • DIPALI RUHELA

Bio-organic Chemistry Laboratory
National Institute of Immunology
New Delhi 110067, India

I. Abstract

The remarkable structure, biosynthesis, cell biology, and membrane biology of glycosylphosphatidylinositol (GPI) anchors of cell surface proteins have generated significant interest in organic chemistry, and number of syntheses of full-length GPI anchors and their fragments/intermediates have been reported. In spite of their structural complexity and microheterogeneity, the GPI anchors of many species can now be assembled in the laboratory from easily accessible starting materials. These major chemical advances have provided many probe molecules and biosynthetic intermediates required for addressing key questions pertaining to the form and function of GPI molecules. In this chapter, the authors have reviewed the synthetic efforts, with focus on the total synthetic routes reported by various synthetic chemistry groups. The relevant work on partial synthesis of GPI intermediates and mimics have also been discussed.

THE ENZYMES, Vol. XXVI   181   ISSN NO: 1874-6047
                                                                DOI: 10.1016/S1874-6047(09)26010-0

II. Introduction

The discovery of glycosylphosphatidylinositol (GPI) molecules, as a unique class of complex glycolipids, which anchor proteins to the plasma membrane of eukaryotic cell, was a landmark in modern biology as it unraveled an alternative mode for the membrane anchoring of surface macromolecules, a mechanism quite distinct from that of the well-known hydrophobic peptide domains. The full chemical structure of GPI anchor was revealed in 1988; first for the GPI anchor of variant surface protein (VSG) of the parasitic protozoa *Trypanosoma brucei* [1] and the second one for Thy-1 glycoprotein of rat brain [2]. Subsequently, several GPI anchors and protein-free GPIs have been isolated all across the eukaryotic species, including humans. The biology of GPI anchors have been periodically reviewed [3–6].

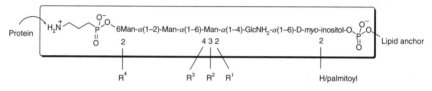

Species	R^4	R^3	R^2	R^1	Lipid	Ref.
T. brucei VSG	H	H	Gal$_2\alpha$(1–2)	H	DAG	5
Rat Thy-1	Manα(1–2)	GalNAcβ(1–4)	H	EA-P	Alkylacylglycerol	5
Yeast	Manα(1–2)	H	H	H	Ceramide DAG	5
Malaria	Manα(1–2)	H	H	H	DAG	5

The common core structure [6-*O*-aminoethylphosphoryl-Man-α(1–2)-Man-α(1–6)-Man-α(1–4)-GlcNH$_2$-α(1–6)-D-*myo*-inositol-1-*O*-phosphate] has been found to be conserved across the species during evolution but lots of species-specific modifications have occurred at various branching points of the GPI core, that is, presence of additional mannose, galactose, galactos-amine, and ethanolamine-phosphate residues. Significant microheterogeneity has been found in the lipid domain: *sn*-1,2-dimyristoylglycerol in *T. brucei*, *sn*-1-*O*-alkyl-2-*O*-acylglycerol in human erythrocyte acetylcholine esterse and the folate receptor, the inositol residue can be palmitoylated at 2-*O*-position (malarial GPI anchors). In addition, for example, in yeast, a ceramide-containing sphingosine occurs in GPI anchor in place of the glycerolipids.

Since the discovery of GPI anchoring mode of membrane association of cell-surface proteins, the chemistry and biology of these complex glycoli-pids have remained in focus. Surprisingly, the GPIs are produced in high abundance by the protozoan parasites (*Trypanosoma*, *Leishmania*, and *Plasmodium falciparum*), compared to that of higher organisms, where they serve as essential virulence factors that allow the parasites to infect, proliferate, and subvert the host immunity. Marked differences in the structure, function, and biosynthesis of GPIs from the parasites and human cells have been identified providing valuable targets for drug and vaccine design. Even among the parasites, various species express GPIs with subtle structural differences that manifest in remarkable and, at times, opposing biological functions in the host. Their structural complexity and biological function have inspired widespread chemical interest.

Immediately after the disclosure of the remarkable chemical structures of GPI anchor of the variant surface glycoprotein of *T. brucei* and Thy-1 protein of rat brain in 1988, chemical synthetic efforts were initiated by several leading organic chemists and a number of synthetic approaches towards GPIs (Thy-1, yeast, *T. brucei*, sperm CD52, *Leishmania*, and *Plasmodium falciparum*) have been reported.

Figure 10.1 shows chemical structures of key GPI anchors of various species.

Chemical synthesis of full-length GPI anchors presents substantial challenges requiring expertise of carbohydrate, lipid, and phosphorus chemistry, which is complicated by the presence of the structural and functional differences among/within the species and microheterogeneity in their lipid and glycan domains. Since the discovery of GPI anchors coincided with the unraveling role of phosphatidylinositol (PI) and phosphoinositides as second messengers in lipid-mediated signal transduction, the access to synthetic PIs and GPIs became indispensable for further advances in membrane and cell biology. Initial effort on the synthesis of a partial GPI structure pentasaccharide core was reported by Fraser-Reid in 1989; however, the total synthesis of a full-length GPI anchor was first reported by Ogawa and coworkers in 1992 [7–11].

III.   Synthesis of GPI Anchor of VSG of *T. brucei*

The synthetic plan adopted by Ogawa group [7–11] used six key building blocks, the mannose-glucosamine-inositol **2**, galactobioside **3**, Z-protected ethanolamine-*H*-phosphonate **4**, glycerol-lipid-*H*-phosphonate **5**, and two mannose intermediates **6** and **7** as shown in Scheme 10.1.

The critical requirement for synthon **2** was based on the access of optically pure *myo*-inositol (1-*O*-*p*-methoxybenzyl-2,3,4,5-tetra-*O*-benzyl-D-*myo*-inositol **16**).This was prepared from racemic 2,3:4,5-bis-cyclohexylidene-*myo*-inositol (**8**) by the following reaction sequence (Scheme 10.2).

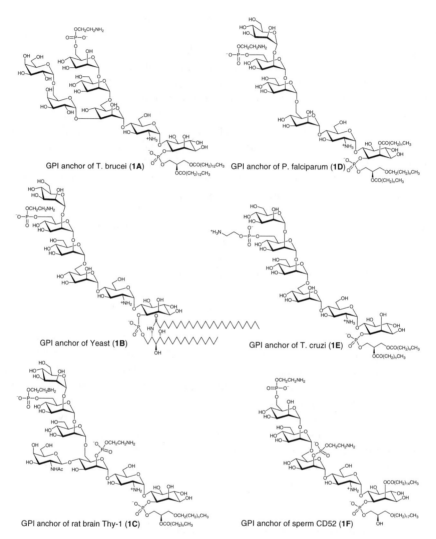

FIG.10.1. The chemical structures of key GPI anchors of various species.

The tin-oxide-mediated regioselective *p*-methoxybenzylation at 6-position (→**9**), allylation (→**10**), acetal deprotection followed by benzylation (→**11**), allyl group removal (→**12**), resolution of required D-isomer through camphanate-derived diastereoisomers (→**13**), replacement of PMB with allyl group (→**14**), hydrolysis of camphanate ester (→**15**), and allyl deprotection.

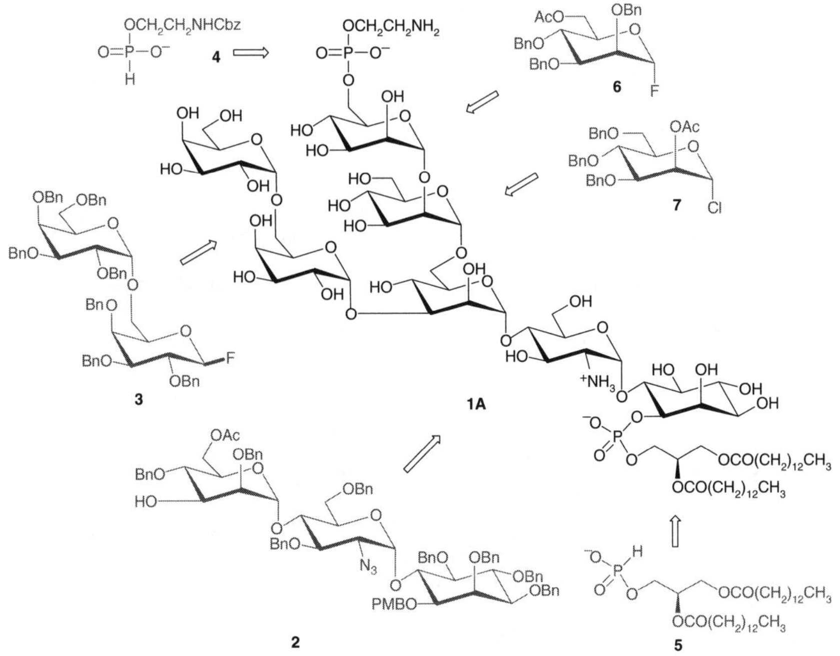

SCHEME 10.1. Synthetic approach of Ogawa for GPI anchor *T. brucei*.

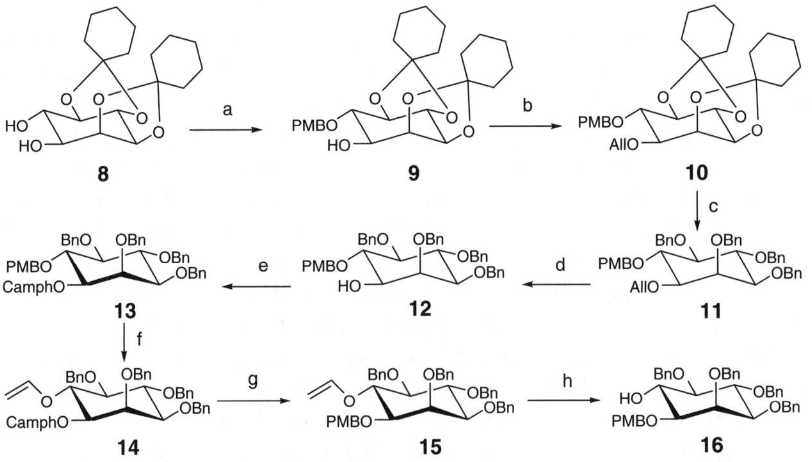

SCHEME 10.2. Synthesis of *myo*-inositol building block. Reagents and conditions: (a) Bu$_2$SnO, CsF, PMBCl, KI; (b) allyl bromide, NaH, DMF; (c) 0.1 M HCl-MeOH, BnBr, NaH, DMF; (d) KOtBu, DMSO, and acetone-1 M HCl, 9:1; (e) camphanic acid chloride, DMAP, Et3N, DCM; (f) CAN, CH$_3$CN-H$_2$O, and ethylvinylether, PTSA; (g) NaOH, MeOH, and PMBCl, NaH, DMF; (h) AcOH-MeOH.

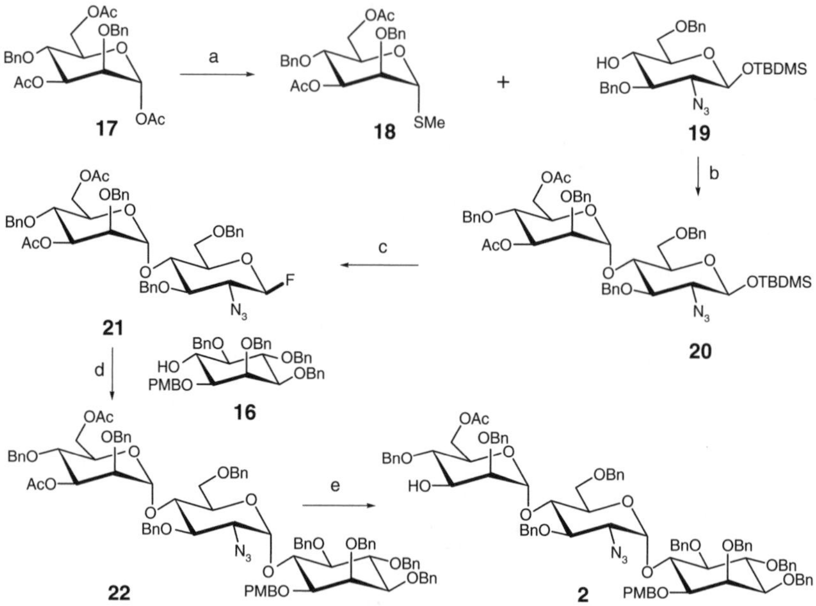

SCHEME 10.3. Synthesis of *myo*-inositol building Man-GlcN-inositol block. Reagents and conditions: (a) methyl tributyltin sulfide, tin(V)chloride; (b) Cu(II) bromide, silver triflate in CH_3NO_2 (c) TBAF, AcOH, and DAST; (d) zirconocene dichloride, silver perchlorate; (e) NaOMe-MeOH and AcCl, pyridine.

The disaccharide **20** (Scheme 10.3) was synthesized by Cu(II)-mediated glycosidation of the known azidoglycosyl acceptor **19** with the mannosyl donor **18** (prepared from 1,3,6-tri-*O*-acetyl-2,4-di-*O*-benzyl-α-D-mannopyranose **17** with methyl tributyltinsulfide and tin (IV) chloride). The intermediate **20** was converted to the corresponding glycosyl donor **21** by desilylation (TBAF, THF) followed by the reaction with diethylamino-sulfurtrifluoride. The crucial glycosidation of **21** with the inositol synthon **16** using zirconocene dichloride and silver perchlorate gave pseudotrisaccharide **22** in α:β (3.7:1) ratio. The removal of two acetyl groups followed by regioselective acetylation at the primary OH gave the building block **2** (Scheme 10.3).

The galactobiosyl building block **3** was prepared (Scheme 10.4) by coupling of *p*-methoxy phenyl-2,3,4-tri-*O*-benzyl-D-galactopyranose (**28**, from penta-acetylmannose) with thiomethyl galactoside **29** by cupric-bromide and tetrabutylammonium-bromide (→**30**), followed by conversion to the galactobiosyl fluoride donor (**3**).

In the further elaboration of synthesis (Scheme 10.5), two key building blocks **2** and **3** were coupled using zirconocene-dichloride and silver-perchlorate to obtain pseudopentasaccharide **32**. The acetyl group from

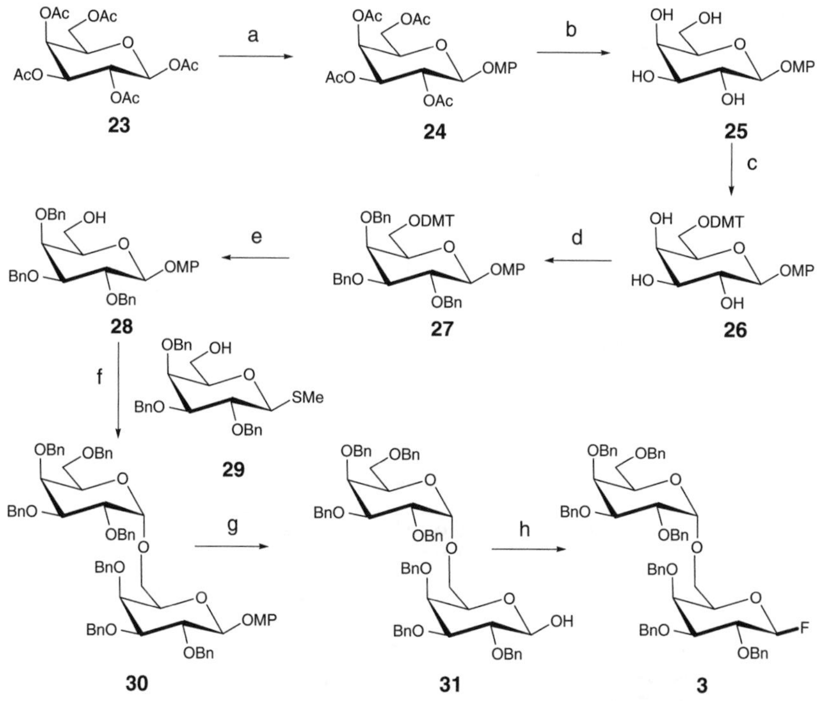

SCHEME 10.4. Synthesis of galactobiosyl donor. Reagents and conditions: (a) *p*-methoxy-phenol, TESOTf; (b) NaOMe-MeOH; (c) DMTCl, pyridine; (d) BnBr, NaH, DMF; (e) PTSA-MeOH; (f) CuBr$_2$, Bu$_4$NBr; (g) ammonium cerium (IV)nitrate, toluene-acetonitrile-water; (h) DAST, DCM.

the middle mannose was removed (→**33**) followed by the glycosidation with 2-*O*-acetyl-3,4,6-tri-*O*-benzyl-mannosyl chloride (**7**) provided pseudohexa-saccharide **34**. The deacetylation at the 2-position of the newly introduced mannose residue (→**35**) followed by one more cycle of glycosidation with 2,3,4-tri-*O*-benzyl-6-*O*-acetyl-mannosyl fluoride (**6**) by zirconocene catalyst yielded heptasaccharide core (→**36**).

For strategic reasons, the terminal acetyl of **36** was replaced with the chloroacetyl group (→**37**) and then the PMB group from 1-position of the inositol residue was removed by TMSOTf method to give pseudoheptasac-charide **38** ready for the phospholipidation with 1,2-di-*O*-acyl-glycero-*H*-phosphonate (**5**) affording the heptasaccharide-lipid-*H*-phosphonate **39** (Scheme 10.6).

The deprotection of chloroacetyl group by thiourea (→**40**) followed by phosphitylation with Cbz-ethanolamine-*H*-phosphonate (**4**) and *in situ* oxi-dation gave fully protected GPI anchor **42**. The final global deprotection by

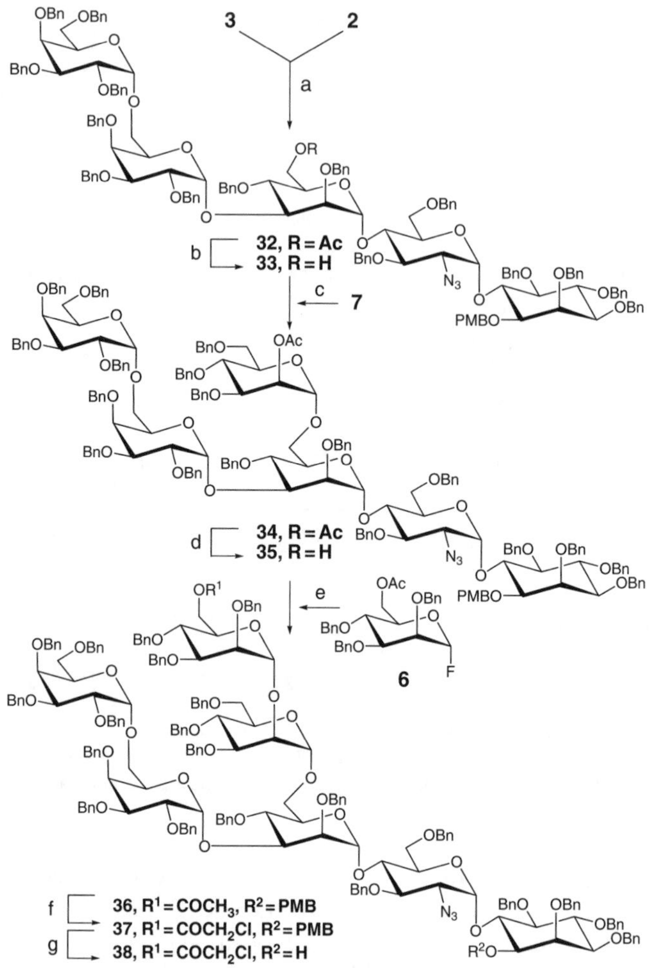

SCHEME 10.5. Coupling of the various building blocks. Reagents and conditions: (a) zico-nocene dichloride, silver perchlorate; (b) NaOMe, MeOH; (c) HgCl$_2$, Hg(CN)$_2$; (d) NaOMe-MeOH; (e) ziconocene dichloride, silver perchlorate; (f) NaOMe, MeOH and chloroacetyl anhydride, pyridine; (g) TMSOTf-DCM.

Pd(OH)$_2$-mediated hydrogenation afforded the target GPI anchor of VSG of *T. brucei*. After the publication of this synthesis, it was discovered that the *myo*-inositol intermediate used by Ogawa group was of unnatural L-*myo*, but not the desired D-*myo* stereochemistry. Despite this mistake, the synthetic design provided the first roadmap to access full-length GPI anchors.

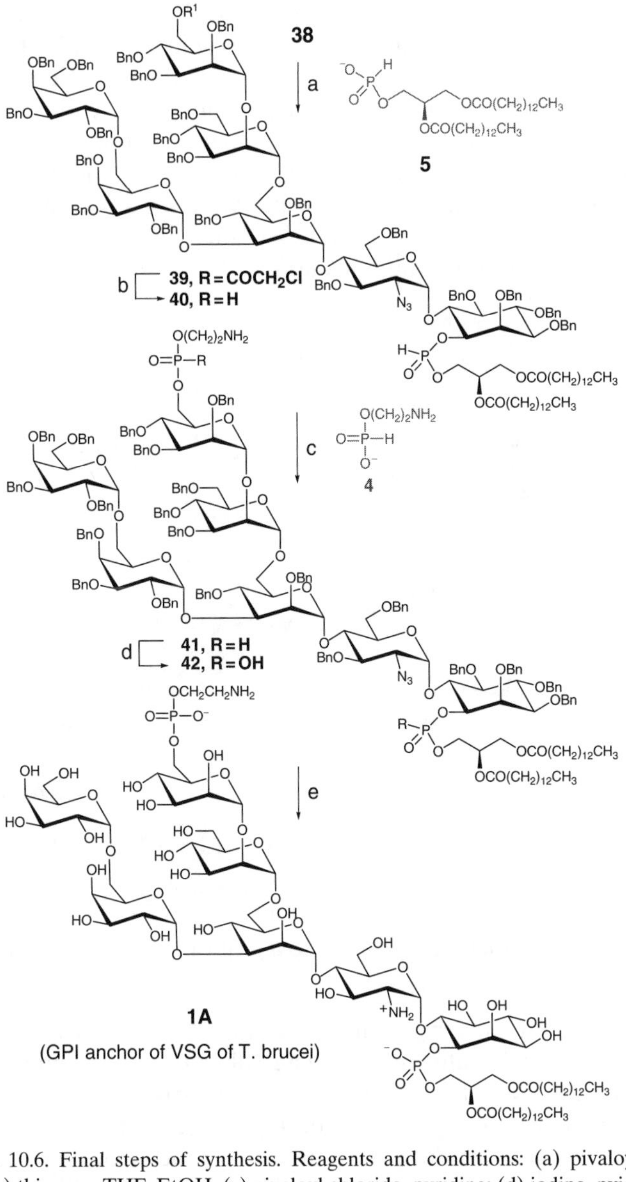

SCHEME 10.6. Final steps of synthesis. Reagents and conditions: (a) pivaloyl chloride, pyridine; (b) thiourea, THF, EtOH; (c) pivaloyl chloride, pyridine; (d) iodine, pyridine; (e) Pd (OH)$_2$, CHCl$_3$-MeOH-H$_2$O.

Steven Ley's group at Cambridge University reported their elegant synthesis of full-length GPI anchor of *T. brucei* in 1998 [12]. The synthetic approach [12, 13] was based on the new 1,2-acetal methodology discovered in their lab showing that (a) bis-(dihydropyran)s form corresponding dis-piroketals with 1,2-diols leading to stereoselective protection and desym-metrization, and (b) diacetal protection can be used to tune the reactivity of glycosyl donors. The highly convergent synthesis reported by Ley and coworkers was an elegant demonstration of the power of 1,2-acetal method (Scheme 10.7 as the retrosynthetic plan).

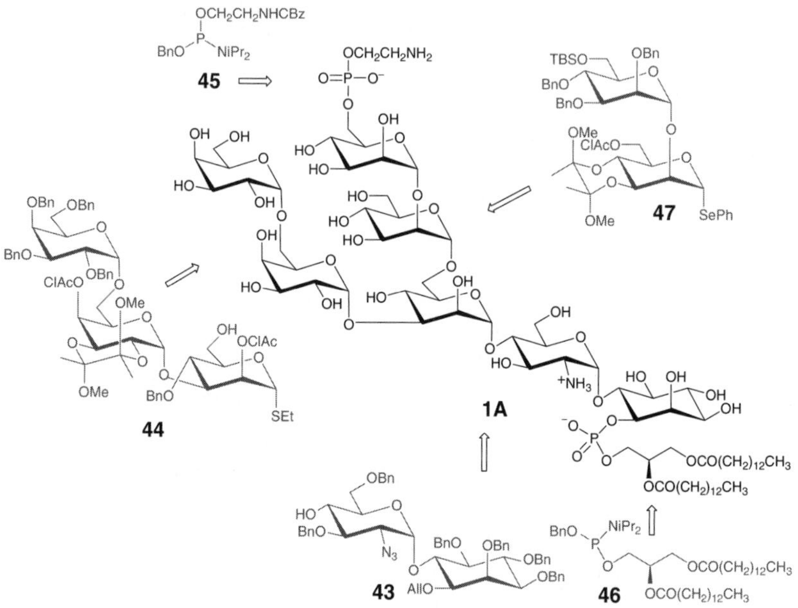

SCHEME 10.7. Synthetic approach of Ley for GPI anchor of VSG of *T. brucei*.

The key feature of this synthesis is the desymmetrization protocol (Scheme 10.8) for inositol intermediate solving an outstanding problem of GPI synthesis, as most of the earlier synthesis of *myo*-inositols used either chiral resolution (chemical or enzymatic methods), low-yielding desymmetrization, or synthesis from D-glucose. One of the two enantiotopic diol pairs in intermediate **48** (Scheme 10.8) was selectively protected with chiral bis-dihydropyran to obtain the dispiroketal **49** as single optically pure diastereoisomer (ee 98%). This key intermediate on deben-zoylation, per-*O*-benzylation, and oxidation led to the intermediate **50**, which on, dispiroketal deprotection provided the *myo*-inositol-diol **51**.

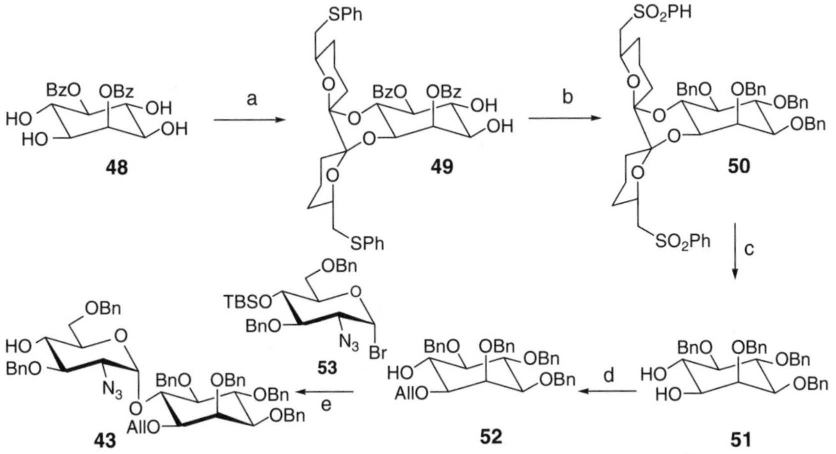

SCHEME 10.8. Synthesis of GlcN-inositol building blocks. Reagents and conditions: (a) bis (dihydropyran), PPh$_3$, HBr, CHCl$_3$; (b) K$_2$CO$_3$, MeOH; NaH, BnBr, DMF; (c) mCPBA, DCM, LiN(TMS)$_2$, THF, 0 °C; (d) Bu$_2$Sn(OMe)$_2$, Tol, AllBr, TBAI; (e) 53, TBABr, DCM, MS 4A, 3d; (f) TBAF, TH.

Selective allylation via tin-acetal complex yielded 1-*O*-allyl-2,3,4,5-tetra-*O*-benzyl-*myo*-D-inositol (52), which on, glycosidation with known azidoglycosyl bromide (53) using Lamieux's inversion protocol, provided the required pseudodisaccharide 43 (Scheme 10.8).

The trisaccharide building block 44 was assembled (Scheme 10.9) through three intermediates 56, 57, and 59. The *trans* diol of readily available galactoside selenide 54 [14] was protected as butane bis-acetal. These steps were then followed by selective silylation, chloroacylation, and desilylation to give 56. The glycosidation of 56 with perbenzylated galactoselenide using NIS and TMSOTf furnished desired α-linked digalactoside 57, along with a minor amount of β-linked isomer. The deactivating effect of BDA and chloroacetate groups in the acceptor 56 allowed selective activation of galactoselenide donor thus preventing homocoupling. The digalactoside 57 was then used as a glycosyl donor by activation with methyltriflate and treated with mannoside 58 to afford a trisaccharide 59. The higher reactivity of selenophenyl group in 58 allowed its activation without disturbing anomeric thioethyl group of acceptor 57. The removal of silyl group by tetrabutylammonium fluoride (TBAF) in THF provided desired trisaccharide building block 44.

The synthesis of mannobiose building block 47 (Scheme 10.10) required mannoside donor 61 and the acceptor 63. The donor was prepared from 1-*O*-phenylselenomannose (54) by selective silylation at 6-position (→60) followed by perbenzylation.

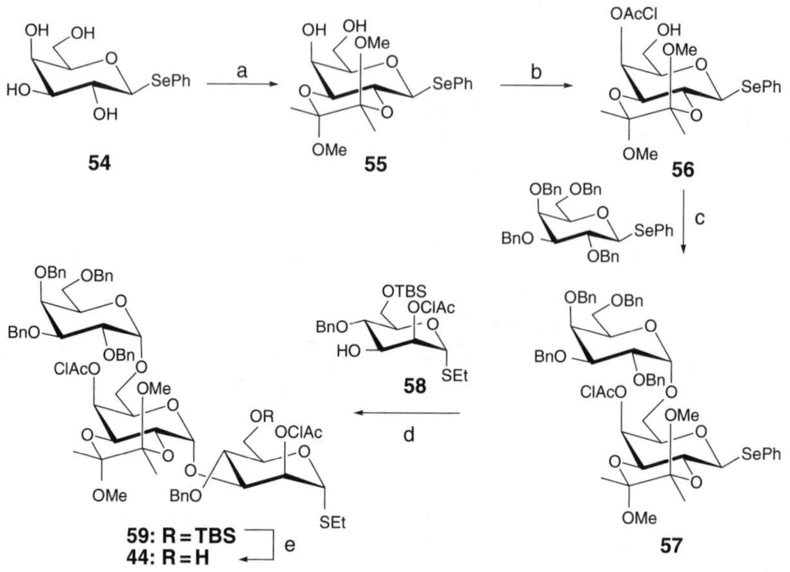

SCHEME 10.9. Synthesis of trisaccharide donor. Reagents and conditions: (a) butane 2,3-dione, HC(OCH$_3$)$_2$, CSA, MeOH; (b) TBSCl, imidazole, THF; (ClAc)$_2$O, pyridine, 0 °C, 48% HF, CH$_3$CN; (c) NIS, TMSOTf, ether/DCM; (d) MeOTf, ether; (e) 48% HF, CH$_3$CN.

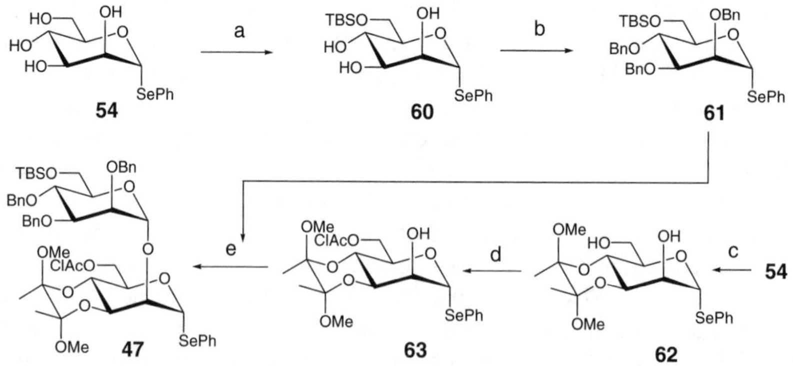

SCHEME 10.10. Synthesis of mannobiose block. Reagents and conditions: (a) TBSCl, imidazole; (b) NaH, BnBr, DMF; (c) butane 2,3 dione, HC(OCH$_3$)$_2$, CSA, MeOH; (d) (ClAc)$_2$O, pyridine; (e) NIS, TMSOTf, Ether.

The acceptor mannose **63** was also prepared from **54** by 3,4-bis-acetylation (→**62**) followed by 6-*O*-chloroacetylation. The coupling of **61** and **63** provided disaccharide **47** in excellent yield. As in the previous case, the

BDA and chloroacetate groups tuned the reactivity of the acceptor and prevented the homocoupling.

For the final assembly of the key pieces, trisaccharide **44** was glycosylated with selenophenyl mannobiose donor **47** under methyltriflate catalysis to give pentasaccharide **64** (Scheme 10.11). The reaction required four equivalent of donor **47** to suppress the formation of anhydrosugar of **44** due to intramolecular glycosidation at 6-OH position. The activation (NIS/ TfOH) of thioethyl of the pentasaccharide **64** and glycosidation with glucosamine-inositol **43** furnished pseudoheptasaccharide core **65** in reasonable (51%) yield. The TBS group from the terminal mannose residue of

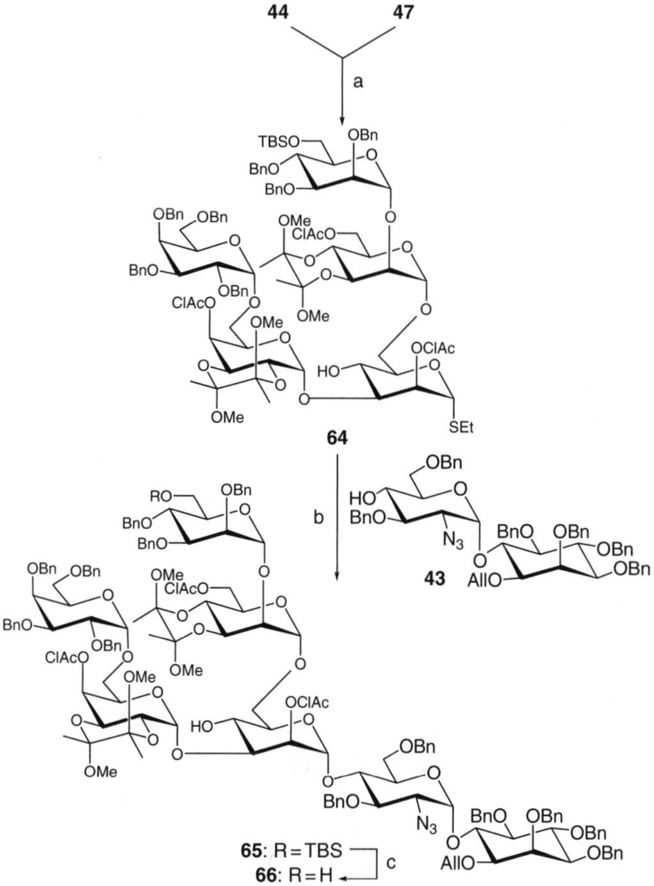

SCHEME 10.11. Coupling of various building blocks. (a) MeOTf, DCM, 4A MS; (b) NIS, TfOH, ether/DCM; (c) 48% HF, CH$_3$CN.

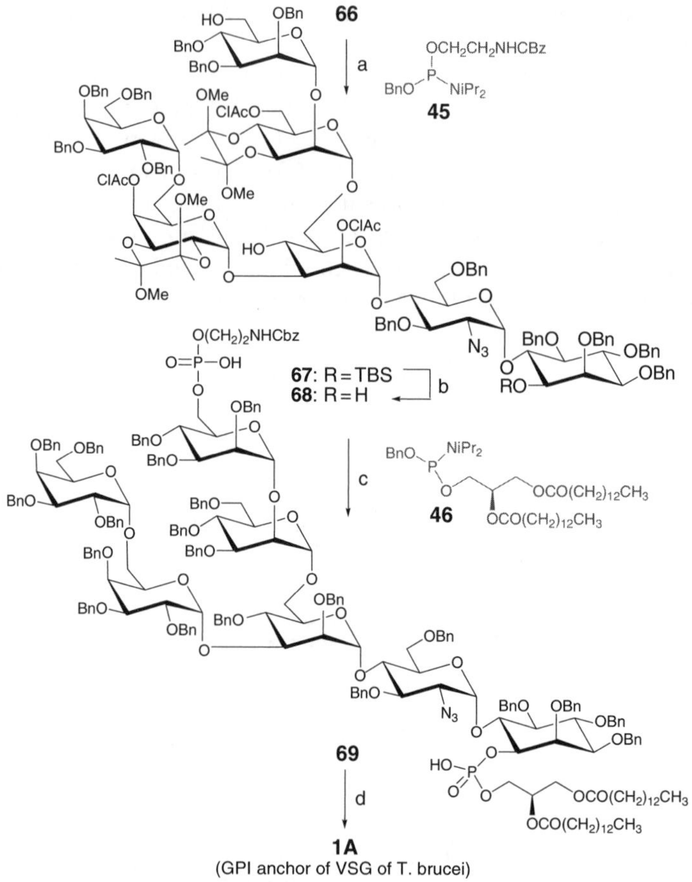

SCHEME 10.12. Final steps of synthesis. Reagents and conditions: (a) 1-*H*-tetrazole, CH$_3$CN/DCM (1:1), then mCPBA (−40 to 20 °C); (b) 48% HF aq., CH$_3$CN; (c) 1-*H*-tetrazole, CH$_3$CN/DCM (1:1), then mCPBA (−40 to 20 °C); (d) PdCl$_2$, H$_2$, CHCl$_3$/MeOH/H$_2$O, 3:3:1, NH$_2$NHC(S)SH, 2.6-lutidine, CH$_3$CN; TFA/H$_2$O, 9:1.

65 was removed (TBAF) to give **66**, exposing the primary alcohol ready for phosphorylation with the ethanolamine phosphate unit.

This carbohydrate core **66** was further elaborated to fully protected GPI anchor by phosphoramidite chemistry (Scheme 10.12). The ethanolamine phosphate unit was installed by coupling of **66** with phosphoramidite **45** followed by oxidation to provide compound **67**. The HF-mediated desilylation at 1-position of *myo*-inositol residue (→**68**) followed by phosphorylation with diacylglycerophosphoramidite **46** and oxidation provided fully protected GPI anchor (**69**).

In the final step, hydrogenation of **69** with Pd/C removed the benzyl ethers, the benzyloxycarbonyl group and converted azide to amine group, followed by the treatment with hydrazine dithiocarbonate (removal of chloroacetyl group) and aqueous TFA (removal of BDA group) gave the GPI anchor **1A** in a remarkably high (90%) yield. All the intermediates were fully characterized by high-resolution NMR and MS analysis.

IV.   Synthesis of GPI Anchor of *Saccharomyces cerevisiae* (Yeast)

The Richard Schmidt group (Germany) reported [15, 16] first the total synthesis of ceramide containing GPI anchor (**1B**) of yeast in 1994. Their synthetic strategy (Scheme 10.13) was convergent and versatile with elegant demonstration of the power of trichloroacetimidate glycosidation method, which was discovered by his group.

The ceramide lipid building block **70**, containing phytosphingosine as nitrogen base, was prepared from azido-lipid intermediate **74** as shown in Scheme 10.14. The protection of bis-diol of azido intermediate **74** with

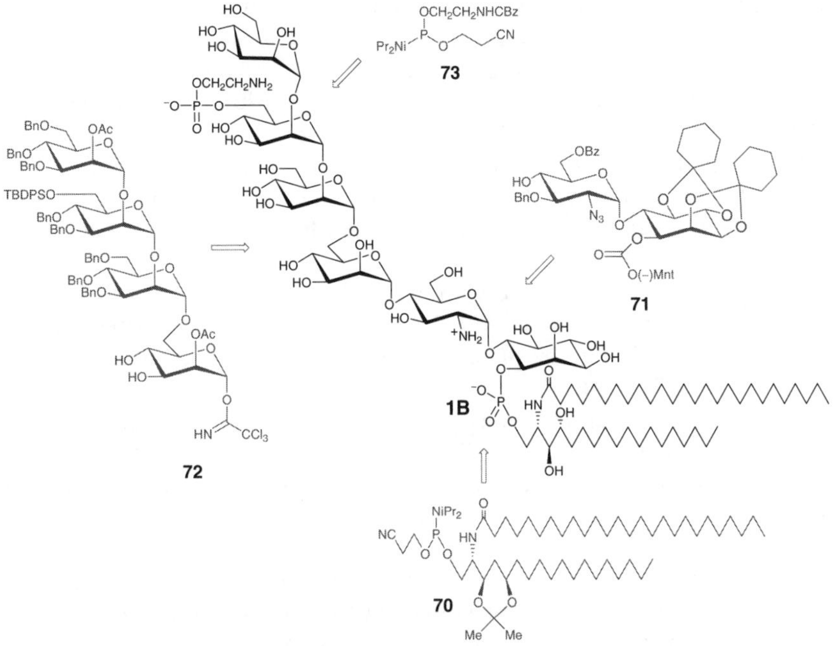

SCHEME 10.13. Synthetic approach of RR Schmidt for GPI anchor of *S. cerevisae* (Yeast).

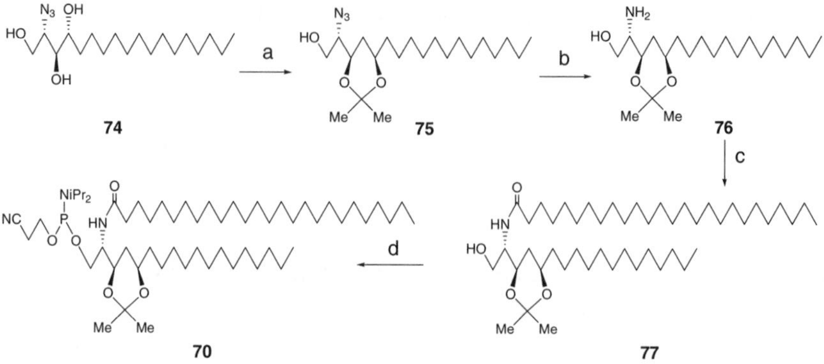

SCHEME 10.14. Synthesis of ceramide lipid intermediate. Reagents and conditions: (a) $Me_2C(OMe)_2$, PTSA; (b) PPh_3, H_2O; (c) $C_{25}H_{51}COOH$, EEDQ; (d) $NCCH_2CH_2OP$ $(NiPr_2)_2$, 1-H-tetrazole, DCM.

acetal group (dimethoxypropane, pTSA) to give **75** followed by the reduction of azide group by Staundinger method ($\rightarrow$**76**) and N-acylation with cerotic acid (hexacosanic acid) using 2-ethoxy-1-ethyoxycarbonyl-1,2-dihydroquinoline (EEDQ) produced 1-O-unprotected ceramide **77**. This was followed by phosphitylation with a bifunctional cyanoethoxy bis-(diisopropylamino)-phosphane in the presence of 1-H-tetrazole activator providing the desired ceramide intermediate **70**.

For the synthesis of α-glucosaminyl-(1–4)-inositol building block **71** (Scheme 10.15), the optically pure *myo*-inositol intermediate **78** was prepared by the resolution of racemic 2,3:4,5-bis-cyclohexylidene-*myo*-inositol (**8**) using menthyl chiral auxiliary. The glycosidation of **78** with 2-azidoglycosyl trichloroacetimidate (**79**) led to the α-linked pseudodisaccharide **80** in excellent yield.

This was followed by deacetylation ($\rightarrow$**81**), regioselective 6-O-benzoylation with benzoyl cyanide at $-70\,^{\circ}$C ($\rightarrow$**82**), and selective 3-O-benzylation with benzyl bromide in the presence of silver oxide providing the required building block **71** in high yield.

The synthesis of tetramannose building block **72** was based on the access to three key mannose intermediates **88**, **93**, and **94** (Scheme 10.16). All these compounds were prepared from a common starting material, the mannose-orthoester **83** readily available from D-mannose.

The synthon **88** was obtained by the following steps: benzylation with benzyl bromide and sodium hydride ($\rightarrow$**84**), opening of the orthoester group with glacial acetic acid ($\rightarrow$**85**), acetylation ($\rightarrow$**86**), selective 1-O-deacetylation with ammonium carbonate in DMF ($\rightarrow$**87**), and conversion to the trichloroacetimidate ($\rightarrow$**88**) by trichloroacetonitrile and

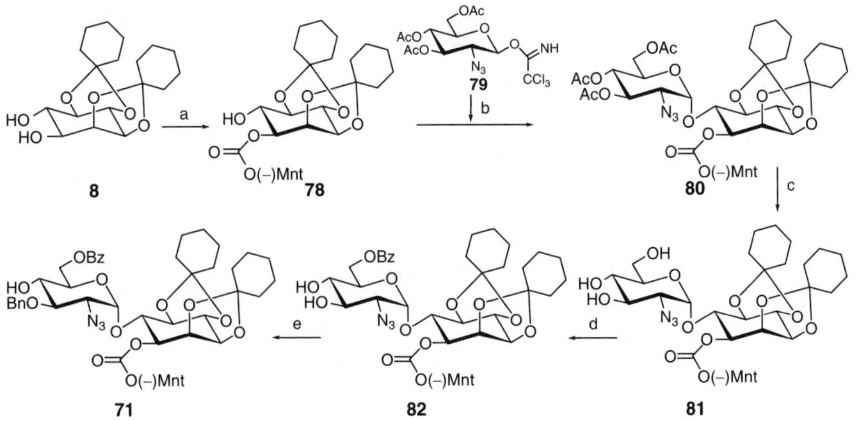

SCHEME 10.15. Synthesis of glucosamine-inositol fragment. Reagents and conditions: (a) Bu$_2$SnO, menthylchloroformate; (b) TMSOTf, ether, MS 3A; (c) NaOMe, MeOH; (d) BzCN, NEt$_3$, −70 °C; (e) Ag$_2$O, BnBr, MS 3A.

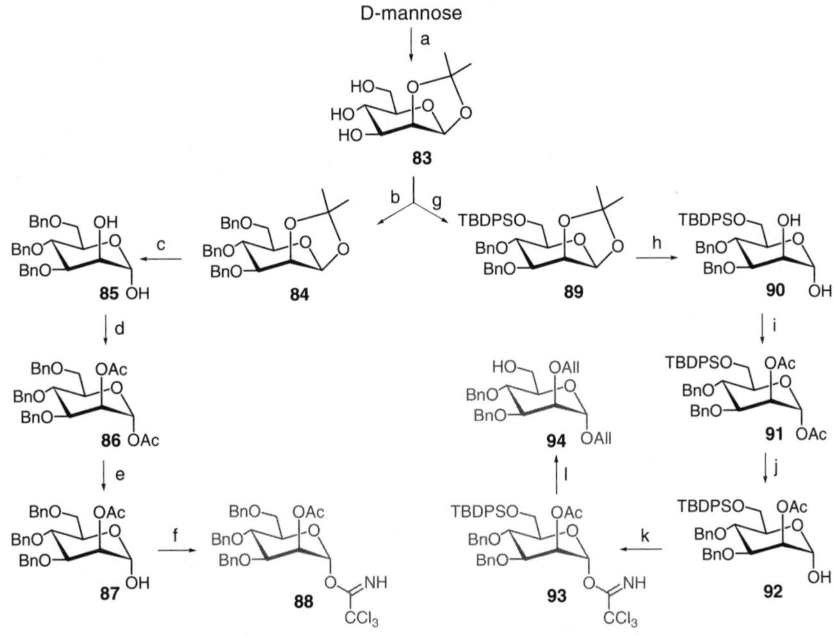

SCHEME 10.16. Synthesis of key building blocks for tetramannose fragment. Reagents and conditions: (a) HC(OMe)$_3$, H$^+$; (b) BnBr, NaH, DMF; (c) AcOH; (d) Ac$_2$O, pyridine; (e) (NH$_4$)$_2$CO$_3$, DMF; (f) CCl$_3$CN, DBU; (g) TBDPSCI, imidazole; (h) NaH, BnBr, AcOH; (i) Ac$_2$O, pyridine; (j) (NH$_4$)$_2$CO$_3$, DMF; (k) CCl$_3$CN, DBU; (l) allyl alcohol, NaOMe, MeOH TMSOTf, ether; TBAF, AcOH.

diazabicycloundecene (DBU). The 2-*O*-acetyl group in **88** was placed to promote α-selectivity during further glycosidations. The synthon **93** was prepared by regioselective 6-*O*-silylation of **83** followed by benzylation (→**89**), which on orthoester hydrolysis (→**90**), acetylation (→**91**), selective 1-*O*-deacetylation (→**92**), and trichloroacetimidation provided the desired donor **93**. The synthon **94** was easily prepared from **93** by reaction with allyl alcohol, 2-*O*-deacetylation, 2-*O*-allylation, and the removal of 6-*O*-silyl group.

The iterative assembly of the hexasaccharide building block **72** has been described in Scheme 10.17. The reaction of the mannosyl donor **88** with the

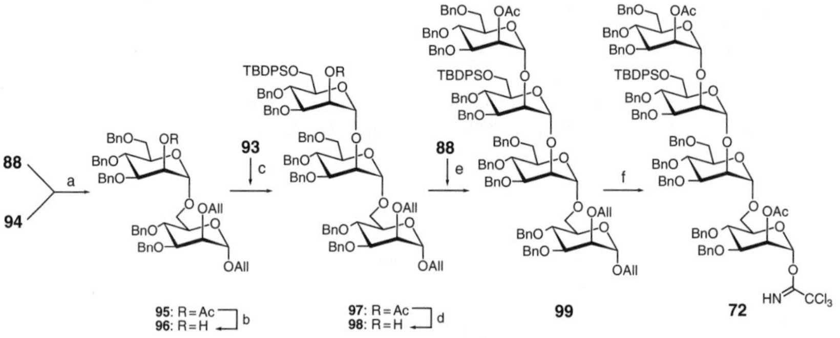

SCHEME 10.17. Synthesis of tetramannose fragment. Reagents and conditions: (a) TMSOTf, ether; (b) NAOMe, MeOH; (c) TMSOTf, ether; (d) NaOMe, MeOH; (e) TMSOTf, ether; (f) (Ph$_3$P)$_2$RhCl, Tol, ethanol, H$_2$O, 100 °C; Ac$_2$O, pyridine; (NH$_4$)$_2$CO$_3$, DMF, 40 °C; CCl$_3$CN, DBU.

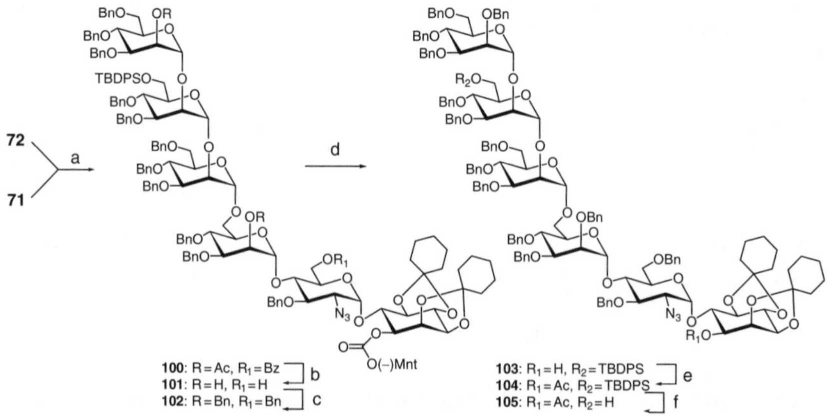

SCHEME 10.18. Assembly of hexasaccharide block. Reagents and conditions: (a) TMSOTf, ether; (b) KCN, MeOH-ether; (c) NaH, BnBr, DMF; (d) K$_2$CO$_3$, MeOH; (e) Ac$_2$O, Pyridine; (f) TBAF, THF.

acceptor **94** in diethyl ether using TMSOTf catalyst gave α-(1–6)-linked product **95** in good yield.

The removal of 2-*O*-acetyl group (→**96**) followed by the glycosidation with the second mannose donor **93** under the same condition gave α-(1–2)-α-(1–6)-linked trisaccharide **97**. In one more iteration, removal of the 2-*O*-acetyl group (→**98**) followed by the glycosidation with mannosyl donor **88** gave α-(1–20)-α-(1–2)-α-(1–6)-linked tetrasaccharide **99**. Now, both the allyl groups from **99** were removed by treatment with Wilkinson catalyst and hydrolysis and the free hydroxyl groups were acetylated. The regioselective anomeric deacetylation followed by trichloroacetimidation provided the desired tetramannose building block **72**.

In the advanced stage of synthesis (Scheme 10.18), the above tetramannose donor **72** was reacted with the glucosamine-inositol acceptor **71** in diethyl ether at RT in the presence of TMSOTf, which provided the pseudohexasaccharide **100** with good α-selectivity in good yield. All the acyl groups (two acetyls and one benzoyl) were now removed (→**101**) and replaced with stable benzyl groups (→**102**). This was followed by the replacement of chiral menthyl carbonate from inositol residue (→**103**) with an acetyl group (→**104**).

The stage was now set for the placement of ethanolamine phosphate at the 6-position of the third mannose residue, which was accomplished by the removal of TBDPS group (→**105**) and phosphitylation with Z-protected ethanolamine phosphoramidite **73** to give the intermediate **106** (Scheme 10.19). The cleavage of cyanoethyl and acetyl groups (NaOMe, MeOH, **107**) followed by phosphitylation with ceramide-lipid-phosphoramidite **70** (1-*H*-tetrazole,

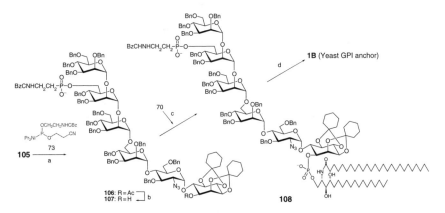

SCHEME 10.19. Final coupling with lipid anchor and deprotection. Reagents and conditions: (a) 1-*H*-tetrazole, t-BuO$_2$H; (b) NaOMe, MeOH; (c) 1-*H*-tetrazole, t-BuO$_2$H; Me$_2$NH; (d) HOCH$_2$CH$_2$OH, CSA; Pd(OH)$_2$/C, H$_2$.

m-chloroperbenzoic acid, and dimethylamine) gave fully protected GPI anchor **108**.

In the final steps, simultaneous deprotection of *O*-isopropylidene and *O*-cyclohexylidene groups with camphor sulphonic acid in glycol followed by hydrogenolysis of the Z-group, the *O*-benzyl groups, and the azido group yielded the target GPI anchor (**1B**) in excellent yield.

The above synthetic methodology was extended by the Schmidt group for the preparation of a variety of GPI anchors and their analogs [17–20].

V.   Synthesis of GPI Anchor of Rat Brain Thy-1

The first chemical synthesis of full-length GPI anchor of rat brain Thy-1 protein was reported by Fraser-Reid and coworkers in 1995 based on the new chemistry of *n*-pentenyl glycosides discovered in their labs [21–26]. The overall synthetic plan envisioned by Fraser-Reid group, illustrated in Scheme 10.20, was based on a triply convergent approach using five building blocks: trimannoside donor **109**, galactomannoside donor **110** (both in

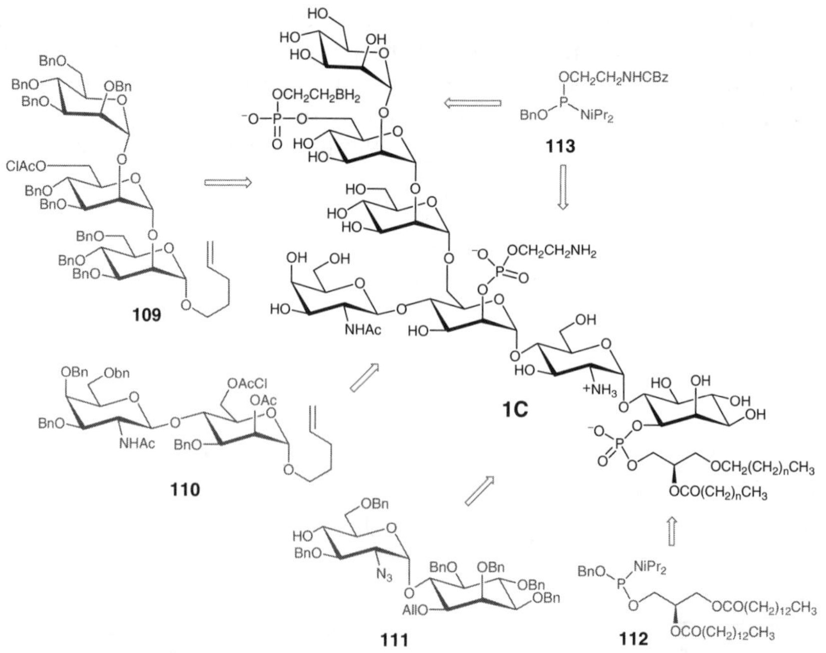

Scheme 10.20. Synthetic approach of Fraser-Reid for GPI anchor of rat brain Thy-1.

the form of *n*-pentenyl glycosides), glucosamine-inositol **111**, glycerolipid phosphoramidite **112**, and the ethanolamine phosphate **113**.

The synthesis of trimannose building block **109** was achieved (Scheme 10.21) from a single starting material *n*-pentenylmannosyl orthoester (**115**),

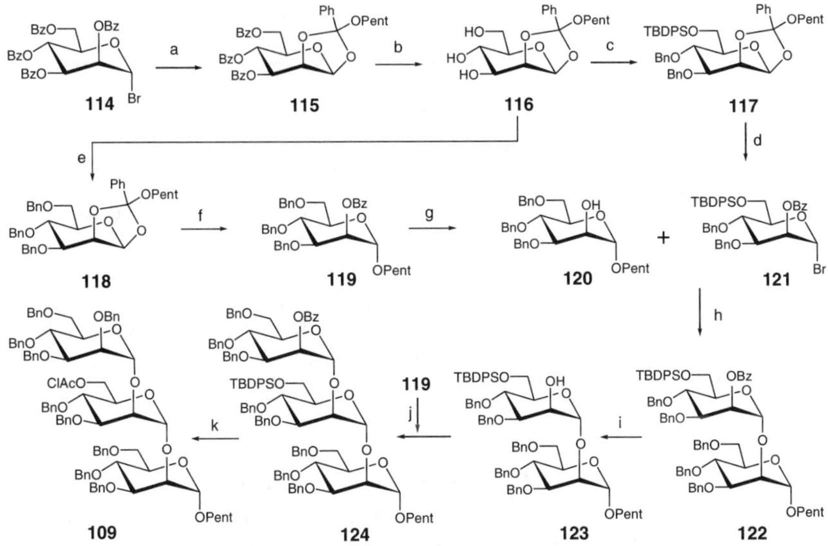

SCHEME 10.21. Synthesis of trimannose building block **109**. Reagents and conditions: (a) 4-penten-1-ol, 2,6-lutidine, DCM; (b) NaOMe, MeOH; (c) TBDPSCl, imidazole, THF; BnBr, NaH, DMF; (d) Br₂, DCM, 0 °C; (e) BnBr, NaH, DMF; (f) CSA, DCM, 50 °C; (g) NaOMe, MeOH; (h) AgOTf, 4A MS DCM; (i) NaOMe, MeOH; (j) NIS, TESTf, DCM; (k) NaOMe, MeOH; BnBr, NaH, DMF; TBAF THF; (ClAc)₂O, Et₃N, DCM.

readily accessible from tetrabenzoyl mannosyl bromide (**114**). The first intermediate **121** was prepared from **115** by selective 6-*O*-silylation, benzylation (→**117**), and treatment with bromine. The acceptor intermediate **120** was prepared from **116** by benzylation (→**118**), H⁺-induced transfer of the pentenyloxy group to the anomeric center (→**119**), and debenzoylation to expose OH group (→**120**) for glycosylation. The silver triflate-mediated coupling of **120** and **121** gave the disaccharide **122**. The removal of the benzoyl group provided the acceptor **123**, which was coupled with the *n*-pentenyl mannose donor **119** to provide trimannoside intermediate **124**. The TBDPS group of **124** was now replaced with a chloroacetyl group to provide the desired trimannoside block **109**.

The galactosamine-mannose building block **110** was prepared (Scheme 10.22) from two intermediates, the galactosamine donor **130** and *n*-pentenyl

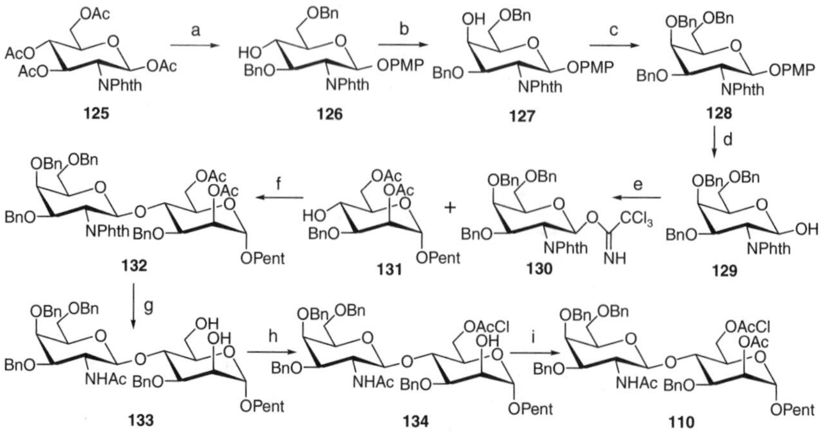

SCHEME 10.22. Synthesis of galacto-manno building block. Reagents and conditions: (a) *p*-methoxyphenol, TMSOTf, 4A MS, DCM; (b) NaOMe, MeOH; (c) BnBr, NaH, DMF; (d) CAN, Tol, MeCN; (e) CCl₃CN, DBU, Tol; (f) TMSOTf, 4A MS, tol, $-20\,^\circ$C; (g) MeNH₂, EtOH; Ac₂O; (h) (ClAc)₂O, Et₃N, DCM; (i) Ac₂O, Et₃N, DMAP, DCM.

mannoside acceptor **131**. Two different methods were reported for the donor **130**, either from galactosamine through phthalimido intermediate or from a cheaper glucosamine utilizing C-4 inversion protocol devised by Chaplin *et al.* [27]. The mannose acceptor **131** was prepared from *n*-pentenyl-α-D-mannopyranoside by 4,6-*O*-benzylidenation, regioselective 3-*O*-benzylation, 2-*O*-acetylation, acetal-hydrolysis, and 6-*O*-acetylation. The coupling of **130** and **131** using TMSOTf provided the disaccharide **132** with good β-stereocontrol. Further steps of the removal of the acetyl and phthalimido groups, *N*-acetylation, 6-*O*-chloroacetylation followed by 2-*O*-acetylation provided the desired intermediate **110**.

The glucosamine-inositol building block **111** was synthesized from two key intermediates, the optically pure suitably protected *myo*-inositol acceptor **141** (Scheme 10.23) and the azidoglycosyl donor **145** (Scheme 10.24). For the synthesis of **141**, the starting bis-cyclohexylidene derivative **8** was prepared from the commercially available *myo*-inositol by the Garegg's procedure for cyclohexylidenation. Regioselective *p*-methoxybenzylation of the diol **8** through tributyltin-oxide complex gave two regioisomers **135** and **136**, both of which were efficiently used to make **141** as described below. In the first sequence, the resolution method using camphanic acid chloride provided the desired diastereoisomer **137**, which on replacement of camphanate ester with allyl ether (→**138**), deprotection of cyclohexylidene-acetals, perbenzylation, and PMB group removal provided

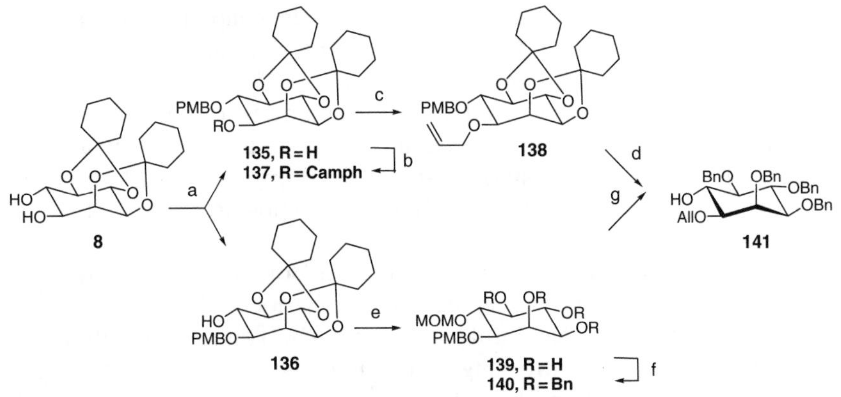

SCHEME 10.23. Synthesis of inositol building block. Reagents and conditions: (a) Bu$_2$SnO, MeOH then PMBCl, CsF, KI, DMF; (b) CamphCl, Et$_3$N, MeOH; (c) LiOH, DME, H$_2$O; Allyl bromide, NaH, TBAl, DMF; (d) pTSA, MeOH, DCM; BnBr, NaH, TBAl, DMF; CAN, DCM, MeCN, H$_2$O; (e) MOMCl, NaH, DMF; AcOH, H$_2$O; (f) BnBr, NaH, DMF; (g) CAN, DCM, MeCN; CamphCl, Et$_3$N; LiOH, DME, H$_2$O; Allyl bromide, NaH, DMF; Aq HCl, THF.

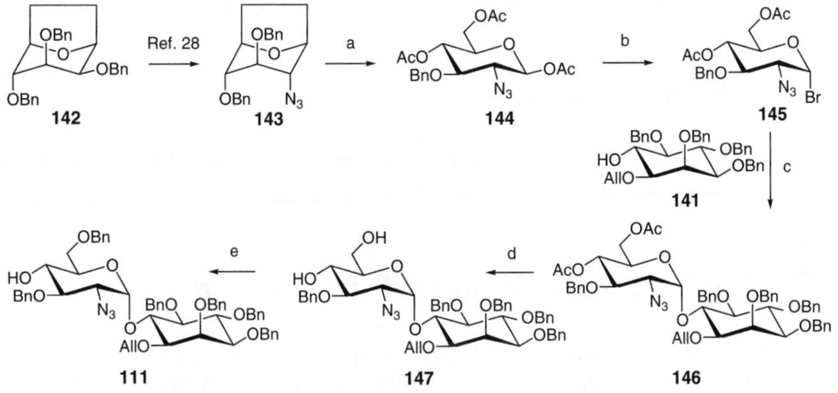

SCHEME 10.24. Synthesis of glucosamine inositol building block. Reagents and conditions: (a) Ac$_2$O, Et$_3$N, DMAP then Ac$_2$O, TFA; (b) TiBr$_4$, DCM, EtOAc; (c) AgClO$_4$, Et$_2$O, −35 °C; (d) NaOMe, MeOH; (e) Ac$_2$O, Et$_3$N, DCM then DHP, PPTS, DCM, then NaOMe, MeOH, then BnBr, TBAl, DMF, then MeOH, PPTS.

the desired intermediate **141**. In the second sequence, the other regioisomer **136** was transformed to **141** as described in Scheme 10.23.

For the synthesis of azidoglycosyl donor **145**, the method of Hori et al. [28] was used to convert 1,6-anhydro-β-D-mannopyranose **142** into 2-azidoanhydroglucose **143** (Scheme 10.24). The latter was then treated with TFA/acetic anhydride to give glucosyl acetate **144**, which was converted to the glycosyl bromide donor **145**. The coupling of **145** and the *myo*-inositol intermediate **141** using silver perchlorate in diethyl ether provided good α-selectivity in the product **146**.

This was followed by deacetylation to **147**; however, the selective 6-*O*-benzylation turned out to be challenging. This had to be done by a longer route (selective 6-*O*-acetylation, 4-*O*-tetrahydropyranylation, deacetylation, benzylation, and removal of THF group) to prepare required building block **111**.

The coupling of galactosamine-mannose donor **110** with glucosamine-inositol acceptor **111** provided the tetrasaccharide **148** (Scheme 10.25), which was followed by dechloroacetylation (→**149**) and glycosidation with trimannoside donor **109** affording pseudoheptasaccharide **150**. The chloroacetyl from the 6-position of third mannose was removed and the product was phosphorylated with Z-protected ethanolamine phosphoramidite (**113**), followed by acetyl deprotection from first mannose and one more cycle of phosphorylation with **113** providing intermediate **152**. The deallylation from 1-position of inositol residue followed by phosphorylation with glycerolipid **112** gave fully protected GPI anchor (**154**). In the final step, all the protecting groups were removed by hydrogenolysis to afford the target GPI (**1C**).

An alternative synthesis of GPI anchor of Rat Brain Thy-1 was reported [17] by Schmidt and coworkers in 1999, by the overall strategy presented in Scheme 10.26.

This synthesis was based on their earlier reported synthesis of Yeast GPI anchor described previously in this review in detail.

VI.   Synthesis of GPI Anchor of *P. falciparum*

Early synthetic efforts on the GPI anchor of malarial parasite *P. falciparum* were reported in 2002, when Seeberger group published synthesis of the glycan domain (without the glycerolipid domain) as malarial vaccine [29, 30]. This was followed by the Seeberger synthesis of a Pf-GPI mimic (short-chain lipid attached to the inositol residue through a phosphodiester [31, 32]. In the same year, Fraser-Reid reported [26] synthesis of a model Pf-GPI anchor (without the fourth mannose residue and with short-chain glycerolipid domain). However, the first total synthesis of fully lipidated Pf-GPI anchor was reported [33] in 2005 by the Seeberger group, as shown in the Scheme 10.27.

The key feature of Pf-GPIs is the presence of a third fatty acid at the 2-position of *myo*-inositol residue. In the Seeberger's synthetic design, the Pf-GPI (**1D**) was derived from a 4+2 coupling of a tetramannoside block

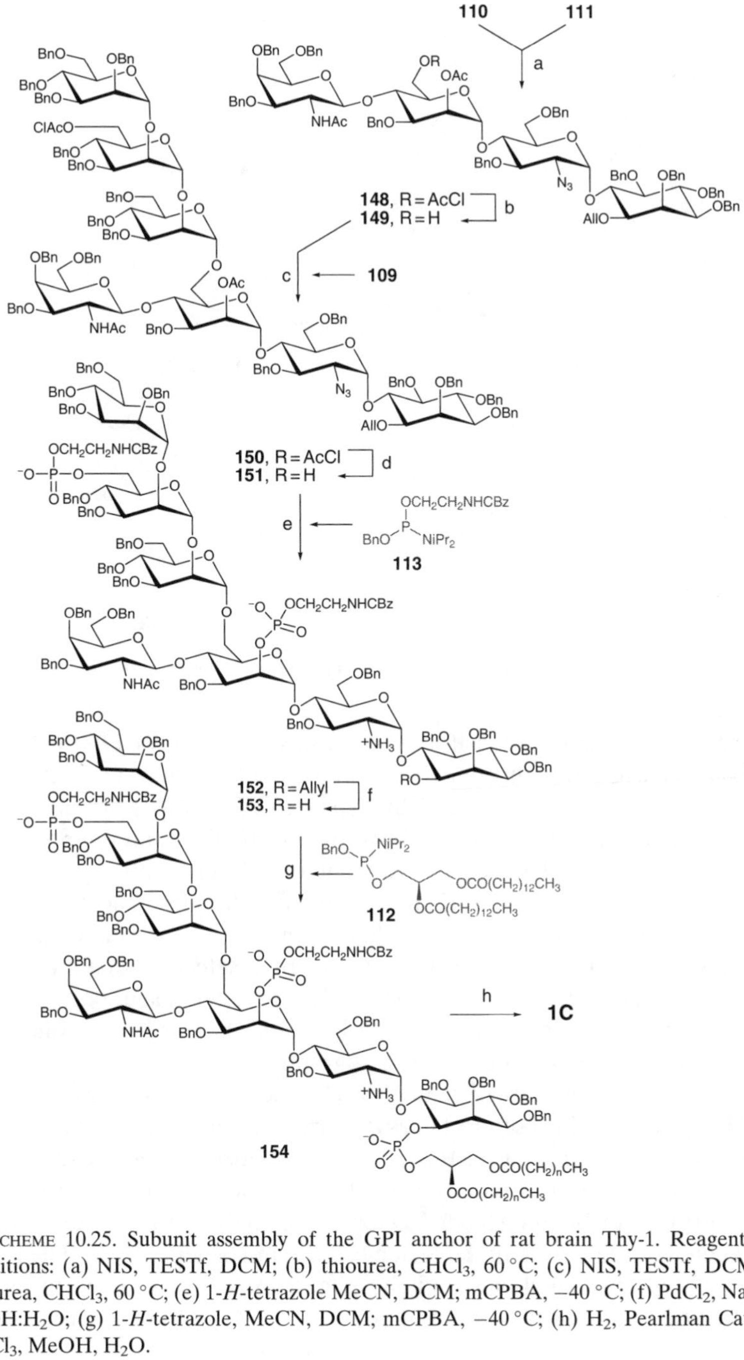

SCHEME 10.25. Subunit assembly of the GPI anchor of rat brain Thy-1. Reagents and conditions: (a) NIS, TESTf, DCM; (b) thiourea, CHCl₃, 60 °C; (c) NIS, TESTf, DCM; (d) thiourea, CHCl₃, 60 °C; (e) 1-*H*-tetrazole MeCN, DCM; mCPBA, −40 °C; (f) PdCl₂, NaOAc, AcOH:H₂O; (g) 1-*H*-tetrazole, MeCN, DCM; mCPBA, −40 °C; (h) H₂, Pearlman Catalyst, CHCl₃, MeOH, H₂O.

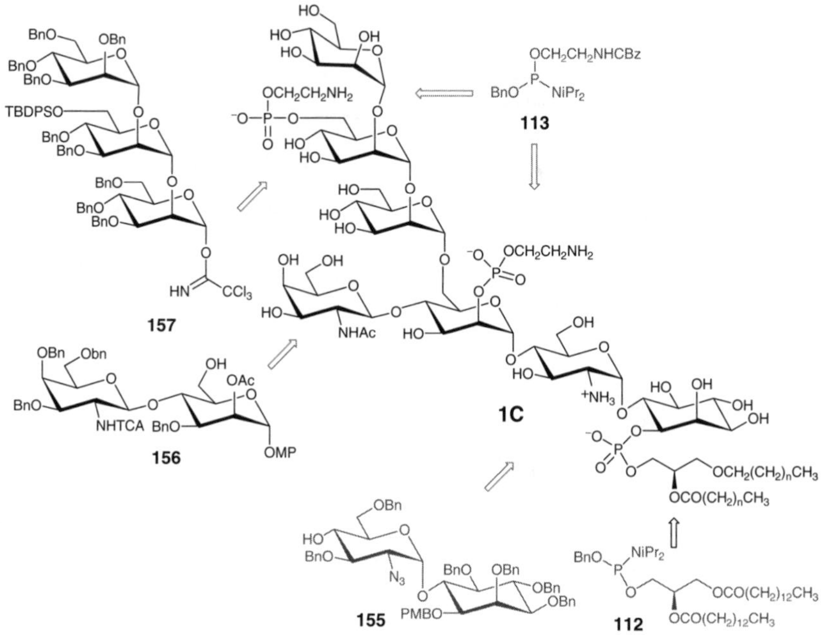

SCHEME 10.26. Synthetic approach of RR Schmidt for GPI anchor of rat brain Thy-1.

(160) and a glucosamine-inositol pseudodisaccharide (159). On the basis of their earlier work, triisopropyl ether and allyl ether groups were chosen to mark two phosphorylation sites. This synthesis started with stereoselective synthesis of α-linked glucosamine-inositol intermediate and the regioselective alkylation with PMBCl resulting in the differentiation of the two hydroxyl groups in inositol (Scheme 10.28). The three acetate groups were removed followed by the installation of 4,6-O-benzylidene and 3-O-benzyl groups. This was followed by the regioselective opening of the 4,6-O-benzylidene to afford differentially protected pseudodisaccharide intermediate. At this stage, both the α- and β-isomers of 166 could be separated by silica column chromatography.

The coupling of the above pseudodisaccharide 166 and the tetramannoside 160 [31] provided the pseudo hexasaccharide 167 in excellent yield under complete stereocontrol due to the neighboring benzoyl ester at C-2 position. The ester group was now replaced with benzyl ether group (167→168→169) to provide the key building block 169 for the elaboration to the fully lipidated GPI anchor, as described in the Scheme 10.29 and Scheme 10.30. The installation of lipids, phospholipids, and phosphoethanolamine were performed in varying order, and most efficient approach turned out to be the removal of

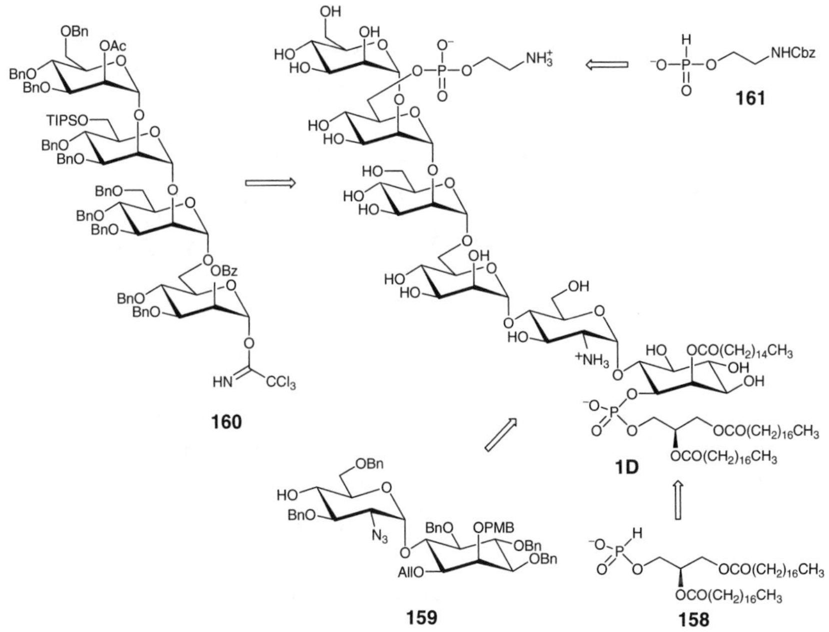

SCHEME 10.27. Synthetic approach of Seeberger for GPI anchor *P. falciparum.*

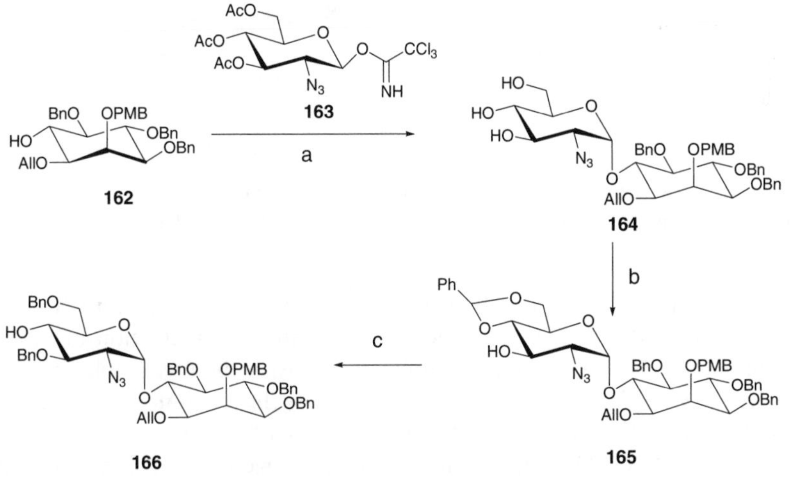

SCHEME 10.28. Reagents and conditions: (a) TMSOTf, CH$_2$Cl$_2$, −30 °C; NaOMe, MeOH; (b) PhCH(OCH$_3$)$_2$, CSA, CH$_3$CN; (c) (i) BnBr, NaH, DMF, 0 °C, 2h; (ii) HCl-Et$_2$O, 0 °C, NaCNBH$_3$.

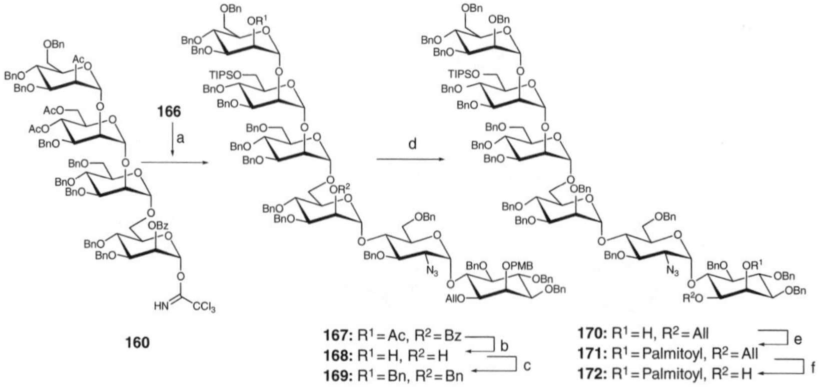

160

166

167: R¹ = Ac, R² = Bz
168: R¹ = H, R² = H
169: R¹ = Bn, R² = Bn

170: R¹ = H, R² = All
171: R¹ = Palmitoyl, R² = All
172: R¹ = Palmitoyl, R² = H

SCHEME 10.29. (a) **166**, TMSOTf, DCM, −40 °C; (b) NaOMe, MeOH; (c) NaH, BnBr, DMF; (d) CAN, MeCN, Tol; (e) $C_{15}H_{31}COOH$, DCC, DMAP, DCM; (f) $PdCl_2$, NaOAc, AcOH, H_2O.

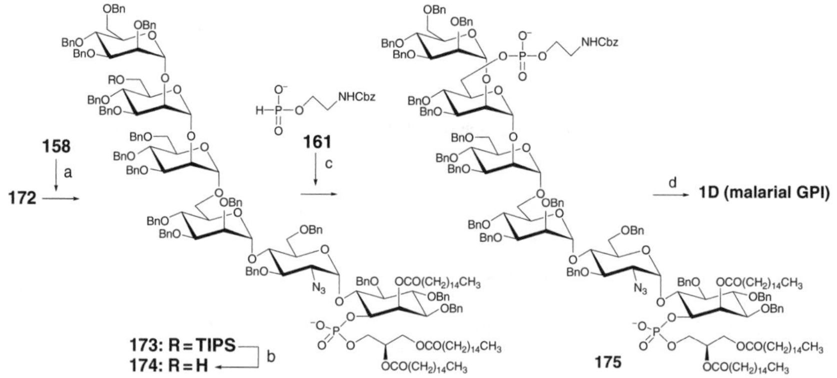

158

172 →

161

173: R = TIPS
174: R = H

175

1D (malarial GPI)

SCHEME 10.30. (a) PivCl, pyridine; iodine, pyridine, H_2O; (b) $Sc(OTf)_3$, MeCN, H_2O; (c) PivCl, pyridine; iodine, pyridine, H_2O; (d) H_2, $Pd(OH)_2$, MeCN, MeOH, H_2O.

PMB group from 2-position of inositol residue (**169→170**) and lipidation with palmitic acid (**170→171**), followed by deallylation at 1-position of inositol residue (**171→172**) and the phospholipidation with the glycerophospholipid moiety (**172→173**). Now the silyl group from the third mannose was removed and the free OH was phosphorylated with Cbz-protected ethanolamine *H*-phosphonate (**161**). In the final step, the hydrogenolysis using Pearlman's catalyst provided fully lipidated Pf-GPI anchor (**1D**).

Subsequently, this methodology was exploited by the Seeberger group to synthesize a series of malarial GPI anchor oligosaccharide probes [34], the

GPI anchor of *Toxoplasma gondii* [35] and the synthetic GPI arrays to study their antitoxic malaria response in animal models [36].

VII.   Synthesis of GPI Anchors of *Trypanosoma cruzi*

Our group reported in 2005 [37], a new convergent [2+2+2] approach (Scheme 10.31) to construct GPI molecules through efficient synthesis of glucosamine-inositol and tetramannose intermediates, which led to a total synthesis of a GPI anchor of *T. cruzi* IG7 antigen, and also afforded key intermediates for synthesis of valuable [4-deoxy-Man-III]-GPI analogs.

Arguably, the most demanding aspect of GPI synthesis has been to access suitably protected glucosamine-inositol intermediates requiring optically pure protected D-*myo*-inositol acceptor and 2-azido-2-deoxyglucosyl donor. This has mainly been done either by the painstaking resolution of bis-*cyclo*-hexylidene-*myo*-inositols using expensive camphanate auxiliaries/ enzymes or through a multistep synthesis from D-glucose by Ferrier reaction. In our ongoing efforts on the chemical biology of the GPI molecules, such an intermediate was required for fluorescent GPI analogs. Instead of following the reported methods based on *a priori* resolution of *myo*-inositol, it was reasoned on the basis of structural modeling, that if sufficient strain was built through a cyclic protective group, the azidoglycosyl unit itself could function as an efficient chiral auxiliary on the way to GPIs, making a number of early steps redundant. This proposition was tested by glycosylation of racemic 1-*O*-PMB-2,3,4, 5-tetra-*O*-benzyl-*myo*-inositol **16** with 2-azido glycosyl donor **163** to obtain pseudodisaccharide **184** (Scheme 10.32). The product on deacylation (**185**) and benzylidenation gave 4,6-cyclic-acetal protected disaccharide as diastereoisomeric pair (**186/187**). This quantitative reaction led to a clean separation (Rf difference of 0.1) of two enantiomeric disaccharides **186** and **187** by a simple silica column. The next two steps, benzylation at 3-OH and regioselective opening of benzylidene acetal by NaCNBH$_3$ provided key building block **176** and undesired isomer **188**.

The spectral and [α]$_D$ data of **186** and **176** were identical to that reported [19, 20] for the compounds prepared by alternative routes. The method also worked with 2-*O*-allyl-1-*O*-PMB-3,4,5-tri-*O*-benzyl- and 1-*O*-allyl-2,3,4,5-tetra-*O*-benzyl inositols showing its generality. Since good separation was obtained for **186/187** and their benzylated pairs, it was envisaged that the benzylidene-protected donors could also be used for stereoselective glycosidation.

The easy access to the key intermediate **176** encouraged us to apply this new method for the total synthesis of structurally and biologically

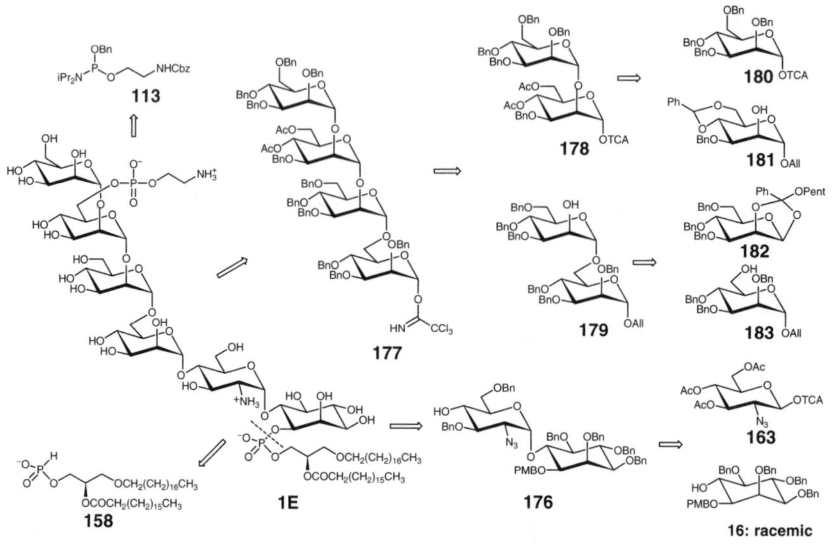

SCHEME 10.31. Synthetic strategy for GPI anchor of *T. cruzi*.

challenging target such as the GPI anchor of *T. cruzi* IG7 antigen (**1E**). The synthetic design (Scheme 10.31) also accommodated a feature that allowed the access to valuable [4-deoxy-Man-III]-GPI analogs to address a key biological question: why proteins are transferred only to 6-OH of Man-III residue and what happens if the conformation of this residue is disturbed by 4-deoxygenation? After the efficient access to **176**, a new and convergent [2+2] approach for construction of the tetramannose building block **177** was designed. This was prepared from two protected mannobiosides, the activated donor **178** and acceptor **179**.

The donor **178** was prepared (Scheme 10.33) by coupling of allyl-3-*O*-benzyl-4,6-*O*-benzylidene-α-D-mannoside (**181**, prepared in four steps from D-mannose via compound **185**) with 2,3,4,6-tetra-*O*-benzyl-α-D-mannosyl trichloroacetimidate (**180**, synthesized from D-mannose via **184**). The glycosylation (TMSOTf, DCM, −20 °C) went smoothly and the product **186** was taken directly to the next simultaneous steps of the removal of anomeric allyl and 4,6-benzylidine groups (KOtBu, DMSO, 80 °C; 1 M HCl-acetone, 1:9, 60 °C) to obtain the intermediate **187**. The peracetylation of resultant triol (**187→188**), selective removal of anomeric acetyl (Me₂NH, MeCN, −20 °C) followed by Schmidt activation (CCl₃CN, DBU) provided the desired mannobiose **178**. It needs mention that the two acetyls at

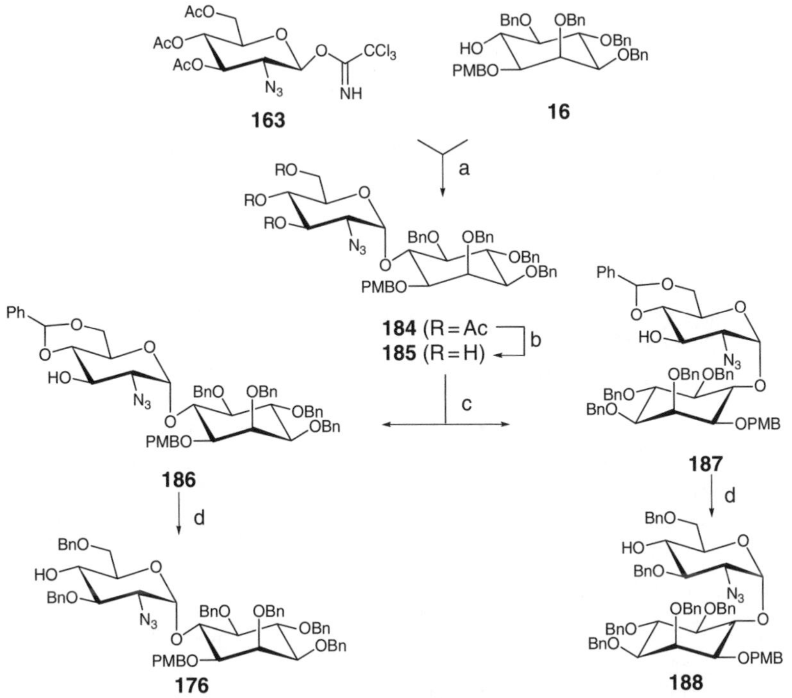

SCHEME 10.32. (a) TMSOTf, CH_2Cl_2, 0 °C; (b) NaOMe, MeOH; (c) PhCH(OCH$_3$)$_2$, CH$_3$CN; (d) (i) BnBr, NaH, DMF, 0 °C, 2 h; (ii) HCl-Et$_2$O, 0 °C, NaCNBH$_3$.

positions 4- and 6-OH in **189** were deliberately placed considering the target [4-deoxy-Man-III]-GPI analog.

Lower mannobiose **179** was prepared (Scheme 10.34) from two key mannose intermediates, allyl-2,3,4-tri-*O*-benzyl-α-D-mannopyranoside (**183**, synthesized from D-mannose in four simple steps, anomeric allylation, selective tritylation at the primary 6-OH, perbenzylation followed by the removal of trityl group) and 3,4,6,-tri-*O*-benzyl-β-D-Man-1,2-pent-4-enylorthobenzo-ate (**182**, prepared from pentabenzoyl mannosyl bromide by the method of Fraser-Reid). The glycosylation of **182** and **183** (TESTf, NIS) provided the desired intermediate **194**, which on removal of benzoyl group from the 2-*O*-position yielded mannobiose **179**.

Having both mannobiose donor **178** and acceptor **179** in hand, further glycosylation, which required considerable optimization, provided a fully protected tetramannose **195** in acceptable 65% yield (Scheme 10.35). This, after anomeric allyl-removal (**196**, KOtBu, DMSO, 80 °C; 1 M HCl-ace-tone, 1:9, 60 °C) and activation (CCl$_3$CN, DBU), afforded the desired tetramannose donor **177**. To our satisfaction, the next critical step of the

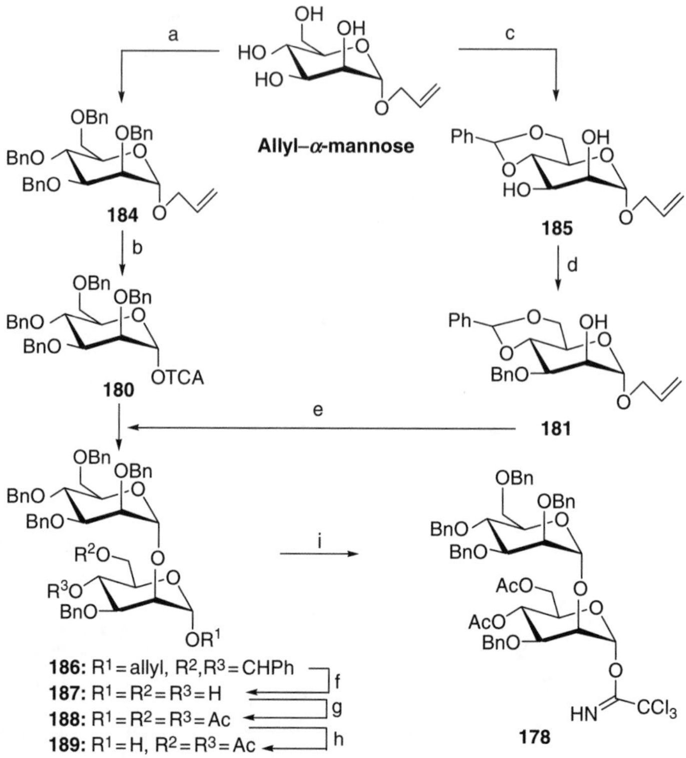

SCHEME 10.33. (a) BnBr, NaH, DMF, 24 h, 83%; (b) t-BuOK, DMSO, 80 °C, 2 h; acetone-1 M HCl (9:1), 40 °C, 0.5 h, 95%; CCl$_3$CN, CH$_2$Cl$_2$, DBU, RT, 1.5 h, 98%; (c) PhCH(OMe)$_2$, CSA, CH$_3$CN, RT, 18 h, 74%; (d) BnBr, NaH, DMF, 0 °C, 65%; (e) TMSOTf, Et$_2$O, RT, 2 h, 81%; (f) t-BuOK, DMSO, 80 °C, 3 h; acetone-1 M HCl (9:1), 55 °C, 2 h, 82%; (g) Ac$_2$O, pyridine, RT, 24 h, 93%; (h) Me$_2$NH, ACN, −20 °C, 1 h, 94%; (i) CCl$_3$CN, CH$_2$Cl$_2$, DBU, RT, 1 h, 96%.

[4+2] glycosylation of glucosamine-inositol **176** with the above tetraman-nose **177** went off smoothly (TMSOTf, DCM, 0 °C, 70%) to provide a pseudohexasaccharide **197** (Scheme 10.35) as the central point for both the GPI anchor as well as the deoxy-GPI analogs. For the synthesis of the GPI anchor, two acetyls from **197** were first removed and the primary 6-OH of the diol **198** was silylated (TBDPSCl, imidazole), followed by the benzy-lation of the 4-OH (BnBr, NaH) to get **199**. The TBDPS group was now removed (TBAF, THF) to obtain the pseudohexasaccharide acceptor **200** ready for phosphorylation with ethanolamine.

A part of the diol **198** (with free 4- and 6-OH of Man-III) was used for 4-deoxygenation by Barton's cyclic-thiocarbonate method for further synthesis of [4-deoxy-Man-III]-GPI probes.

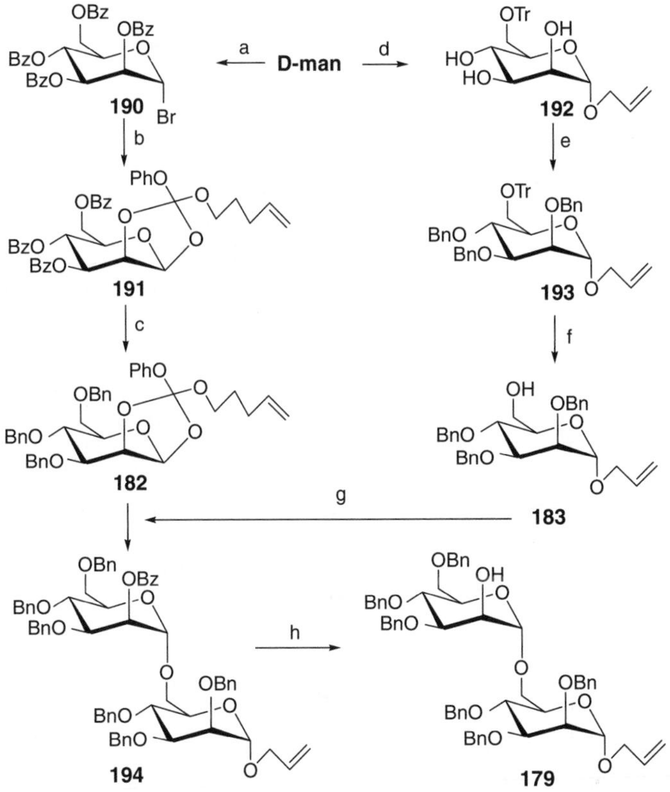

SCHEME 10.34. (a) (i) BzCl, pyridine, RT, 24 h; 30% HBr-AcOH, CH$_2$Cl$_2$, 0 °C, 12 h (b) 4-pentenol, 2,6-lutidine, CH$_2$Cl$_2$, 60 h, 82%; (c) (i) NaOMe, CH$_2$Cl$_2$-CH$_3$OH, 2 h, (ii) BnBr, NaH, DMF, RT, 24 h, 93%; (d) (i) allyl alcohol, BF$_3$-Et$_2$O, 80 °C, 4 h; (ii) TrCl, pyridine, 80 °C, 16 h, 70%; (e) BnBr, NaH, DMF, 4 h, 89%; (f) pTSA, CH$_2$Cl$_2$-CH$_3$OH, RT, 12 h, 93%; (g) TESOTf, NIS, CH$_2$Cl$_2$, 4A MS, RT, 0.5 h, 72%; (h) NaOMe, CH$_2$Cl$_2$-CH$_3$OH, RT, 3 h, 92%.

The coupling of **200** with NHCbz-ethanolamine-phosphoramidite **113** was carried out (Scheme 10.36) with 1-*H*-tetrazole followed by mCPBA oxidation, which provided the intermediate **201** in good yield. Now the PMB group from position-1 of the *myo*-inositol residue of **201** was removed (CAN, MeCN-DCM-H$_2$O) and the product **202** was phospholipidated with 1-*O*-alkyl$_{18:0}$-2-*O*-acyl$_{18:0}$-*sn*-glycero-*H*-phosphonate (**158**) by pivaloyl-chloride and iodine-oxidation method, to provide fully protected GPI anchor (**203**). Final step involved global deprotection and azide-reduction by hydrogenolysis (Pd(OH)$_2$, DCM-MeOH-H$_2$O, H$_2$) to the target GPI anchor (**1E**). This synthesis represented the key advances in our efforts aiming at the related

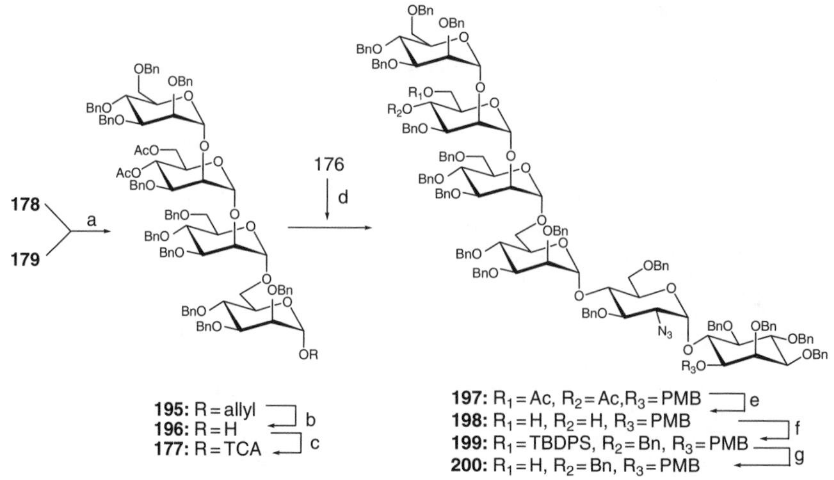

SCHEME 10.35. (a) TMSOTf, Et_2O, 4A MS, 0 °C, 0.5 h; (b) $PdCl_2$, NaOAc, $AcOH-H_2O$, 19:1, 16 h; (c) CCl_3CN, CH_2Cl_2, DBU, RT; (d) TMSOTf, Et_2O, 4A MS, 0 °C, 0.5 h; (e) NaOMe, $CH_3OH-CH_2Cl_2$ (3:1), RT, 1 h; (f) (i) TBDPSCl, imidazole, THF, 5 h; (ii) BnBr, NaH, DMF, 0 °C, 1 h; (g) CAN, $CH_3CN/Tol/H_2O$ (9:5:4), 1.5 h.

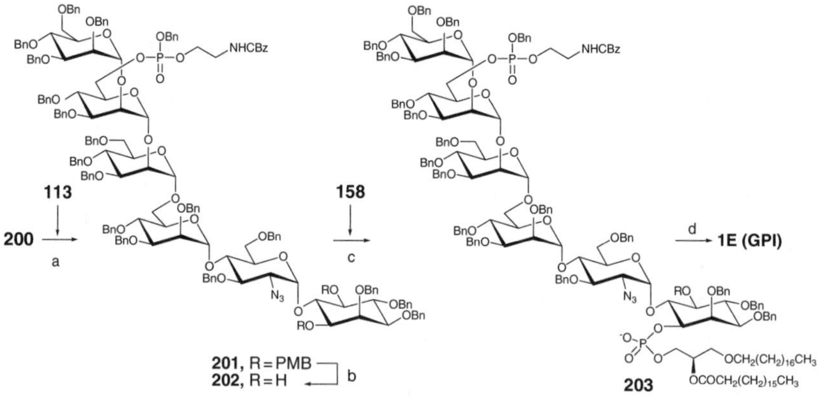

SCHEME 10.36. (a) 113, 1-H-tetrazole, mCPBA oxidation; (b) CAN, DCM-MeOH, 1.5 h, RT; (c) 158, pivaloyl chloride, pyridine; iodine, pyridine-H_2O; (d) H_2, 20% $Pd(OH)_2$, CHCl3, CH_3OH, H_2O.

GPIs (e.g., with unsaturated fatty-acids at sn-2-glycerol and aminoethyl phosphonate on GlcN residue) to probe their cell and membrane biology. The synthesis also afforded key intermediates for valuable [4-deoxy-Man-III]-GPI analogs and the GPIs of malaria parasite. This method was extended to the preparation of a number of GPI and PI probes [38–42].

An elegant synthetic approach towards GPI anchor of *T. cruzi* has been reported by Ferguson and Nikolaev group at Dundee [43]. The key feature of their synthesis was that the method allowed placement of unsaturated fatty acids in the glycerolipid domain. This was a major advance in the area of GPI synthesis because prior to their synthesis, all previous ones were with saturated acids. The ability to place unsaturated fatty acid will help decipher key biological activities of the GPI anchors, particularly the proinflammatory activity of the GPI anchors of *T. cruzi* and *P. falciparum*. To place unsaturated fatty acids, Nikolaev group used ester protection groups instead of the normal benzyl ethers used in all previous syntheses. The overall synthetic design of Dundee groups is depicted in Scheme 10.37. The tetrasaccharide block **207** was

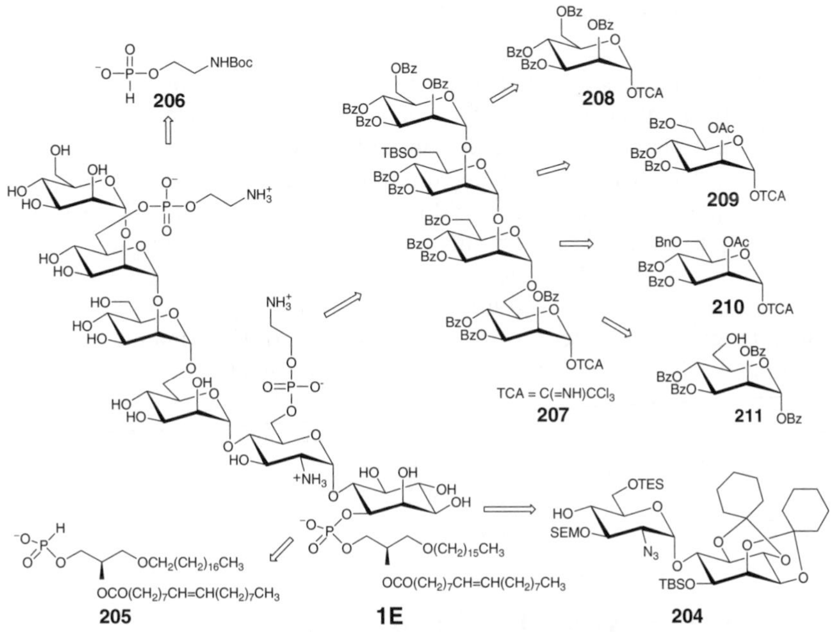

SCHEME 10.37. Synthetic approach of Nikolaev for GPI anchor *T. cruzi*.

prepared from the monosaccharide derivatives **208–211**, which were assembled in a step-by-step manner. The compounds **210** and **211** were synthesized from D-mannose and compound **208** and **209** were prepared via a common intermediate 3,4,6-tri-*O*-acetyl-1,2-*O*-(1-methoxyethylidene)-β-D-mannose.

The azidoglucose-inositol block **204** was synthesized from a precursor described earlier by Schmidt group [15] as a glycosylation product of optically pure D-*myo*-inositol derivative.

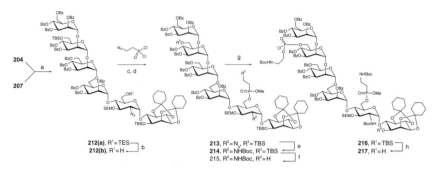

SCHEME 10.38. (a) TMSOTf, DCM, 4A MS; (b) TBAF, AcOH, THF; (c) 1H-tetrazole, iPr$_2$NEt, tol; (d) MeOH; (e) PPh$_3$P, H$_2$O, THF; Boc$_2$O, Et$_3$N, MeOH; (f) HF-Et$_3$N, MeCN, THF; (g) **206**, PivCl, pyridine, iodine, pyridine, H$_2$O; (h) TBAF, AcOH, THF.

In the advanced stage (Scheme 10.38), the glycoinositol backbone was prepared by the glycosylation of the glycosyl acceptor **204** with mannotetrose trichloroacetimidate **207** in the presence of TMSOTf. Subsequent cleavage of TES group with TBAF provided intermediate **212(b)**. The 1-*H*-tetrazole-assisted phosphorylation with azido phosphonodichloridate followed by methanolysis afforded phosphonic diester **213**. This was then subjected to reduction of azide and protection of the amine with Boc group. Selective cleavage of primary TBS group (→**215**), followed by the introduction of ethanolamine phosphate (→**216**) at 6-*O*-position of third mannose residue and final desilylation (→**217**) gave 1-hydroxy glycol derivative.

Now the compound **217** was phospholipidated (Scheme 10.39) to give fully protected oleic ester **219**. Global deprotection was performed in two steps: first the *O*-debenzoylation followed by cleavage of *O*-acetal and *N*-Boc to access the desired GPI anchor (**1E**) with unsaturated fatty acid at *sn*-2 position of the glycerol moiety.

VIII. Synthesis of GPI Anchor of CD52

The synthetic design of Guo's group [44–47] for the GPI anchor of CD52 has been summarized in Scheme 10.40. The overall plan was to assemble the phospholipidated core and then introduce two phosphoethanolamine moieties.

The system of pseudodisaccharide **220** is outlined in Scheme 10.41. The trimannose was prepared according to a procedure developed for similar structure (44).

The glycosylation (Scheme 10.42) of **220** with **221** with TMSOTf gave **233**. The removal of TBS and PMB groups followed by coupling with

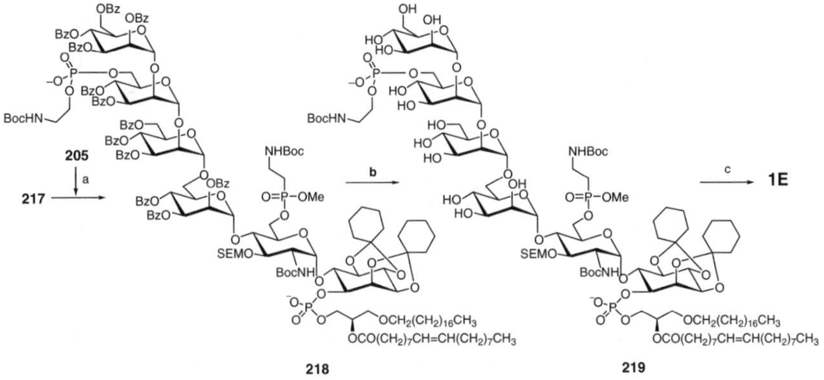

SCHEME 10.39. (a) PivCl, pyridine; iodine, pyridine, H_2O; (b) PhSH, Et_3N, DMF; 0.05 M NaOMe, H_2O; (c) CF_3COOH, H_2O.

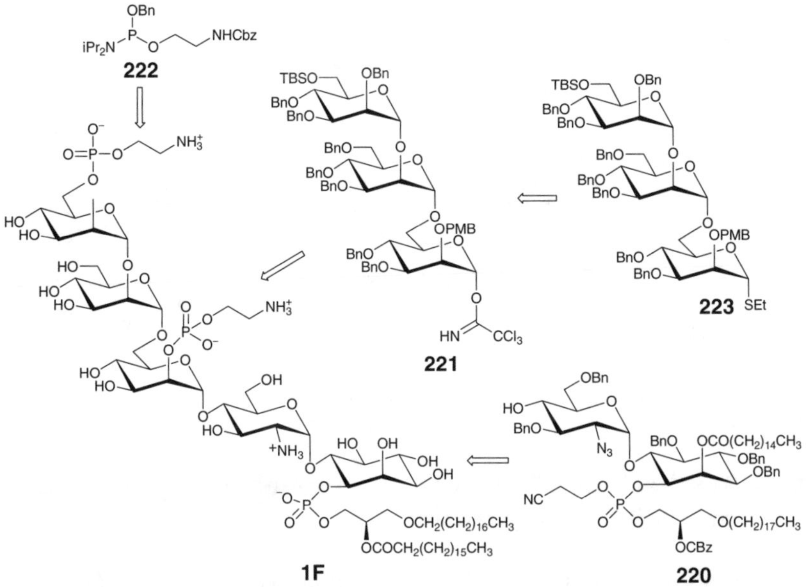

SCHEME 10.40. Synthetic approach of Guo *et al.* for GPI anchor of CD52.

phosphoethanolamine intermediate **222** gave protected GPI anchor **235**. Global deprotection in two steps (DBU treatment to remove cyanoethyl group followed by hydrogenolysis) provided desired GPI anchor CD52 (**1F**).

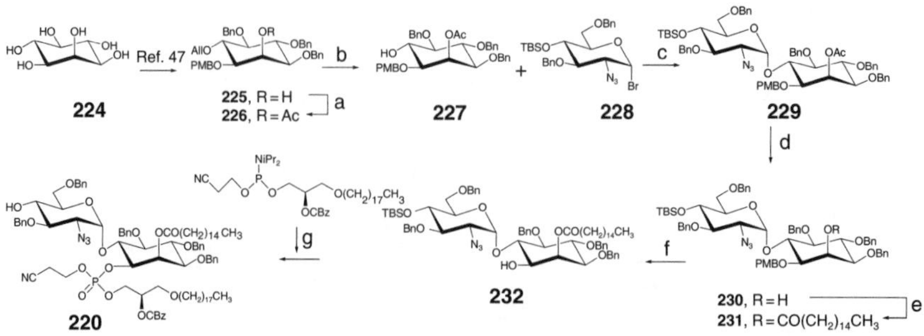

SCHEME 10.41. (a) Ac$_2$O, pyridine; (b) PdCl$_2$, NaOAc, AcOH; (c) TBAl, 4A MS, DCM; (d) NaOMe, MeOH; (e) DCC, Palmitoleic acid, DMAP; (f) CAN, MeCN, H$_2$O; (g) 1-H-tetrazole, DCM; t-BuO$_2$H, −20 °C; BF$_3$Et$_2$O, DCM.

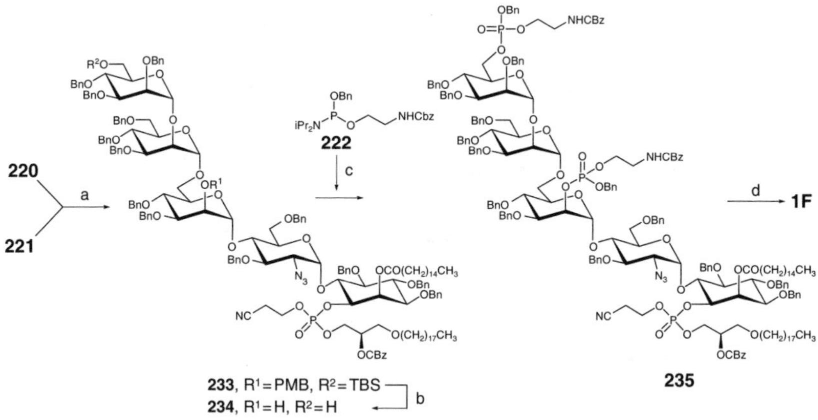

SCHEME 10.42. (a) TMSOTf, DCM, 4A MS; (b) 3% BF$_3$Et$_2$O, DCM; (c) 1-H-tetrazole, DCM; t-BuO$_2$H, −20 °C (d) 10% Pd/C, H$_2$, MeOH, CHCl$_3$, H2O (3:3:1), 3 days.

IX.   Synthesis of GPI Anchor of Lipophosphoglycan (LPG) of *Leishmania* parasite

Synthesis of the unusual GPI anchor of [Gal(α1–6)Gal(α1–3)Gal$_f$(β1–3) [Glc(α1-PO$_4$_6)Man](α1–3)Man(α1–4)GlcN(α1–6)Ins-1-PO$_4$] of LPG of *Leishmania* **1G** was reported [48] by Konradsson *et al.*, as shown in Scheme 10.43 using a convergent block synthetic strategy.

Four appropriately protected building blocks, a glucosyl inositol phosphate **236**, a dimannoside **237**, a trigalactoside **238**, and a glucosyl-α-1-H-phosphonate **239**; were synthesized [49] using multistep sequences. Glycoside **240** was synthesized from ethyl 2-azido-2-deoxy-1-thio-β-glucopyranoside in

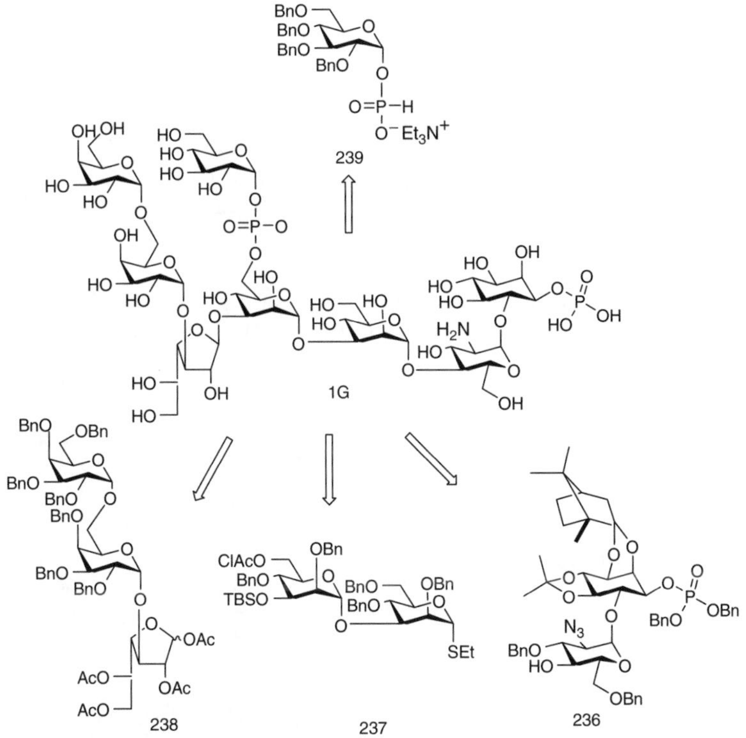

SCHEME 10.43. Synthetic approach of Konradsson for LPG core of *Leishmania*.

six steps. Deprotection of pivaloyl groups and treatment with 2,2-dimethyoxy-propane in the presence of PPTS (→**241**) followed by phosphorylation with dibenzyl *N,N*-diisopropyl phosphoramidite and subsequent oxidation with mCPBA led to the compound **242**. Deallylation of **242** using iridium catalyst, followed by hydrolysis led to the synthesis of building block **236** (Scheme 10.44).

For the synthesis of **237**, the compound **244** (Scheme 10.45) was obtained by benzylation of **243** and was used as a precursor for both mannose residues. Reductive benzylidene opening of **244** followed by benzylation and removal of silyl group or monochloroacetylation of position 6 yielded compounds **245** and **246**, respectively. The treatment of **246** with bromine, followed by AgOTf-mediated coupling with **245** in the presence of *sym*-collidine yielded building block **237**.

Ethyl-1-thio-β-galactopyranoside (**246**) was selectively (Scheme 10.46) protected at position 6 with triphenylmethyl group, followed by benzylation

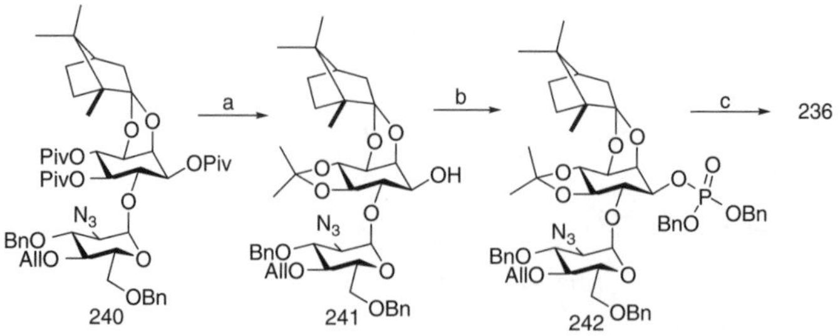

SCHEME 10.44. (a) (i) NaOH, MeOH (ii) 2,2-Dimethoxypropane, PPTS, DMF; (b) (i) (BnO)$_2$PN(i-prop)$_2$, tetrazole, CH$_2$Cl$_2$ (ii) mCPBA; (c) (i) H$_2$, THF, Indium catalyst (ii) NIS, H$_2$O.

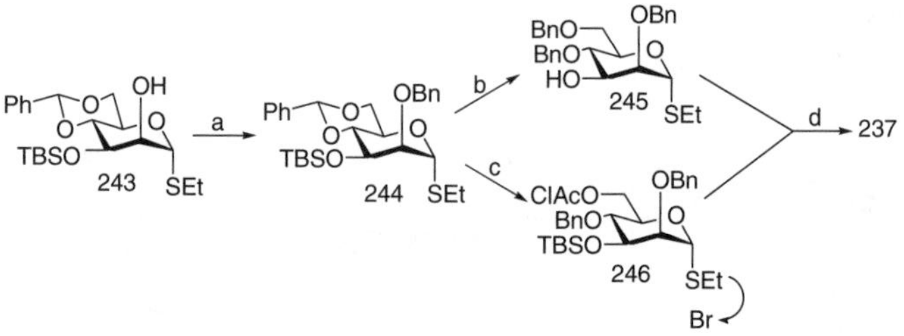

SCHEME 10.45. (a) BnBr, NaH, DMF; (b) (i) Me$_3$N.BH$_3$.AlCl$_3$, CH$_2$Cl$_2$/diethyl ether (ii) BnBr, NaH, DMF (iii) BF$_3$.Et$_2$O, CHCl$_3$; (c) (i) Me$_3$N.BH$_3$.AlCl$_3$, CH$_2$Cl$_2$/diethyl ether (ii) ClAcCl, CH$_2$Cl$_2$/pyridine; (d) AgOTf, collidine, CH$_2$Cl$_2$.

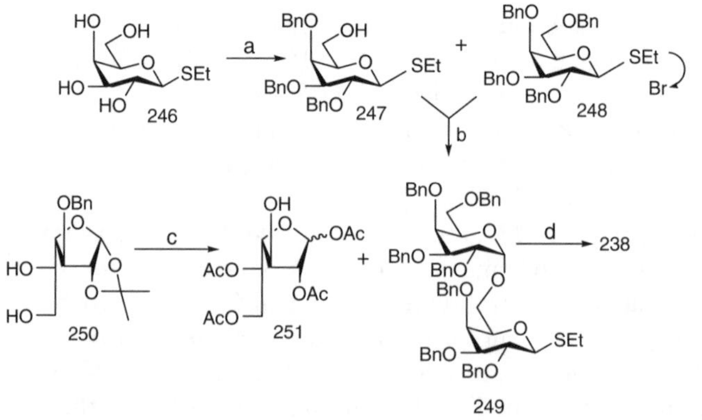

SCHEME 10.46. (a) (i) TrCl, pyridine (ii) BnBr, NaH, DMF (iii) pTsOH in CHCl$_3$/MeOH; (b) (i) Br$_2$, CH$_2$Cl$_2$ (ii) Et$_4$NBr, CH$_2$Cl$_2$/DMF; (c) (i) Ac$_2$O, pyridine (ii) TFA/CHCl$_3$ and Ac$_2$O, pyridine (iii) H$_2$, Pd/C in EtOAc; (D) DMTST.

and subsequent acidic detritylation (→247). Ethyl-2,3,4,6-tetra-O-benzyl-1-thio-β-D-galactopyranoside (248) was converted to the corresponding bromosugar and coupled with 247 using Et₄NBr (→249).

Furanoside derivative 251 was synthesized from 250, obtained from 1,2:5,6-di-O-isopropylidene-α-D-galactofuranose by benzylation and selective removal of the 1,2-isopropylidene acetal in aqueous AcOH. Compound 250 was subjected to acetylation, followed by removal of 1,2-isopropylidene group, acetylation and subsequent catalytic hydrogenolysis (→251). Dimethyl(methylthio)sulfonium trifluoromethane sulfonate (DMTST)-mediated coupling of 251 and 249 afforded trisaccharide building block 238.

The building block 237 was first coupled to block 238 and the resulting pentamer was then joined to building block 236. The fully protected hexasaccharyl-*myo*-inositol intermediate thus obtained was subjected to selective deprotection of chloroacetyl group, and the product was phosphorylated with the building block 239. The anomeric phosphodiester was introduced in the last step before global deprotection, because of the nucleophilic nature of phosphodiesters being incompatible during the glycosylation reactions.

X.   Synthesis of GPI Anchor of GIPL of *T. cruzi*

Konradsson's group reported [50] the first synthesis of heptasaccharyl *myo*-inositol 1H, the inositolphosphoglycan of *T. cruzi* glycoinositolphospholipid (GIPL), formerly called as lipopeptidophosphoglycan (LPPG). This was the first complete 2-aminoethyl phosphonic acid substituted glycan related to GPI anchor family to be synthesized. The synthesis was accomplished (Scheme 10.47) using a convergent block synthetic strategy. Synthesis of block 254 was accomplished by condensation of two Galf-(β1–3)-Man-*p* disaccharides (Scheme 10.48). The first disaccharide 257 was synthesized by SnCl₄-mediated glycosylation of diol 256 with penta-O-acetyl-β-D-galactofuranose 255, with the 2-O-acetyl group of the galactofuranoside directing the reaction to 1,2-*trans* glycosidic linkage. The second disaccharide 258 was obtained from 257 after replacement of all the acetyl groups with nonparticipatory benzyl groups.

Conversion of 258 into a bromo-glycoside, followed by glycosylation with 257 using AgOTf afforded the desired tetrasaccharide 259. Bu₄NIO₄ and TfOH-mediated hydrolysis of 259, followed by treatment with Cl₃CCN and DBU yielded the α-trichloroacetimidate 254.

Synthesis of building block 252 was accomplished by treating ethyl 2-O-benzoyl-3,4,6-tri-O-benzyl-1-α-D-thio-mannopyranoside 260 with NaOMe in CH₂Cl₂/MeOH. Coupling of the tetrasaccharide 254 (Scheme 10.49) with a readily available dimannoside 253 using TMSOTf afforded the protected

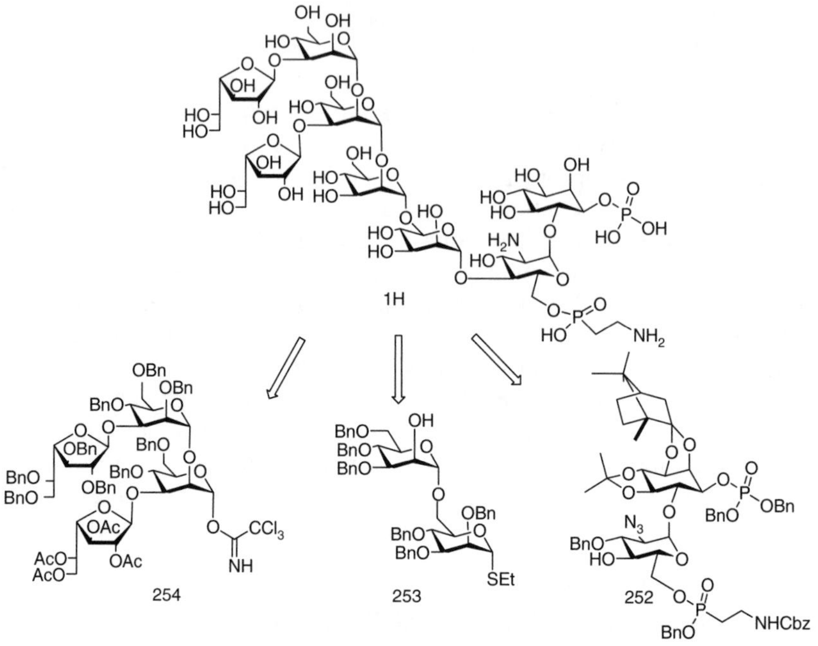

SCHEME 10.47. Synthetic approach of Konradsson for LPPG core of *T. cruzi.*

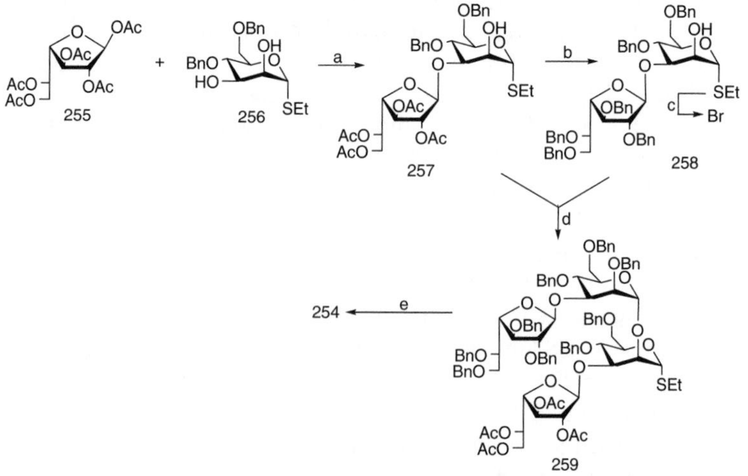

SCHEME 10.48. (a) SnCl$_4$, CH$_2$Cl$_2$; (b) (i) NaOMe, MeOH/CH$_2$Cl$_2$ (ii) BnBr, KI, Ag$_2$O, DMF; (c) Br$_2$, CH$_2$Cl$_2$; (d) AgOTf, CH$_2$Cl$_2$; (e) (i) n-Bu$_4$NIO$_4$, TfOH, H$_2$O, CH$_3$CN (ii) Cl$_3$CCN, DBU, CH$_2$Cl$_2$.

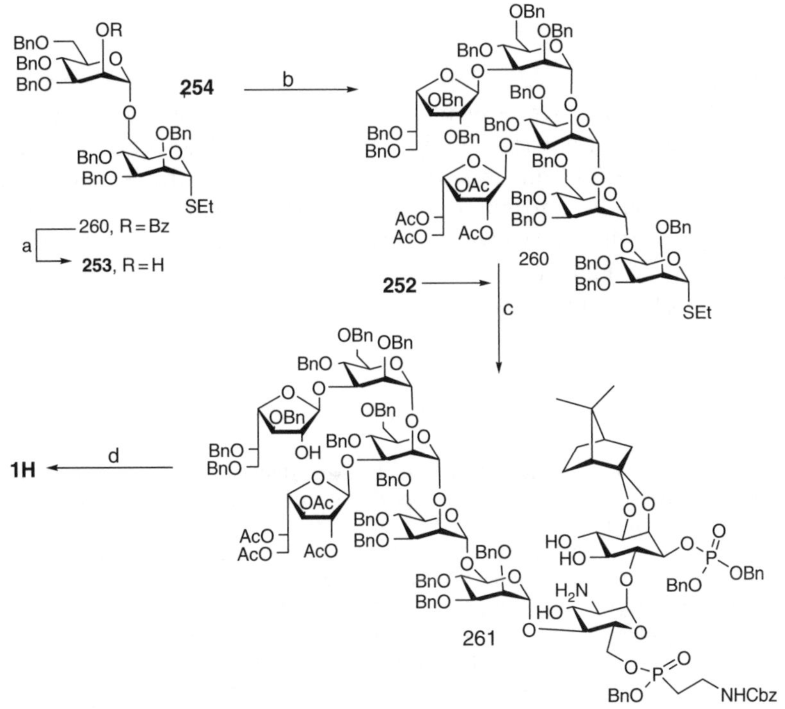

SCHEME 10.49. Assembly of the LPPG heptassaccharyl *myo*-inositol core. (a) NaOMe, CH$_2$Cl$_2$/MeOH; (b) TMSOTf, Et$_2$O; (c) DMTST, Et$_2$O; (d) (i) NaOMe, CH$_2$Cl$_2$/MeOH (ii) Na, NH$_3$ and 0.1 M HCl.

hexasaccharide **260**. DMTST-mediated glycosylation of **260** with **252** yielded the fully protected octasaccharide **261**, followed by sequential deprotection (deacetylation using NaOMe, debenzylation with sodium in liquid ammonia, removal of camphor, and isopropylidene acetals using aqueous HCl) led to the synthesis of the target molecule **1H**.

Since the synthetic efforts on the phosphoglycan repeats of LPG of *Leishmania* species have been recently reviewed [51], we have not included this topic in this chapter.

XI. Other Notable Contributions

The group of Martin Lomas has reported [52–55] the synthesis of a number of inositolphosphoglycans as insulin mimetics and the GPI intermediates on solid support.

XII. Conclusion

As evident from this review, a significant progress has been made towards the chemical synthesis of GPI anchors of various species, and this has led to the major advances in the chemistry of GPI anchors. The access to the full-length GPI anchors, their biosynthetic substrates, and labeled probes have already contributed to our understanding of the biology of this unique class of cell surface glycoconjugates. The stage is now set to build on this progress to elucidate the cell and membrane biology of the GPI anchored proteins and glycans.

REFERENCES

1. Ferguson, M.A.J., Homans, S.W., Dwek, R.A., and Rademacher, T.W. (1988). Glycosyl phosphatidylinositol moiety that anchors *Trypanosoma brucei* variant surface glycoprotein to the membrane. *Science* 239:753–759.
2. Homans, S.W., Ferguson, M.A.J., Dwek, R.A., Rademacher, T.W., Anand, R., and Williams, A.F. (1988). Complete structure of the glycosyl phosphatidylinositol membrane anchor of rat brain Thy-1 glycoprotein. *Nature* 333:269–272.
3. Orlean, P., and Menon, A.K. (2007). Thematic review series: lipid posttranslational modifications. GPI anchoring of protein in yeast and mammalian cells, or: how we learned to stop worrying and love glycophospholipids. *J Lipid Res* 48:993–1011.
4. McConville, M.J., and Menon, A.K. (2000). Recent developments in the cell biology and biochemistry of glycosylphosphatidylinositol. *Mol Membr Biol* 17:1–16.
5. McConville, M.J., and Ferguson, M.A. (1993). The structure, biosynthesis and function of glycosylated phosphatidylinositols in the parasitic protozoa and higher eukaryotes. *Biochem J* 294:305–324.
6. Englund, P.T. (1993). The structure and biosynthesis of glycosyl phosphatidylinositol protein anchors. *Annu Rev Biochem* 62:121–138.
7. Murakata, C., and Ogawa, T. (1990). Synthetic studies on cell-surface glycans. Synthetic study on glycophosphatidyl inositol (GPI) anchor of *Trypanosoma brucei*: glycoheptaosyl core. *Tetrahedron Lett* 31:2439–2442.
8. Murakata, C., and Ogawa, T. (1991). Synthetic studies on cell-surface glycans. Synthetic studies on glycophosphatidylinositol anchor: a highly efficient synthesis of glycobiosyl phosphatidylinositol through H-phosphonate approach. *Tetrahedron Lett* 32:101–104.
9. Murakata, C., and Ogawa, T. (1991). Synthetic studies on cell-surface glycans. A total synthesis of GPI anchor of *Trypanosoma brucei*. *Tetrahedron Lett* 32:671–674.
10. Murakata, C., and Ogawa, T. (1992). Synthetic studies on cell surface glycans.Stereoselective synthesis of glycobiosyl phosphatidylinositol, a part structure of the glycosylphosphatidyl inositol (GPI) anchor of *Trypanosoma brucei*. *Carbohydr Res* 234:75–91.
11. Murakata, C., and Ogawa, T. (1992). Synthetic studies on cell-surface glycans. Stereoselective total synthesis of the glycosyl phosphatidylinositol (GPI) anchor of *Trypanosoma brucei*. *Carbohydr Res* 235:95–114.
12. Baeschlin, D.K., Chaperon, A.R., Charbonneau, V., Green, L.G., Ley, S.V., Lucking, U., and Walther, E. (1999). Rapid assembly of oligosaccharides: total synthesis of a glycosyl phosphatidylinositol anchor of *Trypanosoma brucei*. *Angew Chem Int Ed* 37:3423–3428.

13. Baeschlin, D.K., Chaperon, A.R., Green, L.G., Hahn, M.G., Ince, S.J., and Ley, S.V. (2000). 1,2-Diacetals in synthesis: total synthesis of a glycosylphosphatidylinositol anchor of *Trypanosoma brucei*. *Chem Eur J* 6:172–186.
14. Mallet, A., Mallet, J.M., and Sinay, P. (1994). The use of selenophenyl galactopyrano sides for the synthesis of α and β-(1→4-C-disaccharides. *Tetrahedron Asymmetry* 5:2593–2608.
15. Mayer, T.G., Kratzer, B., and Schmidt, R.R. (1994). Synthesis of a GPI anchor of the yeast *Saccharomyces cerevisiae*. *Angew Chem Int Ed* 33:2177–2181.
16. Mayer, T.G., and Schmidt, R.R. (1999). Glycosylphosphatidylinositol (GPI) anchor synthesis based on versatile building blocks. Total synthesis of a GPI anchor of yeast. *Eur J Org Chem* 5:1153–1165.
17. Tailler, D., Ferrieres, V., Pekari, K., and Schmidt, R.R. (1999). Synthesis of the glycosyl phosphatidylinositol anchor of rat brain Thy-1. *Tetrahedron Lett* 40:679–682.
18. Mayer, T.G., Weingart, R., Munstermann, F., Kawada, T., Kurzchalia, T., and Schmidt, R. R. (1999). Synthesis of labeled glycosylphosphatidylinositol (GPI) anchors. *Eur J Org Chem* 10:2563–2571.
19. Pekari, K., Tailler, D., Weingart, R., and Schmidt, R.R. (2001). Synthesis of the fully phosphorylated GPI anchor pseudohexasaccharide of *Toxoplasma gondii*. *J Org Chem* 66:7432–7442.
20. Pekari, K., and Schmidt, R.R. (2003). A variable concept for the preparation of branched glycosylphosphatidylinositol anchors. *J Org Chem* 68:1295–1308.
21. Mootoo, D.R., Konradsson, P., and Fraser-Reid, B. (1989). n-Pentenyl glycosides facilitate a stereoselective synthesis of the pentasaccharide core of the protein membrane anchor found in *Trypanosoma brucei*. *J Am Chem Soc* 111:8540–8542.
22. Udodong, U.E., Madsen, R., Roberts, C., and Fraser-Reid, B. (1993). A ready convergent synthesis of the heptasaccharide GPI membrane anchor of rat brain Thy-1 glycoprotein. *J Am Chem Soc* 115:7886–7887.
23. Stewart Campbell, A., and Fraser-Reid, B. (1995). First synthesis of a fully phosphorylated GPI membrane anchor: rat brain Thy-1. *J Am Chem Soc* 117:10387–10388.
24. Roberts, C., Madsen, R., and Fraser-Reid, B. (1995). Studies related to synthesis of glycophosphatidylinositol membrane-bound protein anchors: *n*-pentenyl ortho esters for mannan components. *J Am Chem Soc* 117:1546–1553.
25. Madsen, R., Udodong, U.E., Roberts, C., Mootoo, D.R., Konradsson, P., and Fraser-Reid, B. (1995). Studies related to synthesis of glycophosphatidylinositol membrane-bound protein anchors: convergent assembly of subunits. *J Am Chem Soc* 117:1554–1565.
26. Lu, J., Jayaprakash, K.N., Schlueter, U., and Fraser-Reid, B. (2004). Synthesis of a malaria candidate glycosylphosphatidylinositol (GPI) structure: a strategy for fully inositol acylated and phosphorylated GPIs. *J Am Chem Soc* 126:7540–7547.
27. Chaplin, D., Crout, D.H.G., Bornemann, S., Hutchinson, D.W., and Khan, R. (1992). Conversion of 2-acetamido-2-deoxy-β-D-glucopyranose (*N*-acetylglucosamine) into 2-acetamido-2-deoxy-β-D-galactopyranose (*N*-acetylgalactosamine) using a biotransfor mation to generate a selectively deprotected substrate for SN2 inversion. *J Chem Soc Perkin Trans 1* 3:235–237.
28. Hori, H., Nishida, Y., Ohrui, H., and Meguro, H. (1989). Regioselective de-*O*-benzylation with Lewis acids. *J Org Chem* 54:1346–1353.
29. Schofield, L., Hewitt, M.C., Evans, K., Siomos, M.A., and Seeberger, P.H. (2002). Synthetic GPI as a candidate anti-toxic vaccine in a model of malaria. *Nature* 418:785–789.
30. Hewitt, M.C., Snyder, D.A., and Seeberger, P.H. (2002). Rapid synthesis of a glycosyl phosphatidylinositol-based malaria vaccine using automated solid-phase oligosaccharide synthesis. *J Am Chem Soc* 124:13434–13436.

31. Seeberger, P.H., Soucy, R.L., Kwon, Y.U., Snyder, D.A., and Kanemitsu, T. (2004). A convergent, versatile route to two synthetic conjugate anti-toxin malaria vaccines. *Chem Commun* 15:1706–1707.

32. Liu, X., and Seeberger, P.H. (2004). A Suzuki-Miyaura coupling mediated deprotection as key to the synthesis of a fully lipidated malarial GPI disaccharide. *Chem Commun* 15: 1708–1709.

33. Liu, X., Kwon, Y.U., and Seeberger, P.H. (2005). Convergent synthesis of a fully lipidated glycosylphosphatidylinositol anchor of *Plasmodium falciparum*. *J Am Chem Soc* 127:5004–5005.

34. Kwon, Y.U., Soucy, R.L., Snyder, D.A., and Seeberger, P.H. (2005). Assembly of a series of malarial glycosylphosphatidylinositol anchor oligosaccharides. *Chem Eur J* 11: 2493–2504.

35. Kwon, Y.U., Liu, X., and Seeberger, P.H. (2005). Total syntheses of fully lipidated glycosyl phosphatidylinositol anchors of *Toxoplasma gondii*. *Chem Commun* 17: 2280–2282.

36. Kamena, F., Tamborrini, M., Liu, X., Kwon, Y.U., Thompson, F., Pluschke, G., and Seeberger, P.H. (2008). Synthetic GPI array to study antitoxic malaria response. *Nature Chem Biol* 4:238–240.

37. Ali, A., Gowda, D.C., and Vishwakarma, R.A. (2005). A new approach to construct full-length glycosylphosphatidylinositols of parasitic protozoa and [4-deoxy-Man-III]-GPI analogues. *Chem Commun* 4:519–521.

38. Vishwakarma, R.A., and Menon, A.K. (2005). Flip-flop of glycosylphosphatidylinositols (GPI's) across the ER. *Chem Commun* 4:453–455.

39. Ruhela, D., Chatterjee, P., and Vishwakarma, R.A. (2005). 1-Oxabicyclic β-lactams as new inhibitors of elongating MPT–a key enzyme responsible for assembly of cell-surface phosphoglycans of *Leishmania* parasite. *Org Biomol Chem* 3:1043–1048.

40. Vishwakarma, R.A., Vehring, S., Mehta, A., Sinha, A., Pomorski, T., Herrmann, A., and Menon, A.K. (2005). New fluorescent probes reveal that flippase-mediated flip-flop of phosphatidylinositol across the ER membrane does not depend on the stereochemistry of the lipid. *Org Biomol Chem* 3:1275–1283.

41. Ruhela, D., and Vishwakarma, R.A. (2003). Iterative synthesis of *Leishmania* phosphoglycans by solution, solid-phase and polycondensation approaches without involving any glycosylation. *J Org Chem* 68:4446–4456.

42. Chawla, M., and Vishwakarma, R.A. (2003). Alkylacylglycerolipid domain of GPI molecules of *Leishmania* is responsible for inhibition of PKC mediated *c-fos* gene expression. *J Lipid Res* 44:594–600.

43. Yashunsky, D.V., Borodkin, S., Ferguson, M.A.J., and Nikolaev, A.V. (2006). The chemical synthesis of bioactive glycosylphosphatidylinositols from *Trypanosoma cruzi* containing unsaturated fatty acid in the lipid. *Angew Chem Int Ed* 45:468–474.

44. Xue, J., Shao, N., and Guo, Z. (2003). First total synthesis of a GPI-anchored peptide. *J Org Chem* 68:4020–4029.

45. Xue, J., and Guo, Z. (2003). Convergent synthesis of a GPI Containing an acylated inositol. *J Am Chem Soc* 125:16334–16339.

46. Wu, X., and Guo, Z. (2007). Convergent synthesis of a fully phosphorylated GPI anchor of the CD52 antigen. *Org Lett* 9:4311–4313.

47. Xue, J., and Guo, Z. (2002). Convergent synthesis of an inner core GPI Sperm CD52. *Bioorg Med Chem Lett* 12:2015–2018.

48. Ruda, K., Lindberg, J., Garegg, P.J., Oscarson, S., and Konradsson, P. (2000). Synthesis of the *Leishmania* LPG core heptasaccharyl *myo*-inositol. *J Am Chem Soc* 122:11067–11072.

49. Lindberg, J., Ohberg, L., Garegg, P.J., and Konradsson, P. (2002). Efficient routes to glucosamine-*myo*-inositol derivatives, key building blocks in the synthesis of glycosylphosphatidylinositol anchor substances. *Tetrahedron* 58:1387–1398.
50. Hederos, M., and Konradsson, P. (2006). Synthesis of the *Trypanosoma cruzi* LPPG heptasaccharyl *myo*-inositol. *J Am Chem Soc* 128:3414–3419.
51. Nikolaev, A.V., Botvinko, I.V., and Ross, A.J. (2006). Natural phosphoglycans containing glycosyl phosphate units: structural diversity and chemical synthesis. *Carbohydr Res* 342:297–344.
52. Martin-Lomas, M., Khiar, N., Garcia, S., Koessler, J.L., Nieto, P.M., and Rademacher, T.W. (2000). Inositolphosphoglycan mediators structurally related to glycosyl phosphatidylinositol anchors: synthesis, structure and biological activity. *Chem Eur J* 6:3608–3621.
53. Lopez-Prados, J., and Martin-Lomas, M. (2005). Inositolphosphoglycan mediators: an effective synthesis of the conserved linear GPI anchor structure. *J Carbohydr Chem* 24:393–414.
54. Lopez-Prados, J., Cuevas, F., Reichardt, N.C., de Paz, J.L., Morales, E.Q., and Martin Lomas, M. (2005). Design and synthesis of inositol-phospho-glycan putative insulin mediators. *Org Biomol Chem* 3:764–786.
55. Reichardt, H.C., and Martin-Lomas, M. (2003). A practical solid-phase synthesis of glycosylphosphatidylinositol precursors. *Angew Chem Int Ed* 42:4674–4677.

11

GPI-Based Malarial Vaccine: Past, Present, and Future

XINYU LIU • DANIEL VARON SILVA • FAUSTIN KAMENA •
PETER H. SEEBERGER

Laboratory for Organic Chemistry
Swiss Federal Institute of Technology (ETH) Zurich
8093 Zurich, Switzerland

I. Abstract

Vaccine, the most cost-effective approach to control and prevent infectious diseases in human, is not yet available for malaria. The pursuit of DNA- and protein (or peptide)-based malarial vaccines have been pursued for several decades, yet no commercial vaccine against malaria is available. A glycosylphosphatidylinositol (GPI) anchor of *Plamosidum falciparum* origin has recently emerged as the central malarial toxin. Immunization against malaria based on a synthetic GPI glycan proved effective in a rodent model. Further generations of a series of structurally defined GPI glycans with chemically unique tags allowed for the precise correlation of anti-GPI antibody responses in malaria-infected patients and healthy people. In this chapter, we describe the emergence of GPI-based malaria toxin theory, followed by highlighting the several applications of synthetic GPIs to study their biological and physiological relevances to malarial pathogenesis, and conclude with the hope to provide a perspective to future research in this area.

ISSN NO: 1874-6047
DOI: 10.1016/S1874-6047(09)26011-2

II.   Introduction to GPI in Malarial Pathogenesis

Malaria is the major parasitic disease that affects humans and is an enormous public health problem mainly in tropical developing countries [1]. According to health statistics, more than 400 million people are infected and 2–3 million die every year, mainly children below 2 years old of age [2]. Malaria is transmitted by the byte of the female *Anopheles* mosquitoes that are infected with the plasmodium parasites. There are four different species of plasmodium that infect humans; *Plasmodium avole*, *Plasmodium malariae*, *Plasmodium vivax*, and *Plasmodium falciparum* which causes most cases of severe disease and deaths. The parasite initially infects the liver cells and subsequently initiates blood-stage infection, the main cause of the disease [3]. The emergence of resistant forms of parasites and poor eradication of mosquitoes lead to the development of a vaccine, the best alternative to eradicate this disease [4].

In contrast to other infectious diseases, humans are able to develop only partial immunity against malaria even after one or two infections, which render the development of antimalarial vaccines a daunting task [5]. The majority of the vaccine candidates target antigenic proteins or peptides expressed either on the surface or within invasion organelles of the different life stages of the parasites, including the sporozoite, erythrocytic, and transmission-blocking vaccines [6]. However, all these strategies suffer from similar problems in generating an appropriate protection caused by antigenic diversities, immunosuppression, immune evasion strategies, and genetic restrictions in the immune system among others. The antitoxin vaccine represents an alternative approach that targets the main cause of the host pathology [7]. Vaccines against tetanus and diphtheria have been highly successful in humans by targeting the toxins causing the diseases. To adopt such an approach to malarial vaccine development, an understanding of the molecular basis of malarial physiopathology as well as the identification of the malarial toxin that contribute to the disease is essential.

The toxin theory of malarial pathogenesis can be ascribed to Camillo Golgi, who hypothesized in 1886 that the cause of the malarial febrile paroxysm was a released toxin of parasite origin [8]. However, the specific biochemical identity of the toxin had long been remained obscure. The identification of tumor necrosis factor-α (TNF-α) in the late 1970s offered a convenient biomarker to examine the malarial endotoxin activity [9]. Macrophages from malaria-infected mice were found to produce high levels of TNF-α by stimulation with different agonists, along with the observation that crude extracts of rodent malaria parasites induced macrophages to secret TNF-α *in vitro* [10, 11]. Playfair initially suggested that the malarial toxin to be a phospholipid [12]. Schofield, subsequently, extracted and tentatively assigned a structure related to the glycosylphosphatidylinositol (GPI) anchor to the putative toxin from the malarial parasite *P. falciparum* in 1993 [13]. This assignment was based on the fact that administration of the isolated material to mice induced symptoms similar to accute malaria, including the release of proinflammatory cytokines, pyrexia, hypoglycaemia, and eventually caused death [14]. Despite the initial skeptisms concerning the GPI-based malarial toxin theory, additional results published in the area shed more light on the relevance of GPI to malarial pathogenesis, albeit the results are sometimes contradictory [15, 16].

GPI constitutes more than 95% of the total carbohydrate modification of *P. falciparum* schizonts and correlates with the virtual absence of *N*- and *O*-linked glycosylation on the proteins of these parasites [17, 18]. The structural elucidation of the *P. falciparum* GPI vividly illustrates the microheterogeneous character of this class of molecules (Figure 11.1). Initally proposed in 1996 [17], the malarial GPI was identified as an evolutionarily conserved core GPI glycan with three mannosyl residues with phosphoglycerol moiety acylated with palmic acids and C-2 inositol acylated with myristic acid. Another report in 2001 identified the GPI glycan backbone with a terminal fourth mannose residue in an α1–2 linkage to the third mannose in the core glycan and phosphoglycerol moiety dominantly acylated with stearic acid and oleic acid and C-2 inositol acylated with palmitic acid [19]. Interestingly, parasite proteins were found to be anchored exclusively by the Man_4-GPIs, indicating a high degree of selectivity towards the presence of the fourth, terminal mannose residue in the GPI anchor moiety, while it is not clear whether protein-free GPI present in *P. falciparum* is dominantly Man_3-GPI or mixed with Man_4-GPI.

As the chemical synthesis of GPIs is covered extensively in a separate chapter of this series, it is sufficient to point out that two lipidated malarial GPIs have been independently synthesized by the groups of Fraser-Reid and Seeberger (Figure 11.1) [20, 21], although neither approach will allow for the installation of an unsaturated lipid on the GPI backbone.

Based on the GPI-malarial toxin theory [22], the contribution of immunological processes to severe malarial pathogenesis in humans is considered as direct or indirect association with GPI action. Malarial GPI is known to activate macrophages and induce the expression of acute-phase cytokines such as TNFα, interleukin (IL)-1, and IL-6 and a chemokine cascade, leading to local, organ-specific, and systemic inflammation [23–26]. GPIs potently synergize with IFN-γ in regulating gene expression. GPIs are reported to signal through both TLR2 and TLR4 in a CD14-dependent manner, and GPI is the only TLR2/4 agonist described from malaria to date [27, 28].

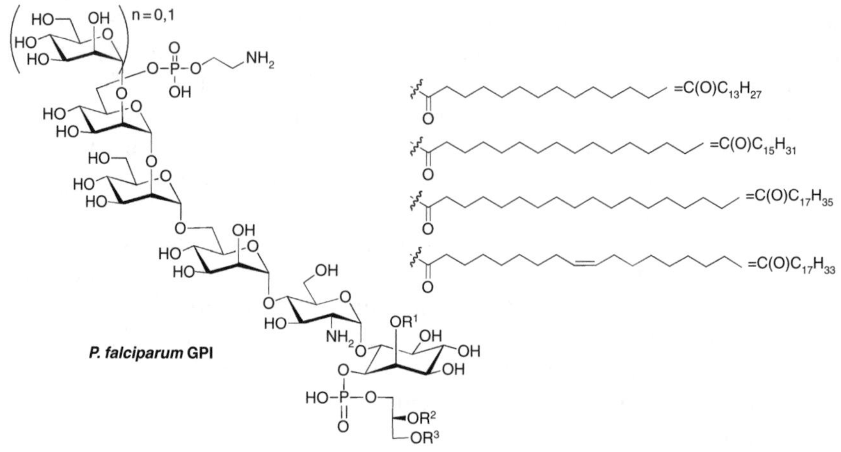

	n	R^1	R^2	R^3
Native structure proposed in 1996	0	$C(O)C_{13}H_{27}$	$C(O)C_{15}H_{31}$	$C(O)C_{15}H_{31}$
Native structure proposed in 2001	1	$C(O)C_{17}H_{35}$	$C(O)C_{17}H_{33}$	$C(O)C_{17}H_{35}$
Chemically synthesized by fraser-reid's group	0	$C(O)C_{13}H_{27}$	$C(O)C_{17}H_{35}$	$C(O)C_{17}H_{35}$
Chemically synthesized by seeberger's group	1	$C(O)C_{15}H_{31}$	$C(O)C_{17}H_{35}$	$C(O)C_{17}H_{35}$

FIG. 11.1. Native *P. falciparum* GPI structures as proposed in 1996 and 2001 (lipid compositions were shown as the dominant ones) [17, 19] and the chemically synthesized *P. falciparum* GPI structures by the groups of Fraser-Reid and Seeberger in 2004 [20, 21].

A major obstacle to clear the doubts around the GPI-based malarial toxin theory have been the difficulties in obtaining sufficient amounts and prove the homogeneity of the isolated GPIs. Standardized methods to purify native malarial GPI to compositional homogeneity are not clearly established and the isolation process is limited to the expertise in several laboratories around the world [17, 19]. As there is no recombinant approach for the production of GPI, the GPI was purified from bulk parasite cultures for initial studies. Obtaining adequate amounts of parasites free of host cell components for detailed biochemical studies and vaccine trials proved to be difficult. To obtain well-defined structures and precisely define the structural requirements for malarial GPI bioactivity and evaluate the potential of *P. falciparum* GPI as malarial vaccine antigen, chemical synthesis emerged as the only option.

III. Synthetic GPI as Antitoxic Malarial Vaccine Candidate in a Rodent Model

Early studies revealed that the isolated malarial GPI could elicit immune responses in both rodents and human, thus suggesting that this molecule may hold potentials for the development of an antitoxic malarial vaccine. A consensus *P. falciparum* GPI glycan, derived from chemical and enzymatic hydrolysis of the native GPI and nontoxic itself, was synthesized (Figure 11.2A) [7]. The initial synthesis was carried out in solution phase,

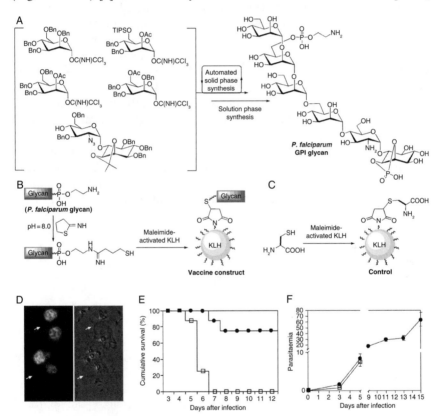

Fig. 11.2. Synthetic GPI glycan–KLH conjugate protects against murine cerebral malaria. (A) Chemical synthesis of a *P. falciparum* GPI glycan as immunogen with the combination of automated solid-phase and solution-phase synthesis. (B) Preparation of *P. falciparum* GPI glycan–KLH conjugate; (C) Preparation of a cysteine–KLH conjugate as control; (D) Left panel, reactivity of antiglycan IgG antibodies with *P. falciparum* trophozoites and schizonts. Right panel, same field under white light; (E) Kaplan-Meier survival plots; and (F) Parasitaemias, to 2 weeks postinfection, of KLH–glycan-immunized (closed circles) and sham-immunized (open squares) mice challenged with malarial parasite.

before the synthetic process was accelerated further using an automated oligosaccharide synthesizer (Figure 11.2A) [29, 30]. The synthetic GPI glycan was modified with 2-iminothiolane and conjugated to maleimide-activated carrier protein, keyhole limpet hemocyanin (KLH), to give the vaccine construct (Figure 11.2B). Cysteine conjugated KLH was prepared in a similar manner as a control (Figure 11.2C) [7].

The GPI glycan–KLH conjugate was immunogenic in rodents [7]. Antibodies from KLH–glycan immunized animals gave positive IgG titers and selectively recognized intact trophozoites and schizonts, but not uninfected erythrocytes, because of the GPI structural differences between human and parasites (Figure 11.2D). Furthermore, mice immunized with the synthetic GPI–KLH conjugate were substantially protected against severe malarial pathology and fatality. Between 60% and 75% of the vaccinated mice survived for 2 weeks after the parasitic challenge, whereas all the sham-immunized mice died within 7 days (Figure 11.3E). More interestingly, immunization of mice did not alter infection rates of the animals and overall parasitemia, indicating that the antibody to the GPI neutralized toxicity without killing the parasites (Figure 11.3F).

This study eventually led to the development of this vaccine candidate to the clinical level, as currently being pursued by Ancora Pharmaceuticals (Medford, MA, USA) [31]. Synthetic strategies to access multigram quantities of the GPI glycan have been devised and preclinical studies are currently underway.

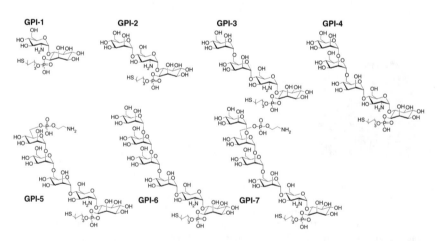

Fig. 11.3. A series of chemically synthesized *P. faliparum* GPI-related glycans equipped with a thiol tag that is suitable for immobilization to various solid supports and probes for biological studies.

This immunization study provided convincing evidence to corroborate that GPI is the dominant toxin of *P. falciparum* origin; however, a central question concerning the association of altered risk of malaria in human with anti-GPI antibodies remains to be clarified. There has been interest in investigating acquired antibody responses to GPI in humans. In several recent studies, Gowda and coworkers have used native *P. falciparum* GPI prepared by a one-step HPLC fractionation of total organic solvent- and water-extractable parasite material, where the GPI-containing peak represents over 20% of the total fractionation gradient [19]. Studies using this material in ELISA have provided contradictory findings [32–38], some reporting a statistically significant association of an anti-GPI IgG response with protection against symptoms of severe malaria [35, 37], and others finding no such relationship [34]. One study found the lipid and other two studies revealed the glycan as the dominant epitope within the GPI. Much of these discrepancies are likely from the compositional impurities associated with the isolation of GPIs from malarial parasite. To precisely address this problem, a series of synthetic GPI that are structurally related to *P. falciparum* GPI were synthesized with unique thiol tag (Figure 11.3) [39].

IV. Synthetic GPI Microarray to Define Antimalarial Antibody Response

To study the anti-GPI antibody responses in a high throughput fashion, an ELISA system based on the microarray technology, which has been used successfully in genomics and proteomics research, was designed [40]. This chip-based format offers many advantages over the conventional methods that allows for the screening of several thousand binding events in parallel with a minimal amount of sera and ligand required [41].

The synthetic *P. falciparum* GPI (1–7) (Figure 11.3) containing a terminal thiol group were covalently attached to maleimide-activated bovine serum albumin (BSA) coated glass slides (Figure 11.4A). Seven synthetic GPIs were arrayed as quadruplicates in a square. Seven squares, each representing one of the GPI epitopes 1–7 and one square containing the buffer control made up the 32 spots that were printed in close proximity. Placement of a 64-well adhesive gasket on the surface of the microarray slides following printing created separated wells, each containing the seven GPIs as well as a control (Figure 11.4A). This setup has helped to define both the minimal length of the glycan structure required to induce an immune response as well as the specific anti-GPI response strictly related to malarial infection. The

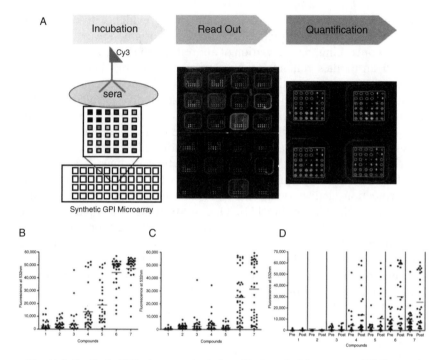

FIG. 11.4. Synthetic GPI microarray to study antitoxic malarial responses. (A) Generation of synthetic GPI glycan microarrays as a convenient tool to study anti-GPI antibody response. (B) IgG levels against synthetic GPI compounds 1–7 in sera from cohorts of African donors living in a malaria endemic area, and (C) in malaria nonexposed Europeans. Each data point represents one serum; bars indicate the mean Ab level. (D) Increase in mean serum IgG levels against synthetic GPI compounds after experimental *P. falciparum* sporozoite challenge of malaria nonexposed Europeans. Sera were taken prior to (pre) or after experimental challenge (post). Each data point represents one serum; bars indicate the mean level.

results obtained with this setup show interesting features that would have been impossible to obtain using natural materials [42].

It was initially observed that the sera from malaria-exposed adult Africans had high-IgG levels against Man$_4$-GPI (6 and 7) and significant reactivity was observed against the Man$_3$-GPIs (4 and 5) although the mean Ab level was lower than against the Man$_4$-GPIs (Figure 11.4B). Surprisingly, most sera of malaria-unexposed Caucasian adults also contained antibodies that bind to the Man$_4$-GPIs 6 and 7 (Figure 11.4C). In contrast to sera from malaria-exposed Africans, no significant reactivity with the Man$_3$-GPIs (4 and 5) was observed (Figure 11.4C). Both Man$_3$-GPI 5 and Man$_4$-GPI

7 have been found on *P. falciparum* blood-stage parasites. The ratio of the two structurally distinct molecules may vary during intraerythrocytic development. Only Man_4-GPI has been found to serve as anchor for surface proteins such as MSP-1 and MSP-2. The *P. falciparum* Man_3-GPI, like some *Toxoplasma gondii* GPI, may primarily exist as the free glycolipid on the parasite cell surface. This finding implicates that the Man_4-GPI response represents a broader anti-GPI response probably involving some nonmalarial pathogenic species as well, whereas the Man_3-GPI response might represent a more specific antimalarial GPI response.

The difference of anti-Man_3-GPI responses between African and European sera may represent a response to nonmalarial parasites found only in Africa. To address this point, it was hypothesized that if anti-Man_3-GPI responses are primarily elicited by malarial parasites, while anti-Man_4-GPI antibodies arise from exposure to a variety of pathogens, a malarial infection in previously unexposed individuals should boost both a preexisting anti-Man_4-GPI B cell response and elicit a *de novo* anti-Man_3-GPI response. In support of this hypothesis, an increase in preexisting anti-Man_4-GPI (6 and 7) IgG levels and the development of a significant anti-Man_3-GPI (4 and 5) Ab response in Caucasian volunteers involved in a sporozoite challenge vaccine trial was observed (Figure 11.4D). This finding is more remarkable when considering that antimalaria chemotherapy started immediately after the first microscopic detection of the blood-stage parasites. These results demonstrate that even one subclinical malarial infection can have a profound effect on the overall anti-GPI Ab level as well as the reactivity pattern.

The approach using synthetic GPI-based microarrays thus provided more detailed insights than those obtained with antigens extracted from parasite preparations. Anti-GPI fine specificity patterns likely reflect exposure to a variety of pathogens and mirror the general epidemiological situation of a population. Therefore, anti-GPI antibody levels measured with GPI isolated from *P. falciparum* parasites as documented in the literature, critically depend on the ratio of Man_3- and Man_4-GPI in the antigen preparations. The ratio of mean anti-Man_3-GPI and anti-Man_4-GPI Ab levels may represent a marker for the relative intensity of malaria exposure. The observation that European control subjects also showed a significant anti-Man_4-response poses an important question as the origin of that response could be of major importance for the design of a GPI-based antitoxic malarial vaccine. A possible explanation is that the anti-Man_4-GPI response in malaria-unexposed subjects may arise as a result of an infection with nonmalarial parasites or even fungi rather than representing an autoimmune response.

V.   Synthetic GPI as Tools to Study Malaria Associated Anemia

Anemia is one of the world's leading causes of disability and represents a serious global public health problem [43]. One major source of anemia is the infection with *P. falciparum* [44]. Severe malaria anemia (SMA) is the most prevalent serious complication of malaria and contributes substantially to the morbidity and mortality from malaria [45]. In malaria-infected patients the rate of infection of the red blood cells (RBCs) mostly varies between <1% and 5%, whereas the degree of RBC's loss rises rather between 25% and 50%. This observation strongly indicates the destruction of great numbers of nonparasitized RBCs (nRBCs) during malarial infection [46], but little is known about the mechanisms leading to the removal of the nRBCs in malaria.

GPIs that retain the fatty acid moiety intact have been shown to be able to leave the membrane and reassociate with other cells. The membrane-form variant surface glycoprotein (mfVSG) of *Trypanosoma brucei* with an intact GPI anchor can transfer from parasite to erythrocytes *in vivo* [47], and human GPI-anchored CD55 can transfer from erythrocytes into schistosomes *in vivo* [48]. When *Leishmania* parasites are grown *in vitro*, the GPI-anchored lipophosphoglycan (LPG) is shed into the culture medium and retains the fatty acid-PI moiety intact, consistent with a nonenzymatic mechanism of release [49]. Furthermore, GPI-anchored proteins are not always associated with membranes. They can also be found under natural conditions in solution with the GPI lipid intact, where they exchange among cells in a normal physiological process [50, 51]. Intact GPI may be released upon schizont rupture to interact with host tissues.

These observations led to a hypothesis that *P. falciparum* GPIs with the intact lipid moiety may inset into the RBC membranes and as a consequence, such tagged cells may be available for humoral and cellular attack leading to lysis of the altered cells (Figure 11.5). The synthetic lipidated *P. falciparum* GPI [21] was therefore explored as the ideal tool to study its association with naive commercially available nRBCs. The detection of the insertion of GPI to RBC was monitored by the anti-GPI antibodies present in a malaria patient, followed by flow-cytometric analysis. Although this detection method was not direct, a series of observations clearly demonstrated that the synthetic *P. falciparum* GPI can insert into the membrane of naive commercially available nRBCs and that the incorporated *P. falciparum* GPI can be recognized by antibodies against the malarial toxin [52]. These observations include: (i) Only nRBCs exposed to complete, lipidated *P. falciparum* GPI were recognized by the antibodies and

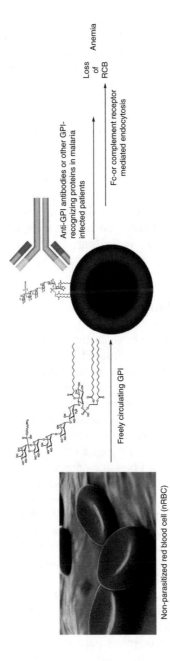

Non-parasitized red blood cell (nRBC)

Freely circulating GPI

Anti-GPI antibodies or other GPI-recognizing proteins in malaria infected patients

Fc-or complement receptor mediated endocytosis

Loss of RCB

Anemia

FIG. 11.5. GPI's involvement in malaria-associated anemia. Freely circulating or released *P. falciparum* GPI inserts into nRBCs and results in the recognition by anti-GPI antibodies that may contribute subsequent RBC eliminations. (See color plate section in the back of the book.)

exhibited fluorescent signals in the flow cytometer, while applying nonlipidated *P. falciparum* GPI led to the failure of the antibody-binding activity. (ii) The fluorescent signals could be demonstrated with serum from a patient with severe malaria, but not with the serum from a nonmalaria-exposed European. (iii) Addition of Triton X-100 during incubation of the nRBCs with *P. falciparum* GPI induced dose-dependent increase of the antibody binding to indicate alleviation of the insertion of *P. falciparum* GPI into the cell membrane. The temperature dependency of the *P. falciparum* GPI effect was connoted by the observation that the fluorescence signals reduced when the nRBC *P. falciparum* GPI incubation was conducted at low temperature (4 °C), allowing only for reduced membrane fluidity. (iv) The *P. falciparum* GPI insertion effect was dose dependent on the concentration of *P. falciparum* GPI and inversely on the numbers of used nRBCs representing the extent of the membrane surface. (v) Preincubation of *P. falciparum* GPI with the patient's serum resulted in significantly reduced recognition indicating neutralization by anti-GPI antibodies.

VI.   Conclusion and Perspectives

Although many doubts have been casted on the initial proposal that the GPI of *P. falciparum* origin constitutes the dominant malarial toxin, the success of a synthetic GPI-based vaccine candidate in a rodent model against malaria [7], as well as the clear distiction of anti-GPI antibody responses in malaria-infected patients and healthy individuals as examined using the synthetic GPI glycan-based microarray [40], clearly demonstrated the central and unequivocal role of the GPI anchor in malarial pathogenesis. With the GPI-based malarial vaccine close to enter the clinical trials, certainly more excitements can be anticipated.

It is obvious that many questions associated with the malarial GPI are still open. Synthetically well-defined GPI molecules will likely continue to contribute to a better understanding of its role in malarial pathogenesis. One key point that needs to be thoroughly investigated is the nature of the molecular receptor of the malarial GPI. TLRs have been often cited, but a direct physical interaction with the GPI molecule is still unproven and are they the only ones? The synthetic GPIs with unique chemical tags will allow for the facile conjugation to solid supports (such as affinity chromatography resins), different crosslinking and fluorescent probes. These tools will be used to unravel the underlying secrets.

In addition, beyond malaria, many parasitic protozoa express abundant amounts of GPIs compared to mammals. In these organisms, besides providing membrane anchorage, GPIs are packed densely to form a "glycocalyx," the protecting "sugar" coat on the cell surface that is critical for their survival within the mammalian host [53]. Like GPI as the primary toxin of *P. falciparum*, the role of parasitic GPI anchors as virulence factors and in modulating the host immune response to infection are also observed in other parasites. Notable examples include the GPIs from *T. gondii* and *Trypanosoma cruzi*, the causative agents of toxoplasmosis and Chagas' disease, respectively. Schwarz and coworkers have shown that highly purified GPIs from *T. gondii* tachyzoites (Figure 11.6A), as well as their core glycan, can induce TNF production in macrophages [54], which implicates the participation of *T. gondii* GPIs as bioactive factors in the production of TNF-α during toxoplasmal pathogenesis. *T. cruzi* GPI has toxin-like proinflammatory activities that may be responsible for the pathology of Chagas' disease. Ferguson and coworkers showed that a highly purified and characterized GPI from *T. cruzi* trypomastigote mucins exhibits the potent TNF-α, IL-12, and nitric oxide-inducing activities that is comparable to the activities of bacterial endotoxin and *Mycoplasma* lipopeptide and constitutes one of the most potent microbial proinflammatory agents known to date (Figure 11.6B) [55]. The synthetic strategies to access the GPIs of *T. gondii* and *T. cruzi* origins have been developed [56–59]. It would be worthwhile to investigate the extent of the suitability of a vaccine prepared on the basis of these GPIs to prevent these infections.

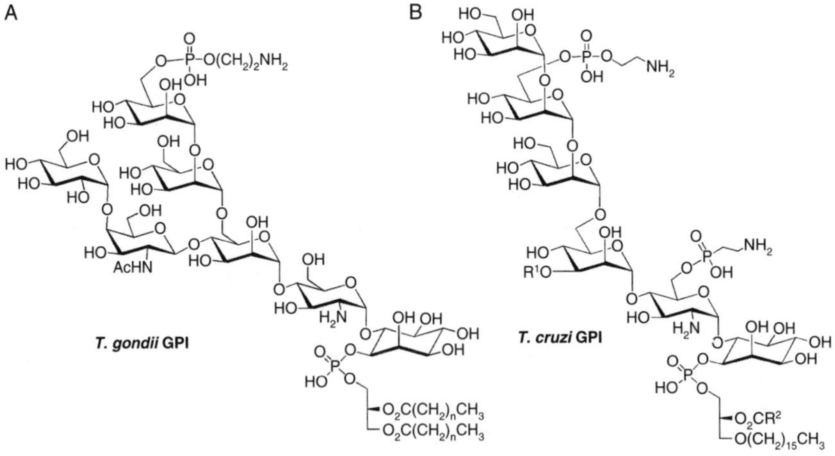

FIG. 11.6. Structures of *T. gondii* and *T. cruzi* GPIs that were found exhibit toxin-like properties. (A) *T. gondii* GPI (n is not specified) and (B) *T. cruzi* GPI (R^1 can be oligogalactose, R^2 is derived from palmitic, or oleic, or linoleic acid).

ACKNOWLEDGMENTS

Research in the Seeberger laboratory on carbohydrate-based vaccines has been supported by ETH Zurich, the Swiss National Science Foundation, Foundation Bay, and a Koerber Award.

REFERENCES

1. Gallup, J.L., and Sachs, J.D. (2001). The economic burden of malaria. *Am J Trop Med Hyg* 64:85–96.
2. http://www.who.int/topics/malaria/en/. Malaria. World Health Organization.
3. Clark, I.A., and Schofield, L. (2000). Pathogenesis of malaria. *Parasitol Today* 16:451–454.
4. Greenwood, B.M., Fidock, D.A., Kyle, D.E., Kappe, S.H., Alonso, P.L., Collins, F.H., and Duffy, P.E. (2008). Malaria: progress, perils, and prospects for eradication. *J Clin Invest* 118:1266–1276.
5. Gupta, S., Snow, R.W., Donnelly, C.A., Marsh, K., and Newbold, C. (1999). Immunity to non-cerebral severe malaria is acquired after one or two infections. *Nat Med* 5:340–343.
6. Sharma, S., and Pathak, S. (2008). Malaria vaccine: a current perspective. *J Vector Borne Dis* 45:1–20.
7. Schofield, L., Hewitt, M.C., Evans, K., Siomos, M.A., and Seeberger, P.H. (2002). Synthetic GPI as a candidate anti-toxic vaccine in a model of malaria. *Nature* 418:785–789.
8. Golgi, C. (1886). Sull' infezione malarica. *Arch Sci Med* 10:109–135.
9. Carswell, E.A., Old, L.J., Kassel, R.L., Green, S., Fiore, N., and Williamson, B. (1975). An endotoxin-induced serum factor that causes necrosis of tumors. *Proc Natl Acad Sci USA* 72:3666–3670.
10. Bate, C.A., Taverne, J., and Playfair, J.H. (1988). Malarial parasites induce TNF production by macrophages. *Immunology* 64:227–231.
11. Bate, C.A., Taverne, J., and Playfair, J.H. (1989). Soluble malarial antigens are toxic and induce the production of tumour necrosis factor *in vivo*. *Immunology* 66:600–605.
12. Bate, C.A.W., Taverne, J., and Playfair, J.H.L. (1992). Detoxified exoantigens and phosphatidylinositol derivatives inhibit tumor necrosis factor induction by malarial exoantigens. *Infect Immun* 60:1894–1901.
13. Schofield, L., and Hackett, F. (1993). Signal transduction in host cells by a glycosylphosphatidylinositol toxin of malaria parasites. *J Exp Med* 177:145–153.
14. Gerold, P., Dieckmann-Schuppert, A., and Schwarz, R.T. (1994). Glycosylphosphatidylinositols synthesized by asexual erythrocytic stages of the malarial parasite, *Plasmodium falciparum*. Candidates for plasmodial glycosylphosphatidylinositol membrane anchor precursors and pathogenicity factors. *J Biol Chem* 269:2597–2606.
15. Nebl, T., De Veer, M.J., and Schofield, L. (2005). Stimulation of innate immune responses by malarial glycosylphosphatidylinositol via pattern recognition receptors. *Parasitology* 130(Suppl.):S45–S62.
16. Schofield, L., and Grau, G.E. (2006). Complexity of immunological processes in the pathogenesis of malaria. *Nat Rev Immunol* 6:423.
17. Gerold, P., Schofield, L., Blackman, M.J., Holder, A.A., and Schwarz, R.T. (1996). Structural analysis of the glycosyl-phosphatidylinositol membrane anchor of the merozoite surface proteins-1 and -2 of *Plasmodium falciparum*. *Mol Biochem Parasitol* 75:131–143.

18. Gowda, D.C., Gupta, P., and Davidson, E.A. (1997). Glycosylphosphatidylinositol anchors represent the major carbohydrate modification in proteins of intraerythrocytic stage *Plasmodium falciparum. J Biol Chem* 272:6428–6439.
19. Naik, R.S., Branch, O.H., Woods, A.S., Vijaykumar, M., Perkins, D.J., Nahlen, B.L., Lal, A.A., Cotter, R.J., Costello, C.E., Ockenhouse, C.F., Davidson, E.A., and Gowda, D.C. (2000). Glycosylphosphatidylinositol anchors of *Plasmodium falciparum*: molecular characterization and naturally elicited antibody response that may provide immunity to malaria pathogenesis. *J Exp Med* 192:1563–1575.
20. Lu, J., Jayaprakash, K.N., Schlueter, U., and Fraser-Reid, B. (2004). Synthesis of a malaria candidate glycosylphosphatidylinositol (GPI) structure: a strategy for fully inositol acylated and phosphorylated gpis. *J Am Chem Soc* 126:7540–7547.
21. Liu, X., Kwon, Y.U., and Seeberger, P.H. (2005). Convergent synthesis of a fully lipidated glycosylphosphatidylinositol anchor of *Plasmodium falciparum. J Am Chem Soc* 127:5004–5005.
22. Schofield, L. (1997). Malaria toxins revisited. *Parasitol Today* 13:275–276: author reply 276–277.
23. Tachado, S.D., Gerold, P., Schwarz, R., Novakovic, S., McConville, M., and Schofield, L. (1997). Signal transduction in macrophages by glycosylphosphatidylinositols of *Plasmodium, Trypanosoma*, and *Leishmania*: activation of protein tyrosine kinases and protein kinase C by inositolglycan and diacylglycerol moieties. *Proc Natl Acad Sci USA* 94:4022–4027.
24. Tachado, S.D., Gerold, P., McConville, M.J., Baldwin, T., Quilici, D., Schwarz, R.T., and Schofield, L. (1996). Glycosylphosphatidylinositol toxin of *Plasmodium* induces nitric oxide synthase expression in macrophages and vascular endothelial cells by a protein tyrosine kinase-dependent and protein kinase C-dependent signaling pathway. *J Immunol* 156:1897–1907.
25. Schofield, L., and Tachado, S.D. (1996). Regulation of host cell function by glycosylphosphatidylinositols of the parasitic protozoa. *Immunol Cell Biol* 74:555–563.
26. Schofield, L., Novakovic, S., Gerold, P., Schwarz, R.T., McConville, M.J., and Tachado, S.D. (1996). Glycosylphosphatidylinositol toxin of *Plasmodium* up-regulates intercellular adhesion molecule-1, vascular cell adhesion molecule-1, and E-selectin expression in vascular endothelial cells and increases leukocyte and parasite cytoadherence via tyrosine kinase-dependent signal transduction. *J Immunol* 156:1886–1896.
27. Campos, M.A., Almeida, I.C., Takeuchi, O., Akira, S., Valente, E.P., Procopio, D.O., Travassos, L.R., Smith, J.A., Golenbock, D.T., and Gazzinelli, R.T. (2001). Activation of Toll-like receptor-2 by glycosylphosphatidylinositol anchors from a protozoan parasite. *J Immunol* 167:416–423.
28. Krishnegowda, G., Hajjar, A.M., Zhu, J.Z., Douglass, E.J., Uematsu, S., Akira, S., Woods, A.S., and Gowda, D.C. (2005). Induction of proinflammatory responses in macrophages by the glycosylphosphatidylinositols of *Plasmodium falciparum*—cell signaling receptors, glycosylphosphatidylinositol (GPI) structural requirement, and regulation of GPI activity. *J Biol Chem* 280:8606–8616.
29. Hewitt, M.C., Snyder, D.A., and Seeberger, P.H. (2002). Rapid synthesis of a glycosylphosphatidylinositol-based malaria vaccine using automated solid-phase oligosaccharide synthesis. *J Am Chem Soc* 124:13434–13436.
30. Seeberger, P.H., Soucy, R.L., Kwon, Y.U., Snyder, D.A., and Kanemitsu, T. (2004). A convergent, versatile route to two synthetic conjugate anti-toxin malaria vaccines. *Chem Commun (Cambridge, UK)*1706–1707.
31. Wolfson, W. (2006). Ancora cooks with carbs synthesizing carbohydrate vaccines. *Chem Biol* 13:689–691.

32. Boutlis, C.S., Fagan, P.K., Gowda, D.C., Lagog, M., Mgone, C.S., Bockarie, M.J., and Anstey, N.M. (2003). Immunoglobulin G (IgG) responses to *Plasmodium falciparum* glycosylphosphatidylinositols are short-lived and predominantly of the igg3 subclass. *J Infect Dis* 187:862–865.

33. Boutlis, C.S., Gowda, D.C., Naik, R.S., Maguire, G.P., Mgone, C.S., Bockarie, M.J., Lagog, M., Ibam, E., Lorry, K., and Anstey, N.M. (2002). Antibodies to *Plasmodium falciparum* glycosylphosphatidylinositols: inverse association with tolerance of parasitemia in Papua New Guinean children and adults. *Infect Immun* 70:5052–5057.

34. de Souza, J.B., Todd, J., Krishegowda, G., Gowda, D.C., Kwiatkowski, D., and Riley, E.M. (2002). Prevalence and boosting of antibodies to *Plasmodium falciparum* glycosylphosphatidylinositols and evaluation of their association with protection from mild and severe clinical malaria. *Infect Immun* 70:5045–5051.

35. Hudson Keenihan, S.N., Ratiwayanto, S., Soebianto, S., Krisin, H., Marwoto, H., Krishnegowda, G., Gowda, D.C., Bangs, M.J., Fryauff, D.J., Richie, T.L., Kumar, S., and Baird, J.K. (2003). Age-dependent impairment of IgG responses to glycosylphosphatidylinositol with equal exposure to *Plasmodium falciparum* among Javanese migrants to Papua, Indonesia. *Am J Trop Med Hyg* 69:36–41.

36. Naik, R.S., Krishnegowda, G., Ockenhouse, C.F., and Gowda, D.C. (2006). Naturally elicited antibodies to glycosylphosphatidylinositols (GPIs) of *Plasmodium falciparum* require intact GPI structures for binding and are directed primarily against the conserved glycan moiety. *Infect Immun* 74:1412–1415.

37. Perraut, R., Diatta, B., Marrama, L., Garraud, O., Jambou, R., Longacre, S., Krishnegowda, G., Dieye, A., and Gowda, D.C. (2005). Differential antibody responses to *Plasmodium falciparum* glycosylphosphatidylinositol anchors in patients with cerebral and mild malaria. *Microbes Infect* 7:682–687.

38. Suguitan, A.L., Jr., Gowda, D.C., Fouda, G., Thuita, L., Zhou, A., Djokam, R., Metenou, S., Leke, R.G., and Taylor, D.W. (2004). Lack of an association between antibodies to *Plasmodium falciparum* glycosylphosphatidylinositols and malaria-associated placental changes in Cameroonian women with preterm and full-term deliveries. *Infect Immun* 72:5267–5273.

39. Kwon, Y.U., Soucy, R.L., Snyder, D.A., and Seeberger, P.H. (2005). Assembly of a series of malarial glycosylphosphatidylinositol anchor oligosaccharides. *Chem Eur J* 11:2493–2504.

40. Kamena, F., Tamborrini, M., Liu, X.Y., Kwon, Y.U., Thompson, F., Pluschke, G., and Seeberger, P.H. (2008). Synthetic GPI array to study antitoxic malaria response. *Nat Chem Biol* 4:238–240.

41. Ratner, D.M., Adams, E.W., Disney, M.D., and Seeberger, P.H. (2004). Tools for glycomics: mapping interactions of carbohydrates in biological systems. *Chembiochem* 5:1375–1383.

42. Ferguson, M.A. (2008). GPIs on a chip. *Nat Chem Biol* 4:223–224.

43. Murray, C.J., and Lopez, A.D. (1997). Global mortality, disability, and the contribution of risk factors: global burden of disease study. *Lancet* 349:1436–1442.

44. Casals-Pascual, C., and Roberts, D.J. (2006). Severe malarial anaemia. *Curr Mol Med* 6:155–168.

45. Lamikanra, A.A., Brown, D., Potocnik, A., Casals-Pascual, C., Langhorne, J., and Roberts, D.J. (2007). Malarial anemia: of mice and men. *Blood* 110:18–28.

46. Ekvall, H. (2003). Malaria and anemia. *Curr Opin Hematol* 10:108–114.

47. Rifkin, M.R., and Landsberger, F.R. (1990). Trypanosome variant surface glycoprotein transfer to target membranes: a model for the pathogenesis of trypanosomiasis. *Proc Natl Acad Sci USA* 87:801–805.

48. Pearce, E.J., Hall, B.F., and Sher, A. (1990). Host-specific evasion of the alternative complement pathway by schistosomes correlates with the presence of a phospholipase C-sensitive surface molecule resembling human decay accelerating factor. *J Immunol* 144:2751–2756.

49. Ilg, T., Etges, R., Overath, P., McConville, M.J., Thomas-Oates, J., Thomas, J., Homans, S.W., and Ferguson, M.A. (1992). Structure of *Leishmania mexicana* lipophosphoglycan. *J Biol Chem* 267:6834–6840.

50. Rooney, I.A., Heuser, J.E., and Atkinson, J.P. (1996). GPI-anchored complement regulatory proteins in seminal plasma. An analysis of their physical condition and the mechanisms of their binding to exogenous cells. *J Clin Invest* 97:1675–1686.

51. Rooney, I.A., Atkinson, J.P., Krul, E.S., Schonfeld, G., Polakoski, K., Saffitz, J.E., and Morgan, B.P. (1993). Physiologic relevance of the membrane attack complex inhibitory protein CD59 in human seminal plasma: CD59 is present on extracellular organelles (prostasomes), binds cell membranes, and inhibits complement-mediated lysis. *J Exp Med* 177:1409–1420.

52. Brattig, N.W., Kowalsky, K., Liu, X., Burchard, G.D., Kamena, F., and Seeberger, P.H. (2008). *Plasmodium falciparum* glycosylphosphatidylinositol toxin interacts with the membrane of non-parasitized red blood cells: a putative mechanism contributing to malaria anemia. *Microbes Infect* 10:885–891.

53. Guha-Niyogi, A., Sullivan, D.R., and Turco, S.J. (2001). Glycoconjugate structures of parasitic protozoa. *Glycobiology* 11:45R–59R.

54. Debierre-Grockiego, F., Azzouz, N., Schmidt, J., Dubremetz, J.F., Geyer, H., Geyer, R., Weingart, R., Schmidt, R.R., and Schwarz, R.T. (2003). Roles of glycosylphosphatidylinositols of *Toxoplasma gondii*—induction of tumor necrosis factor-alpha production in macrophages. *J Biol Chem* 278:32987–32993.

55. Almeida, I.C., Camargo, M.M., Procopio, D.O., Silva, L.S., Mehlert, A., Travassos, L.R., Gazzinelli, R.T., and Ferguson, M.A.J. (2000). Highly purified glycosylphosphatidylinositols from *Trypanosoma cruzi* are potent proinflammatory agents. *EMBO J* 19:1476–1485.

56. Pekari, K., and Schmidt, R.R. (2002). A variable concept for the preparation of branched glycosyl phosphatidyl inositol anchors. *J Org Chem* 68:1295–1308.

57. Ali, A., Gowda, D.C., and Vishwakarma, R.A. (2005). A new approach to construct full-length glycosylphosphatidylinositols of parasitic protozoa and [4-deoxy-Man-III]-GPI analogues. *Chem Commun* 519–521.

58. Kwon, Y.U., Liu, X., and Seeberger, P.H. (2005). Total syntheses of fully lipid glycosylphosphatidylinositol anchors of *Toxoplasma gondii*. *Chem Commun* 2280–2282.

59. Yashunsky, D.V., Borodkin, V.S., Ferguson, M.A.J., and Nikolaev, A.V. (2006). The chemical synthesis of bioactive glycosylphosphatidylinositols from *Trypanosoma cruzi* containing an unsaturated fatty acid in the lipid. *Angew Chem Int Ed* 45:468–474.

12

Inhibitors of GPI Biosynthesis

TERRY K. SMITH

Centre for Biomolecular Sciences
University of St. Andrews, North Haugh
Fife KY16 9ST, Scotland
United Kingdom

I. Abstract

Glycosylphosphatidylinositol (GPI) is a complex glycolipid structure that acts as a membrane anchor for many cell-surface proteins of eukaryotes. The increasing number of completed genomes has allowed considerable progress in the molecular characterization of GPI biosynthesis, not only in terms of human diseases, but also of pathogenic eukaryotic organisms. Although the GPI core glycan is conserved in all organisms, many differences in additional modifications to GPI structures and biosynthetic pathways have been reported. The specificities of these biosynthetic steps must be investigated to allow exploitation of species-specific differences for drug development; however, detailed enzymological studies to date have been limited. Consequently, despite the conserved GPI core structure, the GPI-biosynthetic machinery is different enough between humans and a wide range of eukaryotic pathogens to represent a rich seam of potential therapeutic targets.

Here, we review the recent and promising progress in the field of GPI inhibition.

THE ENZYMES, Vol. XXVI
© 2009 Elsevier Inc. All rights reserved.

ISSN NO: 1874-6047
DOI: 10.1016/S1874-6047(09)26012-4

II. Introduction

Many eukaryotic cell-surface proteins are glycosylphosphatidylinositol
(GPI) anchored [1–5]. This posttranslational modification allows sorting of
apically expressed proteins in epithelial cells [6], and is involved in signal
transduction mechanisms [7], as well as the mechanisms that confer
increased lateral mobility of membrane proteins [8]. Cultured mammalian
cells defective in GPI biosynthesis are viable; however, genetic abrogation
of GPI biosynthesis is embryonically lethal, while certain inherited mutations cause lower levels of GPI biosynthesis leading to thromboses and/or
seizures [9, 10]. Mammalian GPI-anchored proteins also act as receptors for
the bacterial toxins aerolysin [11] and clostridial α-toxin [12]. In yeast and
probably pathogenic fungi, GPI synthesis and/or anchoring are essential for
viability [13], and critical for the maintenance of normal cellular morphology [14], as they are an integral part of fungal cell-wall synthesis and
assembly [15]. GPI biosynthesis in the protozoan parasite *Trypanosoma
brucei*, the causative agent of African sleeping sickness, has also been
genetically and chemically validated as essential in the bloodstream form
of the parasite [16, 17]. This is because their dense cell-surface coat of
GPI-anchored variant surface glycoprotein (VSG) is paramount in their
protection from both the alternative complement pathway, and through
antigenic variation from specific immune responses from the mammalian
host [18, 19]. Several key surface molecules of other trypanosomatids, that
is, *Trypanasoma cruzi* and *Leismania* spp. and apicomplexan parasites, that
is, *Plasmodium falciparum* and *Toxoplasma gondii*, are GPI anchored
(reviewed in Refs. [4, 5]), and it is thought that in some of these pathogens
their GPI biosynthesis is also likely to be essential.

Although GPI anchors are complex glycolipids, the core structure (etha-
nolamine-P-6Manα1–2Manα1–6Manα1–4GlcNH$_2\alpha$1–6-D-*myo*-inositol-1-
phosphate-lipid, where lipid is either diacylglycerol, acyl-alkylglycerol, or
ceramide), is conserved in all organisms (Figure 12.1) [3, 4]. GPI anchors are
synthesized in a step-wise manner in the membrane of the endoplasmic
reticulum (ER) by the concerted action of several glycosyltransferases and
other transferase enzymes. Nearly all of the genes involved in yeast
or mammalian GPI biosynthesis have been identified, mainly by means of
complementation (reviewed in Ref. [20]). However, many subtle but significant differences in additional species-specific modifications to GPI structures (Figure 12.1) and their assembly have been reported, leading to the
notion that these could be exploited as targets for new antipathogenic drugs.

Despite considerable efforts, only a limited number of GPI inhibitors
have been reported to date, most of which are based upon acceptor analogs
and do not possess drug-like properties. Several factors have complicated

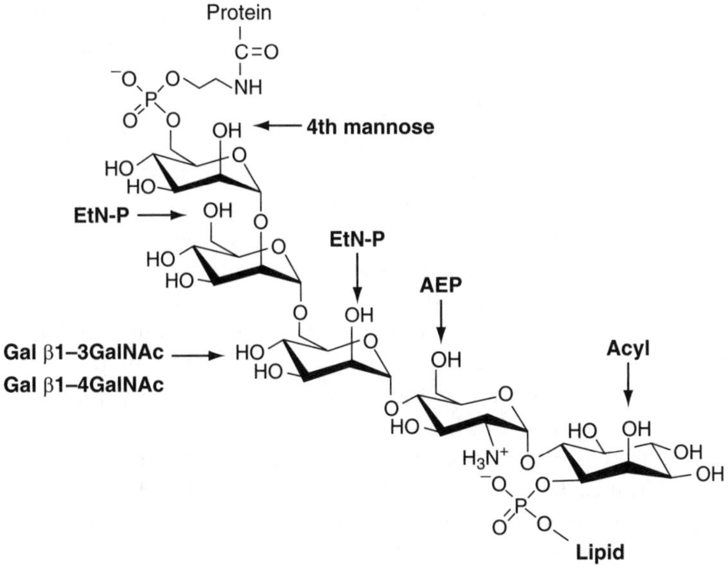

FIG. 12.1. Mature GPI structures found attached to proteins. GPI anchors are complex glycolipids with a conserved core structure of ethanolamine-P-6Manα1–2Manα1–6Manα1–4GlcNH$_2$α1–6-D-myo-inositol-1-phosphate-lipid, where lipid is either diacylglycerol, acyl-alkylglycerol, or ceramide. The arrows show the positions of species-specific additional modifications to the core structure, as described in the text.

the development of GPI inhibitors: the considerable challenge associated with the development of suitable expression systems for the membrane-bound biosynthetic enzymes; the lack of structural data for rational inhibitor design; and a drastic shortage of simple assays.

This last chapter reviews the progress to date in the field of GPI biosynthesis inhibition.

III. GPI Biosynthesis

All GPI anchors have the conserved ethanolamine-P-Manα1–2Manα1–6Manα1–4GlcNH$_2$-PI core (Figure 12.1), which suggests conserved GPI-biosynthetic machinery. GPI-biosynthetic pathways have been elucidated and delineated using various methodologies, including in $vivo$ radiolabeling experiments, but mainly through the use of cell-free systems. It was initially developed in bloodstream form $T.$ $brucei$ [21], but then adapted to other organisms including mammalian cells, yeast, and several other protozoa [22–27]. The cell-free system uses washed membranes with UDP-GlcNAc or synthetic substrate analog and GDP-Man (radiolabeled), which allows step-wise formation of radiolabeled GPI intermediates. The resulting radiolabeled glycolipid intermediates are extracted and separated by high performance thin layer chromatography (HPTLC). Individual intermediates can be identified and structurally characterized using their sensitivities to various chemical and enzymatic digests.

Almost all of the GPI-biosynthetic steps are complexes of polytopic membrane proteins, making expression, purification, and characterization (including three-dimensional structural analysis) a major challenge. Consequently, to date the cell-free systems, and to a minor extent *in vivo* labeling are the main tools to dissect inhibition of GPI-biosynthetic pathways. The substrate specificity of some of the early biosynthetic steps (and recently a later step) has allowed the discovery of several inhibitors (some of which are species specific) but they are still based upon analogs of GPI intermediates. Each of the core biosynthetic steps will now be discussed with reference to their possible exploitable substrate specificities and inhibition.

IV. GlcNAcTransferase

The initial step of GPI biosynthesis is catalyzed by a multiprotein complex, the UDP-GlcNAc:PI α1–6, which transfers GlcNAc from UDP-GlcNAc to phosphatidylinositol [28–33]. In all organisms tested thus far, this first enzymatic step can be inhibited irreversibly by various sulphydryl alkylating reagents, that is, N-ethylmaleimide or iodoacetamide, through the active-site cysteine of PIG-A, the catalytic subunit (Figure 12.2, Ref. [1]) [34]. In *T. brucei*, this inhibitory effect is abolished by pretreatment with either UDP-GlcNAc or UMP or UDP [34]. It has also been shown in *T. brucei* that the PI utilized for GPI biosynthesis is made almost exclusively from the essential *de novo* synthesized inositol in the ER [35, 36].

V. GlcNAc-PI De-*N*-Acetylase

The second step of GPI biosynthesis is the de-N-acetylation of D-GlcNAcα1–6-D-*myo*-inositol-1-HPO$_4$-lipid (GlcNAc-PI), and is a prerequisite for further processing in all GPI-biosynthetic pathways [1]. A wide range of substrate analogs of GlcNAc-PI has allowed specificity's of GlcNAc-PI de-N-acetylases to be explored utilizing cell-free systems of *Leishmania major* [22], *T. brucei* [24, 25, 37–39], *P. falciparum* [40], and human cells [41–43]. These analogs include variations in the lipid,

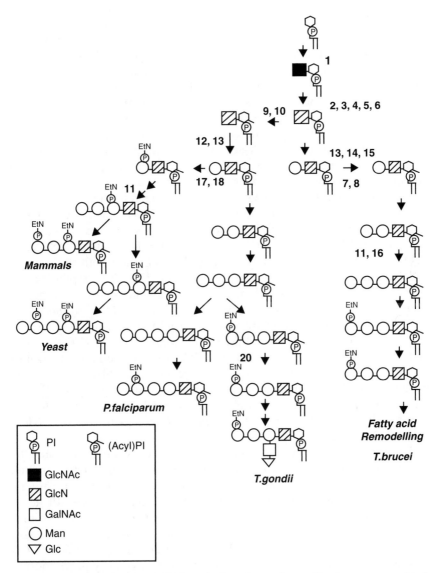

Fig. 12.2. Inhibition of the GPI-biosynthetic pathway of yeast (*Saccharomyces cervisiae*), mammals, *P. falciparum, T. gondii,* and *T. brucei.* The different biosynthetic steps in the GPI pathways of yeast (Y), mammals (M), *P. falciparum* (PF), *T. gondii* (TG), and *T. brucei* (TB) are shown along with reference numbers to specific inhibitors of that enzymatic step in a species-specific manner, as described briefly below and in detail in the main body of the text. (1) (Y, M, PF, TG, TB)—sulphydryl alklating agents; (2) (TB)—GlcN(benzyl)-PI; (3) (TB, M)—GlcNMe$_2$-PI, GlcNCONH$_2$-PI; (4) (TB)—GlcNCONH$_2$-β-PI, GlcNCONH$_2$-(2-*O*-Me)PI; (5) (PF)—GalNCONH$_2$-PI; (6) (TB, M)—1,10-phenanthroline; (7) (TB)—PMSF;

substitution of the 2-hydroxyl of the D-*myo*-inositol with 2-*O*-methyl or 2-*O*-octyl, or variation of the glycosidic linkage, that is, GlcNAc-β-PI or the L-*myo*-inositol analog. All of these were not recognized by the human (HeLa) enzyme, but were weakly processed by the trypanosomal de-*N*-acetylase [37–39]. Mannosylated GlcNAc-PI analogs such as Man-GlcNAc-PI [37], and more recently synthetic Man$_2$-GlcNAc-PI, were neither substrates nor inhibitors for the de-*N*-acetylases tested [44]. Other substrate analogs of the type GlcNR-PI, where R is either acetyl or propionyl were substrates for the de-*N*-acetylases, yet when R is butyrl, isobutyrl, or longer were equally neither substrates nor inhibitors for both *T. brucei* and human systems [37]. However, only *T. brucei* de-*N*-acetylase was able to recognize when R is a benzoyl group, which lead to the corresponding benzyl substitution specifically inhibited the *T. brucei* but not the human enzyme (Figure 12.2, Ref. [2]) [38]. Further elaboration of the R group, that is, GlcNMe$_2$-PI and GlcNCONH$_2$-PI, lead to the first potent GlcNAc-PI de-*N*-acetylases inhibitors, unfortunately they inhibited both the parasite and human de-*N*-acetylases (Figure 12.2, Ref. [3]) [39]. Exploiting the above collective knowledge, lead to the second generation of two potent *T. brucei*-specific inhibitors, 2-deoxy-2-ureido-D-Glcβ1–6D-*myo*-inositol-1-HPO$_4$-*sn*-1,2-dipalmitoylglycerol and 2-deoxy-2-ureido-D-Glcα1–6D-(2-*O*-octyl) *myo*-inositol-1-HPO$_4$-*sn*-1,2-dipalmitoylglycero (GlcNCONH$_2$-β-PI and GlcNCONH$_2$-(2-*O*-octyl)-PI). The IC$_{50}$ of both compounds against *T. brucei* de-*N*-acetylase was <8 nM (Figure 12.2, Ref. [4]), whereas they were not inhibitory to human de-*N*-acetylase up to 100 μM [39, 42]. In a separate study and rather surprisingly, GalNCONH$_2$-PI was found to be a specific inhibitor for the *Plasmodium* GlcNAc-PI de-*N*-acetylase (Figure 12.2, Ref. [5]) [40].

Further studies show the GlcNAc-PI de-*N*-acetylase is inhibited by metal chelators, that is, 1,10-phenanthroline, suggesting a tightly bound divalent metal cation is essential for activity (Figure 12.2, Ref. [6]). Through site-directed mutagenesis and homology modeling the GlcNAc-PI de-*N*-acetylase was postulated to be a zinc-dependent metalloenzyme with a mechanism similar to zinc peptidase [45]. The Ferguson group have now synthesized an *N*-hydroxyurea (–N(H)CON(H)OH) analog; 1-D-6-*O*-[2-(*N*-hydroxyaminocarbonyl) amino-2-deoxy-α-D-glucopyranosyl]-*myo*-inositol

---

(8) (TB)—GlcN-(2-*O*-Me)PI, GlcN-(2-*O*-Oct)PI; (9) (TG)—PMSF; (10) (PF)—GlcN-[L]PI, GlcN-(2-*O*-alkyl)PI; (11) (TB, Y, M)—mannosamine; (12) (PF)—mannosamine; (13) (Y, M, PF, TG, TB)—4-deoxy-GlcN-PI; (14) (TB)—GlcN-(2-*O*-Hexadecyl)PI; (15) (TB)—GlcN-[L] PI; (16) (TB)—ManN-Man-GlcN-PI; (17) (Y, M)—YW3548; (18) (Y, M)—1,10-phenanthroline; (19) (TB)—DFP; (20) (TG)—DFP, PMSF. In all organisms, the transamidation reaction can be inhibited by sulphydryl alkylating agents.

1-(n-octyldecyl phosphate) [46], where the N-hydroxymate moiety should bind tightly to the zinc, leading to inhibition.

VI. Inositol Acyltransferase

Acylation of the 2-hydroxyl of the inositol of a GPI intermediate with usually, but not exclusively, palmitate, and in all organisms that undergo inositol acylation is a prerequisite for further maturation of the GPI intermediates. However, the timing and the subsequent modifications are dependent upon the species, but can be put into two main classes.

The first class is represented by *T. brucei*, where mannosylation precedes inositol acylation, which is subsequently essential for the transfer of ethanolamine phosphate to the 6-hydroxyl of the third mannose [47–49]. Investigations of the substrate specificity of the *T. brucei* third mannosyltransferase (MT) have indirectly revealed that the 4-hydoxyl group of the first mannose group is required for recognition of the inositol acyltransferase, explaining the absolute requirement for mannosylation prior to inositol acylation [44].

T. brucei inositol acylation is mediated by a yet unknown acyl-CoA-independent protein(s) and can be selectively inhibited *in vivo*, as well as *in vitro*, specifically with the serine esterase inhibitor phenylmethylsulphonylfluoride (PMSF) (Figure 12.2, Ref. [7]), leading to the accumulation of the Man$_3$-GlcN-PI intermediate [50]. Inhibition of the trypanosmoal inositol acylation is also observed with the analogs GlcN-(2-O-methyl)PI and GlcN-(2-O-octyl)PI (Figure 12.2, Ref. [8]) [42, 43], where they are mannosylated themselves but failed to receive the ethanolamine phosphate leading to an accumulation of Man$_3$GlcN-(2-O-akyl)PI. This also implies the ethanolamine phosphate transferase was not fooled by Man$_3$-GlcN-(2-O-akyl)PI given that the 2-hydroxyl of the inositol is occupied as it would be by the natural acceptor, Man$_3$-GlcN-(acyl)PI.

Cell-permeable analogs of GlcN-(2-O-methyl)IPC18 were shown to have trypanocidal activity (at 40 µM, 100% within 6 h). These analogs were metabolized (mannosylated) by the *T. brucei* GPI pathway but inhibited inositol acylation of endogenous GPI intermediates. In the human pathway, the analogs were not recognized and were not cytotoxic (up to 100 mM). Closely related analogs deemed not to be metabolized by the trypanosome GPI pathway, showed no trypanocidal activity [51]. To date, this is the only direct chemical validation of the GPI-biosynthetic pathway as a drug target against African human sleeping sickness.

Most other organisms including; mammals, yeast, *P. falciparum*, and *T. gondii* represent the second class of inositol acyltransferase.

Inositol acylation of GlcN-PI is a prerequisite for mannosylation and other subsequent modifications [22, 40, 42, 43]. In human, yeast, and *P. falciparum* inositol acylation unlike that of *T. brucei* is mediated by the acyl-CoA-dependent inositol acyltransferase (PIG-W), which is not inhibited by serine esterase inhibitors or surprisingly by GlcN-(2-*O*-alkyl)PI analogs. Uniquely, the *T. gondii* inositol acyltransferase seems to have characteristics of both classes in other words acyl-CoA dependent, but also inhibited by PMSF (Figure 12.2, Ref. [9]) [22]. In the plasmodium, but not the mammalian cell-free systems the unnatural diastereoisomer GlcN-[L]-PI and GlcN-(2-*O*-alkyl)PI analogs were either competitively inhibiting the inositol acyltransferase or subsequently interacting in a competitive manner with the first MT (Figure 12.2, Ref. [10]) [40].

VII.   Mannosylation of GPI Anchor Intermediates

The trimannosyl core of all GPI anchors is added stepwise involving three distinct dolichol phosphate mannose (Dol-P-Man)-dependent MTs [52–55]. Dol-P-Man is synthesized from dolichol phosphate and GDP-Man by an enzyme called the dolichol phosphate mannose synthase. Amphomycin, a lipopeptide antibiotic that forms a complex with Dol-P in the presence of Ca^{2+} blocking the interaction between the Dol-P-Man synthase and Dol-P [56], was shown to block GPI biosynthesis *in vitro* [57]. Therefore, Dol-P-Man biosynthesis could be an important target for the development of specific inhibitors; particularly of parasites especially those with little or no *N*-glycosylation, such as the apicomplexan parasites *P. falciparum* [58, 59]. Work in the Smith has shown that *T. brucei* dolichol phosphate mannose synthase to be essential and through recombinant expression, identified active-site residues and *T. brucei*-specific competitive inhibitors (Denton and Smith, unpublished results). The GPI MTs, like all glycosyltransferase contain a DXD motif used to bind divalent cations (i.e., Mg^{2+} and Mn^{2+}) and is essential for their activity [60]; therefore, cation scavengers like EDTA are able to prevent *in vitro* GPI mannosylation reactions, leading to the accumulation of the early GPI intermediates (i.e., GlcNAc-PI, GlcN-PI, and GlcN-(acyl)PI) [61].

VIII.   Mannose Analogs

Mannose analogs; 2-deoxy-2-fluoro-D-glucose and 2-deoxy-D-glucose are known to inhibit the formation of Dol-P-Man *in vivo* [62–64], thus preventing GPI formation, that is, *P. falciparum* GPI biosynthesis is inhibited by micromolar concentrations of these mannose analogs [65, 66]. Mannos-amine (2-deoxy-2-amino-D-mannose), a known inhibitor of N-linked oligo-saccharide biosynthesis [67, 68], affect GPI-anchored protein formation in polarized cells like MDCK [69], HeLa cells, and Thy-1 negative lymphoma mutants B and F [67], where they were shown to accumulate in their secretory pathway [70]. Mannosamine also affects GPI biosynthesis in *L. mexicana* [69] and *T. brucei* [71], and thus is rather nonspecies specific. Analysis of the glycan portion of the accumulating GPI intermediate showed mannosamine incorporation in the second mannose position of the conserved trimannosyl glycan core, thereby preventing the addition of the third GPI mannose residue by inhibiting the respective α1–2-MT [71].

Presumably mannosamine (ManN) is converted into GDP-ManN and subsequently Dol-P-ManN (however, there is no direct evidence for either of these mannosamine metabolites) before being utilized by the second MT and forming the intermediate ManN-Man-GlcN-($\pm$acyl)PI (Figure 12.2, Ref. [11]). The effect of mannosamine in *P. falciparum* has been controver-sial due to possible epimerization of mannosamine to glucosamine that similarly observed in *L. mexicana* [72]. However, *in vivo* studies suggest mannosamine causes an accumulation of the GPI intermediate GlcN-(acyl) PI [73], although later cell-free system studies conclude that mannosamine is used instead of GlcN forming the unique GPI-like structures, Man$_3$-ManN-PI, but at a significantly reduced rate, probably causing a block in GPI biosynthesis (Figure 12.2, Ref. [12]) [74]. In *Tetrahymena vorax*, mannosamine inhibition of the GPI pathway drastically impairs cell differentiation and reduces the rate of digestive vacuole formation [75]. While in *Tetrahymena pyriformis*, various sugar analogs seem to interfere with phosphatidylinositol synthesis, thus having a detrimental effect on GPI biosynthesis [76], possibly by interfering with *de novo* synthesis of inositol.

IX. α1–4-Mannosyltransferase (MT-I)

Delinearation of GPI-biosynthetic pathways has shown the relative order of mannosylation and inositol acylation that differ between species as discussed earlier. This has allowed the substrate specificity of the first MT to be explored and exploited using numerous GlcN-PI analogs in *T. brucei, L. major, P. falciparum*, and mammalian cells [21, 40, 42, 43, 77, 78].

Unsurprisingly, the 4-deoxy-GlcN-PI analogs were found to be a com-petitive inhibitor in all systems (Figure 12.2, Ref. [13]); as there is no 4-hydroxyl on the GlcN to accept the first mannose. GlcN-(2-O-hexade-cyl)PI was found to specifically inhibit competitively the first MT in *T. brucei* (Figure 12.2, Ref. [14]), while not being a substrate for either the mammalian or plasmodial MT, whose acceptor substrate would be GlcN-(2-*O*-acyl)PI [42, 43]. The *T. brucei* α1–4 MT was also inhibited competitively by GlcN-[L]-PI but by binding in a different orientation to normal such that the axial 2-hydroxyl of the L-*myo*-inositol residue is orientated above the glycosidic linkage, thus forming a possible stabilizing interaction with the enzyme (Figure 12.2, Ref. [15]) [78]. Recent work in the Smith laboratory has identified the first non-GPI analog-based inhibitors of the *T. brucei* α1–4 MT (Smith, unpublished results).

X.   α1–6-Mannosyltransferase (MT-II)

Synthetic fluorescent analogs of Man-GlcN-PI [79] have been used successfully to monitor the *T. brucei* α1–6-MT activity (Smith, unpublished results). However, simple mannoside acceptor substrates (thiooctyl- and octyl α-mannosides) have also been shown to accept α1–6 mannose residues in a *T. brucei* cell-free system, implying no recognition of the GlcN-α1–6 inositol-1-phosphate moiety as required for earlier enzymes in the pathway [80]. There are as yet no inhibitors for this mannosyltransferase; however, the utilization of ManN instead of Man as discussed earlier, suggests the specificity of the mannose moiety of the donor, that is, Dol-P-Man may be exploitable.

XI.   α1–2-Mannosyltransferase (MT-III)

Activity for the third MTs has been observed directly in the *T. brucei* cell-free system using Man α1–6Man-*O*-octyl as an exogenous acceptor, implying likewise for the second MT, no recognition of the GlcN-α1–6 inositol-1-phosphate moiety is required. As discussed, early *in vivo* studies with mannosamine in many systems shows the formation of ManN-Man-GlcN-(±acyl)PI, presumably the 2-amino group of mannosamine (the position of the hydroxyl accepting the third mannose) interacts with an active-site residue inhibiting the α1–2-MT.

Both of the above specificity observations have now been confirmed in a recent study where various Man_2GlcN-PI analogs were tested in the *T. brucei* cell-free system. The study shows the ManN-Man-GlcN-IPC18 analog has an IC_{50} of ~1.7 μM for the third MT (Figure 12.2, Ref. [16]), while other analogs show no requirement for the 2-amino group of the GlcN moiety [44].

In the mammalian system, the third MT has a specific substrate recognition for the presence of the ethanolamine phosphate group on the first mannose, as shown by the action of the terpenoid lactone YW3548 (see later section on ethanolamine phosphate transfer) [81], as well as the presence of the acylated inositol. All together, the species differences make the third mannosyltransferase a promising potential therapeutic target against *T. brucei* and possibly other protozoa.

XII. Ethanolamine Phosphate Transferases

The last part of the GPI conserved core is the ethanolamine phosphate which links the GPI anchor to the C-terminal of the newly formed protein [82–84]. Uniquely, *T. cruzi* sometimes utilizes the alternative 2-aminoethyl-phosphonate to attach its mucin-like molecules to GPI anchors [85]. The transfer of ethanolamine phosphate from phosphatidylethanolamine to the 6-hydroxyl of the third mannose of Man$_3$-GlcN-(acyl)PI has no known inhibitors in any organism. However, in *T. brucei*, the ethanolamine phosphate addition is prevented indirectly by inhibition of inositol acylation with PMSF [47, 49].

A natural terpenoid lactone isolated from *Codinea simplex* (YW3548) has been shown to block the addition of the ethanolamine phosphate to the first mannose of GPI intermediates in yeast, *Candida albicans* and mammalian cells mannose [81, 82], causing an accumulation of the Man$_2$-GlcN-(acyl)PI intermediate, which is subsequently not a substrate for the third MT (Figure 12.2, Ref. [17]). Similar effects were observed in yeast and mammalian cells using 1,10-phenanthroline [86, 87], where accumulation of GPI intermediates lacking ethanolamine phosphate were identified (Figure 12.2, Ref. [18]). Therefore, both phenanthroline and YW3548 are likely to inhibit the Man-GlcN-(acyl)PI phosphoethanolamine transferase, in mammalian and yeast cells. These inhibitors do not affect GPI biosynthesis in either *T. brucei* or *P. falciparum* because they do not have ethanolamine phosphate addition on the first mannose.

This species selective inhibition is another indication of the fundamental differences between parasite and mammalian/yeast GPI biosynthesis that could be exploited for future drug discovery.

XIII. Inositol Deacylation

Inositol deacylases can also be split into two classes. The first class represents deacylation of the GPI intermediate C′, to form A′, this occurs prior to further species-specific processing of the GPI intermediate. In the case of *T. brucei*, deacyltion takes place prior to fatty acid remodeling and can be inhibited both *in vivo* and *in vitro* selectively with diisopropylfluor-ophosphate (DFP) (Figure 12.2, Ref. [19]) [47]. In the *T. gondii* biosyn-thetic pathway deacylation is a prerequisite for the addition of the GalNAc side chain on the first mannose and is inhibited by both PMSF and DFP (Figure 12.2, Ref. [20]) [22].

The second class of inositol deacylase removes the acyl chain from the mature GPI anchor after protein attachment in both mammalian and yeast systems, as yet there are no known inhibitors of this biosynthetic step [25, 88, 89].

XIV.   GPI Lipid Remodelling

In *T. brucei*, GPI anchor lipid remodeling takes place on the mature GPI precursor, where a step-wise replacement of the acyl chains exclusively with myristate is mediated by specific *sn*-1 and *sn*-2 phospholipases and two myristoyl-CoA-dependant acyltransferases forming a dimyristoylglycerol GPI anchor, which is then ready for protein attachment [90]. Studies using myristic acid analogs such as 10-(propoxy)decanoic acid showed toxicity towards bloodstream *T. brucei* (LD_{50} less than 1 μM), but not to procyclic *T. brucei* (which do not undergo this GPI fatty acid remodeling) or mammalian cells [91, 92], however, the mode of action is still unclear. In *L. mexicana* similar myristate remodeling of the sn-2 acyl chain of the mature sn-1-alkyl-2-acyl GPI anchor precursor has been observed [93].

GPI remodeling in mammalian cells, occurs in the Golgi, postprotein attachment. The usual unsaturated fatty acids are remodeled to saturated fatty acid [94, 95].

In yeast, the diacylglycerol moiety of newly synthesized GPIs are remo-deled to either C26:0/C26:0 diacylglycerols or ceramides consisting of C18:0 phytosphingosine and a hydroxy-C26:0 fatty acid [96, 97]. As yet, there are no known inhibitors of GPI remodeling in any organism, although presum-ably the phospholipases involved contain a catalytic triads, although this does not always mean they are susceptible to inhibitors [98].

XV.   GPI Transamidase

The transamidation reaction attaches the mature GPI precursor to pro-tein on the lumen of the ER [99]. Despite variation between species in the total number and type of individual proteins in this multisubunit membrane-bound complex, they all contain a catalytic component that has an active-site cysteine, which can be modified by general sulphydryl alkylating agents causing loss of activity.

The GPI transamidase cleaves a specific C-terminal GPI addition signal peptide, forming a new C-terminal, activating the carbonyl group of the ω amino acid. Nucleophilic attack on the activated carbonyl by the amino group of the ethanolamine phosphate linked to the third mannose of the mature GPI precursor forming a new amide linkage [100].

Other nucleophiles such as hydrazine or hydroxylamine have been used instead of a GPI anchor for *in vitro* assays resulting in peptide/protein-hydrazide, where the hydrazide linked at the C-terminus of the protein or peptide acceptor instead of the GPI anchor [101].

The distinct species-specific preferences for the amino acid sequence around the omega site, and species-specific differences in the GPI anchor structures, probably reflects the differences in the protein components of the complex. The differences between mammalian and parasite GPI transamidases could be exploited for drug discovery.

XVI. Species-Specific Modifications to the Core GPI Structure

The following section describes some of the species-specific additions made to the GPI core structure (Figure 12.1). These in some circumstances may make obvious targets for species-specific inhibitors; however, for the most part the genes/proteins involved are unknown.

The Smp3 protein adds a fourth mannose (α1–2) to the third mannose of mature GPI precursors before protein attachment in yeast, and in some cases a fifth α1,2- and/or α1,3-linked mannose can be added by unknown Golgi MTs (reviewed in Ref. [20]).

This fourth mannose is also present on some mammalian GPI-anchored proteins as well as extra ethanolamine phosphates on the first and second mannose and/or a (Gal β1–3)±GalNAc β1–4 on the first mannose, although their significance is unknown [102].

All mature plasmodium GPI anchors also contain this fourth α1,2 mannose, along with never removing the acyl chain on the inositol ring [40] and Ref. therein]. While in *T. gondii*, only (Glc β1–4)±GalNAc β1–4 on the first mannose is the only modification to the core structure, and yet these anchors are highly immunogenic in humans [103]. *T. cruzi* is the only organism to date that has a modification (2-aminoethylphosphonate) on the GlcN moiety of the GPI anchor [85].

Uniquely, *T. brucei* add galactopyranose residues to one or more of the core mannoses, depending upon the class/variant of VSG protein [104] and

Ref. therein. The galactosylation is thought to be a space-filling mechanism to maintain the parasites protective armor against the innate immune system [105]. Consequently as, *T. brucei*, unlike humans, are unable to take up galactose, and thus rely totally upon their UDP-Glc 4' epimerase (genetically validated as a drug target) to satisfy their requirement for galactose [106], which they also use heavily in various polylactosamine structures [107]. Unfortunately, initial inhibitors of the *T. brucei* epimerase fail to show significant cytotoxicity selectivity [108].

XVII. Perspectives

In recent years, and with the advent of completed genomes, considerable progress has been made in the molecular biological characterization of GPI biosynthesis, not only in terms of human diseases, but also of pathogenic eukaryotic organisms. Many more genes involved in the GPI-biosynthetic steps are being identified in medically important pathogenic protozoa and fungi, allowing detailed genetic and biochemical validation. However, detailed enzymological studies to date have been limited. The specificities of the reactions involved in GPI biosynthesis must be investigated to allow exploitation of species-specific differences for drug development. Synthetic chemistry advances together with iterative approaches, has overcome some of the challenging complexities of GPI intermediate analog synthesis. Consequently, despite the conserved GPI core structure, the GPI-biosynthetic machinery is different enough between humans and a wide range of eukaryotic pathogens to represent a rich seam of potential therapeutic targets. Hopefully, future chemical–biological analysis of the specificity of the biosynthetic enzymes in humans and pathogens will allow the design and synthesis of lead inhibitors that will stimulate the pharmaceutical sector to develop cheap, effect drugs for some of the most debilitating and neglected diseases on the planet.

ACKNOWLEDGMENTS

Research in the author's laboratory is supported in part by a Wellcome Trust Senior Research Fellowship (067441), and Wellcome Trust grant (086658) and studentships from the Wellcome Trust and the BBSRC.

REFERENCES

1. Nagamune, K., Nozaki, T., Maeda, Y., Ohishi, K., Fukuma, T., Hara, T., Schwarz, R.T., Sutterlin, C., Brun, R., Riezman, H., and Kinoshita, T. (2000). Critical roles of glycosyl-phosphatidylinositol for *Trypanosoma brucei*. *Proc Natl Acad Sci USA* 97:10336–10341.

2. McConville, M.J., and Menon, A.K. (2000). Recent developments in the cell biology and biochemistry of glycosylphosphatidylinositol lipids. *Mol Membr Biol* 17:1–16.

3. Morita, Y.S., Acosta-Serrano, A., and Englund, P.T. (2000). The biosynthesis of GPI anchors, in P. Ernst, P. Sinay and G. Hart (eds.), *Oligosaccharides in Chemistry and Biology—A comprehensive Handbook*, pp. 417–433, VCH, Weinheim, Germany: Wiley.

4. Ferguson, M.A.J., Brimacombe, J.S., Brown, J.R., Crossman, A., Dix, A., Field, R.A., Guther, M.L., Milne, K.G., Sharma, D.K., and Smith, T.K. (1999). The GPI biosynthetic pathway as a therapeutic target for African sleeping sickness. *Biochim Biophys Acta* 1455:327–340.

5. Ferguson, M.A.J. (1999). The structure, biosynthesis and functions of glycosylphosphati-dylinositol anchors, and the contributions of trypanosome research. *J Cell Sci* 112:2799–2809.

6. Lisanti, M.P., Caras, I.W., and Rodriguez-Boulan, E. (1991). Fusion proteins containing a minimal GPI-attachment signal are apically expressed in transfected MDCK cells. *J Cell Sci* 99:637–640.

7. Tachado, S.D., Gerold, P., McConville, M.J., Baldwin, T., Quilici, D., Schwarz, R.T., and Schofield, L. (1996). Glycosylphosphatidylinositol toxin of *Plasmodium* induces nitric oxide synthase expression inmacrophages and vascular endothelial cells by a protein tyrosine kinase-dependent and protein kinasec-dependent signaling pathway. *J Immunol* 156:1897–1907.

8. Noda, M., Yoon, K., Rodan, G.A., and Koppel, D.E. (1987). High lateral mobility of endogenous and transfected alkaline phosphatase: a phosphatidylinositol-anchored mem-brane protein. *J Cell Biol* 105:1671–1677.

9. Nozaki, M., Ohishi, K., Yamada, N., Kinoshita, T., Nagy, A., and Takeda, J. (1999). Developmental abnormalities of glycosylphosphatidylinositol-anchor-deficient embryos revealed by Cre/loxp system. *Lab Invest* 79:293–299.

10. Almeida, A.M., Murakami, Y., Layton, D.M., Hillmen, P., Sellick, G.S., Maeda, Y., Richards, S., Patterson, S., Kotsianidis, I., Mollica, L., Crawford, D.H., Baker, A., *et al.* (2006). Hypomorphic promoter mutation in PIGM causes inherited glycosylphosphati-dylinositol deficiency. *Nat Med* 12:846–851.

11. Diep, D.B., Nelson, K.L., Raja, S.M., Pleshak, E.N., and Buckley, J.T. (1998). Glycosyl-phosphatidylinositol anchors of membrane glycoproteins are binding determinants for the channel-forming toxin aerolysin. *J Biol Chem* 273:2355–2360.

12. Gordon, V.M., Nelson, K.L., Buckley, J.T., Stevens, V.L., Tweten, R.K., Elwood, P.C., and Leppla, S.H. (1999). *Clostridium septicum* alpha toxin uses glycosylphosphatidylinositol-anchored protein receptors. *J Biol Chem* 274:27274–27280.

13. Leidich, S.D., Drapp, D.A., and Orlean, P. (1994). A conditionally lethal yeast mutant blocked at the first step in glycosyl phosphatidylinositol anchor synthesis. *J Biol Chem* 269:10193–10196.

14. Pittet, M., and Conzelmann, A. (2006). Biosynthesis and function of GPI proteins in the yeast *Saccharomyces cerevisiae*. *Biochim Biophys Acta* 1771(3):405–420.

15. Kollar, R., Reinhold, B.B., Petrakova, E., Yeh, H.J., Ashwell, G., Drgonova, J., Kapteyn, J.C., Klis, F.M., and Cabib, E. (1997). Architecture of the yeast cell wall. β 1–6-glucan interconnects mannoprotein, β1–3-glucan, and chitin. *J Biol Chem* 272:17762–17775.

16. Nagamune, K., Nozaki, T., Maeda, Y., Ohishi, K., Fukuma, T., Hara, T., Schwarz, R.T., Sutterlin, C., Brun, R., Riezman, H., and Kinoshita, T. (2000). Critical roles of glycosyl-phosphatidylinositol for *Trypanosoma brucei*. *Proc Natl Acad Sci USA* 97(19): 10336–10341.
17. Smith, T.K., Crossman, A., Brimacombe, J.S., and Ferguson, M.A. (2004). Chemical validation of GPI biosynthesis as a drug target against African sleeping sickness. *EMBO J* 23(23):4701–4708.
18. Cross, G.A.M. (1996). Antigenic variation in trypanosomes: secrets surface slowly. *Bioessays* 18:283–291.
19. Ferguson, M.A.J. (2000). Glycosylphosphatidylinositol biosynthesis validated as a drug target for African sleeping sickness. *Proc Natl Acad Sci USA* 97:10673–10675.
20. Orlean, P., and Menon, A.K. (2007). Thematic review series: lipid posttranslational modifications. GPI anchoring of protein in yeast and mammalian cells, or: how we learned to stop worrying and love glycophospholipids. *J Lipid Res* 48:993–1011.
21. Smith, T.K., Milne, F.C., Sharma, D.K., Crossman, A., Brimacombe, J.S., and Ferguson, M.A. (1997). Early steps in glycosylphosphatidylinositol biosynthesis in *Leishmania major*. *Biochem J* 326:393–400.
22. Smith, T.K., Kimmel, J., Azzouz, N., Shams-Eldin, H., and Schwarz, R.T. (2007). The role of inositol-acylation and inositol-deacylation in the *Toxoplasma gondii* glycosylphospha-tidylinositol biosynthetic pathway. *JBC* 282(44):32032–32042.
23. Tomavo, S., Dubremetz, J.F., and Schwarz, R.T. (1992). Biosynthesis of glycolipid pre-cursors for glycosylphosphatidylinositol membrane anchors in a *Toxoplasma gondii* cell-free system. *J Biol Chem* 267:21446–21458.
24. Gerold, P., Jung, N., Azzouz, N., Freiberg, N., Kobe, S., and Schwarz, R.T. (1999). Biosynthesis of glycosylphosphatidylinositols of *Plasmodium falciparum* in a cell-free incubation system: inositol acylation is needed for mannosylation of glycosylphosphati-dylinositols. *Biochem J* 344:731–738.
25. Chen, R., Walter, E.I., Parker, G., Lapurga, J.P., Millan, J.L., Ikehara, Y., Udenfriend, S., and Medof, M.E. (1998). Mammalian glycophosphatidylinositol anchor transfer to pro-teins and posttransfer deacylation. *Proc Natl Acad Sci USA* 95:9512–9517.
26. Canivenc-Gansel, E., Imhof, I., Reggiori, F., Burda, P., Conzelmann, A., and Benachour, A. (1998). GPI anchor biosynthesis in yeast: phosphoethanolamine is attached to the a1,4-linked mannose of the complete precursor glycophospholipid. *Glycobiology* 8:761–770.
27. Puoti, A., Desponds, C., Fankhauser, C., and Conzelmann, A. (1991). Characterization of glycophospholipid intermediate in the biosynthesis of glycophosphatidylinositol anchors accumulating in the Thy-1-negative lymphoma line SIA-b. *J Biol Chem* 266:21051–21059.
28. Watanabe, R., Inoue, N., Westfall, B., Taron, C.H., Orlean, P., Takeda, J., and Kinoshita, T. (1998). The first step of glycosylphosphatidylinositol biosynthesis is mediated by a complex of PIG-A, PIG-H, PIG-C and GPI1. *EMBO J* 17:877–885.
29. Watanabe, R., Murakami, Y., Marmor, M.D., Inoue, N., Maeda, Y., Hino, J., Kangawa, K., Julius, M., and Kinoshita, T. (2000). Initial enzyme for glycosylphosphatidylinositol biosynthesis requires PIG-P and is regulated by DPM2. *EMBO J* 19:4402–4411.
30. Leidich, S.D., Kostova, Z., Latek, R.R., Costello, L.C., Drapp, D.A., Gray, W., Fassler, J.S., and Orlean, P. (1995). Temperature-sensitive yeast GPI anchoring mutants gpi2 and gpi3 are defective in the synthesis of *N*-acetylglucosaminyl phosphatidylinositol. Cloning of the GPI2 gene. *J Biol Chem* 270:13029–13035.
31. Schonbachler, M., Horvath, A., Fassler, J., and Riezman, H. (1995). The yeast spt14 gene is homologous to the human PIG-A gene and is required for GPI anchor synthesis. *EMBO J* 14:1637–1645.

32. Inoue, N., Watanabe, R., Takeda, J., and Kinoshita, T. (1996). PIG-C, one of the three human genes involved in the first step of glycosylphosphatidylinositol biosynthesis is a homologue of *Saccharomyces cerevisiae* GPI2. *Biochem Biophys Res Commun* 226:193–199.

33. Shams-Eldin, H., Azzouz, N., Kedees, M.H., Orlean, P., Kinoshita, T., and Schwarz, R.T. (2002). The GPI1 homologue from *Plasmodium falciparum* complements a *Saccharomyces cerevisiae* GPI1 anchoring mutant. *Mol Biochem Parasitol* 120:73–81.

34. Milne, K.G., Ferguson, M.A., and Masterson, W.J. (1992). Inhibition of the GlcNAc transferase of the glycosylphosphatidylinositol anchor biosynthesis in African trypanosomes. *Eur J Biochem* 208:309–314.

35. Martin, K., and Smith, T.K. (2006). Phosphatidylinositol synthase is essential in *Trypanosoma brucei*. *Biochem J* 396:287–295.

36. Martin, K., and Smith, T.K. (2006). The role of *de novo* synthesis of *myo*-inositol in *Trypanosoma brucei*. *Mol Microbiol* 61(1):89–105.

37. Sharma, D.K., Smith, T.K., Crossman, A., Brimacombe, J.S., and Ferguson, M.A. (1997). Substrate specificity of the *N*-acetylglucosaminyl-phosphatidylinositol de-*N*-acetylase of glycosylphosphatidylinositol membrane anchor biosynthesis in African trypanosomes and human cells. *Biochem J* 328:171–177.

38. Sharma, D.K., Smith, T.K., Weller, C.T., Crossman, A., Brimacombe, J.S., and Ferguson, M.A. (1999). Differences between the trypanosomal and human GlcNAc-PI de-*N*-acetylases of glycosylphosphatidylinositol membrane anchor biosynthesis. *Glycobiology* 9:415–422.

39. Smith, T.K., Crossman, A., Borissow, C.N., Paterson, M.J., Dix, A., Brimacombe, J.S., and Ferguson, M.A. (2001). Specificity of glcnac-PI de-*N*-acetylase of GPI biosynthesis and synthesis of parasite-specific suicide substrate inhibitors. *EMBO J* 20:3322–3332.

40. Smith, T.K., Gerold, P., Crossman, A., Paterson, M.J., Borissow, C.N., Brimacombe, J.S., Ferguson, M.A.J., and Schwarz, R.T. (2002). Substrate Specificity of the *Plasmodium falciparum* glycosylphosphatidylinositol biosynthetic pathway and inhibition by species specific suicide substrates. *Biochemistry* 41:12395–12406.

41. Smith, T.K., Crossman, A., Paterson, M.J., Borissow, C.N., Brimacombe, J.S., and Ferguson, M.A. (2002). Specificities of enzymes of glycosylphosphatidylinositol biosynthesis in *Trypanosoma brucei* and HeLa cells. *J Biol Chem* 277:37147–37153.

42. Smith, T.K., Sharma, D.K., Crossman, A., Brimacombe, J.S., and Ferguson, M.A. (1999). Selective inhibitors of the glycosylphosphatidylinositol biosynthetic pathway of *Trypanosoma brucei*. *EMBO J* 18:5922–5930.

43. Smith, T.K., Sharma, D.K., Crossman, A., Dix, A., Brimacombe, J.S., and Ferguson, M.A. (1997). Parasite and mammalian GPI biosynthetic pathways can be distinguished using synthetic substrate analogues. *EMBO J* 16:6667–6675.

44. Urbaniak, M.D., Yashunsky, D.V., Crossman, A., Nikolaev, A.V., and Ferguson, M.A.J. (2008). Probing enzymes late in the Trypanosomal GPI biosynthetic pathway with synthetic GPI analogues. *Chem Biol* 3(10):625–634.

45. Urbaniak, M.D., Crossman, A., Chang, T., Smith, T.K., Van Aalten, D., and Ferguson, M. A.J. (2005). The *N*-acetyl-D-glucosaminyl-phosphatidylinositol de-*N*-acetylase of glycosyl-phosphatidylinositol biosynthesis is a zinc metalloenzyme. *JBC* 280 (24):22831–22838.

46. Urbaniak, M.D., Crossman, A., and Ferguson, M.A.J. (2008). Synthesis of 1-D-6-*O*-[2-(*N*-hydroxyaminocarbonyl) amino-2-deoxy-α-D-glucopyranosyl]-*myo*-inositol 1-(*n*-octyldecyl phosphate): a potential metalloenzyme inhibitor of GPI biosynthesis 343(9):1478–1481.

47. Guther, M.L., and Ferguson, M.A. (1995). The role of inositol acylation and inositol deacylation in GPI biosynthesis in *Trypanosoma brucei*. *EMBO J* 14:3080–3093.

48. Field, M.C. (1992). Inositol acylation of glycosylphosphatidylinositol. *Glycoconj J* 9:155–159.

49. Masterson, W.J., and Ferguson, M.A. (1991). Phenylmethanesulphonyl fluoride inhibits GPI anchor biosynthesis in the African trypanosome. *EMBO J* 10:2041–2045.

50. Guther, M.L., Masterson, W.J., and Ferguson, M.A. (1994). The effects of phenylmethyl-sulfonyl fluoride on inositol-acylation and fatty acid remodeling in African trypanosomes. *J Biol Chem* 269:18694–18701.

51. Smith, T.K., Crossman, A., Brimacombe, J.S., and Ferguson, M.A.J. (2004). Chemical validation of GPI biosynthesis as a drug target African sleeping sickness. *EMBO J* 23 (23):4701–4708.

52. Maeda, Y., Watanabe, R., Harris, C.L., Hong, Y., Ohishi, K., Kinoshita, K., and Kinoshita, T. (2001). PIG-M transfers the first mannose to glycosylphosphatidylinositol on the lumenal side of the ER. *EMBO J* 20:250–261.

53. Takahashi, M., Inoue, N., Ohishi, K., Maeda, Y., Nakamura, N., Endo, Y., Fujita, T., Takeda, J., and Kinoshita, T. (1996). PIG-B, a membrane protein of the endoplasmic reticulum with a large lumenal domain, is involved in transferring the third mannose of the GPI anchor. *EMBO J* 15:4254–4261.

54. Sutterlin, C., Escribano, M.V., Gerold, P., Maeda, Y., Mazon, M.J., Kinoshita, T., Schwarz, R.T., and Riezman, H. (1998). *Saccharomyces cerevisiae* GPI10, the functional homologue of human PIG-B, is required for glycosylphosphatidylinositol-anchor synthesis. *Biochem J* 332:153–159.

55. Waechter, C.J., Lucas, J.J., and Lennarz, W.J. (1973). Membrane glycoproteins. I. Enzy-matic synthesis of mannosyl phosphoryl polyisoprenol and its role as a mannosyl donor in glycoprotein synthesis. *J Biol Chem* 248:7570–7579.

56. Banerjee, D.K. (1989). Amphomycin inhibits mannosylphosphoryldolichol synthesis by forming a complex with dolichylmonophosphate. *J Biol Chem* 264:2024–2028.

57. Odenthal-Schnittler, M., Tomavo, S., Becker, D., Dubremetz, J.F., and Schwarz, R.T. (1993). Evidence for N-linked glycosylation in *Toxoplasma gondii*. *Biochem J* 291:713–721.

58. Dieckmann-Schuppert, A., Bender, S., Odenthal-Schnittler, M., Bause, E., and Schwarz, R.T. (1992). Apparent lack of *N*-glycosylation in the asexual intraerythrocytic stages of *Plasmodium falciparum*. *Eur J Biochem* 205:815–825.

59. Kimura, E.A., Couto, A.S., Peres, V.J., Casal, O.L., and Katzin, A.M. (1996). N-linked glycoproteins are related to schizogony of the intraerythrocytic stage in *Plasmodium falciparum*. *J Biol Chem* 271:14452–14461.

60. Oliver, G.J., Harrison, J., and Hemming, F.W. (1975). The mannosylation of dolichol-diphosphate oligosaccharides in relation to the formation of oligosaccharides and glyco-proteins in pig-liver endoplasmic reticulum. *Eur J Biochem* 58:223–229.

61. Vidugiriene, J., and Menon, A.K. (1993). Early lipid intermediates in glycosyl-phosphatidylinositol anchor assembly are synthesized in the ER and located in the cytoplasmic leaflet of the ER membrane bilayer. *J Cell Biol* 121:987–996.

62. Datema, R., Schwarz, R.T., and Jankowski, A.W. (1980). Fluoroglucose-inhibition of protein glycosylation *in vivo*. Inhibition of mannose and glucose incorporation into lipid-linked oligosaccharides. *Eur J Biochem* 109:331–341.

63. Datema, R., Schwarz, R.T., and Winkler, J. (1980). Glycosylation of influenza virus proteins in the presence of fluoroglucose occurs via a different pathway. *Eur J Biochem* 110:355–361.

64. Datema, R., and Schwarz, R.T. (1978). Formation of 2-deoxyglucose-containing lipid-linked oligosaccharides. Interference with glycosylation of glycoproteins. *Eur J Biochem* 90:505–516.

65. Udeinya, I.J., and Van Dyke, K. (1981). 2-Deoxyglucose: inhibition of parasitemia and of glucosamine incorporation into glycosylated macromolecules, in malarial parasites (*Plasmodium falciparum*). *Pharmacology* 23:171–175.

66. Santos de Macedo, C., Gerold, P., Jung, N., Azzouz, N., Kimmel, J., and Schwarz, R.T. (2001). Inhibition of glycosyl-phosphatidylinositol biosynthesis in *Plasmodium falciparum* by C-2 substituted mannose analogues. *Eur J Biochem* 268:6221–6228.

67. Sevlever, D., and Rosenberry, T.L. (1993). Mannosamine inhibits the synthesis of putative glycoinositol phospholipid anchor precursors in mammalian cells without incorporating into an accumulate intermediate. *J Biol Chem* 268:10938–10945.

68. Pan, Y.T., De Gespari, R., Warren, C.D., and Elbein, A.D. (1992). Formation of unusual mannosamine-containing lipid-linked oligosaccharides in Madin-Darby canine kidney cell cultures. *J Biol Chem* 267:8991–8999.

69. Pan, Y.T., Kamitani, T., Bhuvaneswaran, C., Hallaq, Y., Warren, C., Yeh, E.T.H., and Elbein, A.D. (1992). Inhibition of glycosylphosphatidylinositol anchor formation by mannosamine. *J Biol Chem* 267:21250–21255.

70. Lisanti, M.P., Field, M.C., Caras, W., Menon, A.K., and Rodriguez-Boulan, E. (1991). Mannosamine, a novel inhibitor of glycosylphosphatidylinositol incorporation into proteins. *EMBO J* 10:1969–1977.

71. Ralton, J.E., Milne, K.G., Guther, M.L., Field, R.A., and Ferguson, M.A.J. (1993). The mechanism of inhibition of glycosylphosphatidylinositol anchor biosynthesis in *Trypanosoma brucei* by mannosamine. *J Biol Chem* 268:24183–24189.

72. Field, M.C., Medina-Acosta, E., and Cross, G.A.M. (1993). Inhibition of glycosylphosphatidylinositol biosynthesis in *Leishmania mexicana* by mannosamine. *J Biol Chem* 268:9570–9577.

73. Naik, R.S., Davidson, E.A., and Gowda, D.C. (2000). Developmental stage-specific biosynthesis of glycosylphosphatidylinositol anchors in intraerythrocytic *Plasmodium falciparum* and its inhibition in a novel manner by mannosamine. *J Biol Chem* 275:24506–24511.

74. Azzouz, N., Macedo, C.S., Smith, T.K., and Schwarz, R.T. (2005). Mannnosamine can replace glucosamine in GPIs of *Plasmodiumn falciparum in vitro*. *Mol Biochem Parasitol* 142:12–24.

75. Yang, X., and Ryals, P.E. (1994). Cytodifferentiation in *Tetrahymena vorax* is linked to glycosyl-phosphatidylinositol-anchored protein assembly. *Biochem J* 298:697–703.

76. Kovacs, P., and Csaba, G. (1997). Effects of the amino sugars, glucosamine, mannosamine, or the fluorinated derivative 2-deoxy-fluoroglucose on the phosphatidylinositol and glycosyl phosphatidyl inositol systems of *Tetrahymena*. *Microbios* 89:91–104.

77. Smith, T.K., Cottaz, S., Brimacombe, J.S., and Ferguson, M.A. (1996). Substrate specificity of the dolichol phosphate mannose: glucosaminyl phosphatidylinositol alpha1–4-mannosyltransferase of the GPI biosynthetic pathway of African trypanosomes. *J Biol Chem* 271:6476–6482.

78. Smith, T.K., Paterson, M.J., Crossman, A., Brimacombe, J.S., and Ferguson, M.A. (2000). Parasite-specific inhibition of the GPI biosynthetic pathway by stereoisomeric substrate analogues. *Biochemistry* 39:11801–11807.

79. Mayer, T.G., Weingart, R., Munstermann, F., Kawada, T., Kuurzchalia, T., and Scmidt, R.R. (1999). Synthesis of labeled GPI anchors. *Eur J Org Chem* 10:2563–2571.

80. Brown, J.R., Guther, M.L., Field, R.A., and Ferguson, M.A. (1997). Hydrophobic mannosides act as acceptors for trypanosome α-mannosyltransferases. *Glycobiology* 7:549–558.

81. Sutterlin, C., Horvath, A., Gerold, P., Schwarz, R.T., Wang, Y., Dreyfuss, M., and Riezman, H. (1997). Identification of a species-specific inhibitor of glycosylphosphatidylinositol synthesis. *EMBO J* 16:6374–6383.

82. Canivenc-Gansel, E., Imhof, I., Reggiori, F., Burda, P., Conzelmann, A., and Benachour, A. (1998). GPI anchor biosynthesis in yeast: phosphoethanolamine is attached to the alpha1,4-linked mannose of the complete precursor glycophospholipid. *Glycobiology* 8:761–770.

83. Kamitani, T., Menon, A.K., Hallaq, Y., Warren, C.D., and Yeh, E.T. (1992). Complexity of ethanolamine phosphate addition in the biosynthesis of glycosylphosphatidylinositol anchors in mammalian cells. *J Biol Chem* 267:24611–24619.

84. Hirose, S., Prince, G.M., Sevlever, D., Ravi, L., Rosenberry, T.L., Ueda, E., and Medof, M.E. (1992). Characterization of putative GPI anchor precursors in mammalian cells. Localization of phosphoethanolamine. *J Biol Chem* 267:16968–16974.

85. de Lederkremer, R.M. (1990). Structural features of the lipopeptidophosphoglycan from *Trypanosoma cruzi* common with the glycophosphatidylinositol anchors. *Eur J Biochem* 192:337–345.

86. Mann, K.J., and Sevlever, D. (2001). 1,10-Phenanthroline inhibits glycosylphosphatidylinositol anchoring by preventing phosphoethanolamine addition to GPI anchor precursors. *Biochemistry* 40:1205–1213.

87. Sevlever, D., Mann, K.J., and Medof, M.E. (2001). Differential effect of 1,10-phenanthroline on mammalian, yeast, and parasite glycosylphosphatidylinositol anchor synthesis. *Biochem Biophys Res Commun* 288:1112–1118.

88. Chen, R., Walter, E.I., Parker, G., Lapurga, J.P., Millan, J.L., Ikehara, Y., Udenfriend, S., and Medof, M.E. (1998). Mammalian GPI anchor transfer to proteins and post- transfer deacylation. *Proc Natl Acad Sci USA* 95:9512–9517.

89. Tanaka, S., Maeda, Y., Tashima, Y., and Kinoshita, T. (2004). Inositol deacylation of glycosylphosphatidylinositol-anchored proteins is mediated by mammalian PGAP1 and yeast Bst1p. *J Biol Chem* 279:14256–14263.

90. Masterson, W.J., Raper, J., Doering, T.L., Hart, G.W., and Englund, P.T. (1990). Fatty acid remodeling: a novel reaction sequence in the biosynthesis of trypanosome glycosyl phosphatidylinositol membrane anchors. *Cell* 62:73–80.

91. Doering, T.L., Raper, J., Buxbaum, L.U., Adams, S.P., Gordon, J.I., Hart, G.W., and Englund, P.T. (1991). An analog of myristic acid with selective toxicity for African trypanosomes. *Science* 252:1851–1854.

92. Doering, T.L., Lu, T., Werbovetz, K.A., Gokel, G.W., Hart, G.W., Gordon, J.I., and Englund, P.T. (1994). Toxicity of myristic acid analogs toward African trypanosomes. *Proc Natl Acad Sci USA* 91:9735–9739.

93. Ralton, J.E., and McConville, M.J. (1998). Delineation of three pathways of glycosylphosphatidylinositol biosynthesis in *Leishmania mexicana*. Precursors from different pathways are assembled on distinct pools of phosphatidylinositol and undergo fatty acid remodelling. *J Biol Chem* 273:4245–4257.

94. Singh, N., Zoeller, R.A., Tykocinski, M.L., Lazarow, P.B., and Tartakoff, A.M. (1994). Addition of lipid substituents of mammalian protein glycosylphosphoinositol anchors. *Mol Cell Biol* 14:21–31.

95. Maeda, Y., Tashima, Y., Houjou, T., Fujita, M., Yoko-O, T., Jigami, Y., Taguchi, R., and Kinoshita, T. (2007). Fatty acid remodeling of GPI-anchored proteins is required for their raft association. *Mol Biol Cell* 18:1497–1506.

96. Sipos, G., Reggiori, F., Vionnet, C., and Conzelmann, A. (1997). Alternative lipid remodelling pathways for glycosylphosphatidylinositol membrane anchors in *Saccharomyces cerevisiae*. *EMBO J* 16:3494–3505.

97. Reggiori, F., Canivenc-Gansel, E., and Conzelmann, A. (1997). Lipid remodeling leads to the introduction and exchange of defined ceramides on GPI proteins in the ER and Golgi of *Saccharomyces cerevisiae*. *EMBO J* 16:3506–3518.

98. Richmond, G.S., and Smith, T.K. (2007). A novel phospholipase from *Trypanosoma brucei*. *Mol Microbiol* 63(4):1078–1095.
99. Sharma, D.K., Hilley, J., Bangs, J.D., Coombs, G.H., Mottram, J., and Menon, A.K. (2000). Soluble GPI8 restores glycosylphosphatidylinositol anchoring in a trypanosome cell-free system depleted of soluble endoplasmic reticulum proteins. *Biochem J* 351:717–722.
100. Maxwell, S.E., Ramalingam, S., Gerber, L.D., Brink, L., and Udenfriend, S. (1995). An active carbonyl formed during glycosylphosphatidylinositol addition to a protein is evidence of catalysis by a transamidase. *J Biol Chem* 270:19576–19582.
101. Sharma, D.K., Vidugiriene, J., Bangs, J.D., and Menon, A.K. (1999). A cell-free assay for glycosylphosphatidylinositol anchoring in African trypanosomes. Demonstration of a transamidation reaction mechanism. *J Biol Chem* 274:16479–16486.
102. Grimme, S.J., Westfall, B.A., Wiedman, J.M., Taron, C.H., and Orlean, P. (2001). The essential Smp3 protein is required for addition of the side-branching fourth mannose during assembly of yeast glycosylphosphatidylinositols. *J Biol Chem* 276:27731–27739.
103. Debierre-Grockiego, F., Campos, F., Azzouz, M.A., Schmidt, N., Bieker, J., Resende, U., Mansur, M.G., Weingart, D.S., Schmidt, R., Golenbock, R.R.D., Gazzinelli, R.T., and Schwarz, R.T.T. (2007). Activation of TLR2 and TLR4 by glycosylphosphatidylinositols derived from *Toxoplasma gondii*. *J Immunol* 179(2):1129–1137.
104. Mehlert, A., Zitzmann, N., Richardson, J.M., Traumann, A., and Ferguson, M.A.J. (1998). The glycosylation of the variant surface glycoprotein and procyclic acidic repetitive proteins of *T. Brucei*. *Mol Biochem Parasitol* 91:145–152.
105. Homans, S.W., and Ferguson, M.A.J. (1989). Solution structure of the glycosylphosphatidylinositol membrane anchor glycan of trypanosoma-brucei variant surface glycoprotein. *Biochemistry* 28:2881–2887.
106. Roper, J.R., Guther, M.L., Milne, K.G., and Ferguson, M.A. (2002). Galactose metabolism is essential for the African sleeping sickness parasite *Trypanosoma brucei*. *Proc Natl Acad Sci USA* 99:5884–5889.
107. Atrih, A., Richardson, J.M., Prescott, A.R., and Ferguson, M.A.J. (2005). *Trypanosoma brucei* glycoproteins contain novel giant poly-*N*-acetyllactosamine carbohydrate chains. *J Biol Chem* 280:865–871.
108. Urbaniak, M.D., Tabudravu, J.N., Msaki, A., Matera, K.M., Brenk, R., Jaspars, M., and Ferguson, M.A. (2006). Identification of novel inhibitors of UDP-Glc 4'-epimerase, a validated drug target for African sleeping sickness. *Bioorg Med Chem Lett* 16:5744–5747.

13

Transport of GPI-Anchored Proteins: Connections to Sphingolipid and Sterol Transport

GUILLAUME A. CASTILLON • HOWARD RIEZMAN

Department of Biochemistry
University of Geneva
30 quai Ernest Ansermet
CH-1211 Geneva, Switzerland

I. Abstract

Ten to twenty percent of secretory proteins are modified at their C-termini by glycosylphosphatidylinositol (GPI) after their synthesis and translocation into the endoplasmic reticulum (ER). Addition of GPI allows anchoring to the inner leaflet of the ER membrane and subsequently to the extracellular leaflet of the plasma membrane. Most secretory proteins are thought to exit ER by a concentrative process requiring direct or indirect binding to Sec24p or isoforms. This is true for GPI-anchored proteins, but being exclusively luminal, they must interact with a cargo receptor in order to be concentrated into the COPII-coated vesicles. In yeast, GPI-anchored proteins exit the ER in vesicles distinct from other secretory proteins. In mammalian cells, sorting of GPI-anchored proteins most likely occurs upon Golgi exit. The anchor of GPI-anchored proteins undergoes lipid remodeling in yeast and mammalian cells in the ER and Golgi compartments, respectively. It is likely that remodeling allows association with

ISSN NO: 1874-6047
DOI: 10.1016/S1874-6047(09)26013-6

membrane microdomains that are enriched in liquid-ordered phase due to the presence of sphingolipids and sterols. GPI-anchored protein presence in microdomains is likely to be important for their trafficking. Studies in yeast have revealed an interdependency between GPI-anchored protein and ceramide trafficking as mutants in GPI trafficking affect ceramide transport (CERT) and ceramide synthesis mutants affect GPI-anchored protein traffic. Being essentially luminal proteins, but still membrane anchored, misfolded GPI-anchored proteins in the ER seem to follow a specialized pathway for degradation.

II.   ER Exit of GPI-Anchored Proteins

A.   GPI ANCHOR STRUCTURE AND ER EXPORT

Glycosylphosphatidylinositol (GPI) is synthesized in the endoplasmic reticulum (ER) after sequential enzymatic reactions involving more than 20 gene products [1, 2]. Newly synthesized GPI-anchored proteins are attached to GPI by the GPI transamidase complex. Later in the process the lipid moiety of the GPI anchor is modified. In the first step, the acyl chain on the inositol ring is removed. Then, the unsaturated fatty acyl chain at the *sn-2* position of the phosphatidylinositol is removed and replaced by myristic acid (C14:0) in the case of the bloodstream form of *T. brucei* or by saturated C26 fatty acid in yeast. Eventually the diacylglycerol of the GPI anchor can be modified to ceramide in yeast [3–6]. This process, called remodeling, occurs in the ER in yeast and in the Golgi compartment in mammalian cells [3, 7, 8]. In mammalian cells, remodeling results in the production of mature GPI-anchored proteins whose PI moiety contains only saturated fatty acids (C18:0) [9].

A defect in GPI anchor biosynthesis or in GPI anchoring leads to a delay in GPI-anchored protein maturation as a consequence of an ER to Golgi transport defect in yeast and other organisms [10–13]. The replacement of the GPI attachment site of the GPI-anchored protein Gas1p by an artificial transmembrane domain results in about a 50% reduction in ER budding efficiency [14]. The GPI-anchored protein VSG without its GPI attachment peptide accumulates in the ER in *T. brucei* [10, 15]. Moreover, a defect in attachment of GPI to the Prion protein Prp leads to its retention in the ER [16]. These results suggest that the GPI anchor acts as a positive forward transport signal in the early secretory pathway.

It has been observed in yeast that defects in remodeling of GPI-anchored proteins generate an ER to Golgi transport defect. In *gup1Δ* cells, where the addition of a C26 saturated fatty acid to GPI-lyso-PI is compromised,

Gas1p and Cwp2p accumulate in the ER [17, 18; our unpublished data]. The *cwh43* mutation blocks replacement of diacylglycerol with ceramide for the formation of fully remodeled GPI-anchored proteins [19, 20] and delays the maturation of the GPI-anchored protein Gas1p [20]. This result is surprising considering that the Gas1p GPI-anchor is thought to be remodeled into diacylglycerol [6]. Our recent results suggest that remodeling is important for concentration of GPI-anchored proteins prior to budding [93]. Considering that mammalian enzymes responsible for remodeling, excluding PGAP1, are localized in the Golgi apparatus, it is unlikely that remodeling is important for ER exit in mammalian cells. However, PGAP1-deficient cells display a delay in GPI-anchored protein transport, which might be explained by a quality control mechanism if the acyl group is not removed from the inositol [21].

B.   COPII-MEDIATED ER EXIT

ER-to-Golgi vesicular transport is mediated by the assembly of COPII coat. Sec12p is an ER-bound transmembrane protein, which activates the small Ras-like GTPase Sar1p. Upon activation, Sar1-GDP is converted to Sar1-GTP. Sar1-GTP bound to Sec12p at the cytoplasmic side of the ER recruits the Sec23p/Sec24p complex. The Sar1p/Sec23p/Sec24p complex, called prebudding complex, selects and concentrates cargos directly or via the interaction with a cargo receptor. Subsequently, the prebudding complex recruits Sec13p/Sec31p heterotetramer. Sec13p/Sec31p drives membrane deformation and allows COPII vesicle formation [22–27].

The yeast GPI-anchored proteins have been shown to exit the ER in vesicles that are distinct from other exiting proteins, including the transmembrane protein Gap1p, proalpha-factor, and the vacuolar alkaline phosphatase. The initial studies showed this using an *in vitro* assay reconstituting sorting and budding from the ER [28] and the vesicles produced had different physical properties as seen from their behavior in density gradient centrifugation. Subsequent studies were able to confirm that sorting also occurs *in vivo* using biochemical fractionation of pulse-labeled cells where fusion of ER-derived vesicles with the Golgi was blocked [29]. More recently, we have been able to visualize the sorting using live imaging techniques and have demonstrated that sorting seems completed at the level of ER exit sites [93].

In yeast, GPI-anchored proteins behave similarly to other secretory proteins with respect to COPII-mediated ER export. However, the Sar1p requirement is not as strict for GPI-anchored proteins. The Schekman group could produce ER-derived vesicles containing the GPI-anchored protein Gas1p from a microsomal fraction after addition of purified

Sec23p/Lst1p (Lst1p being an isoform of Sec24p), Sec13p/Sec31p, and in absence of Sar1p [30]. Using permeabilized spheroplasts and cytosol derived from cells carrying the thermosensitive allele *sar1–2*, Gas1p vesicles were able to bud from the ER slightly less efficiently (20% reduction) than when compared to a budding assay using WT membrane and cytosol (our unpublished data). Budding of the transmembrane protein Gap1p was completely blocked by the *sar1–2* mutation. Similar results could be observed using the dominant negative form of Sar1p (GTP-locked form) (our unpublished data). Interestingly the concentration of the GPI-anchored protein Cwp2p at ER exit sites was unaffected in *sec12–4* and *sec16–2* mutant cells. These results suggest that even though GPI-anchored proteins may require COPII coat for ER budding in yeast, the mechanism of concentration into COPII-coated vesicles is independent of COPII in contrast to others types of cargo proteins [93]. In HeLa cells, it has been reported that GFP fused with GPI and VSVG fused with GFP displayed different kinetics of accumulation into ER-derived vesicles in presence of a GTP-restricted form of Sar1p [31].

There are three isoforms of Sec24p in yeast (Sec24p, Sfb2p, and Sbf3p/Lst1p) and four in mammalian cells [32]. Maturation of the GPI-anchored protein Gas1p and the carboxypeptidase Y are delayed in the thermosensitive allele of *SEC24, sec24–13*. In *sfb2Δ* and *sfb3Δ* cells, only Gas1p maturation is affected [33]. These results suggest a difference in the way the prebudding complex loads GPI-anchored proteins compared to other secretory proteins.

C. ER EXIT RECEPTORS FOR GPI-ANCHORED PROTEINS?

Exclusively luminal GPI-anchored proteins (see Section II) might interact with the prebudding complex via the interaction with a cargo receptor. In yeast, a member of the p24 family, Emp24p, has been proposed to be the ER exit receptor of GPI-anchored proteins in association with Erv25p, another member of the p24 family. Gas1p maturation is delayed in *emp24Δ* and in *erv25Δ* cells [34]. The ER budding efficiency of the GPI-anchored protein Gas1p is reduced *in vitro* using *emp24* mutant membranes or by using wild-type membranes and antibody against the cytoplasmic tail of Emp24p [34]. ER exit of Cwp2p is also strongly decreased in *emp24Δ* cells [93]. Furthermore, it has been observed that Emp24p can be cross-linked with Gas1p, but not with the transmembrane protein Gap1p upon ER exit [34]. The mammalian homolog of Emp24p, named p23, is required for GPI-anchored protein ER-to-Golgi transport [35], but attempts to show similar cross-linking were unsuccessful. Still, the exact mechanism whereby Emp24p affects incorporation of GPI-anchored proteins into COPII vesicles is not known. Besides the effect on ER exit, Emp24p and Erv25p have been shown to facilitate the COPI coated vesicle formation through the ArfGAP Glo3p [36]. As Golgi-to-ER retrograde transport seems to be required for anterograde GPI-anchored protein transport [37] it is difficult to rule out completely an indirect role of Emp24p in anterograde transport. To date, the best evidence for a direct role of Emp24 in anterograde transport is the antibody inhibition of Gas1p budding [34] together with the cross-linking of the protein to Gas1p and the ability of the cytoplasmic tail to bind directly to COPII components [38].

D. Genetic Interactions with *SEC13*

EMP24, ERV25, and *BST1* were identified in a genetic screen as mutations allowing growth of a strain carrying a *sec13* deletion [39]. The reason for this genetic interaction remains unclear but suggests that a decrease in GPI-anchored proteins ER export allows cells to cope with a COPII-coated vesicle formation defect. Several possible alterations in these mutants may contribute to the suppression of the *sec13* phenotype. The *emp24* mutant induces an unfolded protein response (UPR) [36] and some of the induced proteins may contribute to overcoming the *sec13* defect. In addition, the GPI-anchored protein exit pathway is less dependent on COPII functions (Section II.B). It is possible that the other proteins, like Gap1p, that have a strict dependence on COPII leave the ER through the GPI pathway when the GPI-anchored proteins are no longer efficiently exported. Finally, we have recently seen that ceramides accumulate in some mutants defective in exit of GPI-anchored proteins from the ER (Sharon Epstein, G.A.C, Isabelle Riezman and H.R., unpublished data) and it is possible that alterations in the ER membrane structure by increased ceramides could allow membrane deformation even in the absence of the Sec13/31 complex.

III.   Sorting of GPI-Anchored Proteins upon ER Exit

A. Role of Sorting Signals

Some proteins that exit the ER have signals in their cytoplasmic tails that drive interaction with COPII components, for instance Gap1p [24], or bind to adaptor proteins that connect to COPII components, for example proalpha-factor, which uses Erv29p as an adaptor protein [40]. The sorting signals on the proteins and the adaptor are required for efficient ER exit. GPI-anchored proteins are sorted away from these secretory proteins and the sorting event appears to be completed when cargos are concentrated at the ER exit sites (Section II.B). One obvious signal for sorting of a GPI-anchored protein would be its anchor. Surprisingly, the replacement of the GPI attachment sequence of Gas1p by an artificial transmembrane domain (Gas1TMD) did not drive this mutant form of Gas1p into the Gap1p-containing vesicles [14]. Moreover the addition of a GPI anchor to the soluble cargo protein proalpha factor, which is normally found in the Gap1p-containing ER-derived vesicles, did not modify the sorting of proalpha factor in wild-type cells. However, in *erv29Δ* cells, where budding efficiency is greatly reduced, proalpha factor was not found in the Gap1p-containing vesicles anymore. From these data, two hypotheses were proposed. To be incorporated into the Gap1p-containing vesicles, cargos require ER exit signals and in the case of soluble proteins like proalpha factor, also a transmembrane adaptor protein, like Erv29p. Cargos without an ER exit signal would be excluded from Gap1p-containing vesicles. Alternatively, if the GPI anchor is an ER exit signal, then it would drive the protein into the GPI-anchor protein-containing vesicles, but then the ER exit signal of proalpha factor must be dominant over the GPI anchor.

B.   SORTING FACTORS FOR ER EXIT

During vesicular traffic, proteins are sorted and packaged into vesicles that bud from a donor membrane. The vesicles are transported and tethered to the acceptor membrane prior to SNARE complex formation and subsequent membrane fusion. The tethering events require the function of distinct Rab GTPases. The specificity of membrane targeting is driven by combined actions of Rabs, tethering factors, and SNARES. Because they are spatially and temporally separated, budding and targeting/fusion steps are considered to be independent. Studies of ER-to-Golgi transport of GPI-anchored proteins in yeast led us to revise our view on tethering and fusion of transport vesicles.

It has been shown that the mutant for the tethering factor Uso1p not only prevents tethering of ER-derived vesicles, but displays a sorting defect in which GPI-anchored proteins and non-GPI-anchored proteins are found in the same ER derived vesicles population [29]. In the same study, it has been found that the small Rab-GTPase Ypt1p and the tethering factors Sec34/35 are also necessary for sorting upon ER exit. This sorting defect appears to occur during the budding event, suggesting that Ypt1p and the tethering complex is required for sorting during budding at the level of ER. Additionally, it has been reported that mutants in the v-SNAREs Bos1p, Sec22p, and Bet1p produce *in vitro* ER-derived vesicles that are unable to fuse to the Golgi and in which Gas1p and Gap1p are not sorted [41].

Once vesicles are formed from the ER they tether together [42]; but in yeast, this tethering only occurs homotypically [29]. It is possible that these tethering factors are not directly involved in sorting in the ER, but may function to prevent heterotypic tethering of the two populations of ER-derived vesicles after budding. A summary of GPI-anchored protein ER exit is provided in Figure 13.1.

IV. Defects in GPI-Anchored Protein Trafficking and Folding

A. UNFOLDED GPI-ANCHORED PROTEIN DEGRADATION

Under normal conditions, it has been estimated that up to 30% of all newly synthesized proteins are misfolded [43]. In addition to the fact that unfolded or misfolded proteins are not functional, accumulation of

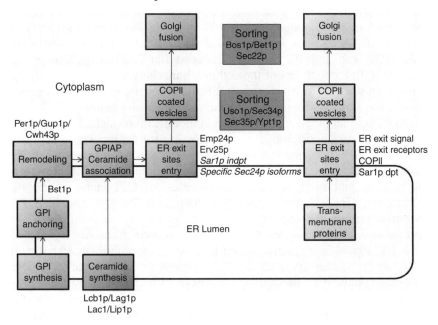

FIG. 13.1. ER exit of GPI-anchored proteins in yeast. After synthesis and anchoring, GPI-anchored proteins are remodeled and associated with DRMs. The DRM association allows concentration in a COPII-independent manner of GPI-anchored proteins into specific ER exit sites, different from those concentrating other cargo molecules. With the direct or indirect assistance of Emp24p and Erv25p, GPI-anchored proteins bud from the ER in a Sar1p-independent manner. Sorting is maintained by SNAREs and tethering factors, which likely prevent heterotypic fusion of the different ER-derived vesicle populations.

unfolded proteins in the ER is toxic for cells [44]. There are two nonexclusive strategies for cells to maintain low levels of unfolded or misfolded proteins. The first one helps proteins to acquire proper folding through the action of chaperones. The second is to destroy the irreversible misfolded proteins. This can occur in two ways: by traffic to the vacuole/lysosomal compartment or through ER-associated protein degradation (ERAD), in which misfolded proteins are translocated out of the ER, ubiquitinated, and degraded by the proteasome [45].

B. UNFOLDED PROTEIN RESPONSE

Upon certain stress conditions or in presence of certain chemicals, the amount of unfolded or misfolded proteins is greatly increased causing cells to activate the UPR [46]. Upon sensing the increase of unfolded proteins, the ER-resident transmembrane protein Ire1p oligomerizes [47–49]. The oligomerization triggers its autophosphorylation, and allows the cleavage of the mRNA of *XBP1* in mammalian cells known as *HAC1* in yeast [50]. The cleaved mRNA encodes an activator of UPR target genes. In mammalian cells, IRE1 can activate the stress-induced Jun *N*-terminal Kinase and interacts with components of the cell-death machinery, such as caspase-12 [51–53]. In yeast, activation of the UPR enhances the expression of genes involved in translocation, protein folding, protein degradation (including ERAD), lipid metabolism, glycosylation/modification (including GPI synthesis, anchoring, and remodeling), and vesicle trafficking with a majority involved in ER and Golgi anterograde and retrograde transport [46]. Even though the response is not specific for GPI-anchored proteins, it is interesting to note that most of the key molecules for GPI-anchored protein maturation (Gpi10p, Arv1p, Gaa1p, Gup1p, and Cwh43p) and for GPI-anchored protein ER export (Lcb1p, Lac1p, Erv25p, and Emp24p) are upregulated, suggesting that GPI-anchored protein transport is induced upon ER stress. In a genetic screen for mutants synthetically lethal with *ire1*Δ, several genes involved in GPI synthesis or GPI-anchored protein remodeling have been identified [54]. These mutants, called *PER* for protein processing in the ER are dependent on the activation of UPR for viability. *PER17* is identical to *BST1* and encodes for the GPI inositol deacylase and *PER1* encodes for the GPI-phospholipase A_2 (see Section II.A). Therefore, the UPR response is probably essential for survival of mutants with defective remodeling and ER exit of GPI-anchored proteins. In fact, it seems that mutants in genes in GPI anchor biosynthesis and GPI anchor remodeling are among the mutations that show the greatest induction of the unfolded protein response, suggesting that they are important for proper folding of GPI anchored proteins [94]. Moreover, the transmembrane protein Tat2p is retained in the ER in *gaa1–1* and *bst1*Δ mutant cells [55], suggesting that these defects may also influence exit of other classes of proteins.

C. ERAD

Even though we have a good molecular understanding of the ERAD mechanism, little is known with respect to GPI-anchored protein degradation upon misfolding. It has been suggested that the GPI inositol deacylation by Bst1p plays an important role in quality control and ER-associated degradation of GPI-anchored proteins in yeast [18] because a misfolded form of Gas1p (Gas1*p) was stabilized in *bst1*Δ mutant cells. There are three types of ERAD pathways: ERAD-L, which recognizes misfolding in the ER lumen; the ERAD-C, which is associated with the misfolded cytoplasmic domains of proteins inserted into the ER; and ERAD-M, which degrades the proteins misfolded in their transmembrane domain [56]. The ERAD-L and ERAD-M pathways require the ubiquitin ligase Hrd1p and ERAD-C requires the ubiquitin ligase Doa10p. It is believed that in the case of ERAD-L the misfolded substrates are transported from the ER to the Golgi and then retrieved to the ER, where they are translocated into the cytoplasm and degraded by the proteasome. Gas1*p degradation requires Sec18p suggesting that Gas1*p degradation follows ERAD-L, but Gas1*p ubiquitination is not dependent on Hrd1p, Doa10p, or Rsp5p [18]. This suggests a novel ERAD pathway, for which the ubiquitin ligase has to be identified. In *emp24*Δ cells, Gas1*p degradation is slowed down. At the moment it is not clear if Bst1p sorts misfolded proteins from folded GPI-anchored proteins or if GPI-anchored protein remodeling and transport is a prerequisite for ERAD.

D. DEGRADATION OF PRION PrP

Prion diseases are fatal neurodegenerative disorders that include Creutzfeldt-Jacob disease in humans and bovine spongiform encephalopathy in animals, for which there is no effective treatment. Some prion proteins, such as PrP are GPI anchored. The toxic, misfolded PrP^{Sc} (Sc for scrapie) cannot replicate in absence of PrP^C (cellular prion protein) and PrP-null mice are resistant to prion infection [57, 58]. Preventing accumulation of more misfolded material by reducing PrP^C expression allows normal metabolic pathways to restore homeostasis by clearing various disease-associated PrP forms [59, 60]. Even a small reduction of PrP^C expression has a significant impact on disease progression [61]. RNAi against PrP^C

rescues early neuronal dysfunction and prolongs survival in mice with prion disease [62]. While still under debate, it is believed that part of the neuronal toxicity observed in prion diseases is due to the accumulation of cytoplasmic aggregated prion. This accumulation has been suggested to be the result of the insensitivity of prion to the proteasome or an inhibition of the proteasome activity by prion itself after retrotranslocation from the ER [63]. Studies of the prion, in addition to the medical contribution, can enlighten us on mechanisms linking GPI-anchored protein ER exit and ERAD. A wide range of mutations and deletions of prion are tolerated and lead to ER export. Many forms of mutated, misfolded, or misprocessed PrP^C are recognized and efficiently degraded by ERAD. A mutant mouse neuroblastoma cell line, termed A4, with a reduced expression of PrP^C has been isolated [16]. In this cell line, PrP^C ER export is defective due to a GPI anchoring defect and the prion protein is translocated out of the ER and degraded by the proteasome. However it seems that the GPI anchoring defect is not sufficient to target GPI-anchored proteins to ERAD. This requires an unknown signal contained in the prion sequence.

V.   GPI-Anchored Protein and Lipid Traffic

The transport of GPI-anchored proteins and ceramides from ER to Golgi in yeast seem to be mutually dependent as defects in ceramide synthesis affect GPI-anchored protein transport and defects in GPI-anchored protein transport cause accumulation of ceramides in the ER [64–66]. Considering that sphingolipid transport and sterol transport are intimately coupled [67], a holistic view of lipid traffic is required to fully understand transport of GPI-anchored proteins (see Figures 13.2 and 13.3).

A.   RELATIONSHIP BETWEEN GPI-ANCHORED PROTEINS, SPHINGOLIPIDS, AND STEROLS

GPI-anchored proteins have been proposed to be enriched in specialized membrane microdomains enriched in sphingolipids and sterols because they are found in small cholesterol-dependent clusters in the plasma membrane and because they are found in detergent-resistant membrane (DRM) fractions [68–70]. GPI-anchored proteins are not associated with DRMs in yeast and mammalian remodeling mutants [8, 18, 20]. The DRM association occurs in the ER in yeast and in the Golgi in mammalian cells coinciding with the localization of the remodeling enzymes. In *lcb1–100* cells, in which the first step of ceramide biosynthesis is impaired, the membrane association of GPI-anchored proteins is severely weakened, and transport of GPI-

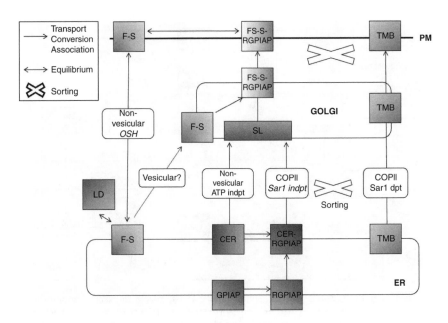

FIG. 13.2. GPI-anchored protein, sphingolipid, and sterol secretory pathway in yeast. After association with DRM (CER-RGPIAP) containing ceramides (CER), remodeled GPI-anchored proteins (RGPIAP) leave the ER in a different ER-derived vesicle population than other secretory proteins (TMB). After arrival in the Golgi apparatus, ceramides are converted to sphingolipids (SL). In the Golgi, GPI-anchored proteins incorporate lipid rafts (SL-S-RGPIAP) enriched in sterols and sphingolipids and are exported to the plasma membrane. At the plasma membrane, concentrations of free sterols (F-S) and sterols contained in lipid rafts are equilibrated. Through a nonvesicular mechanism, an equilibrium is maintained between the plasma membrane and ER-free sterol concentrations. In case of accumulation in the ER, sterols are esterified and stored in lipid droplets (LD). This model does not take into account the contribution of endocytosis to sterol and sphingolipid homeostasis.

anchored proteins from ER to Golgi is strongly affected [66]. Furthermore, ER exit of GPI-anchored proteins is dependent on the ceramide synthases Lag1p and Lac1p, and is inhibited by myriocin, an inhibitor of the first step of ceramide biosynthesis [71, 72]. Altogether, these results suggest that ceramides are essential for membrane association and ER exit of GPI-anchored proteins in yeast. This was not observed in mammalian cells [73]. It is likely that ceramides are building blocks of the ER micro-domains in yeast. One might also expect an important function for sterols in this process, but even severe mutants in ergosterol biosynthesis do not affect ER-to-Golgi transport of GPI-anchored proteins in yeast [74]. This suggests that the presence of sterols within the ER microdomains is dispensable for stable membrane association of GPI-anchored proteins.

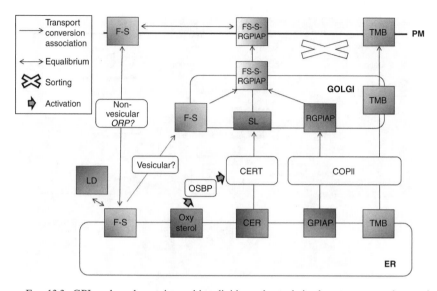

Fig. 13.3. GPI-anchored proteins, sphingolipids, and sterols in the secretory pathway of mammalian cells. Ceramides (CER) are transported to the Golgi by a nonvesicular mechanism involving CERT. This traffic event is stimulated by OSBP upon elevation of oxysterol levels. After arrival to the Golgi apparatus, ceramides are converted to sphingolipids (SL). GPI-anchored proteins (GPIAP) and other cargo molecules (TMB) are likely cotransported from ER to the Golgi apparatus. In the Golgi, GPI-anchored proteins are remodeled (RGPIAP), incorporated into lipid rafts (SL-S-RGPIAP) enriched in sterols and sphingolipids, and exported to the plasma membrane. At the plasma membrane, concentrations of free sterols (F-S) and sterols contained in lipid rafts are equilibrated. Through a nonvesicular mechanism, an equilibrium is maintained between the plasma membrane and ER-free sterol concentrations [67]. In case of accumulation in the ER, sterols are esterified and stored in lipid droplets (LD). This model does not take into account the contribution of endocytosis to sterol and sphingolipid homeostasis.

Later along the secretion, microdomains, in which ceramides are replaced by complex sphingolipids, seem to require ergosterol in yeast. Upon pheromone induction, a GFP-GPI was localized at the mating tip in yeast. Disturbance of ergosterol or sphingolipid synthesis induced the relocalization of GFP-GPI to the cell cortex [75–77]. The erg6 mutation also misdirects the GPI-anchored protein, Yps1p, to the vacuole [78]. In mammalian cells, the DRM association of GPI-anchored proteins occurs in the Golgi apparatus in connection with the GPI remodeling by the sequential action of PGAP3/PERLD1 and PGAP2. Even though not sufficient, association of GPI-anchored proteins with lipid rafts is essential for apical sorting in several epithelial cell lines and to the axonal region of neuronal cells [79–83]. Treatment with an inhibitor of the ceramide synthase, fumonisin B1, disturbs the apical transport of GPI-anchored proteins but does not affect transmembrane proteins [84]. Stabilization of GPI-anchored proteins within microdomains requires oligomerization. The oligomerization of the GPI-anchored proteins is affected by cholesterol depletion.

B. SPHINGOLIPID AND STEROL TRANSPORT REQUIRES GPI-ANCHORED
    PROTEIN TRANSPORT IN YEAST

In yeast and mammalian cells, ceramides have been described to traffic from ER to Golgi via vesicular and nonvesicular mechanisms (see Figures 13.2 and 13.3). In yeast mutants defective in ER-to-Golgi vesicular transport only about one-half of sphingolipid synthesis was inhibited [64, 65]. In mammalian cells most of the CERT is nonvesicular. Most of the nonvesicular ER-to-Golgi transport of ceramides for sphingomyelin synthesis in mammalian cells is ensured by CERT, probably enhanced at membrane contact sites [85]. Interestingly addition of 25-hydroxycholesterol enhances the synthesis of sphingomyelin via stimulation of CERT [86]. The 25-hydroxycholesterol-mediated CERT is regulated through the oxysterol binding protein (OSBP) [87]. CERT and OSBP contain a PH domain for Golgi localization, and a FFAT domain important for ER localization. CERT also contains a START domain, which allows binding of ceramides. OSBP displays a 25-hydroxycholesterol binding motif. Apparently OSBP senses increased cholesterol level through elevation of oxysterol, thus undergoing phosphorylation, which promotes its localization to the ER [88]. This possibly increases the number of membrane contact sites between ER and Golgi, and therefore facilitates the nonvesicular transport of ceramides through CERT. Enhancement of CERT and subsequent increase in complex sphingolipid synthesis might be crucial to keep a constant ratio between sphingolipids and cholesterol. It has not yet been reported that GPI-anchored proteins plays a key role in this process so far.

A possible link between GPI-anchored protein and ceramide transport comes from yeast. In yeast, a CERT homolog has not been identified. This is not surprising because CERT is not able to transport ceramides with very long acyl chains as found in yeast [89]. In addition, as mentioned earlier, less CERT is ensured by a nonvesicular mechanism in yeast. Interestingly the nonvesicular transport of ceramides is ATP independent in contrast to mammalian cells [64]. It has been postulated that membrane contact sites might be sufficient for CERT, but they still require a heat-labile, trypsin-sensitive cytosolic factor.

Synthesis of complex sphingolipids such as IPC is decreased in *arv1Δ* mutant cells. Arv1p is involved in GPI synthesis, somehow regulating the delivery of the glucosaminyl-acyl-PI (GlcN-acylPI) to the GPI-mannosyl-transferase [65]. Therefore GPI-anchored proteins are not efficiently GPI anchored, due to defect in GPI synthesis. Others mutants in GPI synthesis or anchoring exhibit a defect in sphingolipid synthesis, likely through a CERT defect. ATP depletion did not exacerbate the IPC synthesis defect in *arv1Δ* mutant, suggesting that GPI synthesis is important for the vesicular transport of ceramides. These data and the fact that GPI-anchored proteins require ceramides for ER export, strongly suggest that GPI and ceramides cooperate for efficient ER-to-Golgi transport.

Sterols are synthesized in the ER. Cells maintain a gradient of sterol distribution through the secretory pathway to finally display an approximate 10-fold enrichment of sterol in the plasma membrane relative to ER. Newly synthesized sterols traffic from ER to plasma membrane mainly via a nonvesicular pathway because there is no defect in transport rate in secretory mutants such as *sec18* [67]. In mammalian cells, cholesterol transport from ER to plasma membrane uses either a nonvesicular pathway or a vesicular pathway that bypasses the BFA-sensitive route through the Golgi [90]. The actual model of secretion of sterol describes a vesicular pathway where sterols are transported from ER to the Golgi and then to the plasma membrane in association with sphingolipids in microdomains, and a nonvesicular pathway where free sterols reach the plasma membrane directly from the ER through membrane contact sites or/with the help of sterol carriers [91, 92]. Possible sterol carriers are encoded by the OSH (OSBP homologs) genes (7 OSHs) in yeast [92]. In mammalian cells, the OSBP-related proteins (ORP) might contribute to sterol transport, however the individual function of most of them (12 ORPs in human) is not known [92].

In the *arv1Δ* mutant and other GPI synthesis mutants in yeast, sterols accumulate in the ER and in lipid particles [65]. Because in the same mutants sphingolipid synthesis is strongly affected, it has been hypothesized that nonvesicular traffic of sterols is required to maintain an equilibrium between free sterols and sterols complexed with sphingolipids at the plasma membrane [67]. In support of this, in *lcb1–100* mutant cells, ceramides and IPC are not efficiently synthesized and even though there is no increase of total sterols, there is an increase of free sterols at the plasma membrane [67]. To summarize the model, in wild-type cells there is a constant equilibrium between free sterols and sterols complexed with sphingolipids, in a way that the amount of total sterol at the plasma membrane is 10-fold higher that the amount of sterols in the ER and in both compartments while the concentration of free sterol remains equivalent [65, 67]. If the level of sphingolipids is reduced either by disturbing the synthesis of ceramide or blocking

CERT by a GPI synthesis defect, this entails an elevated level of free sterols at the plasma membrane. By equilibration of free sterols between plasma membrane and ER through the nonvesicular transport, the free sterol concentration in the ER is increased. To cope with elevation of sterol levels within the ER, sterols are esterified and accumulated in lipid droplets.

VI. Parallels and Differences between Yeast and Mammalian Cells

Yeast and mammalian cells use similar molecular mechanisms for GPI synthesis and anchoring. After remodeling of GPI, GPI-anchored proteins are incorporated into microdomains, which confer certain particularities to GPI-anchored protein traffic. Crucial differences between yeast and mammalian cells regarding GPI and lipid transport may come from the discrepancy in remodeling enzyme localization. In yeast, GPI-anchored proteins are remodeled in the ER, and incorporated into DRMs with ceramides, ensuring efficient ER exit and arrival to the Golgi apparatus. This property may explain why GPI-anchored proteins are sorted from other secretory proteins upon ER exit. In mammalian cells, after deacylation of PI in the ER, remodeling takes place in the Golgi apparatus. Therefore we anticipate that ER-to-Golgi transport of GPI-anchored proteins is not dependent on ceramide trafficking, and sorting not required for efficient transport. Moreover in yeast vesicular and nonvesicular transport of ceramides coexist, whereas mammalian cells heavily favor nonvesicular transport for which remodeled GPI-anchored protein cotransport is not required. The difference in organization may be a necessity arising from the more hydrophobic very long acyl chains of yeast ceramides and its inherent difficulty to undergo nonvesicular traffic.

ACKNOWLEDGMENTS

The work from the Riezman lab was supported by grants from the Swiss National Science Foundation and the University of Geneva.

REFERENCES

1. Orlean, P., and Menon, A.K. (2007). Thematic review series: lipid posttranslational modifications. GPI anchoring of protein in yeast and mammalian cells, or: how we learned to stop worrying and love glycophospholipids. *J Lipid Res* 48:993–1011.

2. Pittet, M., and Conzelmann, A. (2007). Biosynthesis and function of GPI proteins in the yeast *Saccharomyces cerevisiae*. *Biochim Biophys Acta* 1771:405–420.
3. Sipos, G., Reggiori, F., Vionnet, C., and Conzelmann, A. (1997). Alternative lipid remodelling pathways for glycosylphosphatidylinositol membrane anchors in *Saccharomyces cerevisiae*. *EMBO J* 16:3494–3505.
4. Reggiori, F., Canivenc-Gansel, E., and Conzelmann, A. (1997). Lipid remodeling leads to the introduction and exchange of defined ceramides on GPI proteins in the ER and Golgi of *Saccharomyces cerevisiae*. *EMBO J* 16:3506–3518.
5. Morita, Y.S., Paul, K.S., and Englund, P.T. (2000). Specialized fatty acid synthesis in African trypanosomes: myristate for GPI anchors. *Science* 288:140–143.
6. Fankhauser, C., Homans, S.W., Thomas-Oates, J.E., McConville, M.J., Desponds, C., Conzelmann, A., and Ferguson, M.A. (1993). Structures of glycosylphosphatidylinositol membrane anchors from *Saccharomyces cerevisiae*. *J Biol Chem* 268:26365–26374.
7. Maeda, Y., Tashima, Y., Houjou, T., Fujita, M., Yoko-o, T., Jigami, Y., Taguchi, R., and Kinoshita, T. (2007). Fatty acid remodeling of GPI-anchored proteins is required for their raft association. *Mol Biol Cell* 18:1497–1506.
8. Fujita, M., Umemura, M., Yoko-o, T., and Jigami, Y. (2006). *PER1* is required for GPI-phospholipase A2 activity and involved in lipid remodeling of GPI-anchored proteins. *Mol Biol Cell* 17:5253–5264.
9. Benting, J., Rietveld, A., Ansorge, I., and Simons, K. (1999). Acyl and alkyl chain length of GPI-anchors is critical for raft association *in vitro*. *FEBS Lett* 462:47–50.
10. McDowell, M.A., Ransom, D.M., and Bangs, J.D. (1998). Glycosylphosphatidylinositol-dependent secretory transport in *Trypanosoma brucei*. *Biochem J* 335(Pt 3):681–689.
11. Field, M.C., Moran, P., Li, W., Keller, G.A., and Caras, I.W. (1994). Retention and degradation of proteins containing an uncleaved glycosylphosphatidylinositol signal. *J Biol Chem* 269:10830–10837.
12. Delahunty, M.D., Stafford, F.J., Yuan, L.C., Shaz, D., and Bonifacino, J.S. (1993). Uncleaved signals for glycosylphosphatidylinositol anchoring cause retention of precursor proteins in the endoplasmic reticulum. *J Biol Chem* 268:12017–12027.
13. Doering, T.L., and Schekman, R. (1996). GPI anchor attachment is required for Gas1p transport from the endoplasmic reticulum in COP II vesicles. *EMBO J* 15:182–191.
14. Watanabe, R., Castillon, G.A., Meury, A., and Riezman, H. (2008). The presence of an ER exit signal determines the protein sorting upon ER exit in yeast. *Biochem J* 414:237–245.
15. Triggs, V.P., and Bangs, J.D. (2003). Glycosylphosphatidylinositol-dependent protein trafficking in bloodstream stage *Trypanosoma brucei*. *Eukaryot Cell* 2:76–83.
16. Ashok, A., and Hegde, R.S. (2008). Retrotranslocation of prion proteins from the endoplasmic reticulum by preventing GPI signal transamidation. *Mol Biol Cell* 19:3463–3476.
17. Bosson, R., Jaquenoud, M., and Conzelmann, A. (2006). *GUP1* of *Saccharomyces cerevisiae* encodes an *O*-acyltransferase involved in remodeling of the GPI anchor. *Mol Biol Cell* 17:2636–2645.
18. Fujita, M., Yoko, O.T., and Jigami, Y. (2006). Inositol deacylation by Bst1p is required for the quality control of glycosylphosphatidylinositol-anchored proteins. *Mol Biol Cell* 17:834–850.
19. Ghugtyal, V., Vionnet, C., Roubaty, C., and Conzelmann, A. (2007). *CWH43* is required for the introduction of ceramides into GPI anchors in *Saccharomyces cerevisiae*. *Mol Microbiol* 65:1493–1502.
20. Umemura, M., Fujita, M., Yoko, O.T., Fukamizu, A., and Jigami, Y. (2007). *Saccharomyces cerevisiae CWH43* is involved in the remodeling of the lipid moiety of GPI anchors to ceramides. *Mol Biol Cell* 18:4304–4316.

21. Tanaka, S., Maeda, Y., Tashima, Y., and Kinoshita, T. (2004). Inositol deacylation of glycosylphosphatidylinositol-anchored proteins is mediated by mammalian *PGAP1* and yeast Bst1p. *J Biol Chem* 279:14256–14263.
22. Futai, E., Hamamoto, S., Orci, L., and Schekman, R. (2004). GTP/GDP exchange by Sec12p enables COPII vesicle bud formation on synthetic liposomes. *EMBO J* 23:4146–4155.
23. Gurkan, C., Stagg, S.M., Lapointe, P., and Balch, W.E. (2006). The COPII cage: unifying principles of vesicle coat assembly. *Nat Rev Mol Cell Biol* 7:727–738.
24. Kuehn, M.J., Herrmann, J.M., and Schekman, R. (1998). COPII-cargo interactions direct protein sorting into ER-derived transport vesicles. *Nature* 391:187–190.
25. Lederkremer, G.Z., Cheng, Y., Petre, B.M., Vogan, E., Springer, S., Schekman, R., Walz, T., and Kirchhausen, T. (2001). Structure of the Sec23p/24p and Sec13p/31p complexes of COPII. *Proc Natl Acad Sci USA* 98:10704–10709.
26. Lee, M.C., Miller, E.A., Goldberg, J., Orci, L., and Schekman, R. (2004). Bi-directional protein transport between the ER and Golgi. *Annu Rev Cell Dev Biol* 20:87–123.
27. Stagg, S.M., Gurkan, C., Fowler, D.M., LaPointe, P., Foss, T.R., Potter, C.S., Carragher, B., and Balch, W.E. (2006). Structure of the Sec13/31 COPII coat cage. *Nature* 439:234–238.
28. Muniz, M., Morsomme, P., and Riezman, H. (2001). Protein sorting upon exit from the endoplasmic reticulum. *Cell* 104:313–320.
29. Morsomme, P., and Riezman, H. (2002). The Rab GTPase Ypt1p and tethering factors couple protein sorting at the ER to vesicle targeting to the Golgi apparatus. *Dev Cell* 2:307–317.
30. Miller, E., Antonny, B., Hamamoto, S., and Schekman, R. (2002). Cargo selection into COPII vesicles is driven by the Sec24p subunit. *EMBO J* 21:6105–6113.
31. Stephens, D.J., and Pepperkok, R. (2004). Differential effects of a GTP-restricted mutant of Sar1p on segregation of cargo during export from the endoplasmic reticulum. *J Cell Sci* 117:3635–3644.
32. Miller, E.A., Beilharz, T.H., Malkus, P.N., Lee, M.C., Hamamoto, S., Orci, L., and Schekman, R. (2003). Multiple cargo binding sites on the COPII subunit Sec24p ensure capture of diverse membrane proteins into transport vesicles. *Cell* 114:497–509.
33. Peng, R., De Antoni, A., and Gallwitz, D. (2000). Evidence for overlapping and distinct functions in protein transport of coat protein Sec24p family members. *J Biol Chem* 275:11521–11528.
34. Muniz, M., Nuoffer, C., Hauri, H.P., and Riezman, H. (2000). The Emp24 complex recruits a specific cargo molecule into endoplasmic reticulum-derived vesicles. *J Cell Biol* 148:925–930.
35. Takida, S., Maeda, Y., and Kinoshita, T. (2008). Mammalian GPI-anchored proteins require p24 proteins for their efficient transport from the ER to the plasma membrane. *Biochem J* 409:555–562.
36. Aguilera-Romero, A., Kaminska, J., Spang, A., Riezman, H., and Muniz, M. (2008). The yeast p24 complex is required for the formation of COPI retrograde transport vesicles from the Golgi apparatus. *J Cell Biol* 180:713–720.
37. Sutterlin, C., Doering, T.L., Schimmoller, F., Schroder, S., and Riezman, H. (1997). Specific requirements for the ER to Golgi transport of GPI-anchored proteins in yeast. *J Cell Sci* 110(Pt 21):2703–2714.
38. Belden, W.J., and Barlowe, C. (2001). Distinct roles for the cytoplasmic tail sequences of Emp24p and Erv25p in transport between the endoplasmic reticulum and Golgi complex. *J Biol Chem* 276:43040–43048.

39. Elrod-Erickson, M.J., and Kaiser, C.A. (1996). Genes that control the fidelity of endoplasmic reticulum to Golgi transport identified as suppressors of vesicle budding mutations. *Mol Biol Cell* 7:1043–1058.
40. Belden, W.J., and Barlowe, C. (2001). Role of Erv29p in collecting soluble secretory proteins into ER-derived transport vesicles. *Science* 294:1528–1531.
41. Morsomme, P., Prescianotto-Baschong, C., and Riezman, H. (2003). The ER v-SNAREs are required for GPI-anchored protein sorting from other secretory proteins upon exit from the ER. *J Cell Biol* 162:403–412.
42. Xu, D., and Hay, J.C. (2004). Reconstitution of COPII vesicle fusion to generate a pre-Golgi intermediate compartment. *J Cell Biol* 167:997–1003.
43. Schubert, U., Anton, L.C., Gibbs, J., Norbury, C.C., Yewdell, J.W., and Bennink, J.R. (2000). Rapid degradation of a large fraction of newly synthesized proteins by proteasomes. *Nature* 404:770–774.
44. Bucciantini, M., Giannoni, E., Chiti, F., Baroni, F., Formigli, L., Zurdo, J., Taddei, N., Ramponi, G., Dobson, C.M., and Stefani, M. (2002). Inherent toxicity of aggregates implies a common mechanism for protein misfolding diseases. *Nature* 416:507–511.
45. Sayeed, A., and Ng, D.T. (2005). Search and destroy: ER quality control and ER-associated protein degradation. *Crit Rev Biochem Mol Biol* 40:75–91.
46. Travers, K.J., Patil, C.K., Wodicka, L., Lockhart, D.J., Weissman, J.S., and Walter, P. (2000). Functional and genomic analyses reveal an essential coordination between the unfolded protein response and ER-associated degradation. *Cell* 101:249–258.
47. Cox, J.S., Shamu, C.E., and Walter, P. (1993). Transcriptional induction of genes encoding endoplasmic reticulum resident proteins requires a transmembrane protein kinase. *Cell* 73:1197–1206.
48. Mori, K., Ma, W., Gething, M.J., and Sambrook, J. (1993). A transmembrane protein with a cdc2+/CDC28-related kinase activity is required for signaling from the ER to the nucleus. *Cell* 74:743–756.
49. Shamu, C.E., and Walter, P. (1996). Oligomerization and phosphorylation of the Ire1p kinase during intracellular signaling from the endoplasmic reticulum to the nucleus. *EMBO J* 15:3028–3039.
50. Cox, J.S., and Walter, P. (1996). A novel mechanism for regulating activity of a transcription factor that controls the unfolded protein response. *Cell* 87:391–404.
51. Hetz, C., Bernasconi, P., Fisher, J., Lee, A.H., Bassik, M.C., Antonsson, B., Brandt, G.S., Iwakoshi, N.N., Schinzel, A., Glimcher, L.H., and Korsmeyer, S.J. (2006). Proapoptotic BAX and BAK modulate the unfolded protein response by a direct interaction with IRE1alpha. *Science* 312:572–576.
52. Urano, F., Wang, X., Bertolotti, A., Zhang, Y., Chung, P., Harding, H.P., and Ron, D. (2000). Coupling of stress in the ER to activation of JNK protein kinases by transmembrane protein kinase IRE1. *Science* 287:664–666.
53. Yoneda, T., Imaizumi, K., Oono, K., Yui, D., Gomi, F., Katayama, T., and Tohyama, M. (2001). Activation of caspase-12, an endoplastic reticulum (ER) resident caspase, through tumor necrosis factor receptor-associated factor 2-dependent mechanism in response to the ER stress. *J Biol Chem* 276:13935–13940.
54. Ng, D.T., Spear, E.D., and Walter, P. (2000). The unfolded protein response regulates multiple aspects of secretory and membrane protein biogenesis and endoplasmic reticulum quality control. *J Cell Biol* 150:77–88.
55. Okamoto, M., Yoko-o, T., Umemura, M., Nakayama, K., and Jigami, Y. (2006). Glycosyl-phosphatidylinositol-anchored proteins are required for the transport of detergent-resistant microdomain-associated membrane proteins Tat2p and Fur4p. *J Biol Chem* 281:4013–4023.
56. Carvalho, P., Goder, V., and Rapoport, T.A. (2006). Distinct ubiquitin–ligase complexes define convergent pathways for the degradation of ER proteins. *Cell* 126:361–373.

57. Bueler, H., Aguzzi, A., Sailer, A., Greiner, R.A., Autenried, P., Aguet, M., and Weissmann, C. (1993). Mice devoid of PrP are resistant to scrapie. *Cell* 73:1339–1347.
58. Sailer, A., Bueler, H., Fischer, M., Aguzzi, A., and Weissmann, C. (1994). No propagation of prions in mice devoid of PrP. *Cell* 77:967–968.
59. Mallucci, G., Dickinson, A., Linehan, J., Klohn, P.C., Brandner, S., and Collinge, J. (2003). Depleting neuronal PrP in prion infection prevents disease and reverses spongiosis. *Science* 302:871–874.
60. Safar, J.G., DeArmond, S.J., Kociuba, K., Deering, C., Didorenko, S., Bouzamondo-Bernstein, E., Prusiner, S.B., and Tremblay, P. (2005). Prion clearance in bigenic mice. *J Gen Virol* 86:2913–2923.
61. Mallucci, G., and Collinge, J. (2005). Rational targeting for prion therapeutics. *Nat Rev Neurosci* 6:23–34.
62. White, M.D., Farmer, M., Mirabile, I., Brandner, S., Collinge, J., and Mallucci, G.R. (2008). Single treatment with RNAi against prion protein rescues early neuronal dysfunction and prolongs survival in mice with prion disease. *Proc Natl Acad Sci USA* 105:10238–10243.
63. Deriziotis, P., and Tabrizi, S.J. (2008). Prions and the proteasome. *Biochim Biophys Acta* 1782:713–722.
64. Funato, K., and Riezman, H. (2001). Vesicular and nonvesicular transport of ceramide from ER to the Golgi apparatus in yeast. *J Cell Biol* 155:949–959.
65. Kajiwara, K., Watanabe, R., Pichler, H., Ihara, K., Murakami, S., Riezman, H., and Funato, K. (2008). Yeast *ARV1* is required for efficient delivery of an early GPI intermediate to the first mannosyltransferase during GPI assembly and controls lipid flow from the endoplasmic reticulum. *Mol Biol Cell* 19:2069–2082.
66. Watanabe, R., Funato, K., Venkataraman, K., Futerman, A.H., and Riezman, H. (2002). Sphingolipids are required for the stable membrane association of glycosylphosphatidylinositol-anchored proteins in yeast. *J Biol Chem* 277: 49538–49544.
67. Baumann, N.A., Sullivan, D.P., Ohvo-Rekila, H., Simonot, C., Pottekat, A., Klaassen, Z., Beh, C.T., and Menon, A.K. (2005). Transport of newly synthesized sterol to the sterol-enriched plasma membrane occurs via nonvesicular equilibration. *Biochemistry* 44:5816–5826.
68. Mayor, S., and Riezman, H. (2004). Sorting GPI-anchored proteins. *Nat Rev Mol Cell Biol* 5:110–120.
69. Brown, D.A., and Rose, J.K. (1992). Sorting of GPI-anchored proteins to glycolipid-enriched membrane subdomains during transport to the apical cell surface. *Cell* 68:533–544.
70. Simons, K., and Ikonen, E. (1997). Functional rafts in cell membranes. *Nature* 387:569–572.
71. Horvath, A., Sutterlin, C., Manning-Krieg, U., Movva, N.R., and Riezman, H. (1994). Ceramide synthesis enhances transport of GPI-anchored proteins to the Golgi apparatus in yeast. *EMBO J* 13:3687–3695.
72. Barz, W.P., and Walter, P. (1999). Two endoplasmic reticulum (ER) membrane proteins that facilitate ER-to-Golgi transport of glycosylphosphatidylinositol-anchored proteins. *Mol Biol Cell* 10:1043–1059.
73. Yasuda, S., Kitagawa, H., Ueno, M., Ishitani, H., Fukasawa, M., Nishijima, M., Kobayashi, S., and Hanada, K. (2001). A novel inhibitor of ceramide trafficking from the endoplasmic reticulum to the site of sphingomyelin synthesis. *J Biol Chem* 276:43994–44002.
74. Heese-Peck, A., Pichler, H., Zanolari, B., Watanabe, R., Daum, G., and Riezman, H. (2002). Multiple functions of sterols in yeast endocytosis. *Mol Biol Cell* 13:2664–2680.
75. Bagnat, M., and Simons, K. (2002). Lipid rafts in protein sorting and cell polarity in budding yeast *Saccharomyces cerevisiae*. *Biol Chem* 383:1475–1480.

76. Bagnat, M., and Simons, K. (2002). Cell surface polarization during yeast mating. *Proc Natl Acad Sci USA* 99:14183–14188.
77. Valdez-Taubas, J., and Pelham, H.R. (2003). Slow diffusion of proteins in the yeast plasma membrane allows polarity to be maintained by endocytic cycling. *Curr Biol* 13:1636–1640.
78. Sievi, E., Suntio, T., and Makarow, M. (2001). Proteolytic function of GPI-anchored plasma membrane protease Yps1p in the yeast vacuole and Golgi. *Traffic* 2:896–907.
79. Brown, D.A., Crise, B., and Rose, J.K. (1989). Mechanism of membrane anchoring affects polarized expression of two proteins in MDCK cells. *Science* 245:1499–1501.
80. Lisanti, M.P., Caras, I.W., Davitz, M.A., and Rodriguez-Boulan, E. (1989). A glycophospholipid membrane anchor acts as an apical targeting signal in polarized epithelial cells. *J Cell Biol* 109:2145–2156.
81. Lisanti, M.P., Le Bivic, A., Saltiel, A.R., and Rodriguez-Boulan, E. (1990). Preferred apical distribution of glycosyl-phosphatidylinositol (GPI) anchored proteins: a highly conserved feature of the polarized epithelial cell phenotype. *J Membr Biol* 113:155–167.
82. Benting, J.H., Rietveld, A.G., and Simons, K. (1999). N-Glycans mediate the apical sorting of a GPI-anchored, raft-associated protein in Madin-Darby canine kidney cells. *J Cell Biol* 146:313–320.
83. Paladino, S., Sarnataro, D., Pillich, R., Tivodar, S., Nitsch, L., and Zurzolo, C. (2004). Protein oligomerization modulates raft partitioning and apical sorting of GPI-anchored proteins. *J Cell Biol* 167:699–709.
84. Lipardi, C., Nitsch, L., and Zurzolo, C. (2000). Detergent-insoluble GPI-anchored proteins are apically sorted in fischer rat thyroid cells, but interference with cholesterol or sphingolipids differentially affects detergent insolubility and apical sorting. *Mol Biol Cell* 11:531–542.
85. Fukasawa, M., Nishijima, M., and Hanada, K. (1999). Genetic evidence for ATP-dependent endoplasmic reticulum-to-Golgi apparatus trafficking of ceramide for sphingomyelin synthesis in Chinese hamster ovary cells. *J Cell Biol* 144:673–685.
86. Ridgway, N.D. (1995). 25-Hydroxycholesterol stimulates sphingomyelin synthesis in Chinese hamster ovary cells. *J Lipid Res* 36:1345–1358.
87. Lagace, T.A., Byers, D.M., Cook, H.W., and Ridgway, N.D. (1999). Chinese hamster ovary cells overexpressing the oxysterol binding protein (OSBP) display enhanced synthesis of sphingomyelin in response to 25-hydroxycholesterol. *J Lipid Res* 40:109–116.
88. Storey, M.K., Byers, D.M., Cook, H.W., and Ridgway, N.D. (1998). Cholesterol regulates oxysterol binding protein (OSBP) phosphorylation and Golgi localization in Chinese hamster ovary cells: correlation with stimulation of sphingomyelin synthesis by 25-hydroxycholesterol. *Biochem J* 336(Pt 1):247–256.
89. Kumagai, K., Yasuda, S., Okemoto, K., Nishijima, M., Kobayashi, S., and Hanada, K. (2005). CERT mediates intermembrane transfer of various molecular species of ceramides. *J Biol Chem* 280:6488–6495.
90. Lange, Y., and Matthies, H.J. (1984). Transfer of cholesterol from its site of synthesis to the plasma membrane. *J Biol Chem* 259:14624–14630.
91. Henneberry, A.L., and Sturley, S.L. (2005). Sterol homeostasis in the budding yeast, *Saccharomyces cerevisiae*. *Semin Cell Dev Biol* 16:155–161.
92. Schulz, T.A., and Prinz, W.A. (2007). Sterol transport in yeast and the oxysterol binding protein homologue (OSH) family. *Biochim Biophys Acta* 1771:769–780.
93. Castillon, G.A., Watanabe, R., Taylor, M., Schwabe, T.M.E., and Riezman, H. (2008). Concentration of GPI-anchored proteins upon ER exit in yeast. *Traffic* 10:186–200.
94. Jonikas, M.C., Collins, S.R., Denic, V., Oh, E., Quan, E.M., Schmid, V., Weibezahn, J., Schwappach, B., Walter, P., Weissman, J.S., and Schuldiner, M. (2009). Comprehensive characterization of genes required for protein folding in the endoplasmic reticulum. *Science* 323:1693–1697.

14

Mechanisms of Polarized Sorting of GPI-anchored Proteins in Epithelial Cells

SIMONA PALADINO[a] • CHIARA ZURZOLO[a,b]

[a]*Dipartimento di Biologia e Patologia Cellulare e Molecolare*
Università degli Studi di Napoli Federico II
Italy

[b]*Unité de Trafic Membranaire et Pathogénèse*
Institut Pasteur
25 rue du Docteur Roux
75724 Paris, France

I. Abstract

The asymmetric distribution of proteins and lipids in two distinct domains of the plasma membrane (PM) of polarized epithelial cells is accomplished by continuous sorting of newly synthesized components and their regulated internalization. Here we discuss the mechanisms of apical sorting of glycosylphosphatidilinositol-anchored proteins (GPI-APs) especially focusing on their biosynthetic pathway. Recent evidences indicate that this event depends on both association to specialized lipid domains (rafts) and clustering in high-molecular-weight complexes at the level of the Golgi apparatus. We discuss the evidences present in literature pointing toward the involvement of the protein ectodomain, the GPI anchor, the lipid environment, and the presence of a putative receptor in this process. We also examine the role of the cytoskeleton and other molecular components of the exocytic pathway in post-TGN transport of GPI-APs to the PM.

THE ENZYMES, Vol. XXVI   289   ISSN NO: 1874-6047
DOI: 10.1016/S1874-6047(09)26014-8

II.   Introduction

Most cell types of eukaryotic organisms are polarized. This is particularly evident in epithelial cells, the fundamental polarized cell type in Metazoa, which line both the outside of the body (skin) and the inside cavities of internal organs such as lung, kidney, gastrointestinal tract, and vasculature. Polarization of epithelial cells is a complex process directed by external stimuli (such as cell–cell and cell–extracellular matrix adhesion) devoted to the establishment of a unique cytoarchitecture, which includes a specialized organization of organelles, plasma membrane (PM), and cytoskeleton, and to the acquisition of specialized trafficking pathways [1–4]. This process results in the formation of a selective epithelial barrier with unique vectorial functions (secretion, reabsorption, hormone response, etc.) necessary for the function of the whole organism. The PM of polarized epithelial cells is divided by tight junctions into morphologically and functionally distinct domains (apical and basolateral), which display different protein and lipid compositions required for specialized functions [2, 3, 5]. Once polarity is established, this asymmetric distribution is achieved by continuous sorting of newly synthesized components and their regulated internalization [6–10]. These processes are strictly regulated and, consequently, alterations in trafficking pathways that are used to maintain epithelial polarity can result in different pathologies [11]. Understanding how cells establish and maintain the polarized phenotype and identifying the molecular machinery required for polarization still remains a major challenge in cell biology and it appears to be essential for understanding how these processes are altered in cancer and/or other pathologies [11, 12].

Here, we discuss the mechanisms of polarized sorting in epithelial cells of glycosylphosphatidilinositol-anchored proteins (GPI-APs), a class of membrane proteins, especially focusing on their biosynthetic pathways. GPI-APs are attached to the external leaflet of the PM by a GPI anchor and play a wide variety of physiological roles, including signaling, uptake of ions and other molecules, enzymatic catalysis, cell surface protection, and cell adhesion and migration [13].

III.   Secretory Pathway and Polarized Sorting

To establish and maintain cell polarity, epithelial cells have to ensure proper delivery of apical and basolateral membrane components to the right membrane domains. Along the secretory pathway, proteins are delivered to their correct destinations via the sequential action of several sorting signals and multiple sorting events (see Ref. [5] for review). Proteins contain specific sorting signals, which can be recognized by cellular machineries able to decode the information and can be used, at each intermediate step of their journey, for forward transport, retention in a specific compartment, or retrieval to the compartment of origin.

Sorting signals have been found along the entire length of transmembrane proteins and can be recognized by either luminal or cytosolic receptors. Instead, GPI-APs cannot interact directly with cytosolic components and could contain signals only within the GPI anchor (e.g., in the lipid and/or glycan portion) and/or in the protein moiety [14]. However, GPI-AP trafficking appears to be regulated by cytosolic proteins (as small GTPases of the Ras superfamily) [10, 15, 16]. This could be achieved through membrane-spanning proteins acting as a link between GPI-AP and cytosolic components.

A. ENDOPLASMIC RETICULUM

Attachment of the GPI anchor, which occurs in the endoplasmic reticulum (ER), together with the acquisition of the proper structure (e.g., deacylation of inositol) is required for efficient ER exit of GPI-APs, which are sorted from their misfolded counterparts or from ER residents proteins [17–20]. In yeast, GPI-APs are sorted from other secretory proteins at the level of the ER (see also Chapter 13 in this book). Indeed, GPI-APs and other PM proteins exit the ER in distinct vesicles [21], and this process is mediated by the small GTPase Ypt1p along with specific tethering factors and SNARE proteins [22, 23]. Consistently in mammalian cells, the traffic from the ER to the Golgi of GPI-APs is specifically slowed down by depletion of p23 (which belongs to the p24 family proteins first identified as constituents of the COPI and COPII vesicles) [24, 25]. Furthermore, a GTP-restricted mutant of Sar1, which acts at the ER level, also affects ER exits of GPI-APs [26]. However, it appears that both soluble GPI-anchored and non-GPI-linked proteins are transported from the ER to the Golgi in the same vesicular–tubular clusters [26, 27]. Therefore, it is not clear whether and how GPI-APs are sorted from other secretory proteins at the ER level in mammalian cells (see Chapter 13 in this book).

B. GOLGI APPARATUS AND SORTING TO THE SURFACE

After leaving the ER, all secretory proteins are transported through the Golgi complex, where they are subjected to posttranslational modifications before being sorted to their final destination (PM or endolysosomal compartments) at the level of the trans-Golgi network (TGN). Additionally in polarized epithelial cells, apical and basolateral proteins are also segregated in the TGN into distinct vesicles upon recognition of specific apical or basolateral sorting signals [9, 16] and are separately delivered to the apical or basolateral surface [5, 28–30]. Basolateral sorting is mediated by discrete domains in the cytosolic protein tail, frequently containing tyrosine or dileucine motifs [8, 31, 32], which are recognized by clathrin adaptor complexes [33, 34].

Apical sorting signals are more variable and less well-characterized. Lumen-localized signals such as N- and O-linked glycosylation, transmembrane domains, and certain determinants in the cytoplasmic tail have all been shown to be important for apical sorting [35–39]. In addition, the sorting of many apical proteins appears to be dependent on their inclusion in sphingolipid- and cholesterol-rich microdomains (rafts), which can be extracted as detergent-resistant membranes (DRMs), as they are insoluble in Triton X-100 and other nonionic detergents [40]. Because of their capacity to segregate specific classes of lipids and proteins [41, 42] and their enrichment in apical membranes [43], lipid rafts (Box 14.1) have been postulated to act as apical sorting platforms [40]. This mechanism of sorting is particularly suitable for GPI-APs, which are apically sorted in several epithelial cell lines [44, 45] and associate with DRMs [46–49] during their passage through the Golgi apparatus [46, 50–52]. Moreover, DRM perturbation by cholesterol and/or sphingolipid depletion results in impaired trafficking to the PM or altered polarity [50, 53, 54]. These combined evidences lead to the proposal that the GPI anchor acts as an apical sorting determinant by mediating raft association [40, 41]. However, this hypothesis has been challenged by the fact that in FRT cells the majority of endogenous and some transfected GPI-APs are basolaterally sorted [50, 55] and that some DRM-associated GPI-APs are basolaterally delivered in MDCK cells [51, 52, 55–57]. While association to DRMs appears to be a property of both apical and basolateral GPI-APs, only apical GPI-APs are able to form high molecular weight (HMW) complexes during their transport to the PM [51]. Because impairment of this GPI-AP "oligomerization" capacity impairs their apical sorting, it has been proposed that this is an essential step for their sorting [51]. Interestingly, this mechanism appears to be functional in epithelial cells of different origin (FRT and Caco-2 other than MDCK cells) indicating that oligomerization in the Golgi apparatus might be a general requirement for apical sorting of GPI-APs in epithelial cells [52].

Formation of HMW complexes of apical GPI-APs occurs during their transport to the PM and is concomitant with raft association [51, 52]. This suggests that oligomerization and raft association cooperate in promoting apical sorting of GPI-APs. A multistep model for the sorting of GPI-APs has been proposed (Figure 14.1). Both apical and basolateral GPI-APs partition into rafts because of the natural chemical affinity of the GPI

BOX 14.1

Biological membranes are composed of a variety of lipids and sterols that do not mix heterogeneously, but form compositionally and functionally distinct microdomains. Rafts represent a class of such domain enriched in sphingolipids and cholesterol [168, 169]. Rafts have been proposed to generate the following weak and transient lipid interactions between the sugar head groups and the acyl chains of sphingolipid. Long-chain, saturated sphingolipids are more tightly packed and have decreased rotational mobility in the plane of the bilayer than neighboring unsaturated glycerophospholipids. Hence in model membranes at physiologic temperatures, these differences give rise to phase separation where sphingolipids display gel-like properties (lo phase), while unsaturated lipids are more fluid (ld phase) [170]. Cholesterol, which is intercalated between the hydrocarbon chains of sphingolipids, decreases their flexibility and enhances their packing. Although phase separation can occur in binary lipid mixtures in the absence of cholesterol [171, 172] and despite the existence of cholesterol-independent rafts domains in the brush-border membrane of enterocytes [173], cholesterol content has a major role in determining the properties (as shape and size) of the sphingolipid-rich rafts [172, 174, 175]. Similarly, cholesterol levels can influence the degree of partitioning of raft proteins into these lipid domains [95, 176, 177]. Lipid-anchored proteins such as GPI-APs, palmitoylated and myristoylated proteins such as flotillins, caveolins, heterotrimeric G proteins, transmembrane proteins bearing a transmembrane stretch, particularly hydrophobic (such as neuraminidase, influenza hemagglutinin), preferentially partition in lipid raft at the steady state [40, 178]. Moreover, some membrane receptors (such as EGFR and uPAR) can enter into rafts after binding to their ligand [179, 180] and their association with rafts can modulate their signaling [181, 182].

Although rafts are difficult to study, modern biophysical approaches (SPT, FRET, FRAP and FCS studies) have provided insights into the size, shape, and dynamics of these domains and their molecular components, all suggesting the existence of lipid rafts in living cells [104, 183–188]. However, there is no common view about the characteristics of these membrane domains, about the size, the mechanism of formation, and their heterogeneity [189–191]. An important biochemical property of rafts is their insolubility in cold nonionic detergents like Triton X-100; hence, rafts can be extracted as DRMs. Proteins associated with rafts also are insoluble in nonionic detergents and have been found to be enriched

(*Continues*)

BOX 14.1   (*Continued*)

in DRM fractions. Because detergents can lead to the aggregation of raft domains, it is clear that DRMs do not reflect the native organization of lipid rafts. However, the detergent extraction remains the only biochemical tool to assess the potential affinity of the membrane components to rafts [182, 192].

For their ability to recruit and/or segregate specific molecules, lipid rafts are proposed to function in membrane trafficking, signaling events, and cell polarization [178, 193].

1. Raft partition

2. Stabilization into rafts

3. Raft coalescence

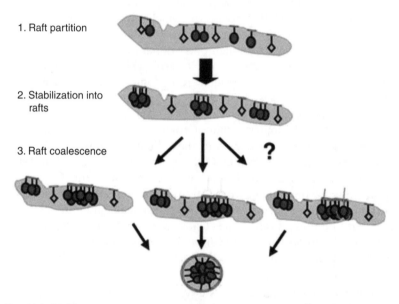

Fig. 14.1. Multistep model for apical sorting of GPI-APs in polarized epithelial cells. Both apical and basolateral GPI-APs partition with rafts due to chemical affinity of GPI-APs for this lipid environment (*1. Raft partition*). In order to be included in a raft-enriched apical vesicle, GPI-APs need to be stabilized into rafts (*2. Stabilization into rafts*). This step is mediated by protein oligomerization, which increases their raft affinity. The clustering of individual small rafts into a larger functional raft should reinforce the segregation among different lipid phases and, consequently, of the components associated with them promoting the selective packaging of GPI-APs in the apical vesicle (*3. Raft coalescence*). Raft coalescence may simply be a consequence of protein oligomerization (left panel); alternatively, a putative apical receptor could be involved in this step. This receptor could interact directly with GPI-APs recognizing specific signals or the three-dimensional structure of the protein (right panel, green receptor), or could bind specific raft lipids, thus, promoting their coalescence (middle panel, yellow receptor). In both these cases, raft clustering should drive the formation of apical vesicles.

anchor for these lipid microdomains [42, 50, 58]. This partition is believed to be very dynamic; therefore in order to be included in a raft-enriched apical vesicle, GPI-APs need to be stabilized into rafts. Oligomerization might promote this process by increasing the affinity of GPI-APs for rafts as previously suggested in the case of sorting in early endosomes of GPI-APs [59]. GPI-AP oligomerization could also lead to the coalescence of small rafts into a larger functional raft from which the apical vesicle could bud (Figure 14.1). Whether it is oligomerization per se or the interaction of GPI-AP oligomers with a putative receptor to drive the coalescence of more rafts into an apical vesicle is still unknown [51, 52]. Clustering of individual small rafts should reinforce the segregation of different lipid phases and, consequently, of membrane components associated with them, and it is has been found to be implicated in different membrane functions such as signaling, protein internalization, pathogen entry, and virus budding [60–64]. Moreover, raft clustering could provide a mechanism for the formation of the apical vesicles by increasing the local line energy at the boundary between raft and nonraft phases, thus enabling the membrane curvature necessary to drive vesicle formation [65–69]. Thus, the mechanism of oligomerization of GPI-APs could have a double role: first it would enable GPI-APs to be segregated by the rest of the proteins in the Golgi membranes by stabilizing the proteins in rafts and, second, it would drive the budding of an apical vesicle by following raft coalescence.

C. MECHANISM OF POLARIZED SORTING

Although oligomerization-driven apical sorting is an attractive hypothesis, the mechanism by which apical GPI-APs oligomerize is still unclear. Specifically, it is not known what oligomerization depends on, and what is the nature of the interactions that determines apical GPI-AP clustering prior to sorting and whether these interactions involve the ectodomain, the GPI anchor, or both [51, 52, 70]. In the section, we will discuss the evidences present in literature pointing toward the involvement of (i) the protein ectodomain, (ii) a putative receptor, and (iii) the GPI anchor and the lipid environment in this process.

1. Protein Ectodomain

It has been shown that GPI-AP oligomers, once formed, are insensitive to depletion of cholesterol and are resistant to conditions in which DRMs are disrupted (such as SDS extraction) [51]. This suggests that GPI-AP oligomers are maintained by protein–protein interactions. Although oligomerization of a chimeric model GPI-AP, GFP-GPI, is stabilized by disulfide bonds between the protein ectodomains [51], it is likely that in the case of native GPI-APs, noncovalent interactions between protein ectodomains are responsible for oligomer formation. Hence, posttranslational modifications of protein ectodomains could be involved in protein oligomerization. For example, glycosylation determinants within protein ectodomains could facilitate weak interactions and could therefore be suitable candidates. The addition of N-linked glycans to the GPI-anchored form of rat growth hormone confers its apical targeting [71]. N-glycosylation is also involved in the apical localization of the native GPI-anchored membrane dipeptidase [72]. In contrast to these data, it has been shown that N-glycans are not required for the sorting of the GPI-anchored form of endolyn [73]. While O-glycosylation is important for the apical delivery of several transmembrane proteins [74–76], there is no evidence for its role in the sorting of GPI-APs with the exception of CEA (carcinoembryonic antigen) shown to accumulate intracellularly after pharmacological inhibition of O-glycosylation [74]. Interestingly, recent evidences demonstrate that mutagenesis of either N- or O-glycosylation sites does not affect DRM association, oligomerization, or apical sorting of both a native and a chimeric GPI-AP (placental alkaline phosphatase, PLAP, and p75GPI) [77]. These data indicate that N- and O-glycosylation do not have a direct role in GPI-AP oligomerization and apical sorting. Thus, our hypothesis is that glycan residues on the protein ectodomain may be important to stabilize the correct conformation of the protein, which is required for TGN export. The fact that mutagenesis of the N-glycosylation sites did not interfere with PLAP oligomerization but that treatment with tunicamycin did [77], indicates that N-linked sugars of the protein ectodomains are not directly involved in its apical sorting. Nonetheless, an N-glycosylated raft-associated interactor could be involved. In support of this hypothesis, Catino et al. also showed that when cells are depleted of cholesterol and treated with tunicamycin, treatments that by themselves do not have any effect, PLAP is missorted to the basolateral surface. Moreover, the finding that in Con A-resistant MDCK cells, defective in high-mannose residues synthesis [78] newly synthesized GPI-APs, arrive at the surface not in clusters as in wild-type MDCK cells [79] is consistent with a general role of N-glycosylation in apical sorting. This supports the hypothesis that basolateral missorting is due to the impairment of the oligomerization machinery.

2. Putative Apical Receptor

To date, a variety of potential raft-associated proteins that might promote oligomerization and, consequently, rafts clustering have been described [64]. One of them, the VIP17/MAL protein, is present in lipid rafts and has been shown to be required for apical delivery of some, but not all, GPI-APs [80, 81]. However, MAL is also necessary for apical transport of transmembrane proteins independently of their association with rafts. Caveolins, flotillins, and stomatin, which are all raft associated and form oligomers, have also been hypothesized to act in promoting GPI-AP clustering [64]. However, caveolin1 knockdown does not affect sorting and kinetics of apical transport of raft-associated proteins (both transmembrane and GPI-linked) [82]. Similarly, transfection of caveolin1 in FRT cells did not restore the apical polarity of basolateral GPI-APs in this cell line [56]. On the other hand, stomatin and flotillins have been found to copatch and coprecipitate with some GPI-APs, and it has been proposed that they might function in concentrating GPI-anchored proteins into rafts [83, 84]. However, it is not clear how these proteins linked to the cytoplasmic leaflet of membranes by their myristoylated or palmitoylated tail could interact with GPI-APs, which associate with the ectoplasmic leaflet. Other possible candidates are proteins belonging to the annexin family (such as annexin II and annexin XIII), which have been found to be enriched in rafts and involved in the apical transport of raft transmembrane proteins [85–88]. Although their mechanistic roles have not been elucidated, it will be very interesting to explore their possible function in apical sorting of GPI-APs. Another candidate could be galectin 4, which interacts with glycosphingolipids, and in particular with the sulfatides carrying long-chain-hydroxylated fatty acids, which are specifically enriched in lipid rafts of HT-29 cells [89]. Interestingly, it has been shown that its depletion by RNAi affects apical trafficking of both transmembrane and GPI-anchored, raft-associated proteins [89]. However, because galectin-4 binds specifically to lipids and its knockdown induces the intracellular accumulation of apical membrane proteins, but not a considerable basolateral missorting [89], it is more likely that galectin-4 is involved in the transport process of all apical proteins, and not as specific sorting receptor for GPI-APs. Similarly, it has been found that the RNAi depletion of FAPP2 (four-phosphate-adaptor protein 2), a cytosolic protein that localizes to the TGN through binding the phospoinositide PtdIns(4)P via a pleckstrin homology domain [90], resulted in delayed delivery and/or intracellular accumulation of apical raft-associated proteins (either transmembrane and GPI-APs) in polarized MDCK cells, whereas the basolateral transport was unaffected [91]. These findings suggest that both galectin-4 and FAPP2 are involved in the transport of apical cargoes rather than to be implicated in their sorting. Because FAPP2 is required for glycosphingolipids synthesis [92], it could participate in the formation of different lipid environments in the Golgi apparatus. On the other hand, galectin-4 would be able to cross-link rafts and lead to the generation of transport carriers through binding to glycosphingolipids. Therefore, both these proteins might be components of a scaffold machinery involved in the TGN to PM transport (Figure 14.2).

Although these are interesting hypothesis, further studies are needed to understand whether a putative apical receptor is necessary to drive oligo-merization and apical sorting of GPI-APs and to allow its identification.

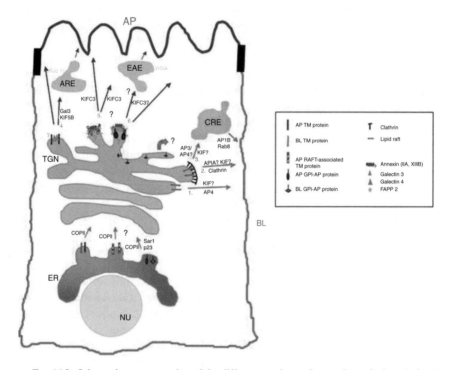

Fig. 14.2. Schematic representation of the different sorting pathways that exist in polarized epithelial cells. Transport of membrane proteins at the cell surface requires sorting at multiple intracellular levels. Newly synthesized membrane proteins might be transported from the ER to the Golgi apparatus in the same tubular vesicle carriers, although some factors appear to be specifically implicated in the ER exit of different secretory proteins (such as p23 and Sar1 for GPI-APs). At the TGN, basolateral signals interact with adaptors of the clathrin (AP1, AP3) or nonclathrin type (AP4) (pathways 1, 2, and 3). Some proteins are directly delivered to the basolateral surface (1, 2) and others move from TGN into recycling endosomes (3) before being delivered to the basolateral membrane. In the latter case, the transport from the common recycling endosomes (CRE) to the basolateral surface is mediated by AP1B (an epithelial-specific variant of AP1) and Rab8, while AP3 and AP4 could regulate the transport from the TGN to CRE. At least two mechanisms, a raft-independent and a raft-dependent, are impli-cated in the sorting of apical proteins (4 and 5/6). In the raft-independent pathway, one mechanism of sorting involves the recognition of glycans by the lectin galectin 3 (Gal3), which by clustering glycoproteins should promote their incorporation into the AP vesicle.

3. Rafts Environment and GPI Anchor

It has been shown that oligomerization occurs concomitantly with raft association and that cholesterol depletion impairs oligomerization in the Golgi apparatus [51]. This suggests that rafts may constitute a favorable environment for the formation of HMW complexes. It is therefore possible that besides the protein ectodomain and a putative interactor, the lipid anchor also has a role in favoring clustering of apical GPI-APs. Furthermore, several reports suggest that differences in the lipid anchor (such as the length of acyl and alkyl chains and the remodeling of glycan portion or the inositol ring) are critical for their raft association or could mediate a different affinity for lipid rafts [58, 93–97]. Interestingly, it has been recently shown that different GPI-attachment signals (derived from an apical or basolateral native GPI-AP, FR and PrP, respectively) affect the ability of the resulting GFP-fusion proteins to oligomerize and to be apically sorted. Indeed, while both GFP-FR and GFP-PrP are associated with DRMs, only GFP-FR was able to oligomerize and was apically sorted, while GFP-PrP did not oligomerize and was basolaterally sorted [70]. Consistent with these data using a FRAP approach, Lebreton and colleagues found that, at the level of the Golgi apparatus, GPI-APs having GFP as ectodomain and different GPI-attachment signals display a different apparent coefficient diffusion (D) [98]. In particular, basolateral GPI-AP (GFP-PrP) displays a higher D as compared to the apical one, GFP-FR. This is in agreement with the fact that GFP-PrP is monomeric, while GFP-FR forms HMW complexes. Interestingly upon cholesterol addition, the apparent diffusion coefficient of GFP-PrP decreased significantly [98], and the protein was then able to oligomerize and was redirected to the apical cell surface [70].

Nonraft transmembrane proteins can be delivered directly or indirectly to the apical surface by passing through Rab11-positive apical recycling endosomes and move along MTs by using specific motors like KIF5B. Some apical transmembrane proteins use a raft-dependent pathway to be sorted/delivered to the apical membranes, similar to GPI-APs. However, they seem to be transported in distinct carriers. Indeed, only GPI-APs need to oligomerize in order to be apically sorted (6). Stabilization into rafts might be promoted by the protein clustering itself (as in the case of GPI-APs) or by the luminal lectin galectin4 (Gal4), which binds glycosphingolipids or by another unknown receptor. FAPP2, regulating the synthesis of glycosphingolipids, could play a role in modulating the lipid environment of the rafts and, consequently, the sorting/delivery of raft-associated proteins. Moreover, proteins belonging to the annexin family (such as annexin II and XIIIb) appear to be part of the scaffold machinery implicated in the apical transport of raft-associated proteins. Thus far, early apical endosomes (EAE) containing WGA seem to be involved in the transport of raft-associated transmembrane proteins and some GPI-APs. Similar to nonraft TM proteins, MTs are involved in the delivery of raft TM proteins and motor KIFC3 is crucial for their transport. How basolateral GPI-APs are delivered to basolateral surface and by which mechanism remains unknown.

These findings could be explained in two possible ways. Either different GPI-attachment signals mediate a different affinity for the same lipid domain or they modulate the association of the resulting GPI-AP with different lipid rafts. This, in turn, influences the oligomerization state of the protein and therefore its sorting behavior. Consistent with these two hypotheses, cholesterol addition can act in two ways: (i) it may stabilize the interaction of basolateral GPI-APs with lipid rafts (e.g., either rigidifying the raft membranes or constituting an obstacle to free diffusion of proteins), thus allowing its oligomerization and apical sorting or (ii) it may change the characteristics of the surrounding lipid environment, thus, enabling the protein originally associated with the "basolateral raft" to oligomerize and to be delivered apically. All together, these data suggest that a specific lipid environment is required for oligomerization and, consequently, for apical sorting of GPI-APs.

However because it has been previously shown that a double cys mutation (S49/71) in the GFP ectodomain of GFP-FR impairs both oligomerization and apical sorting of this protein [51], these data indicate that the protein ectodomain is required to allow or stabilize oligomers occurring in rafts. Interestingly, differently from basolateral GFP-PrP, cholesterol addition does not affect the sorting of the S49/71 mutant, thus suggesting that a specific lipid environment and a "permissive" ectodomain able to oligomerize are necessary to determine the apical sorting of GPI-APs [70].

On the basis of all these findings, we have proposed that in order to be apically sorted, a GPI-AP should have two properties: (i) to partition in a favorable environment that allows oligomerization to occur and (ii) to have an intrinsic capability to form HMW complexes (a "permissive" ectodomain). These two properties are derived by the integrated action of the GPI anchor and the ectodomain of the protein. Analysis of the structural composition of the GPI anchors of differently sorted proteins and of the chemical–physical properties of lipid microdomains surrounding GPI-APs at the level of the Golgi apparatus are required to define the roles of both the GPI anchor and the membrane domains in GPI-AP sorting.

IV.   Site of Sorting and Routes to the Surface

A. THE TGN

Morphological and biochemical studies have shown that the segregation of apical and basolateral cargoes occurs into distinct vesicles upon exit from the TGN [28–30, 99]. The TGN appears to be the site of sorting also for GPI-APs. Indeed although GPI-APs progressively segregate from other cargo molecules through the Golgi stack, they have been shown to be completely segregated in the TGN, from where they exit into distinct vesicles [30, 100]. This is in line with the findings showing that oligomerization of GPI-APs begins in the medial Golgi [51]. In addition, it has been shown that GPI-AP carrying vesicles emerge from large Golgi domains with a spherical appearance, in contrast to the elongated Golgi extensions from which basolateral carriers appear to arise [100, 101]. On the other hand, it is unknown whether and where GPI-APs are segregated from other raft-associated proteins. Growing evidences show that apical transmembrane raft and nonraft-associated proteins are transported to the apical surface in distinct carriers, which display different morphologies. Their formation occurs from distinct subdomains of the Golgi and involve different factors [85, 99, 102, 103]. Thus, it appears that multiple pathways exist from the TGN to the apical surface. Consistently, it has been shown that in FRT cells, two GPI-APs (PLAP and the GPI-anchored form of p75NTR) are delivered to the apical surface independently from their transmembrane and secreted forms (PLAP-Sec, NTR-Sec, p75NTR), respectively, using raft-dependent and raft-independent mechanisms [50]. Furthermore, in contrast to the GPI-anchored form, neither the TM nor the secretory forms of p75NTR oligomerize during their transport to the apical membrane [52] confirming the existence of different mechanisms for apical sorting. Another open question is where apical and basolateral DRM-associated GPI-APs segregate. There are two major possibilities: in the Golgi apparatus or in the post-TGN carriers. By using a FRAP approach, Lebreton et al. have recently shown that differently sorted GPI-APs have a different apparent diffusion coefficient in the Golgi apparatus, while they behave similarly in the PM [98]. Because the apparent diffusion coefficient of a protein is directly related to the surrounding environment, this suggests that segregation between apical and basolateral GPI-APs occurs either in the Golgi apparatus or earlier. Furthermore, the authors' unpublished observations indicate that post-TGN carriers of apical GPI-APs contain threefold more molecules than basolateral ones (Paladino and Zurzolo, unpublished results). This is also consistent with the fact that the apical GPI-AP, but not the basolateral one, oligomerize in the Golgi apparatus [51, 52].

Another open question is whether different apical GPI-APs are in the same or distinct oligomeric complex and/or in the same carrier. Evidences suggest that GPI-APs are organized in distinct homoclusters at the apical surface of MDCK cells (Tivodar et al., unpublished data), thus indirectly suggesting that they might arrive at the PM in distinct vesicles. However, this is in contrast with the data of Sharma et al., suggesting that in non-polarized cells different GPI-APs are in the same nanodomains [104].

More studies will be required to study the organization of GPI-AP at the PM and in the Golgi apparatus. New techniques that are able to overcome the resolution limit of optical microscopy (e.g., PALM, FLIM, STED, and FCS) will be very useful to resolve this question.

B. DIRECT AND INDIRECT DELIVERY

After their segregation in the TGN, GPI-APs are delivered to the apical PM following a direct route from the Golgi apparatus [30, 47, 50]. This hypothesis has been recently confirmed by using the speed and sensitivity of spinning-disc confocal microscopy that allowed the analysis of the transport of different GPI-APs in fully polarized live MDCK cells grown on filters. These studies clearly showed that GPI-APs progressively accumulate at the apical surface after exit from the Golgi [105, 106]. Furthermore, it was established that the rate-limiting step for GPI-AP transport was the exit from Golgi and the surface arrival was largely completed within 20 min [106]. These data support the hypothesis that apical sorting of GPI-APs occurs at the Golgi in contrast to a previous report, which proposed that GPI-AP would pass through the basoalteral surface before reaching the apical membrane (e.g., indirect or transcytotic pathway) [107]. However, the observations of Polishchuk and colleagues that GFP-GPI was following a transcytotic route to the apical surface in unpolarized MDCK cells [107], is consistent with previous data showing that an indirect route could be used by apical proteins at the early stage of polarization in FRT cells [108]. The transitory use of the transcytotic pathway appears to be a protein-specific feature (e.g., in the case of GPI-APs it was observed only for GFP-GPI and NTR-PLAP, but not for PLAP; [105]). Interestingly, it has been found that in hippocampal neurons the GPI-anchored prion protein (PrP) is localized in all neurites early during neuron differentiation, but is restricted to the axon in differentiated neurons [109]. Similarly, PLAP is redistributed from the basolateral to the apical PM during the epithelial morphogenesis of *Drosophila melanogaster* embryo [110]. It will be interesting to understand the functional implication for the use of a transcytotic pathway during the early stages of polarization by GPI-APs.

Endocytic compartments have also been implicated in the biosynthetic traffic of proteins in polarized epithelial cells, similarly to what happens in yeast [111]. This has been clearly shown for proteins delivered to the basolateral surface. Specifically, newly synthesized transferrin receptor (TfR) [112] and the asialoglycoprotein receptor transit through recycling endosomes prior to their basolateral surface delivery [113]. More recently, the basolateral vesicular stomatitis virus glycoprotein (VSV-G) and E-cadherin were shown to traverse Tf-positive and Rab11-positive recycling endosomes, respectively, before their delivery to the basolateral surface [114, 115]. These data indicate that proteins could be sorted a second time in the exocytic pathway after the first sorting event occurring in the Golgi apparatus (Figure 14.2). Beyond their role in basolateral protein sorting, early endosomes seem to have a role in apical trafficking, and multiple pathways to the apical membrane involving these compartments have been recently described [116]. Interestingly, whereas raft-associated HA together with the GPI-anchored form of endolyn appear to transit through early endosomes, delivery of nonraft-associated endolyn proceeds through the Rab11-positive apical recycling endosome [116]. This is consistent with the earlier observation that in live MDCK cells, the chimeric YFP-GPI appeared to pass through subapical intermediates en route to the apical membrane [106]. Although some of these data suggest the existence of an endosomal sorting station for GPI-APs, further evidences for native GPI-APs need to be gathered.

C. CYTOSKELETON

In epithelial cells, polarization is not restricted to the PM, but also spans across organelles and cytoskeleton.

1. Microtubules

Microtubules (MTs) are aligned along the apicobasal axis of epithelial cells and are oriented with the minus end facing the apical surface. In addition a horizontal meshwork composed of short tubulin filaments with mixed orientations underlies the apical and basal membranes [117]. The perturbation of MT organization by MT-depolymerizing compounds (such as nocodazole and colchicine) inhibits both apical and basolateral protein transport ([118–120] and reviewed in Ref. [117]). This is confirmed by recent studies in live cells showing that post-Golgi intermediates delivering both apical and basolateral proteins move along MT tracks [29, 30, 121–124]. Because both apical and basolateral transport require the integrity of MT architecture, it is clear that it is mediated by MT-driven motors with different orientations. Kinesin KIF5B has been recently demonstrated to be specifically involved in the apical trafficking of neurotrophin receptor p75, but not in the trafficking of other apical (such as prominin and a GPI version of YFP, YFP-GPI) or basolateral proteins (such as LDL receptor, E-cadherin) [125]. Interestingly, these authors found that p75 associate with KIF5B exclusively in polarized cells, but not in subconfluent cells, suggesting that different motors could be implicated in protein transport at the earlier and later stages of polarization. This provides an explanation as to why proteins use different routes to reach the surface during polarization as aforementioned [105, 108]. In contrast, apical transport of raft-associated HA is inhibited by a dominant-negative form of KIFC3 and is strongly stimulated by overexpression of its wild-type form [126]. Furthermore, KIFC3 was resistant to Triton X-100 extraction and was copurified with annexin XIIIb-associated vesicles isolated by immunoaffinity chromatography [126]. These data suggest a specific role of KIFC3 in the transport of raft proteins; hence, it will be interesting to assay whether it plays a role also for the apical transport of GPI-APs. The fact that kinesins with a different polarity (e.g., KIF5B plus-end directed and KIFC3 minus-end directed) are both implicated in the apical transport supports the existence of multiple pathways toward this domain. Kinesins have been shown to have a role in the transport of basolateral proteins [118]. However, no specific type of kinesin has been identified yet to be specifically involved in the transport of basolateral proteins from Golgi to PM. KIF5, which colocalize transiently with VSV-G at the level of the Golgi apparatus, appears to be involved exclusively in the retrograde transport between Golgi and ER [127]. The molecular mechanisms of how kinesins select their cargoes have not been elucidated. Although KIF5B seems to interact directly with p75 [125], it is more likely that kinesins select their cargoes through the interaction with adaptor proteins like in the case of KIF13A, which mediates TGN-to-PM transport of the mannose 6-phosphate receptor interacting directly with the ear domain of β1-adaptin [128]. Because inhibition of annexin XIIIb impairs the apical transport of several raft proteins and a functional KIFC3 is required to deliver annexin XIIIb apically, this kinesin could be a good candidate for the apical transport of raft proteins. Hence, annexin XIIIb and KIFC3 could constitute the two faces of the transport machinery, the first by selecting the cargo and the second one by promoting the delivery in the correct apical direction. It will be important to test whether this annexin also plays a role in the transport of the GPI-APs. Kinesins could bind specific lipids surrounding GPI-APs. The fact that KIF1A and KIF16B bind to phosphatidylinositides through their PH domain [129, 130] supports this possibility. On the other hand, other kinesins could interact with galectin-3, which is implicated in the transport of nonraft proteins [131], promoting the selective transport of this class of proteins.

Finally, posttranslation modifications of tubulins, such as acetylation, glutamination, detyrosination, and glycylation, could modulate the interaction between motor proteins and MTs as already described for KIF1, which binds specifically acetylated MT promoting the transport within neurites [132]. Further studies will be necessary to elucidate all these possibilities.

2. Actin Filaments

Besides the central role of MTs, the integrity of actin filaments is also required for the efficient protein transport to the PM as demonstrated by the impairment of both apical and basolateral delivery using actin-depolymerizing agents like cytochalasin D [121, 124, 133]. Increasing evidences show that actin regulatory/binding proteins are implicated in Golgi-to-PM trafficking, suggesting a crucial role of actin filaments in this process [134]. Nonetheless, it remains to be understood whether actin filaments play a role in cargo selection, in membrane budding/fission, and in the polarized transport to the surface. In support of a role for actin dynamics at the Golgi apparatus, it has been shown that in cos-1 cells, actin-stabilizing/depolymerizing agents inhibit the exit of nonraft-associated apical and basolateral proteins (p75 and VSV-G) without affecting the exit of the raft-associated GPI-AP [135]. Consistent with these findings, in unpolarized MDCK cells latrunculin A, which depolymerize actin filaments, decreases the apparent diffusion coefficient of P75 measured by FRAP, while the GPI-AP GFP-FR is unaffected [98]. Furthermore, the depletion of actin-binding kinase LIMK1 or the overexpression of its inactive form slowed down the exit from the TGN of p75-GFP, but did not interfere with the GPI-AP GPI-GFP [136]. However in polarized MDCK cells, all apical proteins, independent of their raft association, display a significant increase in their apparent diffusion coefficients upon addition of latrunculin A, thus providing important evidence that the sorting machinery at the Golgi level could change during polarization [98]. Moreover, the fact that latrunculin A does not have any effect on the mobility of basolateral proteins clearly indicate that apical and basolateral proteins do not share a common membrane environment at the Golgi complex and that actin seems to have a major role in the segregation of apical proteins at this level [98].

The small GTPase cdc42, which regulates the dynamics of actin filaments, might control polarized protein trafficking in epithelial cells [137–139]. Interestingly, while the apical protein transport is accelerated, the basolateral one is drastically delayed after the expression of either a dominant-negative or an activated form of cdc42 [138]. Because the expression of activated cdc42 mutant resulted in the disappearance of actin perinuclear filaments and in a prominent redistribution of actin to the cell cortex [138], it is possible that different actin pools regulate the apical and basolateral pathways. Moreover, it has been reported that the existence of distinct pools of actin isolated from Golgi membranes display different biochemical properties [140]. Alternatively, both apical and basolateral transport is regulated by cdc42 through different actin-binding proteins acting in the two pathways. This is supported by previous findings showing that myosin IIA controls specifically the release of basolateral, but not of apical vesicles [141]. To date, it is unknown if cdc42 is implicated in the sorting and PM delivery of newly synthesized GPI-APs, while its role in the GPI-AP endocytic pathway has been reported [142, 143].

Interestingly, the raft-associated sucrose isomaltase (SI), but not the nonraft-associated lactase phlorizin hydrolase is delivered to the apical surface in an actin-dependent fashion. In particular, after cytochalasin D treatment, SI-containing vesicles accumulated in areas of the cell periphery were costained with monomeric actin, thus suggesting a role of actin filaments in selective protein transport. The fact that the actin motor myosin Ia has been found exclusively in the immunoisolated vesicles carrying SI [121] and that the depletion of the alpha kinase-1, which is able to phosphorylate myosin Ia *in vitro*, decreases the transport of SI to the apical surface [144], supports this hypothesis. Furthermore, a variety of actin-related proteins, such as Arp2/3, cortactin, and spir1, are being detected as to be implicated in post-Golgi trafficking (reviewed in [134]). This indicates that a fine regulation of actin filaments could play a crucial role in polarized protein sorting/transport. Whether these factors will have a role in GPI-AP transport remains to be determined.

V.   Regulation of Membrane Traffic

The complex patterns of membrane traffic are subjected to numerous levels of regulation and several proteins have been implicated. The small GTP-binding proteins belonging to the Rab family are involved in different steps of both exocytic and endocytic pathway [145]. In particular, it has been demonstrated that Rab8 is involved in transport to the basolateral membrane in MDCK cells [146]. However, Rab8 seems to be responsible also for the apical localization of some peptidases and transporters in intestinal cells, which was mislocalized to lysosomes in Rab8-deficient mouse [147]. Therefore, the same Rab protein could act in different pathways. Similarly, Rab10, initially reported to be localized at the common recycling endosomes where it mediates the transport from basolateral to common recycling endosomes [148], has been recently found to be involved with Rab8 in trafficking from the Golgi to the basolateral membrane [149]. This scenario could be even more complicated considering that the same Rab protein could play different roles in distinct epithelial cells. Another interesting candidate in the specific regulation of apical traffic is Rab14, because the expression of a GDP-mutant form affects the apical delivery, but does not disrupt basolateral targeting or recycling [150]. Furthermore, it has been shown that the inactive mutant of Rab14 selectively affects the targeting of raft-associated VIP17/MAL, but has no effect on apical transport of nonraft-associated gp135 and pIgA receptors [150]. Thus, it will be interesting to analyze whether Rab14 could also mediate the trafficking of GPI-APs. Because Rab proteins and GPI-APs are located on opposite leaflets of the membrane bilayer, Rab14 could regulate GPI-AP traffic through the action of VIP17/MAL, which directly interacts with its active form, or through other factors, which could mediate the link with GPI-APs. In support of this hypothesis, Rab14 has been coimmunoprecipitated with annexin A2 in alveolar type II cells [151], thus suggesting that both proteins could be part of the same scaffold machinery.

The last regulation of protein sorting is at the level of the fusion of the cargo vesicles with the target membrane. Membrane fusion is dependent on the formation of complexes between SNARE proteins located at the target membrane (t-SNARE) and on transport vesicles (v-SNARE). A polarized distribution of t-SNAREs has been found in epithelial cells. The t-SNARE syntaxin 3 is localized at the apical PM, whereas syntaxin 4 is expressed predominantly at the basolateral membrane domain of different epithelia [152–155]. Whereas the role of syntaxin 4 in basolateral trafficking has not yet formally been demonstrated, the correct localization of syntaxin 3, which is dependent on intact MT [29], is necessary for targeting of proteins to the apical membrane [156–158]. Indeed, either overexpression of syntaxin 3 or its inhibition by antibody injection decreased strongly the apical targeting of both transmembrane and soluble secretory proteins [156, 158–160]. Interestingly, two recent papers have been shown that specific v-SNAREs are involved in apical and basolateral protein trafficking. Fields et al. have reported that the v-SNARE cellubrevin coimmunoprecipitates with syntaxin 4 and is involved in basolateral trafficking of AP-1B-dependent cargos [161]. By using the RNAi approach Pocard et al. have demonstrated that TI-VAMP mediates direct apical delivery of both raft-associated GPI-AP and nonraft-associated proteins [162]. By contrast, the transcytotic pathway is not affected by TI-VAMP knockdown, but appears to be regulated by VAMP8 [162].

It remains to be understood why the components of the apical t-SNARE complex (syntaxin 3 and SNAP-23) are associated with rafts [156, 158] when they are implicated in the fusion of vesicles carrying both raft and nonraft proteins. One possibility is that perhaps they promote the fusion of each type of cargo in specific regions of the apical membrane, probably through the involvement of other factors. Also in this case further studies are necessary to elucidate this possibility.

VI.   Conclusion/Perspectives

In the last five years, evidences have been accumulated indicating that apical sorting of GPI-APs occurs in the Golgi apparatus and is dependent on two factors: association to rafts and oligomerization in HMW complexes. The next challenge is to understand the molecular mechanisms for this event. Specifically, which are the respective roles of the GPI anchor, the protein ectodomain, and the surrounding membrane environment (e.g., specific lipids or protein chaperones).

Furthermore, it is still unknown whether different raft-associated GPI-APs and TM proteins partition in the same lipid environment, whether TM proteins and GPI-APs segregate at the level of the TGN or travel together toward the apical surface, and whether they use the same or different trafficking routes. In addition, several factors have been implicated in the transport of raft transmembrane proteins. Do they also play a role for GPI-APs?

Many evidences indicate that there are multiple pathways toward the apical surface. Probably this allows the cells to finely regulate protein delivery in response to physiological stimuli. In fact, the apical surface being in direct contact with the outside environment needs to be more plastic and versatile to environmental changes. It will be interesting to dissect these pathways and to put these routes in relationship with the metabolic and the differentiation state of the cell. Several evidences indicate that during polarization various epithelial cells switch between different transport routes, but the underlying molecular mechanisms are still unknown. On the other hand, different pathways could allow the cells to deliver different cargoes in different subdomains of the apical surface, where they could explicate their proper function. It is now evident that many different Rabs and their effectors work together to coordinate sorting, budding transport, and fusion. However, how they control these cellular events temporally and spatially between intracellular compartments is still not clear. Finally in epithelial cells, polarized transport relies on the integrity of MTs and actin filaments and is strictly regulated by motor proteins. Many of the effectors and regulators of cytoskeleton have been found to be associated to the Golgi or with the post-Golgi intermediates, suggesting that they could play a direct role in protein sorting and transport. Because GPI-APs lack a cytosolic tail, it is likely that they interact via a putative raft-associated receptor. Alternatively, a more direct role for MT and actin through their association and modulation of lipid raft could be envisaged. This hypothesis is supported by the fact that interactions between MT and rafts could be regulated by CLIPR-59, which is raft associated [163]. Recent evidences that actin filaments could stabilize lipid rafts [164–167] favor this hypothesis. In conclusion, while overall the basis for the role of rafts in the sorting and trafficking of GPI-APs are quite clear, we await for a more thorough analysis of the molecular machinery controlling these events within the next five years.

REFERENCES

1. Martin-Belmonte, F., and Mostov, K. (2008). Regulation of cell polarity during epithelial morphogenesis. *Curr Opin Cell Biol* 20:227–234.
2. Nelson, W.J. (2003). Epithelial cell polarity from the outside looking in. *News Physiol Sci* 18:143–146.
3. Mostov, K., Su, T., and ter Beest, M. (2003). Polarized epithelial membrane traffic: conservation and plasticity. *Nat Cell Biol* 5:287–293.
4. Bryant, D.M., and Mostov, K.E. (2008). From cells to organs: building polarized tissue. *Nat Rev Mol Cell Biol* 9:887–901.
5. Rodriguez-Boulan, E., Kreitzer, G., and Musch, A. (2005). Organization of vesicular trafficking in epithelia. *Nat Rev Mol Cell Biol* 6:233–247.
6. Mellman, I., and Nelson, W.J. (2008). Coordinated protein sorting, targeting and distribution in polarized cells. *Nat Rev Mol Cell Biol* 9:833–845.
7. Mellman, I. (1996). Endocytosis and molecular sorting. *Ann Rev Cell Dev Biol* 12:575–625.
8. Nelson, W.J., and Yeaman, C. (2001). Protein trafficking in the exocytic pathway of polarized epithelial cells. *Trends Cell Biol* 11:483–486.
9. Matter, K. (2000). Epithelial polarity: sorting out the sorters. *Curr Biol* 10:R39–R42.
10. Folsch, H. (2008). Regulation of membrane trafficking in polarized epithelial cells. *Curr Opin Cell Biol* 20:208–213.
11. Stein, M., Wandinger-Ness, A., and Roitbak, T. (2002). Altered trafficking and epithelial cell polarity in disease. *Trends Cell Biol* 12:374–381.
12. Lee, M., and Vasioukhin, V. (2008). Cell polarity and cancer–cell and tissue polarity as a noncanonical tumor suppressor. *J Cell Sci* 121:1141–1150.
13. Chatterjee, S., and Mayor, S. (2001). The GPI-anchor and protein sorting. *Cell Mol Life Sci* 58:1969–1987.
14. Mayor, S., and Riezman, H. (2004). Sorting GPI-anchored proteins. *Nat Rev Mol Cell Biol* 5:110–120.
15. Altschuler, Y., Hodson, C., and Milgram, S.L. (2003). The apical compartment: trafficking pathways, regulators and scaffolding proteins. *Curr Opin Cell Biol* 15:423–429.
16. Mostov, K.E., Verges, M., and Altschuler, Y. (2000). Membrane traffic in polarized epithelial cells. *Curr Opin Cell Biol* 12:483–490.
17. McDowell, M.A., Ransom, D.M., and Bangs, J.D. (1998). Glycosylphosphatidylinositol-dependent secretory transport in *Trypanosoma brucei*. *Biochem J* 335(Pt. 3):681–689.
18. Vashist, S., Kim, W., Belden, W.J., Spear, E.D., Barlowe, C., and Ng, D.T. (2001). Distinct retrieval and retention mechanisms are required for the quality control of endoplasmic reticulum protein folding. *J Cell Biol* 155:355–368.
19. Tanaka, S., Maeda, Y., Tashima, Y., and Kinoshita, T. (2004). Inositol deacylation of glycosylphosphatidylinositol-anchored proteins is mediated by mammalian PGAP1 and yeast Bst1p. *J Biol Chem* 279:14256–14263.
20. Watanabe, R., and Riezman, H. (2004). Differential ER exit in yeast and mammalian cells. *Curr Opin Cell Biol* 16:350–355.

21. Muniz, M., Morsomme, P., and Riezman, H. (2001). Protein sorting upon exit from the endoplasmic reticulum. *Cell* 104:313–320.
22. Morsomme, P., Prescianotto-Baschong, C., and Riezman, H. (2003). The ER v-SNAREs are required for GPI-anchored protein sorting from other secretory proteins upon exit from the ER. *J Cell Biol* 162:403–412.
23. Morsomme, P., and Riezman, H. (2002). The Rab GTPase Ypt1p and tethering factors couple protein sorting at the ER to vesicle targeting to the Golgi apparatus. *Dev Cell* 2:307–317.
24. Carney, G.E., and Bowen, N.J. (2004). p24 proteins, intracellular trafficking, and behavior: *Drosophila melanogaster* provides insights and opportunities. *Biol Cell* 96:271–278.
25. Takida, S., Maeda, Y., and Kinoshita, T. (2008). Mammalian GPI-anchored proteins require p24 proteins for their efficient transport from the ER to the plasma membrane. *Biochem J* 409:555–562.
26. Stephens, D.J., and Pepperkok, R. (2004). Differential effects of a GTP-restricted mutant of Sar1p on segregation of cargo during export from the endoplasmic reticulum. *J Cell Sci* 117:3635–3644.
27. Mironov, A.A., Mironov, A.A., Jr., Beznoussenko, G.V., Trucco, A., Lupetti, P., Smith, J.D., Geerts, W.J., Koster, A.J., Burger, K.N., Martone, M.E., Deerinck, T.J., Ellisman, M.H., *et al.* (2003). ER-to-Golgi carriers arise through direct en bloc protrusion and multistage maturation of specialized ER exit domains. *Dev Cell* 5:583–594.
28. Wandinger-Ness, A., Bennett, M.K., Antony, C., and Simons, K. (1990). Distinct transport vesicles mediate the delivery of plasma membrane proteins to the apical and basolateral domains of MDCK cells. *J Cell Biol* 111:987–1000.
29. Kreitzer, G., Schmoranzer, J., Low, S.H., Li, X., Gan, Y., Weimbs, T., Simon, S.M., and Rodriguez-Boulan, E. (2003). Three-dimensional analysis of post-Golgi carrier exocytosis in epithelial cells. *Nat Cell Biol* 5:126–136.
30. Keller, P., Toomre, D., Diaz, E., White, J., and Simons, K. (2001). Multicolour imaging of post-Golgi sorting and trafficking in live cells. *Nat Cell Biol* 3:140–149.
31. Bonifacino, J.S., and Traub, L.M. (2003). Signals for sorting of transmembrane proteins to endosomes and lysosomes. *Ann Rev Biochem* 72:395–447.
32. Folsch, H. (2005). The building blocks for basolateral vesicles in polarized epithelial cells. *Trends Cell Biol* 15:222–228.
33. Sugimoto, H., Sugahara, M., Folsch, H., Koide, Y., Nakatsu, F., Tanaka, N., Nishimura, T., Furukawa, M., Mullins, C., Nakamura, N., Mellman, I., and Ohno, H. (2002). Differential recognition of tyrosine-based basolateral signals by AP-1B subunit mu1B in polarized epithelial cells. *Mol Biol Cell* 13:2374–2382.
34. Folsch, H., Ohno, H., Bonifacino, J.S., and Mellman, I. (1999). A novel clathrin adaptor complex mediates basolateral targeting in polarized epithelial cells. *Cell* 99:189–198.
35. Scheiffele, P., Peranen, J., and Simons, K. (1995). *N*-glycans as apical sorting signals in epithelial cells. *Nature* 378:96–98.
36. Scheiffele, P., Roth, M.G., and Simons, K. (1997). Interaction of influenza virus haemagglutinin with sphingolipid-cholesterol membrane domains via its transmembrane domain. *Embo J* 16:5501–5508.
37. Sun, A.Q., Ananthanarayanan, M., Soroka, C.J., Thevananther, S., Shneider, B.L., and Suchy, F.J. (1998). Sorting of rat liver and ileal sodium-dependent bile acid transporters in polarized epithelial cells. *Am J Physiol* 275:G1045–G1055.
38. Sun, A.Q., Salkar, R., Sachchidanand, S., Xu, S., Zeng, L., Zhou, M.M., and Suchy, F.J. (2003). A 14-amino acid sequence with a beta-turn structure is required for apical membrane sorting of the rat ileal bile acid transporter. *J Biol Chem* 278:4000–4009.

39. Chuang, J.Z., and Sung, C.H. (1998). The cytoplasmic tail of rhodopsin acts as a novel apical sorting signal in polarized MDCK cells. *J Cell Biol* 142:1245–1256.
40. Simons, K., and Ikonen, E. (1997). Functional rafts in cell membranes. *Nature* 387:569–572.
41. Simons, K., and van Meer, G. (1988). Lipid sorting in epithelial cells. *Biochemistry* 27:6197–6202.
42. Brown, D.A., and London, E. (1998). Functions of lipid rafts in biological membranes. *Ann Rev Cell Dev Biol* 14:111–136.
43. van Meer, G., Stelzer, E.H., Wijnaendts-van-Resandt, R.W., and Simons, K. (1987). Sorting of sphingolipids in epithelial (Madin-Darby canine kidney) cells. *J Cell Biol* 105:1623–1635.
44. Brown, D.A., Crise, B., and Rose, J.K. (1989). Mechanism of membrane anchoring affects polarized expression of two proteins in MDCK cells. *Science* 245:1499–1501.
45. Lisanti, M.P., Caras, I.W., Davitz, M.A., and Rodriguez-Boulan, E. (1989). A glycophospholipid membrane anchor acts as an apical targeting signal in polarized epithelial cells. *J Cell Biol* 109:2145–2156.
46. Brown, D.A., and Rose, J.K. (1992). Sorting of GPI-anchored proteins to glycolipid-enriched membrane subdomains during transport to the apical cell surface. *Cell* 68:533–544.
47. Arreaza, G., and Brown, D.A. (1995). Sorting and intracellular trafficking of a glycosylphosphatidylinositol-anchored protein and two hybrid transmembrane proteins with the same ectodomain in Madin-Darby canine kidney epithelial cells. *J Biol Chem* 270:23641–23647.
48. Fiedler, K., Kobayashi, T., Kurzchalia, T.V., and Simons, K. (1993). Glycosphingolipid-enriched, detergent-insoluble complexes in protein sorting in epithelial cells. *Biochemistry* 32:6365–6373.
49. Nosjean, O., Briolay, A., and Roux, B. (1997). Mammalian GPI proteins: sorting, membrane residence and functions. *Biochim Biophys Acta* 1331:153–186.
50. Lipardi, C., Nitsch, L., and Zurzolo, C. (2000). Detergent-insoluble GPI-anchored proteins are apically sorted in fischer rat thyroid cells, but interference with cholesterol or sphingolipids differentially affects detergent insolubility and apical sorting. *Mol Biol Cell* 11:531–542.
51. Paladino, S., Sarnataro, D., Pillich, R., Tivodar, S., Nitsch, L., and Zurzolo, C. (2004). Protein oligomerization modulates raft partitioning and apical sorting of GPI-anchored proteins. *J Cell Biol* 167:699–709.
52. Paladino, S., Sarnataro, D., Tivodar, S., and Zurzolo, C. (2007). Oligomerization is a specific requirement for apical sorting of glycosyl-phosphatidylinositol-anchored proteins but not for nonraft-associated apical proteins. *Traffic* 8:251–258.
53. Mays, R.W., Siemers, K.A., Fritz, B.A., Lowe, A.W., van Meer, G., and Nelson, W.J. (1995). Hierarchy of mechanisms involved in generating Na/K-ATPase polarity in MDCK epithelial cells. *J Cell Biol* 130:1105–1115.
54. Ehehalt, R., Krautter, M., Zorn, M., Sparla, R., Fullekrug, J., Kulaksiz, H., and Stremmel, W. (2008). Increased basolateral sorting of carcinoembryonic antigen in a polarized colon carcinoma cell line after cholesterol depletion—Implications for treatment of inflammatory bowel disease. *World J Gastroenterol* 14:1528–1533.
55. Zurzolo, C., Lisanti, M.P., Caras, I.W., Nitsch, L., and Rodriguez-Boulan, E. (1993). Glycosylphosphatidylinositol-anchored proteins are preferentially targeted to the basolateral surface in Fischer rat thyroid epithelial cells. *J Cell Biol* 121:1031–1039.
56. Lipardi, C., Mora, R., Colomer, V., Paladino, S., Nitsch, L., Rodriguez-Boulan, E., and Zurzolo, C. (1998). Caveolin transfection results in caveolae formation but not apical

sorting of glycosylphosphatidylinositol (GPI)-anchored proteins in epithelial cells. *J Cell Biol* 140:617–626.

57. Paladino, S., Sarnataro, D., and Zurzolo, C. (2002). Detergent-resistant membrane micro-domains and apical sorting of GPI-anchored proteins in polarized epithelial cells. *IJMM Int J Med Microbiol* 291:439–445.

58. Benting, J., Rietveld, A., Ansorge, I., and Simons, K. (1999). Acyl and alkyl chain length of GPI-anchors is critical for raft association *in vitro*. *FEBS Lett* 462:47–50.

59. Fivaz, M., Vilbois, F., Thurnheer, S., Pasquali, C., Abrami, L., Bickel, P.E., Parton, R.G., and van der Goot, F.G. (2002). Differential sorting and fate of endocytosed GPI-anchored proteins. *EMBO J* 21:3989–4000.

60. Zhang, J., Pekosz, A., and Lamb, R.A. (2000). Influenza virus assembly and lipid raft microdomains: a role for the cytoplasmic tails of the spike glycoproteins. *J Virol* 74:4634–4644.

61. Parton, R.G., and Richards, A.A. (2003). Lipid rafts and caveolae as portals for endocy-tosis: new insights and common mechanisms. *Traffic* 4:724–738.

62. Pelkmans, L., and Helenius, A. (2002). Endocytosis via caveolae. *Traffic* 3:311–320.

63. Abrami, L., and van Der Goot, F.G. (1999). Plasma membrane microdomains act as concentration platforms to facilitate intoxication by aerolysin. *J Cell Biol* 147:175–184.

64. Schuck, S., and Simons, K. (2004). Polarized sorting in epithelial cells: raft clustering and the biogenesis of the apical membrane. *J Cell Sci* 117:5955–5964.

65. Huttner, W.B., and Zimmerberg, J. (2001). Implications of lipid microdomains for mem-brane curvature, budding and fission. *Curr Opin Cell Biol* 13:478–484.

66. Liu, J., Kaksonen, M., Drubin, D.G., and Oster, G. (2006). Endocytic vesicle scission by lipid phase boundary forces. *Proc Natl Acad Sci USA* 103:10277–10282.

67. Hanzal-Bayer, M.F., and Hancock, J.F. (2007). Lipid rafts and membrane traffic. *FEBS Lett* 581:2098–2104.

68. McMahon, H.T., and Gallop, J.L. (2005). Membrane curvature and mechanisms of dynamic cell membrane remodelling. *Nature* 438:590–596.

69. Kuzmin, P.I., Akimov, S.A., Chizmadzhev, Y.A., Zimmerberg, J., and Cohen, F.S. (2005). Line tension and interaction energies of membrane rafts calculated from lipid splay and tilt. *Biophys J* 88:1120–1133.

70. Paladino, S., Lebreton, S., Tivodar, S., Campana, V., Tempre, R., and Zurzolo, C. (2008). Different GPI-attachment signals affect the oligomerisation of GPI-anchored proteins and their apical sorting. *J Cell Sci* 121:4001–4007.

71. Benting, J.H., Rietveld, A.G., and Simons, K. (1999). *N*-Glycans mediate the apical sorting of a GPI-anchored, raft-associated protein in Madin-Darby canine kidney cells. *J Cell Biol* 146:313–320.

72. Pang, S., Urquhart, P., and Hooper, N.M. (2004). *N*-glycans, not the GPI anchor, mediate the apical targeting of a naturally glycosylated, GPI-anchored protein in polarised epi-thelial cells. *J Cell Sci* 117:5079–5086.

73. Potter, B.A., Ihrke, G., Bruns, J.R., Weixel, K.M., and Weisz, O.A. (2004). Specific *N*-glycans direct apical delivery of transmembrane, but not soluble or glycosylphosphatidylinositol-anchored forms of endolyn in Madin-Darby canine kidney cells. *Mol Biol Cell* 15:1407–1416.

74. Huet, G., Hennebicq-Reig, S., de Bolos, C., Ulloa, F., Lesuffleur, T., Barbat, A., Carriere, V., Kim, I., Real, F.X., Delannoy, P., and Zweibaum, A. (1998). GalNAc-alpha-*O*-benzyl inhibits NeuAcalpha2–3 glycosylation and blocks the intracellular trans-port of apical glycoproteins and mucus in differentiated HT-29 cells. *J Cell Biol* 141:1311–1322.

75. Alfalah, M., Jacob, R., Preuss, U., Zimmer, K.P., Naim, H., and Naim, H.Y. (1999). O-linked glycans mediate apical sorting of human intestinal sucrase-isomaltase through association with lipid rafts. *Curr Biol* 9:593–596.
76. Yeaman, C., Le Gall, A.H., Baldwin, A.N., Monlauzeur, L., Le Bivic, A., and Rodriguez-Boulan, E. (1997). The O-glycosylated stalk domain is required for apical sorting of neurotrophin receptors in polarized MDCK cells. *J Cell Biol* 139:929–940.
77. Catino, M.A., Paladino, S., Tivodar, S., Pocard, T., and Zurzolo, C. (2008). N- and O-glycans are not directly involved in the oligomerization and apical sorting of GPI proteins. *Traffic* 9:2141–2150.
78. Meiss, H.K., Green, R.F., and Rodriguez-Boulan, E.J. (1982). Lectin-resistant mutants of polarized epithelial cells. *Mol Cell Biol* 2:1287–1294.
79. Hannan, L.A., Lisanti, M.P., Rodriguez-Boulan, E., and Edidin, M. (1993). Correctly sorted molecules of a GPI-anchored protein are clustered and immobile when they arrive at the apical surface of MDCK cells. *J Cell Biol* 120:353–358.
80. Cheong, K.H., Zacchetti, D., Schneeberger, E.E., and Simons, K. (1999). VIP17/MAL, a lipid raft-associated protein, is involved in apical transport in MDCK cells. *Proc Natl Acad Sci USA* 96:6241–6248.
81. Martin-Belmonte, F., Puertollano, R., Millan, J., Alonso, M.A., de Marco, M.C., Albar, J.P., and Kremer, L. (2000). The MAL proteolipid is necessary for the overall apical delivery of membrane proteins in the polarized epithelial Madin-Darby canine kidney and fischer rat thyroid cell lines. The MAL proteolipid is necessary for normal apical transport and accurate sorting of the influenza virus hemagglutinin in Madin-Darby canine kidney cells. *Mol Biol Cell* 11:2033–2045.
82. Manninen, A., Verkade, P., Le Lay, S., Torkko, J., Kasper, M., Fullekrug, J., and Simons, K. (2005). Caveolin-1 is not essential for biosynthetic apical membrane transport. *Mol Cell Biol* 25:10087–10096.
83. Snyers, L., Umlauf, E., and Prohaska, R. (1999). Association of stomatin with lipid-protein complexes in the plasma membrane and the endocytic compartment. *Eur J Cell Biol* 78:802–812.
84. Stuermer, C.A., Langhorst, M.F., Wiechers, M.F., Legler, D.F., Von Hanwehr, S.H., Guse, A.H., and Plattner, H. (2004). PrPc capping in T cells promotes its association with the lipid raft proteins reggie-1 and reggie-2 and leads to signal transduction. *FASEB J* 18:1731–1733.
85. Jacob, R., Heine, M., Eikemeyer, J., Frerker, N., Zimmer, K.P., Rescher, U., Gerke, V., and Naim, H.Y. (2004). Annexin II is required for apical transport in polarized epithelial cells. *J Biol Chem* 279:3680–3684.
86. Fiedler, K., Lafont, F., Parton, R.G., and Simons, K. (1995). Annexin XIIIb: a novel epithelial specific annexin is implicated in vesicular traffic to the apical plasma membrane. *J Cell Biol* 128:1043–1053.
87. Lecat, S., Verkade, P., Thiele, C., Fiedler, K., Simons, K., and Lafont, F. (2000). Different properties of two isoforms of annexin XIII in MDCK cells. *J Cell Sci* 113 (Pt. 14):2607–2618.
88. Lafont, F., Lecat, S., Verkade, P., and Simons, K. (1998). Annexin XIIIb associates with lipid microdomains to function in apical delivery. *J Cell Biol* 142:1413–1427.
89. Delacour, D., Gouyer, V., Zanetta, J.P., Drobecq, H., Leteurtre, E., Grard, G., Moreau-Hannedouche, O., Maes, E., Pons, A., Andre, S., Le Bivic, A., Gabius, H.J., *et al.* (2005). Galectin-4 and sulfatides in apical membrane trafficking in enterocyte-like cells. *J Cell Biol* 169:491–501.
90. Godi, A., Di Campli, A., Konstantakopoulos, A., Di Tullio, G., Alessi, D.R., Kular, G.S., Daniele, T., Marra, P., Lucocq, J.M., and De Matteis, M.A. (2004). FAPPs control

Golgi-to-cell-surface membrane traffic by binding to ARF and PtdIns(4)P. *Nat Cell Biol* 6:393–404.

91. Vieira, O.V., Verkade, P., Manninen, A., and Simons, K. (2005). FAPP2 is involved in the transport of apical cargo in polarized MDCK cells. *J Cell Biol* 170:521–526.

92. D'Angelo, G., Polishchuk, E., Di Tullio, G., Santoro, M., Di Campli, A., Godi, A., West, G., Bielawski, J., Chuang, C.C., van der Spoel, A.C., Platt, F.M., Hannun, Y.A., *et al.* (2007). Glycosphingolipid synthesis requires FAPP2 transfer of glucosylceramide. *Nature* 449:62–67.

93. Fujita, M., Umemura, M., Yoko-o, T., and Jigami, Y. (2006). PER1 is required for GPI-phospholipase A2 activity and involved in lipid remodeling of GPI-anchored proteins. *Mol Biol Cell* 17:5253–5264.

94. Bosson, R., Jaquenoud, M., and Conzelmann, A. (2006). GUP1 of *Saccharomyces cerevisiae* encodes an *O*-acyltransferase involved in remodeling of the GPI anchor. *Mol Biol Cell* 17:2636–2645.

95. Legler, D.F., Doucey, M.A., Schneider, P., Chapatte, L., Bender, F.C., and Bron, C. (2005). Differential insertion of GPI-anchored GFPs into lipid rafts of live cells. *FASEB J* 19:73–75.

96. Maeda, Y., Tashima, Y., Houjou, T., Fujita, M., Yoko-o, T., Jigami, Y., Taguchi, R., and Kinoshita, T. (2007). Fatty acid remodeling of GPI-anchored proteins is required for their raft association. *Mol Biol Cell* 18:1497–1506.

97. Jaquenoud, M., Pagac, M., Signorell, A., Benghezal, M., Jelk, J., Butikofer, P., and Conzelmann, A. (2008). The Gup1 homologue of *Trypanosoma brucei* is a GPI glycosylphosphatidylinositol remodelase. *Mol Microbiol* 67:202–212.

98. Lebreton, S., Paladino, S., and Zurzolo, C. (2008). Selective roles for cholesterol and actin in compartmentalization of different proteins in the Golgi and plasma membrane of polarized cells. *J Biol Chem* 283:29545–29553.

99. Jacob, R., and Naim, H.Y. (2001). Apical membrane proteins are transported in distinct vesicular carriers. *Curr Biol* 11:1444–1450.

100. Rustom, A., Bajohrs, M., Kaether, C., Keller, P., Toomre, D., Corbeil, D., and Gerdes, H.H. (2002). Selective delivery of secretory cargo in Golgi-derived carriers of nonepithelial cells. *Traffic* 3:279–288.

101. Luini, A., Ragnini-Wilson, A., Polishchuck, R.S., and De Matteis, M.A. (2005). Large pleiomorphic traffic intermediates in the secretory pathway. *Curr Opin Cell Biol* 17:353–361.

102. Guerriero, C.J., Lai, Y., and Weisz, O.A. (2008). Differential sorting and Golgi export requirements for raft-associated and raft-independent apical proteins along the biosynthetic pathway. *J Biol Chem* 283:18040–18047.

103. Guerriero, C.J., Weixel, K.M., Bruns, J.R., and Weisz, O.A. (2006). Phosphatidylinositol 5-kinase stimulates apical biosynthetic delivery via an Arp2/3-dependent mechanism. *J Biol Chem* 281:15376–15384.

104. Sharma, P., Varma, R., Sarasij, R.C., Ira, Gousset,K., Krishnamoorthy, G., Rao, M., and Mayor, S. (2004). Nanoscale organization of multiple GPI-anchored proteins in living cell membranes. *Cell* 116:577–589.

105. Paladino, S., Pocard, T., Catino, M.A., and Zurzolo, C. (2006). GPI-anchored proteins are directly targeted to the apical surface in fully polarized MDCK cells. *J Cell Biol* 172:1023–1034.

106. Hua, W., Sheff, D., Toomre, D., and Mellman, I. (2006). Vectorial insertion of apical and basolateral membrane proteins in polarized epithelial cells revealed by quantitative 3D live cell imaging. *J Cell Biol* 172:1035–1044.

107. Polishchuk, R., Di Pentima, A., and Lippincott-Schwartz, J. (2004). Delivery of raft-associated, GPI-anchored proteins to the apical surface of polarized MDCK cells by a transcytotic pathway. *Nat Cell Biol* 6:297–307; Epub 2004 March 28.
108. Zurzolo, C., Le Bivic, A., Quaroni, A., Nitsch, L., and Rodriguez-Boulan, E. (1992). Modulation of transcytotic and direct targeting pathways in a polarized thyroid cell line. *EMBO J* 11:2337–2344.
109. Galvan, C., Camoletto, P.G., Dotti, C.G., Aguzzi, A., and Ledesma, M.D. (2005). Proper axonal distribution of PrP(C) depends on cholesterol-sphingomyelin-enriched membrane domains and is developmentally regulated in hippocampal neurons. *Mol Cell Neurosci* 30:304–315.
110. Shiel, M.J., and Caplan, M.J. (1995). Developmental regulation of membrane protein sorting in Drosophila embryos. *Am J Physiol* 269:C207–C216.
111. Luo, W., and Chang, A. (2000). An endosome-to-plasma membrane pathway involved in trafficking of a mutant plasma membrane ATPase in yeast. *Mol Biol Cell* 11:579–592.
112. Futter, C.E., Connolly, C.N., Cutler, D.F., and Hopkins, C.R. (1995). Newly synthesized transferrin receptors can be detected in the endosome before they appear on the cell surface. *J Biol Chem* 270:10999–11003.
113. Leitinger, B., Hille-Rehfeld, A., and Spiess, M. (1995). Biosynthetic transport of the asialoglycoprotein receptor H1 to the cell surface occurs via endosomes. *Proc Natl Acad Sci USA* 92:10109–10113.
114. Ang, A.L., Taguchi, T., Francis, S., Folsch, H., Murrells, L.J., Pypaert, M., Warren, G., and Mellman, I. (2004). Recycling endosomes can serve as intermediates during transport from the Golgi to the plasma membrane of MDCK cells. *J Cell Biol* 167:531–543.
115. Lock, J.G., and Stow, J.L. (2005). Rab11 in recycling endosomes regulates the sorting and basolateral transport of E-cadherin. *Mol Biol Cell* 16:1744–1755.
116. Cresawn, K.O., Potter, B.A., Oztan, A., Guerriero, C.J., Ihrke, G., Goldenring, J.R., Apodaca, G., and Weisz, O.A. (2007). Differential involvement of endocytic compartments in the biosynthetic traffic of apical proteins. *EMBO J* 26:3737–3748.
117. Musch, A. (2004). Microtubule organization and function in epithelial cells. *Traffic* 5:1–9.
118. Lafont, F., Burkhardt, J.K., and Simons, K. (1994). Involvement of microtubule motors in basolateral and apical transport in kidney cells. *Nature* 372:801–803.
119. Grindstaff, K.K., Bacallao, R.L., and Nelson, W.J. (1998). Apiconuclear organization of microtubules does not specify protein delivery from the trans-Golgi network to different membrane domains in polarized epithelial cells. *Mol Biol Cell* 9:685–699.
120. Rindler, M.J., Ivanov, I.E., and Sabatini, D.D. (1987). Microtubule-acting drugs lead to the nonpolarized delivery of the influenza hemagglutinin to the cell surface of polarized Madin-Darby canine kidney cells. *J Cell Biol* 104:231–241.
121. Jacob, R., Heine, M., Alfalah, M., and Naim, H.Y. (2003). Distinct cytoskeletal tracks direct individual vesicle populations to the apical membrane of epithelial cells. *Curr Biol* 13:607–612.
122. Toomre, D., Keller, P., White, J., Olivo, J.C., and Simons, K. (1999). Dual-color visualization of trans-Golgi network to plasma membrane traffic along microtubules in living cells. *J Cell Sci* 112:21–33.
123. Kreitzer, G., Marmorstein, A., Okamoto, P., Vallee, R., and Rodriguez-Boulan, E. (2000). Kinesin and dynamin are required for post-Golgi transport of a plasma-membrane protein. *Nat Cell Biol* 2:125–127.
124. Hirschberg, K., Miller, C.M., Ellenberg, J., Presley, J.F., Siggia, E.D., Phair, R.D., and Lippincott-Schwartz, J. (1998). Kinetic analysis of secretory protein traffic and characterization of Golgi to plasma membrane transport intermediates in living cells. *J Cell Biol* 143:1485–1503.

316   SIMONA PALADINO AND CHIARA ZURZOLO

125. Jaulin, F., Xue, X., Rodriguez-Boulan, E., and Kreitzer, G. (2007). Polarization-dependent selective transport to the apical membrane by KIF5B in MDCK cells. *Dev Cell* 13:511–522.
126. Noda, Y., Okada, Y., Saito, N., Setou, M., Xu, Y., Zhang, Z., and Hirokawa, N. (2001). KIFC3, a microtubule minus end-directed motor for the apical transport of annexin XIIIb-associated Triton-insoluble membranes. *J Cell Biol* 155:77–88.
127. Lippincott-Schwartz, J., Cole, N.B., Marotta, A., Conrad, P.A., and Bloom, G.S. (1995). Kinesin is the motor for microtubule-mediated Golgi-to-ER membrane traffic. *J Cell Biol* 128:293–306.
128. Nakagawa, T., Setou, M., Seog, D., Ogasawara, K., Dohmae, N., Takio, K., and Hirokawa, N. (2000). A novel motor, KIF13A, transports mannose-6-phosphate receptor to plasma membrane through direct interaction with AP-1 complex. *Cell* 103:569–581.
129. Klopfenstein, D.R., Tomishige, M., Stuurman, N., and Vale, R.D. (2002). Role of phosphatidylinositol(4,5)bisphosphate organization in membrane transport by the Unc104 kinesin motor. *Cell* 109:347–358.
130. Hoepfner, S., Severin, F., Cabezas, A., Habermann, B., Runge, A., Gillooly, D., Stenmark, H., and Zerial, M. (2005). Modulation of receptor recycling and degradation by the endosomal kinesin KIF16B. *Cell* 121:437–450.
131. Delacour, D., Koch, A., Ackermann, W., Eude-Le Parco, I., Elsasser, H.P., Poirier, F., and Jacob, R. (2008). Loss of galectin-3 impairs membrane polarisation of mouse enterocytes *in vivo*. *J Cell Sci* 121:458–465.
132. Reed, N.A., Cai, D., Blasius, T.L., Jih, G.T., Meyhofer, E., Gaertig, J., and Verhey, K.J. (2006). Microtubule acetylation promotes kinesin-1 binding and transport. *Curr Biol* 16:2166–2172.
133. Valentijn, K.M., Gumkowski, F.D., and Jamieson, J.D. (1999). The subapical actin cytoskeleton regulates secretion and membrane retrieval in pancreatic acinar cells. *J Cell Sci* 112(Pt. 1):81–96.
134. Egea, G., Lazaro-Dieguez, F., and Vilella, M. (2006). Actin dynamics at the Golgi complex in mammalian cells. *Curr Opin Cell Biol* 18:168–178.
135. Lazaro-Dieguez, F., Colonna, C., Cortegano, M., Calvo, M., Martinez, S.E., and Egea, G. (2007). Variable actin dynamics requirement for the exit of different cargo from the trans-Golgi network. *FEBS Lett* 581:3875–3881.
136. Salvarezza, S.B., Deborde, S., Schreiner, R., Campagne, F., Kessels, M.M., Qualmann, B., Caceres, A., Kreitzer, G., and Rodriguez-Boulan, E. (2008). LIM Kinase 1 and Cofilin Regulate Actin Filament Population Required for Dynamin-dependent Apical Carrier Fission from the TGN. *Mol Biol Cell* 20:438–451. Epub 2008 Nov 5.
137. Cohen, D., Musch, A., and Rodriguez-Boulan, E. (2001). Selective control of basolateral membrane protein polarity by cdc42. *Traffic* 2:556–564.
138. Musch, A., Cohen, D., Kreitzer, G., and Rodriguez-Boulan, E. (2001). cdc42 regulates the exit of apical and basolateral proteins from the trans-Golgi network. *EMBO J* 20:2171–2179.
139. Kroschewski, R., Hall, A., and Mellman, I. (1999). Cdc42 controls secretory and endocytic transport to the basolateral plasma membrane of MDCK cells. *Nat Cell Biol* 1:8–13.
140. Fucini, R.V., Navarrete, A., Vadakkan, C., Lacomis, L., Erdjument-Bromage, H., Tempst, P., and Stamnes, M. (2000). Activated ADP-ribosylation factor assembles distinct pools of actin on Golgi membranes. *J Biol Chem* 275:18824–18829.
141. Musch, A., Cohen, D., and Rodriguez-Boulan, E. (1997). Myosin II is involved in the production of constitutive transport vesicles from the TGN. *J Cell Biol* 138:291–306.

142. Sabharanjak, S., Sharma, P., Parton, R.G., and Mayor, S. (2002). GPI-anchored proteins are delivered to recycling endosomes via a distinct cdc42-regulated, clathrin-independent pinocytic pathway. *Dev Cell* 2:411–423.

143. Chadda, R., Howes, M.T., Plowman, S.J., Hancock, J.F., Parton, R.G., and Mayor, S. (2007). Cholesterol-sensitive Cdc42 activation regulates actin polymerization for endocytosis via the GEEC pathway. *Traffic* 8:702–717.

144. Heine, M., Cramm-Behrens, C.I., Ansari, A., Chu, H.P., Ryazanov, A.G., Naim, H.Y., and Jacob, R. (2005). Alpha-kinase 1, a new component in apical protein transport. *J Biol Chem* 280:25637–25643.

145. Zerial, M., and McBride, H. (2001). Rab proteins as membrane organizers. *Nat Rev Mol Cell Biol* 2:107–117.

146. Ang, A.L., Folsch, H., Koivisto, U.M., Pypaert, M., and Mellman, I. (2003). The Rab8 GTPase selectively regulates AP-1B-dependent basolateral transport in polarized Madin-Darby canine kidney cells. *J Cell Biol* 163:339–350.

147. Sato, T., Mushiake, S., Kato, Y., Sato, K., Sato, M., Takeda, N., Ozono, K., Miki, K., Kubo, Y., Tsuji, A., Harada, R., and Harada, A. (2007). The Rab8 GTPase regulates apical protein localization in intestinal cells. *Nature* 448:366–369.

148. Babbey, C.M., Ahktar, N., Wang, E., Chen, C.C., Grant, B.D., and Dunn, K.W. (2006). Rab10 regulates membrane transport through early endosomes of polarized Madin-Darby canine kidney cells. *Mol Biol Cell* 17:3156–3175.

149. Schuck, S., Gerl, M.J., Ang, A., Manninen, A., Keller, P., Mellman, I., and Simons, K. (2007). Rab10 is involved in basolateral transport in polarized Madin-Darby canine kidney cells. *Traffic* 8:47–60.

150. Kitt, K.N., Hernandez-Deviez, D., Ballantyne, S.D., Spiliotis, E.T., Casanova, J.E., and Wilson, J.M. (2008). Rab14 regulates apical targeting in polarized epithelial cells. *Traffic* 9:1218–1231.

151. Gou, D., Mishra, A., Weng, T., Su, L., Chintagari, N.R., Wang, Z., Zhang, H., Gao, L., Wang, P., Stricker, H.M., and Liu, L. (2008). Annexin A2 interactions with Rab14 in alveolar type II cells. *J Biol Chem* 283:13156–13164.

152. Fujita, H., Tuma, P.L., Finnegan, C.M., Locco, L., and Hubbard, A.L. (1998). Endogenous syntaxins 2, 3 and 4 exhibit distinct but overlapping patterns of expression at the hepatocyte plasma membrane. *Biochem J* 329(Pt. 3):527–538.

153. Li, X., Low, S.H., Miura, M., and Weimbs, T. (2002). SNARE expression and localization in renal epithelial cells suggest mechanism for variability of trafficking phenotypes. *Am J Physiol Renal Physiol* 283:F1111–F1122.

154. Low, S.H., Chapin, S.J., Weimbs, T., Komuves, L.G., Bennett, M.K., and Mostov, K.E. (1996). Differential localization of syntaxin isoforms in polarized Madin-Darby canine kidney cells. *Mol Biol Cell* 7:2007–2018.

155. Delgrossi, M.H., Breuza, L., Mirre, C., Chavrier, P., and Le Bivic, A. (1997). Human syntaxin 3 is localized apically in human intestinal cells. *J Cell Sci* 110(Pt. 18):2207–2214.

156. Sharma, N., Low, S.H., Misra, S., Pallavi, B., and Weimbs, T. (2006). Apical targeting of syntaxin 3 is essential for epithelial cell polarity. *J Cell Biol* 173:937–948.

157. ter Beest, M.B., Chapin, S.J., Avrahami, D., and Mostov, K.E. (2005). The role of syntaxins in the specificity of vesicle targeting in polarized epithelial cells. *Mol Biol Cell* 16:5784–5792.

158. Lafont, F., Verkade, P., Galli, T., Wimmer, C., Louvard, D., and Simons, K. (1999). Raft association of SNAP receptors acting in apical trafficking in Madin-Darby canine kidney cells. *Proc Natl Acad Sci USA* 96:3734–3738.

159. Low, S.H., Chapin, S.J., Wimmer, C., Whiteheart, S.W., Komuves, L.G., Mostov, K.E., and Weimbs, T. (1998). The SNARE machinery is involved in apical plasma membrane trafficking in MDCK cells. *J Cell Biol* 141:1503–1513.

160. Breuza, L., Fransen, J., and Le Bivic, A. (2000). Transport and function of syntaxin 3 in human epithelial intestinal cells. *Am J Physiol Cell Physiol* 279:C1239–C1248.

161. Fields, I.C., Shteyn, E., Pypaert, M., Proux-Gillardeaux, V., Kang, R.S., Galli, T., and Folsch, H. (2007). v-SNARE cellubrevin is required for basolateral sorting of AP-1B-dependent cargo in polarized epithelial cells. *J Cell Biol* 177:477–488.

162. Pocard, T., Le Bivic, A., Galli, T., and Zurzolo, C. (2007). Distinct v-SNAREs regulate direct and indirect apical delivery in polarized epithelial cells. *J Cell Sci* 120:3309–3320.

163. Lallemand-Breitenbach, V., Quesnoit, M., Braun, V., El Marjou, A., Pous, C., Goud, B., and Perez, F. (2004). CLIPR-59 is a lipid raft-associated protein containing a cytoskeleton-associated protein glycine-rich domain (CAP-Gly) that perturbs microtubule dynamics. *J Biol Chem* 279:41168–41178.

164. Chichili, G.R., and Rodgers, W. (2007). Clustering of membrane raft proteins by the actin cytoskeleton. *J Biol Chem* 282:36682–36691.

165. Murase, K., Fujiwara, T., Umemura, Y., Suzuki, K., Iino, R., Yamashita, H., Saito, M., Murakoshi, H., Ritchie, K., and Kusumi, A. (2004). Ultrafine membrane compartments for molecular diffusion as revealed by single molecule techniques. *Biophys J* 86:4075–4093.

166. Morone, N., Fujiwara, T., Murase, K., Kasai, R.S., Ike, H., Yuasa, S., Usukura, J., and Kusumi, A. (2006). Three-dimensional reconstruction of the membrane skeleton at the plasma membrane interface by electron tomography. *J Cell Biol* 174:851–862.

167. Lenne, P.F., Wawrezinieck, L., Conchonaud, F., Wurtz, O., Boned, A., Guo, X.J., Rigneault, H., He, H.T., and Marguet, D. (2006). Dynamic molecular confinement in the plasma membrane by microdomains and the cytoskeleton meshwork. *EMBO J* 25:3245–3256.

168. Edidin, M. (2003). The state of lipid rafts: from model membranes to cells. *Ann Rev Biophys Biomol Struct* 32:257–283.

169. Simons, K., and Vaz, W.L. (2004). Model systems, lipid rafts, and cell membranes. *Ann Rev Biophys Biomol Struct* 33:269–295.

170. Brown, D.A., and London, E. (2000). Structure and function of sphingolipid- and cholesterol-rich membrane rafts. *J Biol Chem* 275:17221–17224.

171. Milhiet, P.E., Giocondi, M.C., and Le Grimellec, C. (2002). Cholesterol is not crucial for the existence of microdomains in kidney brush-border membrane models. *J Biol Chem* 277:875–878.

172. Lawrence, J.C., Saslowsky, D.E., Edwardson, J.M., and Henderson, R.M. (2003). Real-time analysis of the effects of cholesterol on lipid raft behavior using atomic force microscopy. *Biophys J* 84:1827–1832.

173. Hansen, G.H., Niels-Christiansen, L.L., Thorsen, E., Immerdal, L., and Danielsen, E.M. (2000). Cholesterol depletion of enterocytes. Effect on the Golgi complex and apical membrane trafficking. *J Biol Chem* 275:5136–5142.

174. Hao, M., Mukherjee, S., and Maxfield, F.R. (2001). Cholesterol depletion induces large scale domain segregation in living cell membranes. *Proc Natl Acad Sci USA* 98:13072–13077.

175. Dietrich, C., Yang, B., Fujiwara, T., Kusumi, A., and Jacobson, K. (2002). Relationship of lipid rafts to transient confinement zones detected by single particle tracking. *Biophys J* 82:274–284.

176. Saslowsky, D.E., Lawrence, J., Ren, X., Brown, D.A., Henderson, R.M., and Edwardson, J.M. (2002). Placental alkaline phosphatase is efficiently targeted to rafts in supported lipid bilayers. *J Biol Chem* 277:26966–26970.
177. Schroeder, R.J., Ahmed, S.N., Zhu, Y., London, E., and Brown, D.A. (1998). Cholesterol and sphingolipid enhance the Triton X-100 insolubility of glycosylphosphatidylinositol-anchored proteins by promoting the formation of detergent-insoluble ordered membrane domains. *J Biol Chem* 273:1150–1157.
178. Simons, K., and Toomre, D. (2000). Lipid rafts and signal transduction. *Nat Rev Mol Cell Biol* 1:31–39.
179. Cunningham, O., Andolfo, A., Santovito, M.L., Iuzzolino, L., Blasi, F., and Sidenius, N. (2003). Dimerization controls the lipid raft partitioning of uPAR/CD87 and regulates its biological functions. *EMBO J* 22:5994–6003.
180. Mineo, C., Gill, G.N., and Anderson, R.G. (1999). Regulated migration of epidermal growth factor receptor from caveolae. *J Biol Chem* 274:30636–30643.
181. Harder, T., and Engelhardt, K.R. (2004). Membrane domains in lymphocytes–from lipid rafts to protein scaffolds. *Traffic* 5:265–275.
182. Brown, D.A. (2006). Lipid rafts, detergent-resistant membranes, and raft targeting signals. *Physiology (Bethesda)* 21:430–439.
183. Kusumi, A., Nakada, C., Ritchie, K., Murase, K., Suzuki, K., Murakoshi, H., Kasai, R.S., Kondo, J., and Fujiwara, T. (2005). Paradigm shift of the plasma membrane concept from the two-dimensional continuum fluid to the partitioned fluid: high-speed single-molecule tracking of membrane molecules. *Ann Rev Biophys Biomol Struct* 34:351–378.
184. Kusumi, A., Ike, H., Nakada, C., Murase, K., and Fujiwara, T. (2005). Single-molecule tracking of membrane molecules: plasma membrane compartmentalization and dynamic assembly of raftphilic signaling molecules. *Semin Immunol* 17:3–21.
185. Meder, D., Moreno, M.J., Verkade, P., Vaz, W.L., and Simons, K. (2006). Phase coexistence and connectivity in the apical membrane of polarized epithelial cells. *Proc Natl Acad Sci USA* 103:329–334.
186. Kenworthy, A.K., Nichols, B.J., Remmert, C.L., Hendrix, G.M., Kumar, M., Zimmerberg, J., and Lippincott-Schwartz, J. (2004). Dynamics of putative raft-associated proteins at the cell surface. *J Cell Biol* 165:735–746.
187. Mayor, S., and Rao, M. (2004). Rafts: scale-dependent, active lipid organization at the cell surface. *Traffic* 5:231–240.
188. Pralle, A., Keller, P., Florin, E.L., Simons, K., and Horber, J.K. (2000). Sphingolipid-cholesterol rafts diffuse as small entities in the plasma membrane of mammalian cells. *J Cell Biol* 148:997–1008.
189. Pike, L.J. (2004). Lipid rafts: heterogeneity on the high seas. *Biochem J* 378:281–292.
190. Pike, L.J. (2006). Rafts defined: a report on the Keystone Symposium on lipid rafts and cell function. *J Lipid Res* 47:1597–1598.
191. Jacobson, K., Mouritsen, O.G., and Anderson, R.G. (2007). Lipid rafts: at a crossroad between cell biology and physics. *Nat Cell Biol* 9:7–14.
192. Lingwood, D., and Simons, K. (2007). Detergent resistance as a tool in membrane research. *Nat Protoc* 2:2159–2165.
193. Rajendran, L., and Simons, K. (2005). Lipid rafts and membrane dynamics. *J Cell Sci* 118:1099–1102.

15

GPI Proteins in Biogenesis and Structure of Yeast Cell Walls

MARLYN GONZALEZ • PETER N. LIPKE • RAFAEL OVALLE

Department of Biology
Brooklyn College
City University of New York
Brooklyn, NY 11210, USA

I.   Abstract

Glycosyl phosphatidylinositol (GPI)-anchored proteins have an uncommon function in yeasts and other ascomycete fungi: some of them are covalently incorporated into cell walls. Wall-anchored GPI proteins include adhesins, cell wall structural proteins, and enzymes active in cell wall biogenesis and assembly, as well as nutrient acquisition. The wall cross-linking is a result of cleavage of the GPI glycan between the glucosamine residue and the first mannose residue, followed by a transglycosylation reaction that links the reducing end of the GPI glycan remnant to the cell wall $\beta1,6$ glucan. Specific sequences in GPI proteins specify whether a GPI protein remains in the membrane, or is further processed and incorporated into the wall. Some GPI proteins are themselves transglycosidases active in wall assembly. GPI-dependent cell wall anchorage is being exploited for surface display of exogenous peptides and proteins with great stability and high surface density. Such yeast surface displays include single-chain antibodies and antigenic peptides, glycosidases for digestion of polysaccharides including cellulose, and lipases for both hydrolytic and synthetic reactions. Isolated walls with attached enzymes may be used as nonreproductive sources of enzymes or other proteins for bioremediation. Thus, the unique

THE ENZYMES, Vol. XXVI   321   ISSN NO: 1874-6047
DOI: 10.1016/S1874-6047(09)26015-X

ability of yeasts to cross-link GPI-anchored proteins to the cell wall results from a unique metabolism, and has consequences for both fungal life-styles and for use as a surface display platform.

"Oh sweet and lovely wall ..." [1]

II. Fungal GPI-Anchored Proteins and the Cell Wall: General Introduction

Much of our knowledge about Glycosyl phosphatidylinositol (GPI) bio-synthesis and processing derives from genetic and biochemical approaches in bakers' yeast. When compared to other eukaryotes, both the biochemical processes and the sequences of the processing enzymes are conserved. This conservation allows exploration of GPI presence and processing in other organisms by comparative genomics. There are ORFs with apparent GPI addition signals throughout the fungi and at least five other eukaryotic kingdoms: opisthokonts (animals and fungi), plants [2], alveolates (*Plasmodium*) [3], amoebozoa (*Dictyostelium*) [4], discicristates (*Trypanosoma*) [5], and in the domain Archaea [6–8].

A. The Types of GPI-Anchored Proteins in Fungi

Fungal-specific bioinformatic analyses for proteins with N-terminal secretion signals and C-terminal GPI addition sequences have been used to predict 58 GPI-anchored mannoproteins in *Saccharomyces cerevisiae* [9]. In contrast, *Candida glabrata* may have 106 [10], and *Candida albicans* may have about 115 GPI proteins [11]. In the pathogenic yeasts, many of the putative GPI-linked proteins are predicted to mediate interactions of the yeast with mammalian hosts. Known functions include: cell-cell and cell-substrate adhesion [12], modulation of the immune system [13], iron uptake [14, 15], drug resistance [16], cell wall assembly and modification [17, 18], fungal morphogenesis [19–22], substrate invasion [23, 24], and hydrolysis of proteins, carbohydrates, and lipids [25]. Thus, there are many functions of GPI-linked proteins in cell walls. Despite this large number of genes, proteomic analyses detect only a few to a few dozen GPI proteins in the wall of *S. cerevisiae*, *C. albicans*, *Cryptococcus neoformans*, *Schizosaccharomyces pombe*, and *Aspergillus fumigatus* [26–30]. Therefore, expression of different GPI protein genes is under transcriptional regulation [27, 28].

B. GPIs in Cell Walls

In many fungi GPI-linked proteins play a unique role: they are major constituents of cell walls, and they are also active catalysts in wall biogenesis and modulation of structure [17, 31]. Different from most other eukaryote kingdoms, fungi can modify GPI anchors on the cell surface, with the GPI lipid moieties being replaced with covalent cross-links to cell wall polysaccharides [32]. Therefore, some GPI proteins are not membrane associated, but are instead integral parts of a complex organelle exterior to plasma membranes (Figure 15.1).

The GPI wall cross-linked proteins include several classes. There are half a dozen adhesin genes in *S. cerevisiae* [12, 33] and several dozen in *C. albicans* [11]. Proteomic analyses of walls also detect GPI proteins like Cwp1p and Cwp2p, which are major structural components [27]. Several enzyme classes with major roles in wall biogenesis also show evidence of wall cross-linking (Section VII.B), including glycosyl transferases (also called carbohydrate-active enzymes) [34], aspartyl proteases [35–40], and phospholipases [41–43]. Perhaps the greatest numbers of genes for GPI proteins are those encoding "carbohydrate-active enzymes" with known or predicted glycosidase and transglycosidase activity [10, 44, 45].

Both wall- and membrane-associated GPI proteins are key effectors of cell wall synthesis, organization, and assembly. This review will concentrate on GPI protein structures and activities that are unique to fungal cell walls, rather than on their interactions with multicellular host organisms [13]. Additionally, we will summarize the use of covalently cross-linked wall GPI proteins as a platform for surface display of exogenous proteins for medical and industrial uses.

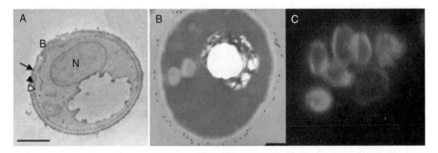

Fig. 15.1. GPI-anchored proteins in yeast cell walls. (A) Transmission electron micrograph of a permanganate-fixed cell showing outer wall proteins, many of which are GPI linked (arrow); glucan layer (filled arrowhead); plasma membrane/periplasmic region (open arrowhead); B bud scar; N, nucleus. (B) Micrograph of a yeast cell stained after sectioning with gold-labeled antibody to a GFP-GPI fusion protein. (C) Direct fluorescence of GFP-GPI protein in isolated, membrane-free washed cell walls. Note the retention of cell shape in the isolated walls. (See color plate section in the back of the book.)

III.   A Brief History of the Discovery of Fungal GPI Proteins in Yeast Cell Walls

The first evidence of GPI-anchored proteins in yeast was published in the late 1980s by several groups [46–48]. This work is well-described in other chapters in this volume. Studies of a surface protein called GP115 (Gas1p) led to the discovery that it was a major yeast GPI protein, and subsequent work has used this protein as a subject for GPI synthesis, processing, and transport [47] (Chapter 13). The association of GPI proteins with cell walls came later in the decade, with the convergence of ideas from several approaches.

Frans Klis' group in Amsterdam focused on the nature of the interactions between proteins and glycans in cell walls. They discovered a novel type of protein-polysaccharide bond between a GPI remnant and β1,6 glucan [49]. Our own work focused on the *S. cerevisiae* sexual agglutinins, and working with Janet Kurjan, we cloned and sequenced genes for two wall-associated sexual agglutinins: *SAG1* (α-agglutinin) and *AGA1* (a subunit of **a**-agglutinin) [50, 51]. At the same time, Widmar Tanner's group cloned *SAG1* and also *AGA2,* (disulfide-linked ligand subunit of **a**-agglutinin) [52, 53]. Sequence gazing led Reza Green to suggest that α-agglutinin is GPI anchored [50]. A series of domain deletion studies by Don Wojciechowicz and Cha Fen Lu in our lab showed that the GPI addition signal was necessary for surface localization, and also delineated the Sag1p binding domain in the N-terminal 325 residues [54, 55]. That left about 300 C-terminal residues of the 650 residue ORF to mediate surface display of the active site. A protein of 300 residues is not long enough to span the region between the membrane and the outer edge of the wall [56], so we proposed that the GPI oligosaccharide must be cleaved to release the protein from the membrane anchor. The reducing terminal of the glycan could then be a transglycosylation donor, with a wall polysaccharide being the receptor, and the transglycosylation reaction would form a covalent association of the protein with the wall matrix.

At a Cold Spring Harbor Yeast meeting, one of the authors (P.N.L.) showed Frans Klis a draft of the article explaining the idea [57]. Frans asked what P.N.L. thought the acceptor was. Having just seen a poster in which Roy Montijn (Frans' student) demonstrated a linkage between protein and cell wall β1,6 glucans, P.N.L. replied that β1,6 glucan was the obvious choice. We arranged that Roy would visit our lab, and bring some antibody to β1,6 glucan. Cha Fen Lu carried out elegant kinetic studies, demonstrating that α-agglutinin followed the classical secretory pathway, being initially N- and O-glycosylated and GPI anchored in the ER, hyperglycosylated in the Golgi, then exported to the exterior surface of the plasma membrane, released with loss of the GPI lipid moiety, and subsequently anchored to wall matrix (Figure 15.2) [54]. The combined efforts of the Klis, Lipke, Kurjan, and Bussey labs resulted in a paper demonstrating that the wall-bound form of α-agglutinin was indeed bound to β1,6 glucan, and was dependent on GPI anchorage of the protein. Mutations that decrease synthesis of β1,6 glucan synthesis decreased cell wall association of α-agglutinin and increased its excretion into the culture medium [54].

This model was subsequently justified by structural studies in Cabib's lab [58], which demonstrated specific linkage structures between GPI-modified wall glycoproteins and the wall polysaccharides. They showed that the remnant of the GPI glycan is transglycosylated to β1,6 glucan, which is in turn linked to the major yeast wall fibrous β1,3 glucan (Figure 15.3B). Small amounts of chitin are also covalently linked to both glucans, to form a cross-linked wall complex surrounding the entire cell [59, 60]. The cross-linking model has since been demonstrated to apply to adhesins, cell wall structural proteins, and many wall-associated enzymes in *S. cerevisiae* and other fungi [12, 17, 61].

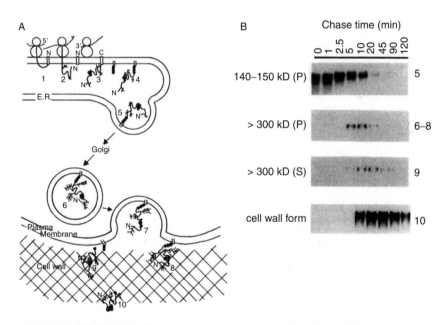

FIG. 15.2. (A) Model for biogenesis and secretion of a wall-anchored GPI protein in yeast ([72]; reprinted with permission). (B) Pulse-chase analysis of cellular fractions of α-agglutinin after induction. The cellular fractions are numbered according to their position in the scheme shown in part A ([54]; reprinted with permission).

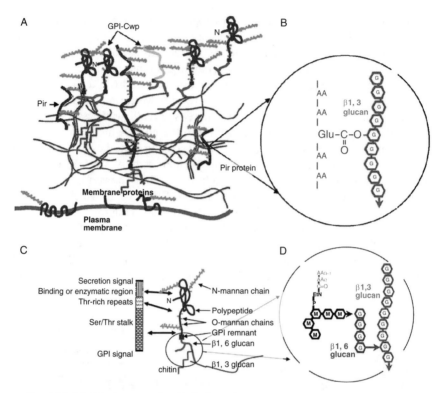

FIG. 15.3. Model of yeast cell wall ([32]; modified with permission). (A) Overall model; (B) structure of the protein-β1,3 glucan cross-link for a PIR protein. (C) Key to structures in Panel A, and the regions of GPI-anchored cell wall proteins. (D) Structure of the cross-link between a GPI-anchored wall protein and cell wall β1,6 glucan.

IV.   Structure of Yeast Cell Walls

A.   INTRODUCTION

A basic structure for yeast cell walls is shown in Figure 15.3. The wall is a covalently and H-bonded complex of polysaccharides and glycoproteins, with minor constituents including glycolipids and cations. Walls are thick, typically 100–200 nm, and can make up 25–30% of the cell wet mass [61]. The major polysaccharides form hydrogen-bonded fibrils that give the walls tensile strength. Proteins are cross-linked to the polysaccharides through two types of bonds: transglycosylation to GPI glycan remnants (Figure 15.3C and D) and esters between amino acid carboxyl groups and polysaccharide hydroxyls (Figure 15.3B) [25, 61, 62]. The ratio of manno-protein, β1,6 glucan, and β1,3 glucan molecules is approximately 1:1:1

[32, 63]. The mannoproteins are also disulfide cross-linked through Cys residues to one another (not shown in Figure 15.3) [64]. The wall therefore forms a discrete organelle that maintains its shape and many of its constituents after isolation and extraction with salts or SDS (Figure 15.1C).

B. POLYSACCHARIDES

1. β1,3 glucan

In *S. cerevisiae* and *C. albicans* wall fibrils are primarily β1,3 glucan molecules of more than 1000 monosaccharide units. These chains form H-bonded single and triple helices, and comprise about 40% of the wall mass. β1,3 Glucan is synthesized at the plasma membrane, and the product is extruded to the extracellular surface. The catalytic and transport activities probably reside in a single protein encoded by *FKS1* in vegetative cells or its paralog *FKS2* in spores [17, 65]. The *FKS* sequences are homologous to plant callose synthetases [66].

Although Fks proteins are necessary for glucan synthesis, and Fks1p is the target of the echinocandins (β1,3 glucan synthesis inhibitors), there has as yet been no demonstration of glucosyl transferase activity in the purified proteins [17]. The small GTP binding protein Rho1p is a regulatory subunit of glucan synthase (GS), and has other activities in regulation of cell polarity and wall synthesis as well [67]. In other fungi, chitin or cellulose serves as the fibrous component.

2. Chitin

In yeasts, chitin helps render the glucan fibrils insoluble, and makes walls tougher and more enzyme-resistant in older cells. It is also a major constituent of the bud scars. Its structure, biosynthesis, and cellular roles have been extensively reviewed [61, 67]. Chitin is synthesized by transfer of N-acetylglucosamine from UDP-GlcNAc to the 4-OH of another GlcNAc to form an oligosaccharide chain. The reaction is catalyzed by the transmembrane proteins Chs1p, Chs2p, or Chs3p, and the oligomer is extruded to the periplasmic space as it is formed. Two paralogous GPI-linked proteins, Crh1p and Crh2p, are critical for chitin cross-linking to cell wall glucans (Figure 15.3A; Section VII.B.5).

3. β1,6 glucan: A Cross-Linking Polysaccharide

A β1,3-branched β1,6 glucan polymer, with a degree of polymerization of a few hundred monosaccharides is a minor but important wall component. This β1,6 glucan cross-links the fibrous β1,3 glucan, GPI-anchored

mannoproteins, and chitin to form the bilayered fungal cell wall
(Figure 15.1). The structures of the cross-links were deduced in *tour de
force* analyses that used NMR and mass spectroscopy on the remnants of
extensive enzymatic digests of isolated cell walls [58, 60]. The reducing ends
of chitin chains are linked to hydroxyls in single sugar β1,3 branches in the
β1,6 glucan. The reducing end of the β1,6 glucan is in turn bonded to the β1,3
glucan (Figure 15.3C and D). The components of β1,6 GS remain mysterious,
but appear to include components of the β1,3 GS system [17, 68, 69].

C. Mannoproteins

About 40% of wall mass consists of mannosylated proteins. Short mannosyl oligosaccharides are linked to Ser and Thr hydroxyl groups (O-link),
and long chains are linked to the γ-amide N of Asn residues (N-link).
N-linked glycans core structures are similar to mammalian N-glycans [70].
In yeasts, the N-glycans of many wall proteins are elongated with up to 200
additional mannosyl groups, including an α1,6-linked mannosyl backbone
and short sidechain branches of α1,2- and α1,3-linked mannose. The structures and biosynthesis of these mannoprotein glycoconjugates have been
recently reviewed [25, 191, 192]. GPI-anchored cell wall proteins are linked
to β1,6 glucan through the reducing end mannose in the GPI glycan remnant [58]. Such GPI-anchored wall proteins form the electron-dense outer
layer of the wall [71] (Figure 15.1A). They may also be cross-linked to β1,3
glucan through ester bonds [61, 62].

1. **GPI-mannoproteins: a common design**. GPI-linked mannoproteins share
   a common domain organization (Figure 15.3C). In order, from the
   N-terminal, they have secretion signals, enzymatically active or ligand
   binding domains, a region of repeat sequences, often rich in β-branched
   amino acids Thr, Val, and Ile, a region rich in glycosylated Asn, Ser, and
   Thr residues, and a C-terminal region signaling GPI anchoring [72]. Some
   wall proteins lack one or more of these regions: wall structural proteins
   such as Cwp1p and Dan1p can lack compactly folded functional domains
   [61], and some other proteins such as α-agglutinin (Sag1p) lack repeat
   regions.
2. **To transglycosylate or not to transglycosylate, that is the question**. Some
   GPI-anchored proteins remain in the plasma membrane, and some are
   transglycosylated to β1,6 glucan in the wall. Our current understanding is
   that the partition between wall and membrane anchorage is not an all-or-
   none decision, and it appears that GPI proteins are localized in wall and
   membrane at ratios between 1:4 and 4:1, depending on the protein and
   the state of the cells [73–75].

The sequence of the "ω minus" region, located just upstream of the GPI addition site, specifies the extent to which proteins are released from the membrane and cross-linked to wall glucan. Most wall-associated proteins have some hydrophobic residues in the first five positions upstream from the ω cleavage site, especially at positions 2, 4, and 5 [9, 76]. In contrast, proteins that are primarily localized to the membrane lack hydrophobic residues and often have multiple basic residues in this same region. The relative proportion of proteins in each fraction has been determined after release from the membrane by PI-PLC, and release from the wall matrix with glucanase. These fractions are altered in predictable ways by site-specific mutations in this region [74, 75, 77].

Freiman and Cormack [73] showed that long Ser/Thr-rich glycosylated segments in the C-terminal part of a protein tend to direct the protein to the wall and can over-ride a basic and hydrophilic sequence upstream of the ω cleavage site [78]. This finding is especially interesting because Gas1p has these characteristics, and shows variable localization. Gas1p is normally membrane-associated, but becomes wall-associated when cells are stressed by mutation in *SPT14/CWH6*, which catalyzes the first step in GPI anchor biosynthesis (Chapters 1 and 2) [79]. Such a change in localization is consistent with a changed role: Gas1p is a transglucosylase, and a member of a widespread family apparently involved in elongating glucan chains (Section VII.B.3) [45, 78]. It is easy to conceive that Gas1p would be useful in wall synthesis in wild-type cells, but also in repairing a weakened wall in a GPI synthesis mutant.

To our knowledge, the functional consequences of changes in cell surface localization have only been tested once. Changing the ω minus region of the GPI-anchored membrane stress sensor Ecm33p [19], to that of the cell-wall-localized proteins Egt2p or Fit1p abrogates its ability to allow growth at high temperature [80]. Thus, the localization to membrane or wall probably affects function of yeast GPI-anchored proteins in many ways.

V. Ordered Cell Wall Assembly and Addition of GPI Proteins

The first evidence for an ordered addition of cell wall components came from observations that *S. cerevisiae* protoplasts growing in osmotically supportive liquid medium generate an incomplete cell wall consisting only of β1,3 glucan and chitin [81]. Nevertheless, the regenerating walls retained their cell shapes [67]. Using temperature-sensitive *rho1* mutants defective in β 1,3, glucan biosynthesis, Roh *et al.* [67] elegantly demonstrated that addition and assembly of cell wall components during cell wall formation proceeds in a step-wise, ordered fashion. Rho1p is involved in multiple cellular functions [82, 83] and is also the regulatory subunit for the β1,3 GS complex that associates with the plasma membrane [84]. Acting as a GTPase, Rho1p can rapidly switch synthesis of β1,3 glucan on and off by interconverting from its GTP to its GDP-bound form, respectively [85]. The *FKS1* and *FKS2* genes probably encode the catalytic subunits of the GS complex and their direct interaction with Rho1p at the plasma membrane has been demonstrated by immunoprecipitation [84]. This Rho1p-GS complex primarily accumulates at sites of active cell wall biosynthesis such as during bud cell wall formation and cell wall assembly, remodeling and repair.

  rho1 Mutants severely impaired in β1,3 glucan synthesis exhibit concomitant reductions in the levels of β1,6 glucan and mannoprotein. In these mutants, the missing protein and glucan could not be found in either concentrated media or trapped within the wall in unbound form, thus, their absence at the wall was thought to reflect a limitation in the number of β1,3 glucan acceptor molecules for cross-linking of the glucan–protein complex [67]. The β1,6 glucan–protein complex was not linked to chitin either, as it is in mutants with reduced β1,6 glucan content [59], since mannoproteins could not be detected in the insoluble fraction of glucanase-treated walls. On the other hand, formation of a β1,3-β1,6 glucan network proceeds undisturbed in the absence of effective incorporation of mannoproteins into the wall. These results show that formation of the wall proceeds as follows: extrusion of β1,3 glucan oligosaccharides as they are synthesized by the plasma membrane Rho1p-regulated GS complex. The β 1,3 glucan oligomers are assembled by Gas/Gel/Phr proteins (Section VII.B.3) into a flexible polysaccharide network onto which β1,6 glucan oligomers are cross-linked [86]. Mannoproteins, the last components added, are cross-linked directly to the β1,3 glucan backbone (PIR proteins; Figure 15.3B) [62], or indirectly by formation of a GPI glycan linkage to β1,6 glucan (Figure 15.3C and D and Figure 15.4; Section VII.B.2), or noncovalently associated with the wall (not shown).

VI.   Phylogenetics of GPI-Cell Wall Transglycosylation

  Genomic analyses predict that GPI-anchored proteins are widespread and may be universal among the fungi. Specifically, a survey of 18 fungal genomes in three phyla showed that there are GPI addition signals in each, and implied conserved roles in wall biogenesis and structure [66] [87]. This result leads to the question of how widespread are GPI proteins that are cross-linked to cell wall glycans by transglycosylation of the anchor. These

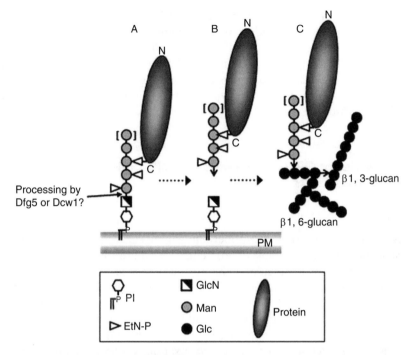

Fɪɢ. 15.4. Synthesis of a transglycosylation link between a GPI protein and wall glucan ([25]; modified with permission). (A) GPI-anchored protein showing membrane association. (B) Cleavage by a glycosidase between the GlcN and the first mannosyl residue. The vertical arrow denoted the reducing end. (C) Cross-link formed between the cleaved GPI remnant and the cell wall β1,6-β1,3 glucan complex.

cross-links are present in each ascomycete budding yeast that has been investigated. In addition, there is biochemical evidence for GPI-glycan cross-linking in the filamentous ascomycete *Aspergillus niger* [88] (but not in *A. fumigatus*, [89]), and there is a cell wall associated GPI protein in the fission yeast *Schizosaccaharomyces pombe* [20]. On the other hand, a recent bioinformatic analysis of adhesins (Figure 15.4), the most common GPI transglycosylated proteins in *S. cerevisiae* and *C. albicans*, shows that they are not GPI linked in *S. pombe* [90]. In the basidiomycetes, *Cryptococcus neoformans* phospholipase Plb1p is released from walls as a complex with β1,6 glucan, good evidence for transglycosylation [42].

BLAST and PSI-BLAST searches for homologs of the putative GPI transglycosylases *DCW1* and *DFG5* (Section VII.B.2) uncover orthologous sequences in 15 tested ascomycete genomes (including *A. fumigatus*) and the basidiomycete smut fungus *Ustilago maydis*, but not in other basidiomycetes, including *C. neoformans*, where there is biochemical evidence for cross-linking [42]. This analysis predicts that GPI-wall transglycosylation is present in ascomycetes and some basidiomycetes. However, there is a conundrum: *A. fumigatus* shows no proteomic evidence for GPI-glucan cross-linking and yet it contains *DCW1* and *DFG5* homologs and *Cryptococcus* lacks close homologs of these genes but apparently has wall cross-linked GPI proteins [88, 89]. We await additional information to resolve this issue.

VII. Roles of GPIs in Biogenesis of GPI-Cell Wall Cross-Links

A. PHOSPHOETHANOLAMINE TRANSFERASES

As in other organisms, phosphoethanolamine (EtN-P) transferases form part of the GPI anchor biosynthetic complex where they function to add ethanolamine phosphate side chains to the mannose residues in the GPI anchor glycan. To date, three EtN-P transferases have been described in *S. cerevisiae*: Mcd4p, Gpi7p/Las21p and Gpi13p. They add EtN-P groups to the first, second and third mannose residues, respectively, of the tetra- and pentamannose chains of the glycan core of the GPI anchor [91, 92] (Figure 15.4). Mcd4p and Gpi13p are essential proteins residing on the lumenal side of the ER membrane. Therefore, addition of EtN-P to the first and third mannoses occurs there. *GPI13* is essential because it adds the EtN-P to the C6 hydroxyl group of the third mannose and links the GPI anchor to the protein via an amide bond [93]. In *gpi13* mutants, GPI anchors are not attached to proteins through other EtN-P groups and reduction of GPI anchoring leads to overall cell wall fragility [94]. Recent unpublished results in our lab show that cells heterozygous for *GPI13* hypersecrete a GFP labeled cell wall GPI glycoprotein and have increased cell surface fluorescence relative to wild-type cells. It is likely that in these cells the synthesis of GPI proteins is upregulated to compensate for the high failure rate of forming the protein-GPI linkage.

In contrast to Mcd4p and Gpi13p, Gpi7p/Las21p is not essential for viability. It resides at the plasma membrane where it adds EtN-P to the second mannose of the GPI anchor liposaccharide. Deletion of *GPI7* compromises cell wall integrity and leads to deficiencies in polarized cell growth and cell separation [95], and to loss of virulence in *C. albicans* [96]. In preparation for cell separation, the cytoplasmic kinase Cbk1p is targeted to the bud neck, where it mediates activation of the transcription factor Ace2p. Ace2p in turn upregulates the expression of daughter-specific cell wall genes that localize to the bud neck and are anchored to the cell wall via

GPI anchors. In *gpi7* cells, the bud neck GPI-proteins Cts1p, Scw11p, Dse1p, Dse3p, Eng1p and Egt2p are mislocalized. Egt2p in particular is a glucanase which, when mislocalized, leads to cell wall damage and activation of the cell wall stress response. Concomitant with a response to cell wall damage is a halt in the cell cycle that prevents further growth of the budding cell [95]. A double deletion of *GPI7* and *EGT2* suppresses daughter cell growth arrest by preventing mislocalization of Egt2p. However, in these cells, a cell separation defect continues, which confirms that addition of EtN-P to the second mannose of the GPI-glycan is required for completion of cell separation [95].

B. GPI-Anchored Protein in Wall Biogenesis

We review here recent evidence on activities of some yeast GPI-anchored proteins important in cell wall assembly and regulation.

1. Dcw1p and Dfg5p

In an effort to identify the enzymes responsible for cleaving the GPI anchor and cross-linking the GPI-remnant protein to the cell wall, Kitagi *et al.* [97] searched the ORFs of the *S. cerevisiae* genome database for homologs of bacterial mannosidases known to participate in similar enzymatic reactions. They identified genes for two putative proteins that share significant homology with bacterial family 75 mannosidases [34]. Dcw1p and Dfg5p are GPI mannoproteins of the plasma membrane, although in *C. albicans* Dfg5p is also localized to the cell wall [98]. A *dcw1* deletion rendered the cells hypersensitive to the cell-wall-digesting glucanase Zymolyase, whereas deletion of *DFG5* did not seem to cause a cell wall defect. Both disruptants exhibited normal morphology and grew normally in rich and complete synthetic media [97]. These two genes are partially functionally redundant: combined deletion of *DCW1* and *DFG5* leads to synthetic lethality that can be rescued by overexpressing a *DFG5* allele from the *GAL1* promoter. The *dcw1Δ dfg5Δ* cells appear round and large, and almost completely revert to wild type size under conditions that activate expression of the *DFG5* allele [97]. Following glucose repression of the rescuing gene, *dcw1Δdfg5Δ* cells quickly develop hypersensitivity to the cell-wall-disrupting agents Calcofluor white and β-glucanase, and secrete the GPI-anchored cell wall protein Cw1p [99]. In *S. cerevisiae, DFG5* and *DCW1* are required for bud formation [99] and *DFG5* is required for agar invasion and for growth at alkaline pH [100]. In *C. albicans*, these genes have similar functions: at least one of them is required for growth [98], and *DGF5* is required for hyphal development.

Combined, these observations strongly implicate Dcw1p and Dfg5p in cell wall biogenesis and as required for cell growth and for GPI-mediated anchoring of mannoproteins to the cell wall (Figure 15.4). However, their specific role in cell wall biogenesis remains to be established.

2. Kre1p

Kre1p is a heavily O-glycosylated GPI-linked glycoprotein that primarily resides in the plasma membrane, although it has also been found cross-linked to wall β1,6 glucan [101]. *kre1* Deletants exhibit a 30–40% reduction in cell wall β1,6 glucan along with shortening and less branching of residual glucan chains compared to wild-type cells [102]. Interestingly, despite the significantly lower levels of β1,6 glucan in *kre1Δ* cell walls, cell wall manno-protein content remains close to that of wild type, although changes occur in their level of expression and distribution at the cell surface [103]. Although the precise function of Kre1p in the synthesis and assembly of cell wall β 1,6 glucan remains to be established, it is predicted to participate in the late stages of β1,6 glucan formation, likely by branching the β1,6 glucan chains to provide transglycosylation acceptors [103].

3. GAS/GEL/PHR Family

This family of enzymes includes the *GEL* genes from *A. fumigatus* [21], *GAS* genes in *S. cerevisiae* [104], and *PHR* genes in *C. albicans* [105]. Deletion of these genes leads to wall defects similar to mutations in glucan synthesis *fks1* mutants [22]. *GAS/GEL/PHR* are GPI-linked β1,3-glucano-syltransferases required for proper cell wall assembly and remodeling. These enzymes appear to function by transglycosylation of β1,3 glucan chains, which are extruded through the plasma membrane as they are synthesized by a GS complex. Short extruded oligomer chains may be concatenated into long strands by the Gas/Gel/Phr proteins. *In vitro* bio-chemical assays suggest that these enzymes work by first cleaving the internal glycosidic linkage of a β1,3 glucan molecule and transferring the newly generated reducing end to the nonreducing end of an acceptor β1,3 glucan chain [45]. The resultant β1,3 glucan strand can then be woven into a lattice that gives the cell wall its tensile strength. Gas/Gel/Phr proteins belong to the GH72 family of glycosidases/transglycosidases, and are fur-ther subdivided into two distinct groups denoted GH72+ and GH72– depending on the presence or absence, respectively, of a cysteine-rich domain or "Cys-box" near the C-terminus [86]. Enzymes of the GH72+ subgroup such as Gas1p and Gas2p require the Cys-box for proper folding and activity of the catalytic domain and share similar substrate specificity. In contrast, Gas3p, Gas4p, and Gas5p lack Cys-boxes, and display different preferences for cleavage sites [86]. This difference in substrate specificity displayed among enzymes of the two subfamilies is hypothesized to help satisfy the cellular needs to remodel β1,3 glucan under differing environmental conditions.

The C-terminal regions of these proteins contain heavily O-glycosylated Ser- rich regions, followed by basic residues and a GPI anchor addition signal [48]. All Gas proteins localize to the cell surface: Gas3p and Gas5p undergo further processing at the GPI anchor to become cross-linked to the wall, whereas Gas1p partitions both to the plasma membrane and the wall [27].

4. Yapsin/SAP Family

The yapsins/SAPs are a conserved family of GPI-linked aspartyl proteases that are synthesized as zymogen precursors [106]. Yapsins cleave after basic residues, and like mammalian β secretases [107] localize to the cell surface, where the proenzyme is activated by cleavage of an inhibitory domain [37]. To date, five or six yapsin genes have been described in *S. cerevisiae*, *YPS1, 2, 3, 5, 6*, and *7*, and homologs of these genes have been identified in *C. albicans* [108], *C. glabrata* [109], and *Aspergillus oryzae* [110]. In *S. cerevisiae*, yapsin genes encode GPI-linked plasma membrane proteins, with the exception of *YPS7*, which has a C-terminal motif that is representative of cell-wall-localized GPI proteins (Section IV.C.2) [76]. Because *S. cerevisiae* yapsin null mutants display hypersensitivity to cell-wall-disrupting agents, and multiple yapsin deletions result in reduced levels of β glucans in the wall [37], yapsins are believed to play important roles in the maintenance of cell wall integrity and stability. Since these yapsin mutants show no deficiencies in their ability to synthesize the glucan polymers [37], the defect appears to be in cell wall assembly, rather than in synthesis of the components. Additional studies implicate yapsins in cell surface mannoprotein turnover [35], regulation of the cell wall stress response [37, 111], and as key virulence factors in *C. albicans* [18].

S. cerevisiae yapsins 1 and 2 were first isolated as suppressors of mutations in *KEX2*, which encodes an intracellular serine protease that processes secretory proteins such as pro-α factor and the cell-wall-associated exoglucanase *Exg1p* [112]. The remaining yapsins were identified through similarity searches [111]. Like Kex2p, yapsins cleave C terminal to basic motifs, although Kex2p is specific for Lys-Arg pairs and yapsins recognize both dibasic and monobasic residues. Obvious potential cleavage sites are in the clustered basic residues in the ω-region of GPI proteins such as the *GAS/GEL/PHR* transglucosidases [35, 113] and in the yapsins themselves [37]. Gas1p is indeed a substrate of *ScYps1* [37].

In the case of *S. cerevisiae* Yps1p, there are basic amino acids flanking a loop insertion that separates its two catalytic subunits. Cleavage and removal of the loop insertion is required for the enzyme to achieve its proper catatytic conformation (i.e., autocatalytic zymogen activation) [37]. Because yapsins are active at the cell surface and harbor the ability to process basic motifs common to cell surface mannoproteins, it is hypothesized that they act to regulate the activity of enzymes involved in cell wall assembly and remodeling and as "sheddases" involved in manno-protein turnover [35].

5. CRH Family

The *S. cerevisiae* CRH family consists of three paralogous genes encoding GPI-anchored surface glycoproteins with similarity to bacterial family 16 glycosidases and plant xyloglucan transglycosidases [17, 34, 114]. The sequences are conserved across ascomycetes and basidiomycetes [66]. Deletions of *CRH1* and *CRH2* cause changes in chitin deposition and prevalence, and *crr1* deletion mutants have defects in biogenesis of spore walls, which have a chitosan (poly-$\beta1,4$ glucosamine) layer [114]. The proteins are localized to both wall and plasma membrane fractions. In dividing cells, Crh1p localizes at the tips of emerging buds and then the septum area later in the cell cycle. Chr2p is at the maternal side of the bud neck and in bud scars. Both show some localization to lateral walls, with great increases at high temperature [115]. Cabib showed that Crh1p and Crh2p are functionally overlapping proteins necessary for cross-linking of nascent chitin to glucan. In *chr* mutants, the amount of chitin that is not cross-linked to glucan rises, and cells become slightly more susceptible to Congo red. The strains *crh2Δ* and *crh1Δ crh2Δ* fail to compensate for cell wall stress caused by mutations in $\beta1,3$ glucan synthetase gene *FKS1* or the transglycosidase gene *GAS1*. Such cells remain hypersensitive to Congo red and show high levels of spontaneous lysis. In contrast, these glucan mutants compensate by increasing lateral wall chitin if they have functional *CRH1* and/or *CRH2* genes. These properties are consistent with Chr1p and Chr2p being functionally redundant GPI-anchored proteins necessary for cross-linking of nascent chitin to pre-existing $\beta1,6$ glucan. Recently, Cabib *et al.* [44] showed that fluorescently labeled oligomers of $\beta1,6$ glucan are acceptors for chitin transglycosylation in a cell-free system. The reaction is enzymatic and is absolutely dependent on presence of Crh1p or Crh2p (Figure 15.3C). Thus, there is now an example where GPI-anchored proteins are essential enzymes for wall assembly in yeast.

VIII.   Surface Display in Yeast

The understanding of how proteins are anchored in the fungal cell wall has now advanced to a stage where expression of heterologous proteins on the fungal cell wall can be exploited for commercial, medicinal, and scientific uses (Figure 15.5) [116–122]. Yeast transformed to display one or more heterologous proteins on their cell walls are called Surface Display (SD) or "Arming" yeast (SDY). Wall-anchored enzymes (i) have access to molecules that cannot cross the plasma membrane; (ii) remain in high concentration in the wall; (iii) are stabilized by their linkage to the wall matrix, perhaps because they are less vulnerable to proteases [118, 122–124]; and (iv) can be extracted without lysis of the cell, thus minimizing the need for extensive protein purification [118, 122].

Most research into heterologous surface display of proteins has been conducted in *S. cerevisiae*. However, several papers have explored transgenic expression of wall-anchored proteins in other species including *Pichia*

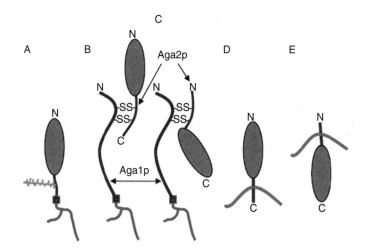

FIG. 15.5. Attachment and cell wall anchorage for surface displayed heterologous proteins in yeast. The heterologous displayed proteins are shown as gray ovals. Symbols and linkages are as in Figure 15.3: black lines: polypeptide; dark gray, β1,6 glucan; light gray, β1,3: glucan; light gray brushes, high mannose glycoprotein N-glycan; -SS-, disulfide bond. N- and C-termini of polypeptides are labeled. (A) Surface-displayed protein fused to the C-terminal and GPI region of an endogenous cell wall protein such as Sag1p or Cwp1p; (B) Surface-displayed protein fused to the N terminal of Aga2p, which is anchored to the GPI cell wall protein Aga1p. (C) Surface-displayed protein fused to C terminal of Aga2p. (D) and (E) Surface-displayed proteins fused to the N or C terminal, respectively, of a cell wall Pir protein fragment, which mediates cross-linking to β1,3 glucan.

pastoris [125], *Kluyveromyces lactis* [126], *Hansenula polymorpha* [127], and *Yarrowia lipolytica* [128]. Ectomycorhizae are also being considered as hosts for display of proteins that trap heavy metals [118].

A. Yeast vs. Bacterial Expression Systems

Surface display on *S. cerevisiae* has the following advantages over bacterial expression systems: (i) eukaryote processing mechanisms for correct folding and glycosylation of proteins, (ii) abundant surface area, and (iii) a greater protein/area ratio for expression of protein. In a head-to-head comparison of bacteria and yeast surface expression systems, Mischo *et al.* [129] determined that conformation-dependent monoclonal antibodies recognized a colorectal cancer protein, NY-ESO-1, more often and/or more robustly when expressed on *S. cerevisiae* cell walls than when expressed in phage particles. In another comparative study, Bowley *et al.* [123] used a cDNA library of IgG light chain sequences from an HIV patient testing for gp120-specific antibodies. The library, expressed as Aga2p-Fv fusion chimeras in *S. cerevisiae* (Figure 15.5B) was compared against a library expressed in *E. coli* from the phage display vector, pComb3X. There were threefold more specific antibody molecules on the surface of yeast cells than on *E. coli* cells. Both of these studies convincingly demonstrate that protein folding of eukaryotic proteins proceeds more efficiently in yeast than in bacteria [130].

B. Anchorage Modes for Surface Display

The modes of anchorage of heterologous proteins to cell walls include: (i) GPI linkage to β1,6 glucan (Figure 15.5A) [131]; (ii) indirect GPI linkage via disulfide bonding to an auxiliary protein (Figure 15.5B and C) [132]; (iii) PIR linkage to β1,3 glucan (Figure 15.5D and E) [133]; and (iv) non-covalent sequestration in the cell wall [134, 135], either by protein–protein or protein–carbohydrate interactions.

Pioneering work on anchorage via direct and indirect GPI-, PIR-, and noncovalent linkage in the wall was performed in the laboratories of Toh-e [94], Klis [136], Lipke [50, 54, 137], and Tanner [138, 139]. These systems were first investigated for heterologous expression in the labs of Klis [131, 140], Jigami [133], and Wittaker [141].

1. Direct GPI Anchorage

Heterologous proteins can be linked directly to wall polysaccharides by inserting the sequence of interest between the secretion signal sequence and the GPI anchorage signal sequence of a surface protein like α-agglutinin (Sag1p) (Figure 15.5) [131, 142, 143]. A great advantage of such anchoring through GPI-linked proteins is that their expression at the outer surface of the wall promotes interaction with large molecules such as proteins [144]. The downsides of direct GPI anchorage are that (i) recovery of the chimera requires proteolytic cleavage [145], and (ii) the method does not work for C-terminal active enzymes [134]. GPI anchorage sequences of different cell wall proteins show differential expression in the wall when promoter, signal sequence, and heterologous protein sequences were held constant; Cwp2p, Sag1p, and Sed1p C-terminal sequences were the highest expressing constructs [146].

2. Indirect GPI Anchorage

Indirect GPI linkage to cell wall using a fusion of a heterologous protein to the N- or C-terminus of S. cerevisiae Aga2p is the next most popular method of cell wall anchorage (Figure 15.5B and C) [132]. The Aga2-hetero chimera is processed, linked to Aga1p through a pair of disulfide bridges, and the protein complex is anchored to the outer cell wall through the Aga1p GPI anchor [51, 138, 141, 147–149]. This expression vehicle has several advantages over direct GPI-wall linkage in that the Aga2-hetero fusion can be extracted from the wall by disulfide reduction; the chimera is exposed on the outer wall surface; and commercial plasmids for N- and C-terminal fusions Aga2-hetero fusions are available (Figure 15.5B and C, respectively).

3. PIR Anchorage

Anchorage to β1,3 glucan of the inner cell wall using fusions to the PIR internal repeats is another Surface Display method [133, 150, 151]. The chimeric protein is anchored to β 1,3 glucan chains of the inner cell wall through an alkali-sensitive linkage [62, 136]. Similar to Aga2 chimeras, the heterologous sequence can be expressed as C-terminal or N-terminal fusions (Figure 15.5D and E, respectively) [151] and can be liberated from the cell wall by treatment with mild base [139, 152]. Because the Pir-glucan linkage occurs on the internal layer of the cell wall, Pir chimeras may only react with wall-permeant molecules [64, 153], but can be useful as supports for lipases [154], glycosidases [133, 150], and lyases [143].

4. Other Anchorage Systems

Aside from direct and indirect covalent anchorage to wall carbohydrates, SDS-extractable Flo1p-hetero fusions use the ability of the N-terminal region of Flo1p to bind to wall carbohydrates noncovalently [134, 155].

5. Promoters and Signal Sequences

Appropriate timing for expression of heterologous proteins is controlled by choice of the promoter sequence, which allows either constitutive expression [133, 156] or expression in selective environments or conditions (*GAL1, AOX1,* or *ICL1*) [134, 148, 157, 158]. Secretion signal sequences from *SUC1, KRE1, FLO1, AGA2,* and *CWP2* are often used to target constructs to the extracellular cell-wall–plasma membrane compartment in *S. cerevisiae* and other fungi [130, 135, 149, 158, 159]. Conversely, native signal sequences of heterologous fungal proteins are also effective in *S. cerevisiae* [120, 160]. Therefore, signal sequences appear to be recognized across species lines.

C. Uses of Surface Display of Heterologous Proteins

The expression of heterologous proteins by yeast surface display falls into the following general categories of uses: (i) substrate modification; (ii) peptide manufacture, maturation, and presentation; (iii) directed evolution; (iv) structural mapping of protein–protein interactions; and (v) biosensors.

1. Substrate Modification

A heterologous enzyme catalyzes a reaction leading to the creation of useful or the destruction of undesirable compounds. In these cases, the yeast wall serves as a platform for the reactants to interact with the catalyst.

a. Conversion of Plant Polysaccharides and Oils to Biofuel

The production of ethanol by anaerobic fermentation of glucose by *S. cerevisiae* is an example of degradation of substrate to form useful products. Previously fermentation has been used to convert sugar cane sap (sucrose) into ethanol for fuel [122]. However, since plant biomass is primarily a mix of glucans and other sugar polysaccharides, research is ongoing to integrate genes encoding enzymes necessary for total degradation of plant structural and storage polysaccharides. The hydrolysis of cellulose and starch are two different problems that can be solved with the same approach.

i. Starch Hydrolysis. *S. cerevisiae* cannot hydrolyze external starch (α1,4 glucan) even though it manufactures cytoplasmic amylases, because neither the reactants nor the enzymes cross the plasma membrane. Murai *et al.* sequentially transformed a single strain of *S. cerevisiae* with Sag1p fusions of glucoamylase (*Rhizopus oryzae*) and α-amylase (*Bacillus stearothermophilus*) (Figure 15.5A) [161]. The transformed yeast gained the ability to hydrolyze and use starch as sole carbon source.

ii. Cellulose Hydrolysis. The hydrolysis of cellulose is a greater problem. Wood is composed of crystalline cellulose (β1,4 glucan), amorphous cellulose, amorphous pectins (miscellaneous carbohydrate polymers), and amorphous lignin (polyphenols). Crystalline cellulose can only be attacked at the crystal surface, and solubilized pectins and polyphenols can act as inhibitor of cellulases; thus the saccharification of wood must be approached in stages. Fujita *et al.* [162] cotransformed *S. cerevisiae* with Sag1p fusions of surface-expressed β-endoglucanase (which hydrolyzes crystalline cellulose to oligosaccharides), cellobiohydrolase (which converts the oligosaccharides to cellobiose and glucose), and β-glucosidase (cellobiase) from *Trichoderma reesei*. Transformed yeast simultaneously expressing the three enzymes could use phosphoric acid-treated cellulose as the sole carbon source, and produce ethanol from cellulose.

b. Lipases

Expression of wall-anchored lipases allows yeast to hydrolyze undesirable fats and oils into fatty acids and alcohols under aqueous conditions. The products can then be fermented or otherwise metabolized. Such yeast can ferment agriculturally and industrially produced oils to ethanol inexpensively [120].

Lipases can also be used to synthesize fatty acid esters. In conditions where the concentration of alcohol and fatty acid exceed the concentration of water, wall-anchored lipases become biosynthetic and produce novel esters that can be used as raw materials for fuels or plastic [130, 134, 135, 155]. Enzyme-catalyzed biosyntheses are desirable because enzymes (i) are inexpensive, (ii) catalyze difficult reactions to make specific products, (iii) minimize side reactions, and (iv) function under gentle reaction conditions that minimize degradation of products. For these reasons, enzymatically generated products are of high purity.

2. Bioremediation

Surface display of heterologous enzymes is also useful to destroy undesirable organic compounds or to sequester toxic metal ions [118, 122].

a. Degradation of Xeno-Organic Compounds

Organophosphate hydrolases (OPH) cleave carbon–phosphate bonds to destroy pesticides and nerve agents. These hydrolyases have a narrow range *in vivo* because cytoplasmic enzymes can only degrade molecules that can cross membranes. External presentation of organophosphate hydrolases extends their range *in vivo* [143]. Degradation of a toxic phenol, Bisphenol-A, was accomplished by surface expression of a *Kluyveromyces lactis* oxidase-Sag1p fusion expressed in *Pichia pastoris* [154]. Future candidate proteins for surface expression include laccases [163] and lignin peroxidases for oxidation of chlorophenols [164].

b. Sequestration of Toxic Metals

Toxic metal ions cannot be destroyed, they must be sequestered [118]. Cells that surface display histidine-rich peptides [165] or cysteine-rich peptides [166] are able to sequester transition metals thus decreasing their availability in soils. Peelle *et al.* tested surface expression of homopolymeric and mixed amino acid hexapeptides for heavy metal trapping potential. They found that histidine homopolymers and co-peptides were most effective for selective precipitation of heavy metals [167].

A downside of using genetically modified organisms (GMO) for bioremediation is the concern that release of GMOs into the environment will have unforeseen consequences [122]. Therefore enzyme expression on sterile spores [122], or heat-killed whole cells [168] have been proposed as answers to the release of GMOs. Other acceptable answers to GMO release include trapping cells within porous polymers [169], or beads [170], and timed expression of flocculation proteins [157]. Nongenetic methods for bioremediation are also under investigation such as cross-linking sulfur amino acids [171] or cross-linking enzymes [172] to yeast surfaces.

3. De novo *Biosyntheses*

Similar to *de novo* production of lipid esters, wall-anchored enzymes can manufacture specific linkages between reactant molecules, especially important in oligosaccharide synthesis. Abe *et al.* showed directed synthesis of manno-oligosaccharides by co-expression of two mannosyltransferases on *S. cerevisiae* cell surfaces [133], whereas Jacob *et al.* using a *Trypanosome cruzi* trans-sialidase-Aga2p fusion, demonstrated the transfer of sialic acid residues from sialyl 2,3 lactose to other carbohydrates. Sugar-coating of various biomaterials can lower immune response and speed healing.

Facile biotinylation of peptides is useful in the production of single-chain monoclonal antibody peptides (scFv). Parthasarathy *et al.* surface-expressed a biotin ligase BirA-Aga2p chimera that could biotinylate soluble proteins [173] (Figure 15.5C). Schollar *et al.* [174] expressed BirA in Golgi using a Kex2signal-BirA-Kex2 construct, and successfully cotransformed the strain to excrete biotinylated scFv. Boder and Wittrup [132] expressed Aga2-scFv chimeras and released the fusion proteins from cell wall with reducing agents, then used Factor Xa to separate the scFv sequence from Aga2p (Figure 15.5B).

4. Peptide and Protein Presentation

Eukaryotic proteins displayed on yeast wall can be useful as vaccines. Schreuder et al. [159] first considered the medical applications of surface display of medicinally important proteins by displaying hepatitis B coat protein on S. cerevisiae cell walls. Tamaru et al. [175] investigated using surface display of iridovirus viral coat proteins as edible vaccines for fisheries.

5. Peptide Libraries

Surface expression of short peptide sequences has been used to determine motif function. Bidlingmaier and Liu [176] inserted cDNAs from a fragmented human testis mRNA library into an Aga2p fusion cassette; they used fluorescent peptides and cell sorting to identify yeast expressing peptide motifs that bind phosphotyrosine residues and then sequenced the plasmids to identify active sequences. Similar procedures were used to identify peptide motifs that conferred tolerance to n-nonane [177] or carbohydrate binding motifs [178].

6. Directed Evolution

A library of peptide sequences can be constructed by random mutation of a single gene using error-prone PCR, either using modified nucleosides [179] or error-prone DNA polymerase [180]. After transformation, a population of different sequences is expressed on the fungal surface as fusion proteins. Fluorescent labeling and flow cytometry is then used to acquire functional transformants, and those transformants that improve upon the properties of the original sequence are selected for mutation in later cycles.

Directed evolution of surface-displayed proteins has been used to (i) accelerate the reaction rates of enzymes and (ii) create mimics with higher binding affinities for receptor molecules. Shiraga et al. surface expressed variants of Rhizopus oryzae lipase-Aga1p chimeras where the lipase sequences were mutated by error prone PCR; they were able to isolate a line that expressed 10^4-fold greater activity than the original construct [180]. Springer's group used SDY to increase affinity of integrin I domain for ICAM-1; such high affinity binding might be useful to block ICAM-1 and reduce lymphocyte migration and invasion [181]. Wild type or PCR-mutated integrin I domains were inserted in-frame between HA and c-Myc tags at the Aga2 C-terminal. The resulting SD proteins could bind anti-c-Myc and anti-HA monoclonals, as well as the integrin ligand ICAM-1. Yeast lines that bound c-Myc and HA antibodies were screened for ICAM-Fc fragment binding by flow cytometry with fluorescent secondary antibody. The best binding sequences were sequenced and expressed from *E. coli*, and their binding to ICAM-1 was analyzed using surface plasmon resonance. Finally, two point mutations were combined to form an optimized sequence with an 8-fold increase in k_{on} a 3×10^4-fold decrease in k_{off}, combining to give a 2×10^5-fold increase in affinity for ICAM-1. Other ligand/receptor studies using directed evolution of SD proteins include scFv/carcinoembryonic antigen [182], MHC/T cell receptor protein [183], and TNF-α/DR5 receptors [184].

7. Structural Mapping of Protein–Protein Interactions

When error-prone PCR is used to generate mutations in the sequence of an antigen, some mutations will result in substitutions that diminish binding between the antigen and a monoclonal antibody. Binding specificity can be determined by mapping the effects of these "loss of binding" mutations. The antigen sequence is sandwiched between two epitope tags in an SD vector and subjected to error-prone PCR. After transformation and expression, flow cytometry is used to select for yeast expressing sequences that have intact flanking epitopes, but no longer bind a mAb directed to the original sequence; this method screens out frameshifts and nonsense mutations. The plasmids are recovered and mapped to identify the epitopes recognized by monoclonal antibodies. Examples of structural mapping include identifying epitope sequences of proteinaceous toxins [185] and cancer antigens [186]. Loss of binding of monoclonals to yeast surfaces after denaturation by heat, DTT, or SDS has been used to differentiate sequence-specific from structure-specific antibodies [187].

8. Detection (Biosensors)

Living organisms can respond to minute changes in their environment, and respond by transcribing new proteins. Surface-displayed fluorescent proteins (GFP and RFP) report environmental changes. Schofield *et al.* [188] created a SDY biosensor for detection of organophosphate by placing GFP behind an *OPH* promoter and RFP behind a promoter for a secondary enzyme that is activated by organophosphate breakdown products. In the absence of substrate, the yeast is not fluorescent; with the addition of organophosphate, the yeast fluoresces green, gradually shifting to red, and finally losing fluorescence when the compounds are completely destroyed.

Surface display organisms can also be glued onto electrodes. Surface-anchored oxidoreductases can be detected directly as they generate measurable changes in pH or ion concentration when they catalyze reactions.

Bacteria [189] and yeast [190] have been directly incorporated into electrodes for quantitative measurements.

The expression of heterologous proteins on yeast and bacteria has far-ranging commercial, medical, and environmental applications. As our understanding of wall biosynthetic pathways increases, new applications for surface display in yeast will materialize as well.

IX. Summary

We have reviewed the discovery of wall cross-linked GPI proteins in yeast, and the biochemistry that has evolved to catalyze this unique modification to GPIs. The GPI-wall cross-linked proteins are present at high spatial density on yeast cell surfaces. Many of the carbohydrate-active enzymes and proteases that create the cross-links can themselves be cross-linked into the cell wall through the GPI; the final localization is dependent on the sequence upstream of the GPI addition site and on physiological condition of the cell. This machinery is being exploited for high density surface display of antigens, single-chain antibodies, and enzymes for scientific and commercial exploitation. This versatile and exploitable system for binding proteins to a stable extracellular carbohydrate-rich matrix shows that it is indeed a "sweet and lovely wall" [1].

REFERENCES

1. Shakespeare, W. (1600). In T. Fisher (ed. and publisher). *A Midsummer Night's Dream*, London.
2. Brady, S.M., Song, S., Dhugga, K.S., Rafalski, J.A., and Benfey, P.N. (2007). Combining expression and comparative evolutionary analysis. The COBRA gene family. *Plant Physiol* 143:172–187.
3. Li, G., Basagoudanavar, S.H., and Gowda, D.C. (2008). Effect of GPI anchor moiety on the immunogenicity of DNA plasmids encoding the 19-kDa C-terminal portion of Plasmodium falciparum MSP-1. *Parasite Immunol* 30:315–322.
4. Yoshida, M., Sakuragi, N., Kondo, K., and Tanesaka, E. (2006). Cleavage with phospholipase of the lipid anchor in the cell adhesion molecule, csA, from Dictyostelium discoideum. *Comp Biochem Physiol B Biochem Mol Biol* 143:138–144.
5. Sakurai, T., Sugimoto, C., and Inoue, N. (2008). Identification and molecular characterization of a novel stage-specific surface protein of Trypanosoma congolense epimastigotes. *Mol Biochem Parasitol* 161:1–11.
6. Kobayashi, T., Nishizaki, R., and Ikezawa, H. (1997). The presence of GPI-linked protein(s) in an archaeobacterium, Sulfolobus acidocaldarius, closely related to eukaryotes. *Biochim Biophys Acta* 1334:1–4.
7. Eisenhaber, B., Bork, P., and Eisenhaber, F. (2001). Post-translational GPI lipid anchor modification of proteins in kingdoms of life: analysis of protein sequence data from complete genomes. *Protein Eng* 14:17–25.
8. Baldauf, S.L., Roger, A.J., Wenk-Siefert, I., and Doolittle, W.F. (2000). A kingdom-level phylogeny of eukaryotes based on combined protein data. *Science* 290:972–977.

9. Caro, L.H., Tettelin, H., Vossen, J.H., Ram, A.F., van den Ende, H., and Klis, F.M. (1997). *In silicio* identification of glycosyl-phosphatidylinositol-anchored plasma-membrane and cell wall proteins of Saccharomyces cerevisiae. *Yeast* 13:1477–1489.

10. Weig, M., Jansch, L., Gross, U., De Koster, C.G., Klis, F.M., and De Groot, P.W. (2004). Systematic identification in silico of covalently bound cell wall proteins and analysis of protein-polysaccharide linkages of the human pathogen Candida glabrata. *Microbiology* 150:3129–3144.

11. Richard, M.L., and Plaine, A. (2007). Comprehensive analysis of glycosylphosphatidyli-nositol-anchored proteins in Candida albicans. *Eukaryot Cell* 6:119–133.

12. Dranginis, A.M., Rauceo, J.M., Coronado, J.E., and Lipke, P.N. (2007). A biochemical guide to yeast adhesins: glycoproteins for social and antisocial occasions. *Microbiol Mol Biol Rev* 71:282–294.

13. Casadevall, A., and Pirofski, L.A. (2006). A reappraisal of humoral immunity based on mechanisms of antibody-mediated protection against intracellular pathogens. *Adv Immunol* 91:1–44.

14. Weissman, Z., and Kornitzer, D. (2004). A family of Candida cell surface haem-binding proteins involved in haemin and haemoglobin-iron utilization. *Mol Microbiol* 53:120920.

15. Bai, C., Chan, F.Y., and Wang, Y. (2005). Identification and functional characterization of a novel Candida albicans gene CaMNN5 that suppresses the iron-dependent growth defect of Saccharomyces cerevisiae aft1Delta mutant. *Biochem J* 389:27–35.

16. Terashima, H., Yabuki, N., Arisawa, M., Hamada, K., and Kitada, K. (2000). Up-regula-tion of genes encoding glycosylphosphatidylinositol (GPI)-attached proteins in response to cell wall damage caused by disruption of FKS1 in Saccharomyces cerevisiae. *Mol Gen Genet* 264:64–74.

17. Lesage, G., and Bussey, H. (2006). Cell wall assembly in Saccharomyces cerevisiae. *Microbiol Mol Biol Rev* 70:317–343.

18. Albrecht, A., Felk, A., Pichova, I., Naglik, J.R., Schaller, M., de Groot, P., Maccallum, D., Odds, F.C., Schafer, W., Klis, F., Monod, M., and Hube, B. (2006). Glycosylphosphati-dylinositol-anchored proteases of Candida albicans target proteins necessary for both cellular processes and host-pathogen interactions. *J Biol Chem* 281:688–694.

19. Martinez-Lopez, R., Park, H., Myers, C.L., Gil, C., and Filler, S.G. (2006). Candida albicans Ecm33p is important for normal cell wall architecture and interactions with host cells. *Eukaryot Cell* 5:140–147.

20. Morita, T., Tanaka, N., Hosomi, A., Giga-Hama, Y., and Takegawa, K. (2006). An alpha-amylase homologue, aah3, encodes a GPI-anchored membrane protein required for cell wall integrity and morphogenesis in Schizosaccharomyces pombe. *Biosci Biotechnol Biochem* 70:1454–1463.

21. Mouyna, I., Fontaine, T., Vai, M., Monod, M., Fonzi, W.A., Diaquin, M., Popolo, L., Hartland, R.P., and Latge, J.P. (2000). Glycosylphosphatidylinositol-anchored glucano-syltransferases play an active role in the biosynthesis of the fungal cell wall. *J Biol Chem* 275:14882–14889.

22. Popolo, L., Vai, M., Gatti, E., Porello, S., Bonfante, P., Balestrini, R., and Alberghina, L. (1993). Physiological analysis of mutants indicates involvement of the Saccharomyces cerevisiae GPI-anchored protein gp115 in morphogenesis and cell separation. *J Bacteriol* 175:1879–1885.

23. Lo, W.S., and Dranginis, A.M. (1996). FLO11, a yeast gene related to the STA genes, encodes a novel cell surface flocculin. *J Bacteriol* 178:7144–7151.

24. Lo, W.S., and Dranginis, A.M. (1998). The cell surface flocculin Flo11 is required for pseudohyphae formation and invasion by Saccharomyces cerevisiae. *Mol Biol Cell* 9:161–171.

25. Gonzalez, M., de Groot, P.W.J., Klis, F.M., and Lipke, P.N. (2009). Glycoconjugate structure and function in fungal cell walls, In A. Moran, P. Brennan, O. Holst and M. von Itzstein (eds.). *Microbial Glycobiology: Structures, Relevance and Applications*, in press.
26. Eigenheer, R.A., Jin Lee, Y., Blumwald, E., Phinney, B.S., and Gelli, A. (2007). Extracellular glycosylphosphatidylinositol-anchored mannoproteins and proteases of Cryptococcus neoformans. *FEMS Yeast Res* 7:499–510.
27. Yin, Q.Y., de Groot, P.W., de Jong, L., Klis, F.M., and De Koster, C.G. (2007). Mass spectrometric quantitation of covalently bound cell wall proteins in Saccharomyces cerevisiae. *FEMS Yeast Res* 7:887–896.
28. Yin, Q.Y., de Groot, P.W., de Koster, C.G., and Klis, F.M. (2008). Mass spectrometry-based proteomics of fungal wall glycoproteins. *Trends Microbiol* 16:20–26.
29. de Groot, P.W., de Boer, A.D., Cunningham, J., Dekker, H.L., de Jong, L., Hellingwerf, K.J., de Koster, C., and Klis, F.M. (2004). Proteomic analysis of Candida albicans cell walls reveals covalently bound carbohydrate-active enzymes and adhesins. *Eukaryot Cell* 3:955–965.
30. de Groot, P.W., Yin, Q.Y., Weig, M., Sosinska, G.J., Klis, F.M., and de Koster, C.G. (2007). Mass spectrometric identification of covalently bound cell wall proteins from the fission yeast Schizosaccharomyces pombe. *Yeast* 24:267–278.
31. Klis, F.M., Mol, P., Hellingwerf, K., and Brul, S. (2002). Dynamics of cell wall structure in Saccharomyces cerevisiae. *FEMS Microbiol Rev* 26:239–256.
32. Lipke, P.N., and Ovalle, R. (1998). Cell wall architecture in yeast: new structure and new challenges. *J Bacteriol* 180:3735–3740.
33. Verstrepen, K.J., and Klis, F.M. (2006). Flocculation, adhesion and biofilm formation in yeasts. *Mol Microbiol* 60:5–15.
34. Henrissat, B., and Davies, G. (1997). Structural and sequence-based classification of glycoside hydrolases. *Curr Opin Struct Biol* 7:637–644.
35. Gagnon-Arsenault, I., Parise, L., Tremblay, J., and Bourbonnais, Y. (2008). Activation mechanism, functional role and shedding of glycosylphosphatidylinositol-anchored Ypslp at the Saccharomyces cerevisiae cell surface. *Mol Microbiol* 69:982–993.
36. Komano, H., Rockwell, N., Wang, G.T., Krafft, G.A., and Fuller, R.S. (1999). Purification and characterization of the yeast glycosylphosphatidylinositol-anchored, monobasic-specific aspartyl protease yapsin 2 (Mkc7p). *J Biol Chem* 274:24431–24437.
37. Krysan, D.J., Ting, E.L., Abeijon, C., Kroos, L., and Fuller, R.S. (2005). Yapsins are a family of aspartyl proteases required for cell wall integrity in Saccharomyces cerevisiae. *Eukaryot Cell* 4:1364–1374.
38. Olsen, V., Cawley, N.X., Brandt, J., Egel-Mitani, M., and Loh, Y.P. (1999). Identification and characterization of Saccharomyces cerevisiae yapsin 3, a new member of the yapsin family of aspartic proteases encoded by the YPS3 gene. *Biochem J* 339(Pt 2):407–411.
39. Schaller, M., Januschke, E., Schackert, C., Woerle, B., and Korting, H.C. (2001). Different isoforms of secreted aspartyl proteinases (Sap) are expressed by Candida albicans during oral and cutaneous candidosis *in vivo*. *J Med Microbiol* 50:743–747.
40. Stringaro, A., Crateri, P., Pellegrini, G., Arancia, G., Cassone, A., and De Bernardis, F. (1997). Ultrastructural localization of the secretory aspartyl proteinase in Candida albicans cell wall *in vitro* and in experimentally infected rat vagina. *Mycopathologia* 137:95–105.
41. Djordjevic, J.T., Del Poeta, M., Sorrell, T.C., Turner, K.M., and Wright, L.C. (2005). Secretion of cryptococcal phospholipase B1 (PLB1) is regulated by a glycosylphosphatidylinositol (GPI) anchor. *Biochem J* 389:803–812.
42. Siafakas, A.R., Sorrell, T.C., Wright, L.C., Wilson, C., Larsen, M., Boadle, R., Williamson, P.R., and Djordjevic, J.T. (2007). Cell wall-linked cryptococcal

phospholipase B1 is a source of secreted enzyme and a determinant of cell wall integrity. *J Biol Chem* 282:37508–37514.

43. Siafakas, A.R., Wright, L.C., Sorrell, T.C., and Djordjevic, J.T. (2006). Lipid rafts in Cryptococcus neoformans concentrate the virulence determinants phospholipase B1 and Cu/Zn superoxide dismutase. *Eukaryot Cell* 5:488–498.

44. Cabib, E., Farkas, V., Kosik, O., Blanco, N., Arroyo, J., and McPhie, P. (2008). Assembly of the yeast cell wall: Crh1p and Crh2p act as transglycosylases *in vivo* and *in vitro*. *J Biol Chem* 283:29859–29872.

45. Mouyna, I., Monod, M., Fontaine, T., Henrissat, B., Lechenne, B., and Latge, J.P. (2000). Identification of the catalytic residues of the first family of beta(13)glucanosyltransferases identified in fungi. *Biochem J* 347(Pt 3):741–747.

46. Conzelmann, A., Riezman, H., Desponds, C., and Bron, C. (1988). A major 125-kd membrane glycoprotein of Saccharomyces cerevisiae is attached to the lipid bilayer through an inositol-containing phospholipid. *EMBO J* 7:2233–2240.

47. Orlean, P. (1990). Dolichol phosphate mannose synthase is required *in vivo* for glycosyl phosphatidylinositol membrane anchoring, O mannosylation, and N glycosylation of protein in Saccharomyces cerevisiae. *Mol Cell Biol* 10:5796–5805.

48. Vai, M., Popolo, L., Grandori, R., Lacana, E., and Alberghina, L. (1990). The cell cycle modulated glycoprotein GP115 is one of the major yeast proteins containing glycosylpho-sphatidylinositol. *Biochim Biophys Acta* 1038:277–285.

49. Montijn, R.C., van Rinsum, J., van Schagen, F.A., and Klis, F.M. (1994). Glucomanno-proteins in the cell wall of Saccharomyces cerevisiae contain a novel type of carbohydrate side chain. *J Biol Chem* 269:19338–19342.

50. Lipke, P.N., Wojciechowicz, D., and Kurjan, J. (1989). AG alpha 1 is the structural gene for the Saccharomyces cerevisiae alpha-agglutinin, a cell surface glycoprotein involved in cell-cell interactions during mating. *Mol Cell Biol* 9:3155–3165.

51. Roy, A., Lu, C.F., Marykwas, D.L., Lipke, P.N., and Kurjan, J. (1991). The AGA1 product is involved in cell surface attachment of the Saccharomyces cerevisiae cell adhesion glycoprotein a-agglutinin. *Mol Cell Biol* 11:4196–4206.

52. Cappellaro, C., Hauser, K., Mrsa, V., Watzele, M., Watzele, G., Gruber, C., and Tanner, W. (1991). Saccharomyces cerevisiae a- and alpha-agglutinin: characterization of their molecular interaction. *EMBO J* 10:4081–4088.

53. Hauser, K., and Tanner, W. (1989). Purification of the inducible alpha-agglutinin of S. cerevisiae and molecular cloning of the gene. *FEBS Lett* 255:290–294.

54. Lu, C.F., Montijn, R.C., Brown, J.L., Klis, F., Kurjan, J., Bussey, H., and Lipke, P.N. (1995). Glycosyl phosphatidylinositol-dependent cross-linking of alpha-agglutinin and beta 1,6-glucan in the Saccharomyces cerevisiae cell wall. *J Cell Biol* 128:333–340.

55. Wojciechowicz, D., and Lipke, P.N. (1989). Alpha-agglutinin expression in Saccharomy-ces cerevisiae. *Biochem Biophys Res Commun* 161:46–51.

56. Jentoft, N. (1990). Why are proteins O-glycosylated?. *Trends Biochem Sci* 15:291–294.

57. de Nobel, H., and Lipke, P.N. (1994). Is there a role for GPIs in yeast cell-wall assembly? *Trends Cell Biol* 4:42–45.

58. Kollar, R., Reinhold, B.B., Petrakova, E., Yeh, H.J., Ashwell, G., Drgonova, J., Kapteyn, J.C., Klis, F.M., and Cabib, E. (1997). Architecture of the yeast cell wall. Beta (1→6)glucan interconnects mannoprotein, beta(1→3)-glucan, and chitin. *J Biol Chem* 272:17762–17775.

59. Kapteyn, J.C., Ram, A.F., Groos, E.M., Kollar, R., Montijn, R.C., Van Den Ende, H., Llobell, A., Cabib, E., and Klis, F.M. (1997). Altered extent of cross-linking of beta1,6-glucosylated mannoproteins to chitin in Saccharomyces cerevisiae mutants with reduced cell wall beta1,3-glucan content. *J Bacteriol* 179:6279–6284.

60. Kollar, R., Petrakova, E., Ashwell, G., Robbins, P.W., and Cabib, E. (1995). Architecture of the yeast cell wall. The linkage between chitin and beta(1→3)-glucan. *J Biol Chem* 270:1170–1178.

61. Klis, F.M., Boorsma, A., and De Groot, P.W. (2006). Cell wall construction in Saccharomyces cerevisiae. *Yeast* 23:185–202.

62. Ecker, M., Deutzmann, R., Lehle, L., Mrsa, V., and Tanner, W. (2006). Pir proteins of Saccharomyces cerevisiae are attached to beta-1,3-glucan by a new protein-carbohydrate linkage. *J Biol Chem* 281:11523–11529.

63. Klis, F.M., Caro, L.H., Vossen, J.H., Kapteyn, J.C., Ram, A.F., Montijn, R.C., Van Berkel, M.A., and Van den Ende, H. (1997). Identification and characterization of a major building block in the cell wall of Saccharomyces cerevisiae. *Biochem Soc Trans* 25:85660.

64. de Nobel, J.G., Klis, F.M., Priem, J., Munnik, T., and van den Ende, H. (1990). The glucanase-soluble mannoproteins limit cell wall porosity in Saccharomyces cerevisiae. *Yeast* 6:491–499.

65. Ishihara, S., Hirata, A., Nogami, S., Beauvais, A., Latge, J.P., and Ohya, Y. (2007). Homologous subunits of 1,3-beta-glucan synthase are important for spore wall assembly in Saccharomyces cerevisiae. *Eukaryot Cell* 6:143–156.

66. Coronado, J.E., Mneimneh, S., Epstein, S.L., Qiu, W.G., and Lipke, P.N. (2007). Conserved processes and lineage-specific proteins in fungal cell wall evolution. *Eukaryot Cell* 6:2269–2277.

67. Roh, D.H., Bowers, B., Riezman, H., and Cabib, E. (2002). Rho1p mutations specific for regulation of beta(1→3)glucan synthesis and the order of assembly of the yeast cell wall. *Mol Microbiol* 44:1167–1183.

68. Dijkgraaf, G.J., Abe, M., Ohya, Y., and Bussey, H. (2002). Mutations in Fks1p affect the cell wall content of beta-1,3- and beta-1,6-glucan in Saccharomyces cerevisiae. *Yeast* 19:671–690.

69. Vink, E., Rodriguez-Suarez, R.J., Gerard-Vincent, M., Ribas, J.C., de Nobel, H., van den Ende, H., Duran, A., Klis, F.M., and Bussey, H. (2004). An *in vitro* assay for (1→6)beta-D-glucan synthesis in Saccharomyces cerevisiae. *Yeast* 21:1121–1131.

70. Gemmill, T.R., and Trimble, R.B. (1999). Overview of N- and O-linked oligosaccharide structures found in various yeast species. *Biochim Biophys Acta* 1426:227–237.

71. Zlotnik, H., Fernandez, M.P., Bowers, B., and Cabib, E. (1984). Saccharomyces cerevisiae mannoproteins form an external cell wall layer that determines wall porosity. *J Bacteriol* 159:1018–1026.

72. Lipke, P.N., and Kurjan, J. (1992). Sexual agglutination in budding yeasts: structure, function, and regulation of adhesion glycoproteins. *Microbiol Rev* 56:180–194.

73. Frieman, M.B., and Cormack, B.P. (2004). Multiple sequence signals determine the distribution of glycosylphosphatidylinositol proteins between the plasma membrane and cell wall in Saccharomyces cerevisiae. *Microbiology* 150:3105–3114.

74. Hamada, K., Terashima, H., Arisawa, M., and Kitada, K. (1998). Amino acid sequence requirement for efficient incorporation of glycosylphosphatidylinositol-associated proteins into the cell wall of Saccharomyces cerevisiae. *J Biol Chem* 273:26946–26953.

75. Mao, Y., Zhang, Z., Gast, C., and Wong, B. (2008). C-terminal signals regulate targeting of the GPI-anchored proteins to the cell wall or the plasma membrane in Candida albicans. *Eukaryot Cell*.

76. Hamada, K., Terashima, H., Arisawa, M., Yabuki, N., and Kitada, K. (1999). Amino acid residues in the omega-minus region participate in cellular localization of yeast glycosylphosphatidylinositol-attached proteins. *J Bacteriol* 181:3886–3889.

77. Frieman, M.B., and Cormack, B.P. (2003). The omega-site sequence of glycosylphosphatidylinositol-anchored proteins in Saccharomyces cerevisiae can determine distribution between the membrane and the cell wall. *Mol Microbiol* 50:88396.

78. Vai, M., Gatti, E., Lacana, E., Popolo, L., and Alberghina, L. (1991). Isolation and deduced amino acid sequence of the gene encoding gp115, a yeast glycophospholipidanchored protein containing a serine-rich region. *J Biol Chem* 266:12242–12248.

79. Vossen, J.H., Ram, A.F., and Klis, F.M. (1995). Identification of SPT14/CWH6 as the yeast homologue of hPIG-A, a gene involved in the biosynthesis of GPI anchors. *Biochim Biophys Acta* 1243:549–551.

80. Terashima, H., Hamada, K., and Kitada, K. (2003). The localization change of Ybr078w/ Ecm33, a yeast GPI-associated protein, from the plasma membrane to the cell wall, affecting the cellular function. *FEMS Microbiol Lett* 218:175–180.

81. Kreger, D.R., and Kopecka, M. (1976). On the nature and formation of the fibrillar nets produced by protoplasts of Saccharomyces cerevisiae in liquid media: an electronmicroscopic, X-ray diffraction and chemical study. *J Gen Microbiol* 92:207–220.

82. Errede, B., Cade, R.M., Yashar, B.M., Kamada, Y., Levin, D.E., Irie, K., and Matsumoto, K. (1995). Dynamics and organization of MAP kinase signal pathways. *Mol Reprod Dev* 42:477–485.

83. Drgonova, J., Drgon, T., Tanaka, K., Kollar, R., Chen, G.C., Ford, R.A., Chan, C.S., Takai, Y., and Cabib, E. (1996). Rho1p, a yeast protein at the interface between cell polarization and morphogenesis. *Science* 272:277–279.

84. Qadota, H., Python, C.P., Inoue, S.B., Arisawa, M., Anraku, Y., Zheng, Y., Watanabe, T., Levin, D.E., and Ohya, Y. (1996). Identification of yeast Rho1p GTPase as a regulatory subunit of 1,3-beta-glucan synthase. *Science* 272:279–281.

85. Szaniszlo, P.J., Kang, M.S., and Cabib, E. (1985). Stimulation of beta(1–3)glucan synthetase of various fungi by nucleoside triphosphates: generalized regulatory mechanism for cell wall biosynthesis. *J Bacteriol* 161:1188–1194.

86. Ragni, E., Fontaine, T., Gissi, C., Latge, J.P., and Popolo, L. (2007). The Gas family of proteins of Saccharomyces cerevisiae: characterization and evolutionary analysis. *Yeast* 24:297–308.

87. Xu, Y., Takvorian, P., Cali, A., Wang, F., Zhang, H., Orr, G., and Weiss, L.M. (2006). Identification of a new spore wall protein from Encephalitozoon cuniculi. *Infect Immun* 74:239–247.

88. Damveld, R.A., Arentshorst, M., VanKuyk, P.A., Klis, F.M., van den Hondel, C.A., and Ram, A.F. (2005). Characterisation of CwpA, a putative glycosylphosphatidylinositolanchored cell wall mannoprotein in the filamentous fungus Aspergillus niger. *Fungal Genet Biol* 42:873–885.

89. Latge, J.P., Mouyna, I., Tekaia, F., Beauvais, A., Debeaupuis, J.P., and Nierman, W. (2005). Specific molecular features in the organization and biosynthesis of the cell wall of Aspergillus fumigatus. *Med Mycol* 43(Suppl 1):S15–S22.

90. Linder, T., and Gustafsson, C.M. (2008). Molecular phylogenetics of ascomycotal adhesins–a novel family of putative cell-surface adhesive proteins in fission yeasts. *Fungal Genet Biol* 45:485–497.

91. Gaynor, E.C., Mondesert, G., Grimme, S.J., Reed, S.I., Orlean, P., and Emr, S.D. (1999). MCD4 encodes a conserved endoplasmic reticulum membrane protein essential for glycosylphosphatidylinositol anchor synthesis in yeast. *Mol Biol Cell* 10:627–648.

92. Benachour, A., Sipos, G., Flury, I., Reggiori, F., Canivenc-Gansel, E., Vionnet, C., Conzelmann, A., and Benghezal, M. (1999). Deletion of GPI7, a yeast gene required for addition of a side chain to the glycosylphosphatidylinositol (GPI) core structure, affects GPI protein transport, remodeling, and cell wall integrity. *J Biol Chem* 274:15251–15261.

93. Tiede, A., Bastisch, I., Schubert, J., Orlean, P., and Schmidt, R.E. (1999). Biosynthesis of glycosylphosphatidylinositols in mammals and unicellular microbes. *Biol Chem* 380:503–523.

94. Toh-e, A., Yasunaga, S., Nisogi, H., Tanaka, K., Oguchi, T., and Matsui, Y. (1993). Three yeast genes, PIR1, PIR2 and PIR3, containing internal tandem repeats, are related to each other, and PIR1 and PIR2 are required for tolerance to heat shock. *Yeast* 9:481–494.

95. Fujita, M., Yoko-o, T., Okamoto, M., and Jigami, Y. (2004). GPI7 involved in glycosyl-phosphatidylinositol biosynthesis is essential for yeast cell separation. *J Biol Chem* 279:51869–51879.

96. Richard, M., Ibata-Ombetta, S., Dromer, F., Bordon-Pallier, F., Jouault, T., and Gaillardin, C. (2002). Complete glycosylphosphatidylinositol anchors are required in Candida albicans for full morphogenesis, virulence and resistance to macrophages. *Mol Microbiol* 44:841–853.

97. Kitagaki, H., Wu, H., Shimoi, H., and Ito, K. (2002). Two homologous genes, DCW1 (YKL046c) and DFG5, are essential for cell growth and encode glycosylphosphatidylinositol (GPI)-anchored membrane proteins required for cell wall biogenesis in Saccharomyces cerevisiae. *Mol Microbiol* 46:1011–1022.

98. Spreghini, E., Davis, D.A., Subaran, R., Kim, M., and Mitchell, A.P. (2003). Roles of Candida albicans Dfg5p and Dcw1p cell surface proteins in growth and hypha formation. *Eukaryot Cell* 2:746–755.

99. Kitagaki, H., Ito, K., and Shimoi, H. (2004). A temperature-sensitive dcw1 mutant of Saccharomyces cerevisiae is cell cycle arrested with small buds which have aberrant cell walls. *Eukaryot Cell* 3:1297–1306.

100. Mosch, H.U., and Fink, G.R. (1997). Dissection of filamentous growth by transposon mutagenesis in Saccharomyces cerevisiae. *Genetics* 145:671–684.

101. Roemer, T., and Bussey, H. (1995). Yeast Kre1p is a cell surface O-glycoprotein. *Mol Gen Genet* 249:209–216.

102. Boone, C., Sommer, S.S., Hensel, A., and Bussey, H. (1990). Yeast KRE genes provide evidence for a pathway of cell wall beta-glucan assembly. *J Cell Biol* 110:1833–1843.

103. Breinig, F., Schleinkofer, K., and Schmitt, M.J. (2004). Yeast Kre1p is GPI-anchored and involved in both cell wall assembly and architecture. *Microbiology* 150:3209–3218.

104. Nuoffer, C., Jeno, P., Conzelmann, A., and Riezman, H. (1991). Determinants for glyco-phospholipid anchoring of the Saccharomyces cerevisiae GAS1 protein to the plasma membrane. *Mol Cell Biol* 11:27–37.

105. Muhlschlegel, F.A., and Fonzi, W.A. (1997). PHR2 of Candida albicans encodes a functional homolog of the pH-regulated gene PHR1 with an inverted pattern of pH-dependent expression. *Mol Cell Biol* 17:5960–5967.

106. Khan, A.R., Khazanovich-Bernstein, N., Bergmann, E.M., and James, M.N. (1999). Structural aspects of activation pathways of aspartic protease zymogens and viral 3C protease precursors. *Proc Natl Acad Sci USA* 96:10968–10975.

107. Huse, J.T., Pijak, D.S., Leslie, G.J., Lee, V.M., and Doms, R.W. (2000). Maturation and endosomal targeting of beta-site amyloid precursor protein-cleaving enzyme. The Alzheimer's disease beta-secretase. *J Biol Chem* 275:33729–33737.

108. Monod, M., Hube, B., Hess, D., and Sanglard, D. (1998). Differential regulation of SAP8 and SAP9, which encode two new members of the secreted aspartic proteinase family in Candida albicans. *Microbiology* 144(Pt 10):2731–2737.

109. Dujon, B., Sherman, D., Fischer, G., Durrens, P., Casaregola, S., Lafontaine, I., De Montigny, J., Marck, C., Neuveglise, C., Talla, E., Goffard, N., Frangeul, L., *et al.* (2004). Genome evolution in yeasts. *Nature* 430:35–44.

110. Kunihiro, S., Kawanishi, Y., Sano, M., Naito, K., Matsuura, Y., Tateno, Y., Gojobori, T., Yamagata, Y., Abe, K., and Machida, M. (2002). A polymerase chain reaction-based method for cloning novel members of a gene family using a combination of degenerate and inhibitory primers. *Gene* 289:177–184.

111. Gagnon-Arsenault, I., Tremblay, J., and Bourbonnais, Y. (2006). Fungal yapsins and cell wall: a unique family of aspartic peptidases for a distinctive cellular function. *FEMS Yeast Res* 6:966–978.

112. Bourbonnais, Y., Ash, J., Daigle, M., and Thomas, D.Y. (1993). Isolation and characterization of S. cerevisiae mutants defective in somatostatin expression: cloning and functional role of a yeast gene encoding an aspartyl protease in precursor processing at monobasic cleavage sites. *EMBO J* 12:285–294.

113. Kaur, R., Ma, B., and Cormack, B.P. (2007). A family of glycosylphosphatidylinositol-linked aspartyl proteases is required for virulence of Candida glabrata. *Proc Natl Acad Sci USA* 104:7628–7633.

114. Rodriguez-Pena, J.M., Cid, V.J., Arroyo, J., and Nombela, C. (2000). A novel family of cell wall-related proteins regulated differently during the yeast life cycle. *Mol Cell Biol* 20:3245–3255.

115. Cabib, E., Blanco, N., Grau, C., Rodriguez-Pena, J.M., and Arroyo, J. (2007). Crh1p and Crh2p are required for the cross-linking of chitin to beta(1–6)glucan in the Saccharomyces cerevisiae cell wall. *Mol Microbiol* 63:921–935.

116. Kondo, A., and Ueda, M. (2004). Yeast cell-surface display: applications of molecular display. *Appl Microbiol Biotechnol* 64:28–40.

117. Pepper, L.R., Cho, Y.K., Boder, E.T., and Shusta, E.V. (2008). A decade of yeast surface display technology: where are we now? *Comb Chem High Throughput Screen* 11:12734.

118. Saleem, M., Brim, H., Hussain, S., Arshad, M., Leigh, M.B., and Zia ul, H. (2008). Perspectives on microbial cell surface display in bioremediation. *Biotechnol Adv* 26:151–161.

119. Schreuder, M.P., Mooren, A.T., Toschka, H.Y., Verrips, C.T., and Klis, F.M. (1996). Immobilizing proteins on the surface of yeast cells. *Trends Biotechnol* 14:115–120.

120. Ueda, M., and Tanaka, A. (2000). Cell surface engineering of yeast: construction of arming yeast with biocatalyst. *J Biosci Bioeng* 90:125–136.

121. Ueda, M., and Tanaka, A. (2000). Genetic immobilization of proteins on the yeast cell surface. *Biotechnol Adv* 18:121–140.

122. Wu, C.H., Mulchandani, A., and Chen, W. (2008). Versatile microbial surface-display for environmental remediation and biofuels production. *Trends Microbiol* 16:181–188.

123. Bowley, D.R., Labrijn, A.F., Zwick, M.B., and Burton, D.R. (2007). Antigen selection from an HIV-1 immune antibody library displayed on yeast yields many novel antibodies compared to selection from the same library displayed on phage. *Protein Eng Des Sel* 20:81–90.

124. Jostock, T., and Dubel, S. (2005). Screening of molecular repertoires by microbial surface display. *Comb Chem High Throughput Screen* 8:127–133.

125. Wang, Q., Li, L., Chen, M., Qi, Q., and Wang, P.G. (2007). Construction of a novel system for cell surface display of heterologous proteins on Pichia pastoris. *Biotechnol Lett* 29:1561–1566.

126. Uccelletti, D., De Jaco, A., Farina, F., Mancini, P., Augusti-Tocco, G., Biagioni, S., and Palleschi, C. (2002). Cell surface expression of a GPI-anchored form of mouse acetylcholinesterase in Klpmr1Delta cells of Kluyveromyces lactis. *Biochem Biophys Res Commun* 298:559–565.

127. Kim, S.Y., Sohn, J.H., Pyun, Y.R., and Choi, E.S. (2002). A cell surface display system using novel GPI-anchored proteins in Hansenula polymorpha. *Yeast* 19:1153–1163.

128. Yue, L., Chi, Z., Wang, L., Liu, J., Madzak, C., Li, J., and Wang, X. (2008). Construction of a new plasmid for surface display on cells of Yarrowia lipolytica. *J Microbiol Methods* 72:116–123.
129. Mischo, A., Wadle, A., Watzig, K., Jager, D., Stockert, E., Santiago, D., Ritter, G., Regitz, E., Jager, E., Knuth, A., Old, L., Pfreundschuh, M., *et al.* (2003). Recombinant antigen expression on yeast surface (RAYS) for the detection of serological immune responses in cancer patients. *Cancer Immunol* 3:5.
130. Breinig, F., Diehl, B., Rau, S., Zimmer, C., Schwab, H., and Schmitt, M.J. (2006). Cell surface expression of bacterial esterase A by Saccharomyces cerevisiae and its enhancement by constitutive activation of the cellular unfolded protein response. *Appl Environ Microbiol* 72:7140–7147.
131. Schreuder, M.P., Brekelmans, S., van den Ende, H., and Klis, F.M. (1993). Targeting of a heterologous protein to the cell wall of Saccharomyces cerevisiae. *Yeast* 9:399–409.
132. Boder, E.T., and Wittrup, K.D. (2000). Yeast surface display for directed evolution of protein expression, affinity, and stability. *Methods Enzymol* 328:430–444.
133. Abe, H., Shimma, Y., and Jigami, Y. (2003). *In vitro* oligosaccharide synthesis using intact yeast cells that display glycosyltransferases at the cell surface through cell wall-anchored protein Pir. *Glycobiology* 13:87–95.
134. Matsumoto, T., Fukuda, H., Ueda, M., Tanaka, A., and Kondo, A. (2002). Construction of yeast strains with high cell surface lipase activity by using novel display systems based on the Flo1p flocculation functional domain. *Appl Environ Microbiol* 68:4517–4522.
135. Jiang, Z.B., Song, H.T., Gupta, N., Ma, L.X., and Wu, Z.B. (2007). Cell surface display of functionally active lipases from Yarrowia lipolytica in Pichia pastoris. *Protein Expr Purif* 56:35–39.
136. Kapteyn, J.C., Van Egmond, P., Sievi, E., Van Den Ende, H., Makarow, M., and Klis, F.M. (1999). The contribution of the O-glycosylated protein Pir2p/Hsp150 to the construction of the yeast cell wall in wild-type cells and beta 1,6-glucan-deficient mutants. *Mol Microbiol* 31:1835–1844.
137. Terrance, K., Heller, P., Wu, Y.S., and Lipke, P.N. (1987). Identification of glycoprotein components of alpha-agglutinin, a cell adhesion protein from Saccharomyces cerevisiae. *J Bacteriol* 169:475–482.
138. Cappellaro, C., Baldermann, C., Rachel, R., and Tanner, W. (1994). Mating type-specific cell-cell recognition of Saccharomyces cerevisiae: cell wall attachment and active sites of a- and alpha-agglutinin. *EMBO J* 13:4737–4744.
139. Mrsa, V., Seidl, T., Gentzsch, M., and Tanner, W. (1997). Specific labelling of cell wall proteins by biotinylation. Identification of four covalently linked O-mannosylated proteins of Saccharomyces cerevisiae. *Yeast* 13:1145–1154.
140. van Berkel, M.A., Caro, L.H., Montijn, R.C., and Klis, F.M. (1994). Glucosylation of chimeric proteins in the cell wall of Saccharomyces cerevisiae. *FEBS Lett* 349:135–138.
141. Boder, E.T., and Wittrup, K.D. (1997). Yeast surface display for screening combinatorial polypeptide libraries. *Nat Biotechnol* 15:553–557.
142. Kaya, M., Ito, J., Kotaka, A., Matsumura, K., Bando, H., Sahara, H., Ogino, C., Shibasaki, S., Kuroda, K., Ueda, M., Kondo, A., and Hata, Y. (2008). Isoflavone aglycones production from isoflavone glycosides by display of beta-glucosidase from Aspergillus oryzae on yeast cell surface. *Appl Microbiol Biotechnol* 79:51–60.
143. Takayama, K., Suye, S., Kuroda, K., Ueda, M., Kitaguchi, T., Tsuchiyama, K., Fukuda, T., Chen, W., and Mulchandani, A. (2006). Surface display of organophosphorus hydrolase on Saccharomyces cerevisiae. *Biotechnol Prog* 22:939–943.

144. Nakamura, Y., Shibasaki, S., Ueda, M., Tanaka, A., Fukuda, H., and Kondo, A. (2001). Development of novel whole-cell immunoadsorbents by yeast surface display of the IgGbinding domain. *Appl Microbiol Biotechnol* 57:500–505.

145. Kato, M., Maeda, H., Kawakami, M., Shiraga, S., and Ueda, M. (2005). Construction of a selective cleavage system for a protein displayed on the cell surface of yeast. *Appl Microbiol Biotechnol* 69:423–427.

146. Van der Vaart, J.M., te Biesebeke, R., Chapman, J.W., Toschka, H.Y., Klis, F.M., and Verrips, C.T. (1997). Comparison of cell wall proteins of Saccharomyces cerevisiae as anchors for cell surface expression of heterologous proteins. *Appl Environ Microbiol* 63:615–620.

147. de Nobel, H., Pike, J., Lipke, P.N., and Kurjan, J. (1995). Genetics of a-agglutunin function in Saccharomyces cerevisiae. *Mol Gen Genet* 247:409–415.

148. Huang, D., and Shusta, E.V. (2005). Secretion and surface display of green fluorescent protein using the yeast Saccharomyces cerevisiae. *Biotechnol Prog* 21:349–357.

149. Wang, Z., Mathias, A., Stavrou, S., and Neville, D.M., Jr. (2005). A new yeast display vector permitting free scFv amino termini can augment ligand binding affinities. *Protein Eng Des Sel* 18:337–343.

150. Shimma, Y., Saito, F., Oosawa, F., and Jigami, Y. (2006). Construction of a library of human glycosyltransferases immobilized in the cell wall of Saccharomyces cerevisiae. *Appl Environ Microbiol* 72:7003–7012.

151. Sumita, T., Yoko-o, T., Shimma, Y., and Jigami, Y. (2005). Comparison of cell wall localization among Pir family proteins and functional dissection of the region required for cell wall binding and bud scar recruitment of Pir1p. *Eukaryot Cell* 4:1872–1881.

152. Moukadiri, I., Jaafar, L., and Zueco, J. (1999). Identification of two mannoproteins released from cell walls of a Saccharomyces cerevisiae mnn1 mnn9 double mutant by reducing agents. *J Bacteriol* 181:4741–4745.

153. De Nobel, J.G., Klis, F.M., Munnik, T., Priem, J., and van den Ende, H. (1990). An assay of relative cell wall porosity in Saccharomyces cerevisiae, Kluyveromyces lactis and Schizosaccharomyces pombe. *Yeast* 6:483–490.

154. Mergler, M., Wolf, K., and Zimmermann, M. (2004). Development of a bisphenol A-adsorbing yeast by surface display of the Kluyveromyces yellow enzyme on Pichia pastoris. *Appl Microbiol Biotechnol* 63:418–421.

155. Jiang, Z., Gao, B., Ren, R., Tao, X., Ma, Y., and Wei, D. (2008). Efficient display of active lipase LipB52 with a Pichia pastoris cell surface display system and comparison with the LipB52 displayed on Saccharomyces cerevisiae cell surface. *BMC Biotechnol* 8:4.

156. Murai, T., Ueda, M., Yamamura, M., Atomi, H., Shibasaki, Y., Kamasawa, N., Osumi, M., Amachi, T., and Tanaka, A. (1997). Construction of a starch-utilizing yeast by cell surface engineering. *Appl Environ Microbiol* 63:1362–1366.

157. Zou, W., Ueda, M., and Tanaka, A. (2001). Genetically controlled self-aggregation of cell-surface-engineered yeast responding to glucose concentration. *Appl Environ Microbiol* 67:2083–2087.

158. Tanino, T., Fukuda, H., and Kondo, A. (2006). Construction of a Pichia pastoris cell-surface display system using Flo1p anchor system. *Biotechnol Prog* 22:989–993.

159. Schreuder, M.P., Deen, C., Boersma, W.J., Pouwels, P.H., and Klis, F.M. (1996). Yeast expressing hepatitis B virus surface antigen determinants on its surface: implications for a possible oral vaccine. *Vaccine* 14:383–388.

160. Kato, M., Fuchimoto, J., Tanino, T., Kondo, A., Fukuda, H., and Ueda, M. (2007). Preparation of a whole-cell biocatalyst of mutated Candida antarctica lipase B (mCALB) by a yeast molecular display system and its practical properties. *Appl Microbiol Biotechnol* 75:549–555.

161. Murai, T., Ueda, M., Shibasaki, Y., Kamasawa, N., Osumi, M., Imanaka, T., and Tanaka, A. (1999). Development of an arming yeast strain for efficient utilization of starch by codisplay of sequential amylolytic enzymes on the cell surface. *Appl Microbiol Biotechnol* 51:65–70.

162. Fujita, Y., Ito, J., Ueda, M., Fukuda, H., and Kondo, A. (2004). Synergistic saccharification, and direct fermentation to ethanol, of amorphous cellulose by use of an engineered yeast strain codisplaying three types of cellulolytic enzyme. *Appl Environ Microbiol* 70:1207–1212.

163. Necochea, R., Valderrama, B., Diaz-Sandoval, S., Folch-Mallol, J.L., Vazquez-Duhalt, R., and Iturriaga, G. (2005). Phylogenetic and biochemical characterisation of a recombinant laccase from Trametes versicolor. *FEMS Microbiol Lett* 244:235–241.

164. Ryu, K., Kang, J.H., Wang, L., and Lee, E.K. (2008). Expression in yeast of secreted lignin peroxidase with improved 2,4-dichlorophenol degradability by DNA shuffling. *J Biotechnol* 135:241–246.

165. Kuroda, K., and Ueda, M. (2003). Bioadsorption of cadmium ion by cell surface-engineered yeasts displaying metallothionein and hexa-His. *Appl Microbiol Biotechnol* 63:182–186.

166. Kuroda, K., and Ueda, M. (2006). Effective display of metallothionein tandem repeats on the bioadsorption of cadmium ion. *Appl Microbiol Biotechnol* 70:458–463.

167. Peelle, B.R., Krauland, E.M., Wittrup, K.D., and Belcher, A.M. (2005). Design criteria for engineering inorganic material-specific peptides. *Langmuir* 21:6929–6933.

168. Machado, M.D., Santos, M.S., Gouveia, C., Soares, H.M., and Soares, E.V. (2008). Removal of heavy metals using a brewer's yeast strain of Saccharomyces cerevisiae: the flocculation as a separation process. *Bioresour Technol* 99:2107–2115.

169. Krol, S., Nolte, M., Diaspro, A., Mazza, D., Magrassi, R., Gliozzi, A., and Fery, A. (2005). Encapsulated living cells on microstructured surfaces. *Langmuir* 21:705–709.

170. Navarro, V.M., Walker, S.L., Badali, O., Abundis, M.I., Ngo, L.L., Weerasinghe, G., Barajas, M., Zem, G., and Oppenheimer, S.B. (2002). Analysis of surface properties of fixed and live cells using derivatized agarose beads. *Acta Histochem* 104:99–106.

171. Yu, J., Tong, M., Sun, X., and Li, B. (2007). Cystine-modified biomass for Cd(II) and Pb(II) biosorption. *J Hazard Mater* 143:277–284.

172. D'Souza, S.F., and Melo, J.S. (1991). A method for the preparation of coimmobilizates by adhesion using polyethylenimine. *Enzyme Microb Technol* 13:508–511.

173. Parthasarathy, R., Bajaj, J., and Boder, E.T. (2005). An immobilized biotin ligase: surface display of *Escherichia coli* BirA on *Saccharomyces cerevisiae*. *Biotechnol Prog* 21:1627–1631.

174. Scholler, N., Garvik, B., Quarles, T., Jiang, S., and Urban, N. (2006). Method for generation of *in vivo* biotinylated recombinant antibodies by yeast mating. *J Immunol Methods* 317:132–143.

175. Tamaru, Y., Ohtsuka, M., Kato, K., Manabe, S., Kuroda, K., Sanada, M., and Ueda, M. (2006). Application of the arming system for the expression of the 380R antigen from red sea bream iridovirus (RSIV) on the surface of yeast cells: a first step for the development of an oral vaccine. *Biotechnol Prog* 22:949–953.

176. Bidlingmaier, S., and Liu, B. (2006). Construction and application of a yeast surface-displayed human cDNA library to identify post-translational modification-dependent protein-protein interactions. *Mol Cell Proteomics* 5:533–540.

177. Zou, W., Ueda, M., Yamanaka, H., and Tanaka, A. (2001). Construction of a combinatorial protein library displayed on yeast cell surface using DNA random priming method. *J Biosci Bioeng* 92:393–396.

178. Ryckaert, S., Callewaert, N., Jacobs, P.P., Dewaele, S., Dewerte, I., and Contreras, R. (2008). Fishing for lectins from diverse sequence libraries by yeast surface display: an exploratory study. *Glycobiology* 18:137–144.

179. Cochran, J.R., Kim, Y.S., Lippow, S.M., Rao, B., and Wittrup, K.D. (2006). Improved mutants from directed evolution are biased to orthologous substitutions. *Protein Eng Des Sel* 19:245–253.

180. Shiraga, S., Kawakami, M., Ishiguro, M., and Ueda, M. (2005). Enhanced reactivity of *Rhizopus oryzae* lipase displayed on yeast cell surfaces in organic solvents: potential as a whole-cell biocatalyst in organic solvents. *Appl Environ Microbiol* 71:4335–4338.

181. Jin, M., Song, G., Carman, C.V., Kim, Y.S., Astrof, N.S., Shimaoka, M., Wittrup, D.K., and Springer, T.A. (2006). Directed evolution to probe protein allostery and integrin I domains of 200,000-fold higher affinity. *Proc Natl Acad Sci USA* 103:5758–5763.

182. Graff, C.P., Chester, K., Begent, R., and Wittrup, K.D. (2004). Directed evolution of an anti-carcinoembryonic antigen scFv with a 4-day monovalent dissociation half-time at 37 degrees C. *Protein Eng Des Sel* 17:293–304.

183. Kieke, M.C., Sundberg, E., Shusta, E.V., Mariuzza, R.A., Wittrup, K.D., and Kranz, D.M. (2001). High affinity T cell receptors from yeast display libraries block T cell activation by superantigens. *J Mol Biol* 307:1305–1315.

184. Lee, H.W., Lee, S.H., Park, K.J., Kim, J.S., Kwon, M.H., and Kim, Y.S. (2006). Construction and characterization of a pseudo-immune human antibody library using yeast surface display. *Biochem Biophys Res Commun* 346:896–903.

185. Levy, R., Forsyth, C.M., LaPorte, S.L., Geren, I.N., Smith, L.A., and Marks, J.D. (2007). Fine and domain-level epitope mapping of botulinum neurotoxin type A neutralizing antibodies by yeast surface display. *J Mol Biol* 365:196–210.

186. Piatesi, A., Howland, S.W., Rakestraw, J.A., Renner, C., Robson, N., Cebon, J., Maraskovsky, E., Ritter, G., Old, L., and Wittrup, K.D. (2006). Directed evolution for improved secretion of cancer-testis antigen NY-ESO-1 from yeast. *Protein Expr Purif* 48:232–242.

187. Cochran, J.R., Kim, Y.S., Olsen, M.J., Bhandari, R., and Wittrup, K.D. (2004). Domain-level antibody epitope mapping through yeast surface display of epidermal growth factor receptor fragments. *J Immunol Methods* 287:147–158.

188. Schofield, D.A., Westwater, C., Barth, J.L., and DiNovo, A.A. (2007). Development of a yeast biosensor-biocatalyst for the detection and biodegradation of the organophosphate paraoxon. *Appl Microbiol Biotechnol* 76:1383–1394.

189. Swers, J.S., Kellogg, B.A., and Wittrup, K.D. (2004). Shuffled antibody libraries created by *in vivo* homologous recombination and yeast surface display. *Nucleic Acids Res* 32:e36.

190. Akyilmaz, E., Yasa, I., and Dinckaya, E. (2006). Whole cell immobilized amperometric biosensor based on Saccharomyces cerevisiae for selective determination of vitamin B1 (thiamine). *Anal Biochem* 354:78–84.

191. Gemmill, T.R. and Trimble, R.B. (1999). Overview of N- and O-linked oligosaccharide structures found in various yeast species. *Biochim Biophys Acta* 1426(2):227–237.

192. Lehle, L., Strahl, S., and Tanner, W. (2006). Protein glycosylation, conserved from yeast to man: A model organism helps elucidate congenital human diseases. *Angew Chem Int Ed Engl* 45(41):6802–6818.

16

Inherited GPI Deficiency

ANTONIO ALMEIDA[a] • MARK LAYTON[b] •
ANASTASIOS KARADIMITRIS[b]

[a]*Departamento de Hematologia, CIPM*
Instituto Portugês de Oncologia Francisco Gentil
Lisbon, Portugal

[b]*Department of Hematology*
Imperial College Healthcare NHS Trust
Hammersmith Hospital
Imperial College London
London W12 0NN
United Kingdom

1. Abstract

Cell surface expression through attachment to glycosylphosphatidylinositol (GPI) is a mode of protein expression highly conserved in eukaryotes. GPI-anchored proteins (GPI-AP) serve a variety of functions that include adhesion, receptors, signal transduction, and complement activation. Paroxysmal nocturnal hemoglobinuria, a rare acquired disorder of hematopoiesis caused by somatic mutations in the X-linked *PIG-A* gene, was until recently the only genetic disorder affecting GPI biosynthesis. The strict requirement for intact GPI biosynthesis during embryonic and fetal development accounted for the paucity of reported cases of inherited forms of GPI deficiency. Here, we review the clinical spectrum, biochemical defect, and genetic pathogenesis of the first cases of inherited GPI deficiency (IGD), an autosomal recessive disorder caused by a hypomorphic mutation in the promoter of *PIG-M*, a gene which like *PIG-A*, is indispensable for GPI biosynthesis. Further, we discuss the evidence suggesting that IGD is a disorder of gene-specific histone hypoacetylation and that pharmacological manipulation targeted to histone acetylation is of therapeutic benefit in IGD.

II. Introduction

Glycosylphosphatidylinositol-anchored proteins (GPI-AP) are found on the surface of all eukaryotic cells and the GPI-biosynthetic pathway is ubiquitously active. The GPI-AP have diverse functions and inactivation of any of the several genes essential for GPI biosynthesis would result in the loss of cell surface expression of dozens of proteins. The nosology of GPI deficiency caused by mutations in *PIG-A* and *PIG-M*, two genes essential for GPI assembly, is reviewed in this and the preceding chapter.

A.   STRUCTURE, BIOSYNTHESIS, AND MODIFICATIONS OF THE GPI ANCHOR

The core structure EtNP-6Manα1–2Manα1–6Manα1–4GlcNα1–6-*myo*-Inositol-phospholipid (EtNP, ethanolamine phosphate; Man, mannose; and GlcN, glucosamine) of GPI is conserved from parasites and fungi to mammals [1–4], although modifications of this core structure may differ between species and tissues. Here, we will only briefly discuss the structure and biosynthesis of the mammalian GPI as the subject has been extensively reviewed in previous chapters of this volume.

1. Biosynthesis of Mature GPI

GPI biosynthesis (recently reviewed in Refs. [2, 3]) begins on the cytoplasmic side of endoplasmic reticulum (ER) with the transfer of GlcNAc from UDP-GlcNAc to PI (Step 1). A multisubunit GPI-GlcNAc transferase (GPI-GnT) comprising six proteins (PIG-A, PIG-C, PIG-H, PIG-P, PIG-Q, and PIG-Y) catalyses this step [2, 3]. All six proteins are essential for GPI biosynthesis. PIG-A provides the catalytic function of GPI-GnT but the role of the other five proteins is not known (see Figure 16.1 for overview of the GPI-biosynthetic pathway).

In step 2, GlcNAc-PI is de-*N*-acetylated to GlcN-PI by PIG-L, a GlcNAc-de-*N*-acetylase; while in step 3, GlcN-PI is flipped across to the luminal side of ER where the synthesis of GPI continues [2, 3].

In step 4, PIG-W, an acyltransferase, mediates the transfer of acyl chain from acyl-CoA to the inositol ring of PI to form GlcN-(acyl)PI.

In the next two steps, Man residues are added sequentially to GlcN-(acyl)PI generating Man-Man-GlcN-(acyl)PI. In step 5, addition of Man-1

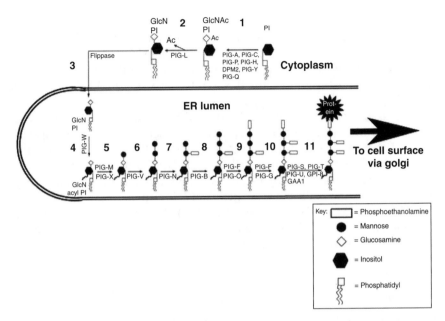

Fig. 16.1. The biosynthetic pathway of GPI. Biosynthesis of GPI starts on the cytoplasmic side of ER but at step 3 GPI is flipped across to the luminal side of ER. Step 1 is blocked in PNH as a result of somatic mutations in *PIG-A*. Step 5, that is, addition of Man-1 is blocked in IGD due to a mutation in *PIG-M*. Dol-P-Man, addition of Man-4, and lipid modification of GPI are not shown.

is catalyzed by the PIG-M/PIG-X complex, a GPI-MTI α1–4 mannosyl-transferase [2, 3]. PIG-M is the catalytic subunit of GPI-MTI while PIG-X, although devoid of catalytic activity, through its stabilizing effect on PIG-M is essential for mannosylation. PIG-V is the GPI-MTII α1–6mannosyltransferase responsible for the addition of Man-2, that is, in step 6 [2, 3].

The first EtNP modification of the GPI core takes place in step 7, in which, PIG-N, a GPI-ethanolamine phosphate transferase I (GPI-ETI), mediates EtNP modification of Man-1 generating Man-(EtNP)Man-GlcN-(acyl)PI [2, 3].

Subsequently, in step 8, PIG-B, an α1–2 mannosyltransferase (GPI-MTIII), catalyses the addition of Man-3 to form Man-Man-(EtNP)Man-GlcN-(acyl)PI [2, 3].

In all three mannosylation steps, Dol-P-Man is the mannose donor [2, 3]. Synthesis of Dol-P-Man requires Dol-P-Man synthase, a three subunit complex consisting of DPM1, DPM2, and DPM3 [5]. DPM1 is the catalytic subunit while DPM2 and DPM3 serve to stabilize DPM1.

In the final-step 9, GPI-ETIII, a complex of PIG-O and PIG-F, adds to Man-3 the crucial for protein attachment to GPI terminal EtNP thus, generating the mature GPI structure EtNP-Man-Man-(EtNP)Man-GlcN-(acyl)PI. Both PIG-F and PIG-O are indispensable for the function of GPI-ETIII, with PIG-F required for the stability of the complex and PIG-O providing the catalytic activity [2, 3].

In mammalian GPI, additional EtNP- and Man-modifications of the core EtNP-Man-Man-(EtNP)Man-GlcN-(acyl)PI structure are possible. Specifically, EtNP modification of Man-2 (step 10) by the PIG-G/PIG-F GPI-ETII complex is common [2, 3], with PIG-G (GPI7) providing the catalytic subunit which is stabilized by PIG-F. A second modification found mainly in the brain, is the addition, after step 8, of a fourth mannose mediated by the α1–2 mannosyltransferase PIG-Z (SMP3). The resulting four mannose intermediates are then EtNP-modified at Man-3 (step 9') generating, thus, another form of mature GPI anchor [2, 3].

2. Posttranslational Attachment of Proteins to GPI

Two structural features are required for the attachment of a protein to any of the three mature forms of GPI: the presence of an N-terminus ER-targeting motif and of a C-terminus domain. The latter provides all the information required for the attachment to GPI through an amidic bond between the ω amino acid in the protein C-terminus and the EtNP attached to Man-3 [2, 3]. This transamidation reaction (step 11) is catalyzed by GPI transamidase, a membrane-bound, multisubunit complex. The transamidase components, PIG-K which provides the catalytic activity, GAA1, PIG-S, PIG-T, and PIG-U, are all essential for the function of the transamidase [2, 3].

3. Lipid Modification of the GPI Anchor

In the ER and after attachment of the protein to the GPI, the acyl chain attached to the inositol moiety is removed [6] while after the GPI-AP complex moves to the Golgi, fatty acid (FA) remodeling of PI comprising conversion at sn1 and sn2 positions of unsaturated to saturated FA takes place. This modification is required for stable association of the GPI anchor with the outer leaflet of the cell membrane and its incorporation into lipid rafts [7]. Exceptionally, in human erythrocyte, in GPI-AP such as CD59 neither the inositol ring acyl chain is removed nor lipid modification takes place at sn2 [2, 8]. This ensures the presence of three FA chains stabilizing GPI-AP on the surface of erythrocytes during their long life.

B. BIOLOGICAL SIGNIFICANCE OF THE GPI ANCHOR

GPI-AP serve a variety of functions that include receptors, adhesion, enzymes, complement regulation, and signal transduction. The evolutionary purpose served by GPI linkage for cell surface protein expression is not clear but its importance is underscored by the fact that complete disruption of GPI biosynthesis is embryonic or conditional lethal in mice [9, 10] and in yeast [11, 12], respectively.

GPI along with glycosphingolipids and cholesterol are enriched in rafts, the relatively rigid membrane microdomain platforms on which surface-cytoplasmic protein interactions necessary for signal transduction activation and vesicular trafficking are organized [13–15]. It is possible that GPI linkage is required for many proteins in order to assume their fully functional conformation [16], while in some cases GPI linkage might serve apical targeting of proteins in polarized cells such as those of the intestinal epithelium [17]. Another potentially important function of GPI might be related to its susceptibility to phospholipase C and D (PI-PLC and -PLD) cleavage and release of the anchored protein as in the case of Notum, a Wnt pathway inhibitor possessing PI-PLC activity. Notum can cleave and release the GPI-anchored glypicans which in their soluble form may function as inhibitors of Wnt interaction with its receptor [18].

III. Disorders of GPI Deficiency

Until recently, the paucity of reports of inherited defect(s) of GPI biosynthesis was in contrast to the several heritable conditions caused by mutations affecting genes involved in other ubiquitous biochemical pathways, such as glycolysis and protein glycosylation. This was not surprising because complete disruption of GPI biosynthesis in $piga^{-/-}$ mice results in embryonic lethality [9, 10], indicating the importance of GPI and/or GPI-linked proteins in early development. However, as cultured cells lacking expression of GPI as a result of various biosynthetic blocks are not growth restricted [19], it could be predicted that complete GPI deficiency can be only tolerated when it affects certain somatic cells or tissues at a time beyond critical developmental stages and mutations that allow reduced but sufficient for survival synthesis of GPI can be tolerated throughout embryonal and fetal development. This explains why until recently, paroxysmal nocturnal hemoglobinuria (PNH), an acquired disorder caused by somatic mutations specific to hematopoietic stem cell (HSC) was the only condition associated with partial or complete deficiency in GPI synthesis and inherited GPI deficiency (IGD), the only congenital disorder of GPI synthesis described recently, is the result of a hypomorphic mutation which permits a degree of GPI production, presumably sufficient for embryonic and fetal viability [20].

A. Acquired GPI Deficiency: PNH

PNH is an acquired clonal disorder characterized by complement-mediated intravascular hemolysis, thrombosis, and bone marrow failure [21–23]. The defining genetic lesions in PNH are somatic mutations affecting the X-linked gene *PIG-A* in a single HSC [21–23] and its progeny, thus disrupting the first step in the GPI biosynthesis. A second process, believed to be a T cell mediated autoimmune attack against normal but not GPI-deficient HSC, allows survival and expansion of the latter resulting in the production of mature blood cells lacking synthesis of GPI and expression of GPI-AP [24, 25]. The clinical consequences are intravascular hemolysis and anemia as a result of erythrocytes lacking expression of the GPI-linked and complement activation inhibitor CD59, propensity to splanchnic (notably abdominal and cerebral) vein thromboses, also possibly linked to CD59 deficiency in erythrocytes and platelets, and finally bone marrow failure and cytopenia as a result of the GPI + HSC being targeted by autoaggressive T cells [26, 27].

B. Inherited GPI Deficiency

1. Clinical Spectrum

In the process of investigating spontaneous and unexplained hepatic vein thrombosis, we identified three children from two consanguineous families (Family 1 of Middle Eastern and Family 2 of Turkish origin) with IGD [20]. The index cases were two infant girls who were diagnosed with hepatic and intraabdominal vein thrombosis with ensuing portal hypertension. The third affected infant, a younger brother from Family 2, was also diagnosed with IGD, but thrombotic complications were averted by prophylactic anticoagulation. Epilepsy in the form of absence seizures, not associated with obvious cerebral vascular pathology, was another clinical feature shared by all three affected children.

Notably, and at variance with PNH, none of the affected children exhibited clinical evidence of significant intravascular hemolysis or bone marrow failure. The parents and unaffected siblings were asymptomatic. Some form of GPI deficiency was suspected initially on the basis of an intermittently low-positive Ham test (a serological test that demonstrates the presence of erythrocytes sensitive to lysis caused by complement activation), suggesting mild deficiency of the GPI-linked complement inhibitor CD59 [28]. This was subsequently confirmed by flow-cytometric analysis of GPI-linked protein expression on the surface of blood cells.

2. Surface GPI Expression in IGD

Expression on blood cells of GPI-linked proteins as determined by GPI-AP-specific mAb and of GPI itself as assessed by staining with FLAER, an inactivated form of aerolysin, revealed a pattern in IGD that is distinct from that in PNH [20] (Figure 16.2). The erythrocytes of children with IGD had relatively normal expression of CD59 with a small proportion of cells (<5%) being completely deficient. This is different from the bi- or trimodal pattern of CD59 expression in PNH-representing populations with normal complete or intermediate loss of GPI biosynthesis, respectively. By contrast, expression of GPI on the surface of granulocytes in IGD was severely reduced in a unimodal fashion whilst it is usually bimodal in PNH.

Monocytes had mostly normal GPI expression. T cells were predominantly positive for GPI whereas B lymphocytes displayed a predominantly deficient pattern of expression but unlike in granulocytes the pattern of GPI deficiency was clearly bimodal resembling that of PNH. Finally, fibroblasts from IGD patients demonstrated moderate reduction of FLAER staining (Figure 16.2).

In summary, the flow-cytometric pattern of GPI expression in IGD revealed a partial yet severe disruption of GPI synthesis. It was interesting to observe that the level of expression of GPI was not uniform in different cells and tissues, suggesting tissue-specific control of GPI synthesis.

3. Identification of PIG-M as the Disease Gene in IGD

A dual approach of homozygosity mapping targeted to the genes known to be involved in GPI biosynthesis and biochemical analysis of GPI intermediate structures was used to identify the disease gene in IGD [20]. Homozygosity mapping using microsatellite markers flanking almost all genes involved in GPI biosynthesis identified PIG-M as one of the possible candidate genes in the affected children of both families. In parallel, by the use of substrate labeling of patient-derived EBV-transformed B cell lines, glycolipid extraction and high-performance thin layer chromatography, the biochemical defect in IGD was pinpointed to the addition of Man-1 to GlcN-(acyl)PI. As mentioned earlier, this step is catalyzed by the GPI-MTI comprising PIG-M and PIG-X.

364

ANTONIO ALMEIDA, ET AL.

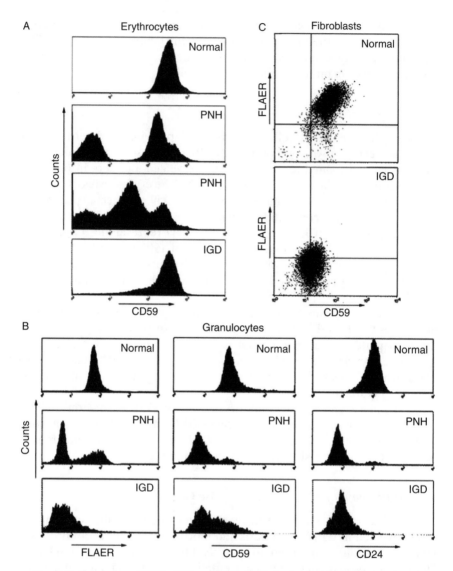

Fig. 16.2. Expression of GPI in IGD and PNH. (A) In patients with IGD, expression of the GPI-AP CD59 is near normal contrasting with the bi- or trimodal pattern in PNH. Expression of CD59 on erythrocytes from the parents of patients with IGD is normal. (B) Expression of GPI and GPI-AP (CD59 and CD24) in granulocytes. Severe deficiency of GPI expression in patients with IGD follows a unimodal pattern whilst it is typically bimodal in patients with PNH. (C). GPI and GPI-AP expression on fibroblasts from a normal donor and a patient with IGD. Expression of GPI-AP in IGD is reduced but not completely deficient.

Transfection of *PIG-M* cDNA into GPI-deficient B cell lines from affected children restored biosynthesis of GPI and expression of GPI-AP on the cell surface, thus confirming its role as the disease gene. PIG-M is a multiple domain membrane protein containing a luminal DXD motif characteristic of many glycosyltransferases [29] and PIG-X.

4. Hypomorphic Promoter Mutation Causes IGD

Direct sequencing of the single exon of *PIG-M* at genomic DNA level revealed no mutations in the coding region in any of the three affected children with IGD [20]. However, it was not possible to amplify *PIG-M* from cDNA that is derived from affected samples, suggesting a defect in transcription. Screening for mutations in the regulatory regions of *PIG-M* revealed that the same C > G substitution at position −270 from ATG in the promoter of *PIG-M* was present in both families and segregated with the phenotype. The same mutation was found in blood cells and fibroblasts in patients with IGD but not in any of control individuals from the patients' ethnic background.

Q-PCR revealed a pattern of *PIG-M* expression consistent with the observed genotypes. In the affected family members (−270 GG), *PIG-M* expression was approximately 1% of that of normal individuals (−270 CC). An intermediate level of expression, about half that of normal, was seen in heterozygotes (parents and siblings) for the −270C > G mutation thus conforming to a typical recessive trait.

Bioinformatics analysis predicted that the −270C > G transversion disrupted a GC-rich binding motif for the generic transcription factor (TF) Sp1. It is notable that this GC-box is the most highly conserved amongst all other predicted TF-binding motifs in the promoter of *PIG-M*, attesting thus to its functional importance. By a combination of transactivation and electromobility shift assays, these predictions were confirmed and it was found that the C > G transversion severely reduced but not completely abrogated the transcriptional activity of the *PIG-M* promoter [20]. These findings, along with the fact that in some cells and tissues in patients with IGD expression of GPI was not completely absent is consistent with −270C > G being a hypomorphic mutation whose impact on transcription varies among different tissues. In addition, it was determined that full activity of the *PIG-M* promoter requires a region of 2 kb upstream of ATG.

5. The Transcription Factor Sp1

Sp1 is a ubiquitously expressed TF that belongs to a wider family of Sp proteins, structurally characterized by the presence of three carboxyterminal, DNA binding zinc finger (ZnF) motifs [30–33].

A large number of *in vitro* studies have shown that Sp1 can be critical for efficient transcription of many house-keeping and inducible genes. Sp1 has been shown to enhance transcription through a variety of mechanisms that include direct interactions with elements of the TFIID complex of the basic transcriptional complex [32, 34], interactions with other TF such as GATA-1 [35–37], and recruitment of histone modifying enzymes such as histone acetyltransferases [38] and deacetylases [39] (HAT and HDAC, respectively). Additionally, the function of Sp1 depends on its ability to form tetramer homopolymers arranged in stacks and can be altered by its covalent modification, for example, phosphorylation and acetylation (reviewed in Refs. [31–33]).

All these interactions require binding of Sp1 through its ZnF motifs to a GC-rich DNA element (motif) found in the proximal promoters of all genes; GC-rich boxes are also found in distal promoters, enhancers, and locus control regions [31–33]. Sp3, another Sp TF binds to the same motif with almost equal affinity [31–33].

6. Sp1 and Gene-Specific Histone Hypoacetylation in IGD

One feature of the function of Sp1 that has attracted considerable interest is its ability to control transcriptional activity through recruitment to the promoter of histone modifying enzymes, especially HAT and HDAC: in most cases, the HAT activity prevails and favors transcription.

Another interesting aspect of Sp1 and its binding element is their role in mediating the effects of HDAC inhibition by butyrate and other HDAC inhibitors. Although HDAC inhibition leads to widespread increase in histone acetylation, it only impacts on the transcription rates of a minority (<5–10%) of genes [40–42]; and for a few amongst these genes, enhanced transcription as a result of HDAC inhibition and histone hyperacetylation, requires binding of Sp1 to its proximal promoter elements (i.e., GC-box) [43–50]. However, it is not known whether these butyrate responsive elements (BRE) are also critical for the maintenance of histone acetylation during baseline physiological transcription.

Using chromatin immunoprecipitation assays, it was shown that in the presence of the homozygous C > G mutation, histone 4 (H4) at the native promoter of *PIG-M* in patient B cell lines was hypoacetylated but, as would be expected of a house-keeping gene, acetylated in the parental B cell lines [51], implying that the Sp1-binding motif mutated in IGD is important for maintenance of histone acetylation at the promoter of *PIG-M* and thus its transcriptional activity.

However, despite disruption of the critical proximal promoter GC-rich motif, butyrate could still enhance the activity of the mutated promoter (although less so in comparison to the WT promoter), suggesting the presence of additional BRE in the promoter of *PIG-M*. These, although not adequate to sustain normal level baseline transcription, they could be required for the increased activity of the mutated promoter after HDAC inhibition [51]. Software analysis indicated the presence of another three GC-rich boxes upstream of the mutated motif, suggesting that these motifs might be required for the butyrate effect on the mutated promoter. Consistent with this prediction, when binding of Sp1 to DNA was disrupted by mithramycin, the butyrate-mediated enhancing effect was dramatically reduced for the mutated as well as the WT promoter [51]. Furthermore, exposure of patient B cell lines to butyrate resulted in restoration of H4 acetylation in the native mutated *PIG-M* promoter, a 20-fold increase in the *PIG-M* mRNA levels and complete restoration of GPI biosynthesis as shown by normal expression levels of GPI-linked proteins on the cell surface [51].

7. HDAC Inhibition as a Treatment for IGD

In view of these results, we treated a child with IGD who suffered chronic intractable seizures and severe disability with butyrate. This resulted in the progressive increase of blood cell *PIG-M* RNA levels from ~1% to 60% and also a progressive increase in the expression of GPI-linked molecules on the surface of granulocytes; more importantly, 2 weeks after the commencement of butyrate, the child became seizure-free for the first time in 12 years [51].

Therefore, the *PIG-M* mutation causing IGD results in significant perturbation of the epigenetic landscape in the promoter of *PIG-M* and eventually in a clinical phenotype directly dependent on the biosynthesis of GPI and cell surface expression of GPI-AP.

IV. A Tentative Model of Transcriptional Control of PIG-M by Sp1 and its Dysregulation in IGD

Considering these findings, we propose a model (Figure 16.3) of transcriptional control of *PIG-M* by Sp1 whereby during baseline transcription of *PIG-M*, binding of Sp1 to the proximal promoter GC-box is associated with corecruitment of HAT and HDAC. In this context, HAT prevails over HDAC activity ensuring histone acetylation and normal transcriptional output. Similar arrangements might occur in the upstream GC-boxes but their role may not be significant in baseline transcription. In IGD, disruption of the proximal GC-box by the C > G mutation significantly

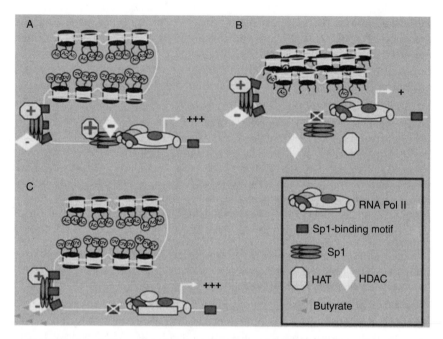

FIG. 16.3. A tentative model of transcriptional control of *PIG-M* by Sp1 and its dysregulation in IGD. (A) The RNA polymerase II transcriptional machinery and the open configuration of acetylated chromatin ensure baseline transcriptional activity of the *PIG-M* promoter. RNA polymerase II activity and histone acetylation are under the control of Sp1 through the ability of the latter to interact with components of the Pol II complex and recruit histone modifying enzymes, that is, acetyltransferases (HAT) which favor transcription and deacetylases (HDAC) which repress transcription. Under basal conditions, the balance favors HAT activity and thus acetylation of histones and active transcription. (B) In IGD, as a result of the $-270C > G$ mutation, Sp1 binding to its binding motif in the core promoter of PIG-M is abolished, resulting in the interruption of interaction of Sp1 with the Pol II complex and loss of its ability to recruit HAT/HDAC. The ensuing chromatin deacetylation and condensation dramatically, but not completely, reduces transcriptional activity of the *PIG-M* promoter. (C) HDAC recruited to the upstream Sp1 binding motifs are inhibited in the presence of the HDAC inhibitor butyrate, leaving activity of HAT unopposed, thus allowing restoration of histone acetylation and transcriptional activity of *PIG-M*. Reproduced by permission from Almeida, A, Mark Layton, M, Karadimitris, A (2009). Inherited glycosylphosphatidylinositol deficiency: a treatable CDG. *BBA—Molecular Basis of Disease*, Jan 9. [Epub ahead of print].

reduces transcriptional output in a cell- and tissue-dependent manner while Sp1 binding to upstream GC-boxes is preserved. In the mutated promoter, during butyrate treatment, selective HDAC inhibition leaves HAT activity unopposed resulting in significant restoration of histone acetylation and thus of transcriptional activity and GPI biosynthesis.

V. IGD: Questions and Perspectives

A. A COMMON ANCESTRAL MUTATION?

An interesting question from the genetic viewpoint is whether the disruption of the same biosynthetic step involving the same gene with exactly the same mutation in ethnically different families with IGD has occurred by chance or it reflects a mutation in a common, distant ancestor. Arguing against the latter, microsatellite- and SNP-based haplotypes encompassing *PIG-M* were clearly disparate [20], suggesting that the mutation arose in the two families independently and by chance.

B. MECHANISM OF CELL- AND TISSUE-DEPENDENT GPI EXPRESSION VARIABILITY

A second question pertains to the mechanism responsible for the cell- and tissue-dependent variability in GPI synthesis in IGD as exemplified by the relatively normal expression of GPI in erythrocytes and its severe deficiency in granulocytes. One possibility is that transcription of *PIG-M* might be regulated by tissue-specific TF which can maintain adequate transcriptional activity even in the presence of the mutated core promoter Sp1-binding motif. Preliminary work suggests that GATA-1, an erythroid/megakaryocyte-specific TF [52] might play a critical role in this process (Almeida *et al.*, unpublished data). This mechanism may also be at play in T lymphocytes and neurons, where other members of the GATA TF family are expressed, and could explain the flow-cytometric profile of GPI expression of T cells and the lack of neurodevelopmental abnormalities in these patients.

C. MECHANISM OF THROMBOSIS AND EPILEPSY

At the moment, we can only speculate on the pathogenesis of seizures and thrombosis in IGD and the role of specific GPI-AP in these processes. The generation of a *pigm* C > G knock-in mouse model will help greatly in this direction.

D. IGD CAUSED BY MUTATIONS IN OTHER GENES OF THE GPI-BIOSYNTHETIC PATHWAYS?

The paradigm of PIG-M-related IGD suggests that probably every gene essential for GPI assembly suffering hypomorphic (e.g., promoter or missense) mutations could lead to inherited deficiency of GPI. However, this would be expected to be a very rare event occurring solely in the context of consanguinity.

E. HDAC INHIBITION AS A MEANS TO ENHANCE TRANSCRIPTION
AND ACTIVITY OF HOUSE-KEEPING GENES ASSOCIATED WITH
HUMAN PATHOLOGY

Because *PIG-M* is a house-keeping gene and part of a ubiquitous bio-synthetic pathway, it can be expected that other genes required for GPI biosynthesis may be subject to the same type of Sp1-dependent transcriptional control as *PIG-M*. Extending this notion further, it could be surmised that Sp1 and its control of histone acetylation are also important for the transcription of genes that are part of other house-keeping, enzyme-based pathways. If so, characterization of Sp1-binding motifs/BRE required for transcriptional control of genes with these characteristics could offer new therapeutic opportunities for inherited disorders of ubiquitous metabolic pathways such as glycolysis [53] and disorders of glycosylation [54]. Often in these autosomal recessive disorders, the molecular pathology comprises missense mutations which allow for residual enzymatic activity. Increased transcription and protein production induced by HDAC inhibitors might translate into increased enzymatic activity and thus amelioration of the cellular and clinical phenotype.

ACKNOWLEDGMENT

Work in the authors' lab has been supported by Leukaemia Research Fund.

REFERENCES

1. Ferguson, M.A.J., and Williams, A.F. (1988). Cell-surface anchoring of proteins via glycosyl-phosphatidylinositol structures. *Annu Rev Biochem* 57:285.
2. Kinoshita, T., Fujita, M., and Maeda, Y. (2008). Biosynthesis, remodeling and functions of mammalian GPI-anchored proteins: recent progress. *J Biochem* 144:287.
3. Orlean, P., and Menon, A.K. (2007). Thematic review series: lipid posttranslational modifications. GPI anchoring of protein in yeast and mammalian cells, or: how we learned to stop worrying and love glycophospholipids. *J Lipid Res* 48:993.
4. Paulick, M.G., and Bertozzi, C.R. (2008). The glycosylphosphatidylinositol anchor: a complex membrane-anchoring structure for proteins. *Biochemistry* 47:6991.
5. Maeda, Y., and Kinoshita, T. (2008). Dolichol-phosphate mannose synthase: structure, function and regulation. *Biochim Biophys Acta* 1780:861.
6. Tanaka, S., Maeda, Y., Tashima, Y., and Kinoshita, T. (2004). Inositol deacylation of glycosylphosphatidylinositol-anchored proteins is mediated by mammalian PGAP1 and yeast Bst1p. *J Biol Chem* 279:14256.
7. Maeda, Y., Tashima, Y., Houjou, T., Fujita, M., Yoko-o, T., Jigami, Y., Taguchi, R., and Kinoshita, T. (2007). Fatty acid remodeling of GPI-anchored proteins is required for their raft association. *Mol Biol Cell* 18:1497.

8. Rudd, P.M., Morgan, B.P., Wormald, M.R., Harvey, D.J., van den Berg, C.W., Davis, S.J., Ferguson, M.A., and Dwek, R.A. (1998). The glycosylation of the complement regulatory protein, human erythrocyte CD59. *Adv Exp Med Biol* 435:153.

9. Kawagoe, K., Kitamura, D., Okabe, M., Taniuchi, I., Ikawa, M., Watanabe, T., Kinoshita, T., and Takeda, J. (1996). Glycosylphosphatidylinositol-anchor-deficient mice: implications for clonal dominance of mutant cells in paroxysmal nocturnal hemoglobinuria. *Blood* 87:3600.

10. Tremml, G., Dominguez, C., Rosti, V., Zhang, Z., Pandolfi, P.P., Keller, P., and Bessler, M. (1999). Increased sensitivity to complement and a decreased red blood cell life span in mice mosaic for a nonfunctional Piga gene. *Blood* 94:2945.

11. Leidich, S.D., Drapp, D.A., and Orlean, P. (1994). A conditionally lethal yeast mutant blocked at the first step in glycosyl phosphatidylinositol anchor synthesis. *J Biol Chem* 269:10193.

12. Orlean, P., Leidich, S.D., Drapp, D.A., and Colussi, P. (1994). Isolation of temperature-sensitive yeast GPI-anchoring mutants. *Braz J Med Biol Res* 27:145.

13. Munro, S. (2003). Lipid rafts: elusive or illusive? *Cell* 115:377.

14. Mukherjee, S., and Maxfield, F.R. (2004). Membrane domains. *Annu Rev Cell Dev Biol* 20:839.

15. Sharma, P., Varma, R., Sarasij, R.C., Ira, K., Gousset, K., Krishnamoorthy, G., Rao, M., and Mayor, S. (2004). Nanoscale organization of multiple GPI-anchored proteins in living cell membranes. *Cell* 116:577.

16. Butikofer, P.E.T.E., Malherbe, T.A.T.I., Boschung, M.O.N.I., and Roditi, I.S.A.B. (2001). GPI-anchored proteins: now you see 'em, now you don't. *FASEB J* 15:545.

17. Schuck, S., and Simons, K. (2006). Controversy fuels trafficking of GPI-anchored proteins. *J Cell Biol* 172:963.

18. Traister, A., Shi, W., and Filmus, J. (2007). Mammalian Notum induces the release of glypicans and other GPI-anchored proteins from the cell surface. *Biochem J* 410:503.

19. Maeda, Y., Ashida, H., and Kinoshita, T. (2006). Cho glycosylation mutants: GPI anchor, in F.Minoru (ed.), *Methods in Enzymology Glycomics* Academic Press, pp. 182–205.

20. Almeida, A.M., Murakami, Y., Layton, D.M., Hillmen, P., Sellick, G.S., Maeda, Y., Richards, S., Patterson, S., Kotsianidis, I., Mollica, L., Crawford, D.H., Baker, A., *et al.* (2006). Hypomorphic promoter mutation in PIGM causes inherited glycosylphosphatidylinositol deficiency. *Nat Med* 12:846.

21. Luzzatto, L., Bessler, M., and Rotoli, B. (1997). Somatic mutations in paroxysmal nocturnal hemoglobinuria: a blessing in disguise? *Cell* 88:1.

22. Rosse, W.F., and Ware, R.E. (1995). The molecular basis of paroxysmal nocturnal hemoglobinuria. *Blood* 86:3277.

23. Takeda, J., Miyata, T., Kawagoe, K., Iida, Y., Endo, Y., Fujita, T., Takahashi, M., Kitani, T., and Kinoshita, T. (1993). Deficiency of the GPI anchor caused by a somatic mutation of the PIG-A gene in paroxysmal nocturnal hemoglobinuria. *Cell* 73:703.

24. Karadimitris, A., and Luzzatto, L. (2001). The cellular pathogenesis of paroxysmal nocturnal haemoglobinuria. *Leukemia* 15:1148.

25. Luzzatto, L., and Bessler, M. (1996). The dual pathogenesis of paroxysmal nocturnal hemoglobinuria. *Curr Opin Hematol* 3:101.

26. Hillmen, P., Lewis, S.M., Bessler, M., Luzzatto, L., and Dacie, J.V. (1995). Natural history of paroxysmal nocturnal hemoglobinuria. *N Engl J Med* 333:1253.

27. Socie, G., Mary, J.Y., de Gramont, A., Rio, B., Leporrier, M., Rose, C., Heudier, P., Rochant, H., Cahn, J.Y., and Gluckman, E. (1996). Paroxysmal nocturnal haemoglobinuria: long-term follow-up and prognostic factors. French Society of Haematology. *Lancet* 348:573.

28. Holguin, M.H., Fredrick, L.R., Bernshaw, N.J., Wilcox, L.A., and Parker, C.J. (1989). Isolation and characterization of a membrane protein from normal human erythrocytes that inhibits reactive lysis of the erythrocytes of paroxysmal nocturnal hemoglobinuria. *J Clin Invest* 84:7.

29. Maeda, Y., Watanabe, R., Harris, C.L., Hong, Y., Ohishi, K., Kinoshita, K., and Kinoshita, T. (2001). PIG-M transfers the first mannose to glycosylphosphatidylinositol on the lumenal side of the ER. *EMBO J* 20:250.

30. Wierstra, I. (2008). Sp1: emerging roles—beyond constitutive activation of TATA-less housekeeping genes. *Biochem Biophys Res Commun* 372:1.

31. Li, L., He, S., Sun, J.M., and Davie, J.R. (2004). Gene regulation by Sp1 and Sp3, Biochem. *Cell Biol* 82:460.

32. Bouwman, P., and Philipsen, S. (2002). Regulation of the activity of Sp1-related transcription factors. *Mol Cell Endocrinol* 195:27.

33. Suske, G. (1999). The Sp-family of transcription factors. *Gene* 238:291.

34. Emili, A., Greenblatt, J., and Ingles, C.J. (1994). Species-specific interaction of the glutamine-rich activation domains of Sp1 with the TATA box-binding protein. *Mol Cell Biol* 14:1582.

35. Merika, M., and Orkin, S.H. (1995). Functional synergy and physical interactions of the erythroid transcription factor GATA-1 with the Kruppel family proteins Sp1 and EKLF. *Mol Cell Biol* 15:2437.

36. Furusawa, M., Taira, T., Iguchi-Ariga, S.M., and Ariga, H. (2003). Molecular cloning of the mouse AMY-1 gene and identification of the synergistic activation of the AMY-1 promoter by GATA-1 and Sp1. *Genomics* 81:221.

37. Chin, K., Oda, N., Shen, K., and Noguchi, C.T. (1995). Regulation of transcription of the human erythropoietin receptor gene by proteins binding to GATA-1 and Sp1 motifs. *Nucleic Acids Res* 23:3041.

38. Xiao, H., Hasegawa, T., and Isobe, K. (2000). p300 collaborates with Sp1 and Sp3 in p21 (waf1/cip1) promoter activation induced by histone deacetylase inhibitor. *J Biol Chem* 275:1371.

39. Doetzlhofer, A., Rotheneder, H., Lagger, G., Koranda, M., Kurtev, V., Brosch, G., Intersberger, E.W., and Seiser, C. (1999). Histone deacetylase 1 can repress transcription by binding to Sp1. *Mol Cell Biol* 19:5504.

40. Gray, S.G., Qian, C.N., Furge, K., Guo, X., and Teh, B.T. (2004). Microarray profiling of the effects of histone deacetylase inhibitors on gene expression in cancer cell lines. *Int J Oncol* 24:773.

41. Mariadason, J.M., Corner, G.A., and Augenlicht, L.H. (2000). Genetic reprogramming in pathways of colonic cell maturation induced by short chain fatty acids: comparison with trichostatin A, sulindac, and curcumin and implications for chemoprevention of colon cancer. *Cancer Res* 60:4561.

42. Van, L.C., Emiliani, S., and Verdin, E. (1996). The expression of a small fraction of cellular genes is changed in response to histone hyperacetylation. *Gene Expr* 5:245.

43. Camarero, N., Nadal, A., Barrero, M.J., Haro, D., and Marrero, P.F. (2003). Histone deacetylase inhibitors stimulate mitochondrial HMG-coA synthase gene expression via a promoter proximal Sp1 site. *Nucleic Acids Res* 31:1693.

44. Davie, J.R. (2003). Inhibition of histone deacetylase activity by butyrate. *J Nutr* 133:2485S.

45. Di, P.E., Cappellini, M.D., Mazzucchelli, R., Moriondo, V., Mologni, D., Zanone, P.B., and Riva, A. (2005). A point mutation affecting an SP1 binding site in the promoter of the ferrochelatase gene impairs gene transcription and causes erythropoietic protoporphyria. *Exp Hematol* 33:584.

46. Gan, Y., Shen, Y.H., Wang, J., Wang, X., Utama, B., Wang, J., and Wang, X.L. (2005). Role of histone deacetylation in cell-specific expression of endothelial nitric-oxide synthase. *J Biol Chem* 280:16467.

47. Guyot, B., Murai, K., Fujiwara, Y., Valverde-Garduno, V., Hammett, M., Wells, S., Dear, N., Orkin, S.H., Porcher, C., and Vyas, P. (2006). Characterization of a megakaryocyte-specific enhancer of the key hemopoietic transcription factor GATA1. *J Biol Chem* 281:13733.

48. Han, L., Lu, J., Pan, L., Wang, X., Shao, Y., Han, S., and Huang, B. (2006). Histone acetyltransferase p300 regulates the transcription of human erythroid-specific 5-aminolevulinate synthase gene. *Biochem Biophys Res Commun* 348:799.

49. Steiner, E., Holzmann, K., Pirker, C., Elbling, L., Micksche, M., and Berger, W. (2004). SP-transcription factors are involved in basal MVP promoter activity and its stimulation by HDAC inhibitors. *Biochem Biophys Res Commun* 317:235.

50. Ye, J., Shedd, D., and Miller, G. (2005). An Sp1 response element in the Kaposi's sarcoma-associated herpesvirus open reading frame 50 promoter mediates lytic cycle induction by butyrate. *J Virol* 79:1397.

51. Almeida, A.M., Murakami, Y., Baker, A., Maeda, Y., Roberts, I.A., Kinoshita, T., Layton, D.M., and Karadimitris, A. (2007). Targeted therapy for inherited GPI deficiency. *N Engl J Med* 356:1641.

52. Cantor, A.B., and Orkin, S.H. (2002). Transcriptional regulation of erythropoiesis: an affair involving multiple partners. *Oncogene* 21:3368.

53. van Wijk, R., and van Solinge, W.W. (2005). The energy-less red blood cell is lost: erythrocyte enzyme abnormalities of glycolysis. *Blood* 106:4034.

54. Jaeken, J., and Matthijs, G. (2007). Congenital disorders of glycosylation: a rapidly expanding disease family. *Annu Rev Genomics Hum Genet* 8:261.

Author Index

Numbers in regular font are reference numbers and indicate that an author's work is referred to although the name is not cited in the text. Numbers in italics refer to the page numbers on which the complete reference appears.

A

Aalten, D. M. F. V., 52–54, 56, 58, 60, *63*
Abdian, P. L., 35, *45*
Abe, H., 338–340, 342, *353*
Abe, K., 335, *351*
Abe, M., 328, *349*
Abeijon, C., 151, *157*, 323, 335–336, *347*
Abrami, L., 295, *312*
Abundis, M. I., 342, *355*
Ackermann, W., 304, *316*
Acosta-Serrano, A., 3, *22*, 93, *110*, 248, 252, *261*
Adams, E. W., 235, *244*
Adams, S. P., 20, *30*, 258, *266*
Aebi, M., 15, *28*, 104, *112*, 137–139, *147–148*, 153, 154, *158*
Aguet, M., 277, *286*
Aguilera-Romero, A., 273, *285*
Aguzzi, A., 277, *286*, 302, *315*
Aharonowitz, Y., 52, *63*
Ahktar, N., 306, *317*
Ahmed, S. N., 293, *319*
Akimov, S. A., 295, *312*
Akira, S., 173, *176*, *178*, 231, *243*
Akyilmaz, E., 344, *356*
Albar, J. P., 297, *313*
Alberghina, L., 322, 324, 329, 334–335, *346*, *349*
Albrecht, A., 322, 335, *346*
Albright, C., 66, *84*, 167, *177*
Alderwick, L. J., 108, 109, *115*
Alessi, D. R., 297, *313*
Alfalah, M., 295, 303, 305–306, *313*, *315*
Ali, A., 209, *226*, 241, *245*

Almeida, A. M., 2, *21*, 104, *113*, 248, 254, *261*, 362–363, 365–367, 369, *373*
Almeida, I. C., 173, *176*, 231, *243–245*, 363
Almond, A., 7, *23*, 66, *85*
Alonso, M. A., 297, *313*
Alonso, P. L., 230, *242*
al-Qahtani, A., 154, *158*
Alton, G., 104, *112*
Altschul, S. F., 50, 55, *63*
Altschuler, Y., 291–292, *309*
Alves, M. J., 20, *30*
Alzari, P. M., 20, *30*
Amachi, T., 339–340, *354*
Amaya, M. F., 20, *30*
Amin, A. G., 109, *115*
Amizuka, N., 79, *90*
Anand, M., 93, *110*, 153, *158*
Anand, R., 3, *22*, 97, *111*, 118, 127, *128*, 182, *224*
Ananthanarayanan, M., 292, *310*
Anderson, D., 107, *114*
Anderson, R. G., 293, *319*
Anderson, V., 141, *149*
Andre, S., 297, *313*
Andrews, B., 16, *29*
Ang, A. L., 303, 306, *315*, *317*
Anraku, Y., 330, *350*
Ansari, A., 306, *317*
Ansorge, I., 82, *90*, 270, *284*, 299, *312*
Anstey, N. M., 235, *244*
Anton, L. C., 276, *286*
Antonny, B., 272, *285*
Antonsson, B., 276, *286*
Antony, C., 292, 300, *318*

Anzick, S. L., 144, *149*
Apodaca, G., 303, *315*
Arad, D., 52, *63*
Arancia, G., 323, *347*
Araujo, F. G, 165, *177*
Arentshorst, M., 331–332, *350*
Ariga, H., 366, *372*
Arisawa, M., 134, *146*, 322–330, 335, *346*, *349–350*
Arnold, K., 52, *63*
Arpagaus, M., 66, *84*
Arreaza, G., 292–302, *311*
Arroyo, J., 323, 336, *347*, *352*
Arshad, M., 337–338, 341–342, *352*
Ash, J., 335, *351*
Ashida, H., 8, 11, 12, 14, 15, 18, *24–29*, 70, *87*, 93, 94, 102, *110*, *111*, 121, 123–127, *130*, 136–140, 143, *147–148*, 168, *178*, 361, *371*
Ashok, A., 270–271, *284*
Ashwell, G., 3, *21*, 248, 256, *261*, 325, 328, *348*
Astrof, N. S., 343, *355*
Atkinson, J. P., 238, *245*
Atomi, H., 339–340, *354*
Atrih, A., 260, *267*
Augenlicht, L. H., 366, *372*
Augusti-Tocco, G., 338, *352*
Autenried, P., 277, *286*
Av-Gay, Y., 52, *63*
Avrahami, D., 307, *317*
Ayala, F. J., 167, *177*
Azuma, M., 123–124, *130–131*
Azzouz, M. A., 259, *267*
Azzouz, N., 71, 72, *88*, 103, *112*, 163, 165–167, 169–170, 173, *176–179*, 241, *245*, 250, 254–255, 258, *262*, *263*, *265*

B

Baba, M., 134, *146*
Babbey, C. M., 306, *317*
Bacallao, R. L., 303, *315*
Bacic, A., 3, *22*
Badali, O., 342, *355*
Baeschlin, D. K., 190, *224*, *225*
Bagnat, M., 17, *29*, 279, *287*
Bai, C., 322, *346*
Bai, Y., 109, *115*
Bailie, N. M., 104, *112*
Baird, J. K., 235, *244*
Bajaj, J., 342, *355*

Bajohrs, M., 301, *314*
Baker, A., 2, *21*, 104, *113*, 362–363, 365–367, 369, *371*, *373*
Baker, E. N., 52, 54, *63*
Balch, W. E., 271, *284*, *285*
Baldauf, S. L., 322, *345*
Baldermann, C., 338–339, *353*
Baldwin, A. N., 295, *313*
Baldwin, T., 170, *180*, 231, *243*, 248, 253, *261*
Balestrini, R., 322, 334, *346*
Ballantyne, S. D., 307, *317*
Ballou, C. E., 104–106, *113*, *114*
Bando, H., 338, *353*
Banerjee, D. K., 254, *264*
Banerjee, S., 162, *178*
Bangs, J. D., 136, 139, *147–148*, 258, 259, *267*, 270, *284*, 291, 309
Bangs, M. J., 235, *244*
Barajas, M., 342, *355*
Baral, T. N., 176, *179*
Barbat, A., 295, *312*
Barlowe, C., 78, *89*, 273, *285*, 291, *309*
Barnwell, J., 168, *178*
Baroni, F., 276, *286*
Barrero, M. J., 366, *372*
Barrio, E., 167, *177*
Barth, J. L., 344, *356*
Barton, G. J., 20, *30*
Barz, W. P., 278, *287*
Basagoudanavar, S. H., 168, *176*, 322, *345*
Bassik, M. C., 276, *286*
Bastisch, I., 332, *350*
Bate, C. A. W., 230, 231, *242*
Bauer, G., 66, *84*
Baulard, A., 107, *114*
Baumann, N. A., 278, *287*
Bause, E., 254, *264*
Beatty, P. G., 167, *177*
Beauvais, A., 327, 331–332, *349*, *350*
Beck, P. J., 66, *84*, 167, *177*
Becker, D., 162, *179*, 254, *264*
Beddoe, T., 108, *114*
Beest Chapin, S. J., 307, *317*
Begent, R., 343, *355*
Beh, C. T., 278, 282, *287*
Beilharz, T. H., 272, *285*
Belcher, A. M., 342, *355*
Belden, W. J., 78, *89*, 273, *285*, 291, *309*
Belisle, J. T., 107, *114*
Bell, R. M., 152, *157*

Benachour, A., 6, 8, 11, 12, 15, *23*, *26–28*, 119–
122, 124–126, *129*, 137–139, *147*, 154, *158*,
250, 257, *262*, *266*, 332, *350*
Bender, F. C., 293, 299, *314*
Bender, S., 254, *264*
Benfey, P. N., 322, *345*
Benghezal, M., 11, 12, 14, 15, 19, *26*, *28*, *30*,
119–120, 122, 124–125, *129*, 137–139,
147–148, 154, *158*, 299, *314*, 332, *350*
Ben-Hur, A., 36, 38, 41, 42, *46*
Bennett, M. K., 292, 300, 307, *317*, *318*
Bennink, J. R., 276, *286*
Benting, J., 82, *90*
Benting, J. H., 270, 280, *284*, *288*, 295, 299, *312*
Berg, S., 105, 109, *113*, *115*
Berger, E. G., 104, *112*
Berger, J., 134, *145*
Berger, W., 366, *373*
Bergmann, E. M., 335, *351*
Berhe, S., 163, *176*
Berkel, M. A., 327, *348*
Bernasconi, P., 276, *286*
Bernshaw, N. J., 363, *372*
Bertello, L. E., 20, *30*
Bertolotti, A., 276, *286*
Bertozzi, C. R., 358, *370*
Besra, G. S., 105, 107, 108, *113–115*
Bessler, M., 2, *21*, 361–362, *371*
Beznoussenko, G. V., 291, *318*
Bhamidipati, A., 16, *29*
Bhandari, R., 344, *356*
Bhatt, A., 108, 109, *115*
Bhuvaneswaran, C., 255, *265*
Bi, E., 15, *28*, 136, 142, 144, *148*
Biagioni, S., 338, *352*
Bickel, P. E., 295, *312*
Bidlingmaier, S., 343, *355*
Bieker, J., 259, *267*
Bieker, U., 165, 173, *176*
Bielawski, J., 297, *314*
Bifani, P., 107, *114*
Bihl, F., 105, *113*
Billman-Jacobe, H., 107, 108, *114*
Birch, H. L., 109, *115*
Black, P. N., 76, *89*
Blackman, M. J., 3, *22*, 87, 163, *177*,
231–232, *242*
Blanco, N., 323, 336, *347*, *352*
Blank, M., 93, *110*
Blasi, F., 293, *319*

Blasius, T. L., 304, *316*
Bloom, G. S., 304, *316*
Blumwald, E., 322, *346*
Blundell, P., 3, *22*, 139–140, *148*
Boadle, R., 323, 331–332, *347*
Bockarie, M. J., 235, *244*
Böcskei, Z., 134, 142, *146*, *149*
Boder, E. T., 337, 339, 342, *352–353*
Boersma, W. J., 340, 343, *354*
Bogyo, M., 136, *146*
Boned, A., 308, *318*
Bonfante, P., 322, 334, *346*
Bonifacino, J. S., 93, *110*, 270, *284*,
292, *310*
Boone, C., 16, *29*, 334, *351*
Boorsma, A., 326–328, *348*
Bordon-Pallier, F., 332, *350*
Borissow, C. N., 58, 60–62, *64*, 72, *88*, 167, *179*,
250, 252, *263*
Bork, P., 134, 140, *146*, *149*, 322, *345*
Bornemann, S., 202, *225*
Borner, G. H., 2, 3, *21*
Borodkin, V. S., 215, *226*, 241, *245*
Borvak, J., 173, *180*
Boschung, M. O. N. I., 361, *371*
Bosson, R., 17, *29*, 271, *284*, 299, *314*
Botha, T., 105, *113*
Botvinko, I. V., 223, *227*
Bounery, J. D., 105, *113*
Bourbonnais, Y., 323, 335–336, *347*, *351*
Boutlis, C. S., 235, *244*
Bouwman, P., 365–366, *372*
Bouzamondo-Bernstein, E., 277, *287*
Bowen, N. J., 291, *318*
Bowers, B., 327–330, *349*
Bowley, D. R., 337–338, *352*
Brady, S. M., 322, *345*
Brammananth, R., 108, *114*
Branch, O. H., 71, *88*, 163, *178*, 231–232,
235, *243*
Brandner, S., 277, *286*, *287*
Brandt, G. S., 276, *286*
Brandt, J., 323, *347*
Brattig, N. W., 175, *180*, 238, *245*
Braun, V., 308, *318*
Breinig, F., 334, 338, 340–341, *351–352*
Brekelmans, S., 338–339, 342, *353*
Brenk, R., 260, *267*
Brennan, P. J., 104–109, *113–115*
Bressler, M., 42, *47*

Breuza, L., 307, *317–318*
Brewis, I. A., 3, *22*
Briken, V., 105, *113*
Brim, H., 337–338, 341, 342, *352*
Brimacombe, J. S., 3, 10, 19, *22*, *24*, *29*, 55, 56, 58, 60–62, *63–64*, 67, 69, 70, 72, *86–88*, 102, *112*, 167, *179*, 248–250, 252–257, *261–265*
Brininstool, G., 3, *21*
Brink, L., 136, *147*, 259, *267*
Briolay, A., 292, *311*
Brodsky, R. A., 33, *44*
Bron, C., 293, 299, *314*, 324, *347*
Brosch, G., 366, *372*
Brown, D., 238, *244*
Brown, D. A., 17, *29*, 278, 280, *287*, *288*, 292–295, 302, *311*, *318*, *319*
Brown, J. L., 125, *131*, 324, 325, 338, *348*
Brown, J. R., 19, *29*, 69, 87, 248, 252, 256, *261*, *265*
Brown, K. N., 163, *177*
Brul, S., 322, *347*
Brun, R., 3, *22*, 93, *110*, 248, 250, 256, *261*, *262*
Bruns, J. R., 295, 301, *312*, *314*
Bryant, D. M., 290, *309*
Brys, L., 176, *179*
Bucciantini, M., 276, *286*
Buckley, J. T., 248, 255, *261*
Bueler, H., 277, *286*
Bulleid, N. J., 139, *148*
Bulone, V., 35, *45*
Burchard, G. D., 175, *180*, 238, *245*
Burda, P., 6, 8, *23*, 104, *112*, 119, *129*, 250, 257, *262*, *266*
Burger, K. N., 291, *318*
Burkhardt, J. K., 303, *315*
Burton, D. R., 337, 338, *352*
Buschiazzo, A., 20, *30*
Bussey, H., 125, *131*, 322–325, 327–328, 334, 336, 338, *346*, *348*, *349*, *351*
Bütikofer, P., 19, *30*, 140, *149*, 299, *314*, 361, *371*
Buxbaum, L. U., 20, *30*, 71, 87, 258, *266*
Byers, D. M., 281, *288*

C

Cabezas, A., 304–305, *316*
Cabib, E., 3, *21*, 248, 256, *261*, 323, 325, 327–330, 336, *347*, *349–350*, *352*

Cade, R. M., 330, *350*
Cahn, J. Y., 362, *371*
Cahuzac, B., 105, *113*
Cai, D., 304, *316*
Cali, A., 330, *350*
Califano, J. A., 144, *149*
Callewaert, N., 343, *355*
Calvo, M., 304, *316*
Camarero, N., 366, *372*
Camargo, M. M., 241, *245*
Camoletto, P. G., 302, *315*
Campagne, F., 305, *316*
Campana, V., 295, *312*
Campbell, J. A., 35, *45*
Campbell, R. E., 35, *45*
Campos, F., 259, *267*
Campos, M. A., 173, *176*, 231, *243*
Canivenc-Gansel, E., 6–8, 11, 12, *23*, *26*, 119–120, 122, 124–125, *129*, 250, 257, 258, *262*, *266*, *267*, 270, *283*, 332, *350*
Cantor, A. B., 369, *373*
Caplan, M. J., 302, *315*
Cappellaro, C., 324, 338–339, *348*, *353*
Cappellini, M. D., 366, *372*
Caras, I. W., 134, 136, *145–147*, 248, 252, 255, *261*, *265*, 270, 280, *284*, *288*, 292, *311*
Carman, C. V., 343, *355*
Carney, G. E., 291, *318*
Caro, L. H., 134, *146*, 322, 327, 329, 338, *345*, *348*, *353*
Carragher, B., 271, *285*
Carriere, V., 295, *312*
Carswell, E. A., 230, *242*
Carvalho, P., 277, *286*
Casadevall, A., 322, 323, *345*
Casal, O. L., 254, *264*
Casals-Pascual, C., 238, *244*
Casanova, J. E., 307, *317*
Casaregola, S., 335, *351*
Cassone, A., 323, *347*
Castillon, G. A., 270, *284*
Catino, M. A., 295, 302, 304, *313–314*
Cawley, N. X., 323, *347*
Cazares, B. X., 93, *110*
Cebon, J., 344, *356*
Cesbron-Delauw, M. F., 173, *180*
Chadda, R., 306, *317*
Chae, Y. K., 144, *149*
Chait, B. T., 167, *178*
Chan, C. S., 330, *350*

Chan, F. Y., 322, *346*
Chang, A., 302, *315*
Chang, H. M., 11, *25*, 36, *46*, 66, *84*, *85*, 126, *131*, 134, *146*, 167, *177*
Chang, K. P., 154, *158*
Chang, T., 14, *27*, 50, 52–54, 56, 58, 60, 62, *63*, 252, *263*
Chang, X., 144, *149*
Channon, J., 173, *180*
Chantret, I., 94, *111*
Chapatte, L., 293, 299, *314*
Chaperon, A. R., 190, *224*, *225*
Chapin, S. J., 307, *317*, *318*
Chaplin, D., 202, *225*
Chapman, J. W., 339, *353*
Charbonneau, V., 190, *224*
Chatterjee, A., 144, *149*
Chatterjee, D., 104, 105, 107–109, *113–115*
Chatterjee, P., 214, *226*
Chatterjee, S., 290, *309*
Chavrier, P., 307, *317*
Chawla, M., 214, *226*
Chen, C. C., 306, *317*
Chen, G. C., 330, *350*
Chen, M., 338, *352*
Chen, R., 14, *27*, 67, *86*, 136, 137, 139, 141, *147*, *149*, 250, 258, *262*, *266*
Chen, W., 337–342, *352*, *353*
Cheng, Y., 271, *285*
Cheong, K. H., 297, *313*
Cherney, M. M., 52, *63*
Chester, K., 343, *355*
Chi, Z., 338, *352*
Chichili, G. R., 308, *318*
Chin, K., 366, *372*
Chintagari, N. R., 307, *317*
Chiti, F., 276, *286*
Chizmadzhev, Y. A., 295, *312*
Cho, Y. K., 337, *352*
Choi, E. S., 338, *352*
Chu, H. P., 306, *317*
Chuang, C. C., 297, *314*
Chuang, J. Z., 292, *311*
Chung, P., 276, *286*
Cid, E., 35, *45*
Cid, V. J., 336, *352*
Clark, I. A., 230, *242*
Clayton, C. E., 66, *84*
Cochran, J. R., 343, 344, *355*, *356*
Codogno, P., 94, *111*

Cohen, D., 305, 306, *316*
Cohen, F. S., 295, *312*
Cohen, G., 52, *63*
Cohen, S., 167, *178*
Cohen, Y., 144, *149*
Cole, N. B., 304, *316*
Coleman, R., 66, *83*
Colli, W., 20, *30*
Collinge, J., 277, *286*, 287
Collins, F. H., 230, *242*
Collins, S. R., 16, *29*
Colomer, V., 297, *311*
Colonna, C., 304, *316*
Colussi, P. A., 6, *23*, 37, *46*, 93, 97, *110*, *111*, 118–119, 127, *129*, *132*, 361, *371*
Conchonaud, F., 308, *318*
Connolly, C. N., 302, *315*
Conrad, P. A., 304, *316*
Conraths, F. J., 166, *179*
Contreras, R., 343, *355*
Conzelmann, A., 2, 3, 6–8, 10–12, 14, 15, 17–19, *21–23*, *26–30*, 67, 78, *85*, *86*, *89*, 94, *111*, 118–126, *128–129*, *131*, 136–140, 143, *147*, *148*, 154, *158*, 248, 250, 255, 257, 258, *261*, *262*, *266*, 270, 271, *283*, *284*, 299, *314*, 324, 332, 334, *347*, *350*, *351*
Cook, H. W., 281, *288*
Cooke, B. M., 162, *180*
Coombs, G. H., 3, 14, *22*, *28*, 138–140, *148*, *149*, 258, *267*
Coppel, R. L., 107, 108, *114*, 162, *180*
Corbeil, D., 301, *314*
Cormack, B. P., 328, 329, 335, *349*, *351*
Corner, G. A., 366, *372*
Coronado, J. E., 322, 323, 325, 327, 336, *345*, *349*
Corrado, K., 15, *28*, 136, 142, 144, *148*
Cortegano, M., 304, *316*
Costello, C. E., 71, *88*, 163, *178*, 231, 232, 235, *243*
Costello, L. C., 12, *26*, 35, 36, *45*, 67, *85*, 167, *176*, 250, *262*
Cottaz, S., 10, *24*, 55, 56, 58, 62, *63–64*, 69, *87*, 255, *265*
Cotter, R. J., 71, *88*, 121, *130*, 163, *178*, 231, 232, 235, *243*
Courtin, O., 125, *131*
Coutinho, P. M., 11, 12, *25*, 35, *44*
Couto, A. S., 254, *264*
Cox, J. S., 276, *286*

Crabb, B. S., 165, *177*
Cramer, J. P., 175, *180*
Cramm-Behrens, C. I., 306, *317*
Crateri, P., 323, *347*
Crawford, D. H., 2, 21, 104, *113*, 362, 363, 365, 369, *371*
Crellin, P. K., 107, 108, *114*
Cremona, M. L., 20, *30*
Cresawn, K. O., 303, *315*
Crick, D. C., 104, 108, 109, *113*, *115*
Crise, B., 280, *288*, 292, *311*
Cross, G. A. M., 14, *27*, 66, 69, 70, *85–87*, 134, 136, *146*, *147*, 248, 255, 257, *262*, *265*
Crossman, A., 3, 10, 14, 19, *22*, *24*, *27*, *29*, 52–56, 58, 60–62, *63–64*, 67, 69, 70, 72, *86–88*, 102, *112*, 167, *179*, 248–250, 252–258, *261–265*
Crout, D. H. G., 202, *225*
Csaba, G., 255, *265*
Cuevas, F., 223, *227*
Cui, J., 162, *178*
Cunningham, J., 322, *346*
Cunningham, O., 293, *319*
Curiel, T. J., 173, *180*
Cutler, D. F., 302, *315*

D

Dacie, J. V., 362, *371*
Dagkessamanskaia, A., 18, *29*
Daigle, M., 335, *351*
Dalley, J. A., 139, *148*
Dallner, G., 167, *178*
Dame, J. B., 168, *178*
Damm, J. B., 165, *179*
Damveld, R. A., 331, 332, *350*
D'Angelo, G., 297, *314*
Daniele, T., 297, *313*
Danielsen, E. M., 293, *318*
Dasgupta, S., 144, *149*
da Silva, A. M., 66, *84*
Datema, R., 254, *264*
Daum, G., 279, *287*
Davidson, E. A., 3, *21*, 71, *88*, 162, 163, *177–178*, 231, 232, 235, *243*, 255, *265*
Davie, J. R., 365, 366, *372*
Davies, G. J., 11, 12, *25*, 35, *44*, *45*, 323, 333, *347*
Davies, J., 52, *63*
Davis, C., 52, *63*

Davis, D. A., 333, *351*
Davis, S. J., 3, 6, *22*, 66, *84*, 360, *371*
Davison, D. B., 72, *88*
Davitz, M. A., 134, *145*, 280, *288*, 292, *311*
de Almeida, M. L., 71, *87*
Dean, N., 42, *46*
De Antoni, A., 272, *285*
Dear, N., 366, *373*
De Armond, S. J., 277, *287*
De Baetselier, P., 176, *179*
Debeaupuis, J. P., 331, 332, *350*
De Bernardis, F., 323, *347*
Debierre-Grockiego, F., 165, 170, 173–175, *176–177*, *180*, 241, *245*, 259, *267*
de Boer, A. D., 322, *346*
de Bolos, C., 295, *312*
Deborde, S., 305, *316*
Debus, V., 104, *112*
Deeg, M. A., 6, 7, *22*, 66, *84*, 118, *128*
Deen, C., 340, 343, *354*
Deerinck, T. J., 291, *318*
Deering, C., 277, *287*
De Gasperi, R., 66, *84*, 85, 126, *131*, 134, *146*, 167, *177*, 255, *265*
de Gramont, A., 362, *371*
De Groot, P. W., 3, 18, *21*, *29*
De Groot, P. W. J., 125, *131*, 134, *146*, 322, 323, 326, 328, 331, 335, *345–348*
De Jaco, A., 338, *352*
de Jong, L., 322, 323, 335, *346*
Dekker, H. L., 322, *346*
De Koster, C. G., 322, 323, 335, *345–347*
Delacour, D., 297, 304, *313*, *316*
de la Cruz, J., 125, *131*
Delahunty, M. D., 93, *110*, 270, *284*
Delannoy, P., 295, *312*
Delauw, M. F., 165, *176*
Delcardayre, S. B., 52, *63*
de Lederkremer, R. M., 20, *30*, 257, 259, *266*
Deleury, E., 11, 12, *25*, 35, *44*
Delgrossi, M. H., 307, *317*
Dell, A., 105, 107, *113*, *114*, 137, *148*
Delorenzi, M., 128, *130*
Del Poeta, M., 323, 329, *347*
De Luca, A. W., 103, *112*
de Macedo, C. S., 163, 167, *176*, *178*, *179*
de Marco, M. C., 297, *313*
De Matteis, M. A., 297, *313*, *314*
De Montigny, J., 335, *351*
Denecke, J., 104, *112*

Denic, V., 16, *29*
Denny, P. W., 82, *90*
De Nobel, H., 324, 328, 339, *348, 349, 353*
De Nobel, J. G., 327, 339, *348, 354*
de Paz, J. L., 223, *227*
Deriziotis, P., 278, *287*
Desmarais, C., 36, 38, 41, 42, *46*
de Souza, J. B., 235, *244*
Desponds, C., 3, 7, *22, 23*, 250, *262*, 270, 271, *284*, 324, *347*
Deutzmann, R., 326, 328, 330, 339, *348*
De Veer, M. J., 231, *242*
Dewaele, S., 343, *355*
Dewerte, I., 343, *355*
de Winde, J. H., 40, *46*
Dhugga, K. S., 322, *345*
Di, P. E., 366, *372*
Diaquin, M., 322, 334, *346*
Diaspro, A., 342, *355*
Diatta, B., 235, *244*
Diaz, E., 292, 300–303, *310*
Diaz-Sandoval, S., 342, *354*
Di Campli, A., 297, *313, 314*
Dickinson, A., 277, *286*
Didorenko, S., 277, *287*
Dieckmann-Schuppert, A., 71, *87*, 163, *177*, 231, *242*, 254, *264*
Diehl, B., 338, 340, 341, *352*
Diep, D. B., 248, 255, *261*
Dietrich, C., 293, *318*
Dieye, A., 235, *244*
Dijkgraaf, G. J., 328, *349*
Dinadayala, P., 109, *115*
Dinckaya, E., 344, *356*
Di Novo, A. A., 344, *356*
Di Pentima, A., 302, *315*
Di Russo, C. C., 76, *89*
Disney, M. D., 235, *244*
Di Tullio, G., 297, *313, 314*
Dix, A., 19, *29*, 60–62, *64*, 67, 69, *86, 87*, 102, *112*, 167, *179*, 248, 250, 252–254, 256, *261, 263*
Djokam, R., 235, *244*
Djordjevic, J. T., 323, 329, 331, 332, *347*
Dobson, C. M., 276, *286*
Doering, T. L., 7, 10, 20, *23, 24, 30*, 32, *44*, 50, 59, 62, *62, 64*, 66, 67, 69, 70, *85–87*, 119, *129*, 258, *266*, 270, 273, *284, 285*
Doerrler, W. T., 7, *24*, 67, *85, 86*
Doetzlhofer, A., 366, *372*

Dohmae, N., 304, *316*
Dominguez, C., 361, *371*
Doms, R. W., 335, *351*
Donnelly, C. A., 230, *242*
Doolittle, W. F., 322, *345*
Dorn, C., 167, *179*
Dotti, C. G., 302, *315*
Doucey, M. A., 93, *110*, 153, *158*, 293, 299, *314*
Douglass, E. J., 173, *178*, 231, *243*
Douwes, J. E., 125, *131*
Dover, L. G., 107, *114*
Dragonová, J., 3, *21*
Dranginis, A. M., 322, 323, 325, *345, 346*
Drapp, D. A., 12, *26*, 35, 36, *45*, 248, 250, 255, *261, 262*, 361, *371*
Drennan, M., 105, *113*, 176, *179*
Dreyfuss, M., 123, *131*, 257, *265*
Drgon, T., 330, *350*
Drgonova, J., 248, 256, *261*, 325, 328, 330, *348, 350*
Drobecq, H., 297, *313*
Dromer, F., 332, *350*
Drubin, D. G., 295, *312*
D'Souza, S. F., 342, *355*
Dubel, S., 337, *352*
Dubremetz, J. F., 71, *88*, 162, 165, 169, 170, *176, 178–180*, 241, *245*, 250, 254, *262, 264*
Duffy, P. E., 230, *242*
Dujon, B., 335, *351*
Dunn, K. W., 306, *317*
Dupree, P., 2, 3, *21*
Duran, A., 328, *349*
Durrens, P., 335, *351*
Duszenko, M., 14, *28*, 69, 137, 139, 140, *147*
Dwek, R. A., 3, 6, 22, 66, *84*, 97, *111*, 118, 119, 127, *128–129*, 182, *224*, 360, *371*

E

Eagle, T., 144, *149*
Echner, H., 14, *28*, 137, 139, 140, *147*
Ecker, M., 326, 328, 330, 339, *348*
Eckert, V., 93, *110*
Edidin, M., 293, 295, *313, 318*
Edwardson, J. M. ., 293, *318, 319*
Egea, G., 304, 305, *316*
Egel-Mitani, M., 323, *347*
Egerton, M., 15, *28*, 137, 140, *147*
Eggeling, L., 107–109, *114, 115*

Egger, D., 6, 7, *23*, 118, 122, *128*
Ehehalt, R., 292, *311*
Ehrhardt, S., 175, *180*
Eigenheer, R. A., 322, *346*
Eikemeyer, J., 297, 301, *313*
Eisenberger, C. F., 144, *149*
Eisenhaber, B., 11, *25*, 34–37, *45*, 134, 140, 142, *146*, *149*, 322, *345*
Eisenhaber, F., 11, *25*, 34–37, *45*, 134, 140, 142, *146*, *149*, 322, *345*
Elbein, A. D., 255, *265*
Elbling, L., 366, *373*
Ellenberg, J., 305, *315*
Ellis, M., 138, *148*
Ellisman, M. H., 291, *318*
El Marjou, A., 308, *318*
Elrod-Erickson, M. J., 78, *89*, 273, *285*
Elsasser, H. P., 304, *316*
Elwood, P. C., 248, 255, *261*
Emili, A., 366, *372*
Emiliani, S., 366, *372*
Emr, S. D., 12, *27*, 92, *109*, 120–122, *130*, 153, *158*, 332, *350*
Endo, M., 167, *177*
Endo, Y., 2, 12, *21*, *26*, 33, 37, 38, 42–43, *44*, *46*, *47*, 92, 104, *109*, *113*, 254, *264*, 362, *371*
Engelhardt, K. R., 293, *319*
Engles, J. M., 144, *149*
Englund, P. T., 7, 10, 19, 20, *23*, *24*, *30*, 32, *44*, 50, 62, *62*, 66, 67, 69, 70, *85–87*, 119, *129*, 182, *224*, 248, 252, 258, *261*, *266*, 270, *284*
Englund, P. W., 59, *64*
Engstler, M. J., 140, *149*
Eppinger, M., 119, *130*, 162, *179*
Epstein, S. L., 327, 330, 336, *349*
Erard, F., 105, *113*
Erdjument-Bromage, H., 305, *316*
Ericsson, L. H., 6, *23*
Erlekotte, A., 104, *112*
Erlich, H. A., 165, *177*
Errede, B., 330, *350*
Escalante, A. A., 167, *177*
Escribano, M. V., 12, *27*, 92, *109*, 123, *131*, 254, *264*
Etges, R., 238, *245*
Eude-Le Parco, I., 304, *316*
Evans, K., 175, *179*, 204, *225*, 230, 233, 234, 240, *242*

F

Fabre, A. L., 12, *27*, 95, *111*, 120, 122, 123, *130*
Fagan, P. K., 235, *244*
Fahey, R. C., 52, *63*
Falck, J. R., 7, *24*, 67, *85*
Fankhauser, C., 3, 7, *22*, 250, *262*, 270, 271, *284*
Farewell, A. E., 15, *28*, 136, 142, 144, *148*
Farina, F., 338, *352*
Farkas, V., 323, 336, *347*
Farmer, M., 277, *287*
Fassler, J. S., 12, *26*, 35, 36, *45*, 250, *262*
Felk, A., 322, 335, *346*
Feng, X., 168, *176*
Ferguson, M. A. J., 2–4, 6, 7, 10, 14, 19–20, *21–24*, *27*, *29*, *30*, 35, *45*, 66, 67, 69, 70, 72, 79, *83–88*, *90*, 93, 97, 102, *110–112*
Ferguson, M. A. J., 3, 7, *22*, 50, 52–56, 58, 60–62, *63–64*, 103, *112*, 118, 119, 122, 127, *128–130*, 135, *145*, *149*, 162, 167, *177*, *179*, 182, 215, *224*, *226*, 236, 238, 241, *244*, *245*, 248–250, 252–258, 260, *264–265*, *267*, 270–271, *284*, 358, 360, *370*, *371*
Ferguson, M. J., 55, 56, 58, *63*
Ferguson, T. R., 6, 7, *22*, 66, *83*, 118, *128*
Fernandez, F., 153, 154, *158*
Fernandez, M. P., 328, *349*
Ferrer, J. C., 35, *45*
Ferrieres, V., 200, 204, *225*
Fery, A., 342, *355*
Fidock, D. A., 230, *242*
Fiedler, K., 292, 297, *311*, *313*
Field, M. C., 3, *22*, 69, 70, 82, *86*, 87, *90*, 136, 139, 140, *147*, *148*, 253, 255, *264*, *265*, 270, *284*
Field, R. A., 19, *29*, 55, 56, 58, *63*, 69, *87*, 103, *112*, 248, 252, 255, 256, *261*, *265*
Fields, I. C., 307, *318*
Fields, S., 36, 38, 41, 42, *46*
Filiano, J., 104, *112*
Filler, S. G., 322, 329, *346*
Filmus, J., 361, *371*
Finean, J. B., 66, *83*
Fink, G. R., 333, *351*
Fink, J. R., 40, *46*
Finnegan, C. M., 307, *317*
Fiore, N., 230, *242*
Fischer, G., 335, *351*
Fischer, M., 277, *286*

Fisher, J., 276, *286*
Fivaz, M., 295, *312*
Florin, E. L., 293, *319*
Flury, I., 6, 7, 11, 12, *23*, *26*, *27*, 118–122, 124–126, *128*, *129*, 332, *350*
Folch-Mallol, J. L., 342, *354*
Folsch, H., 290–293, 306, 307, *310*, *315*, *317*
Fomenkov, A., 144, *149*
Fontaine, T., 322, 323, 330, 334, *346*, *347*, *350*
Fonzi, W. A., 322, 334, *346*, *351*
Ford, R. A., 330, *350*
Formigli, L., 276, *286*
Forsee, W. T., 93, *110*
Forsyth, C. M., 344, *356*
Foss, T. R., 271, *285*
Fouda, G., 235, *244*
Fowler, D. M., 271, *285*
Fraering, P., 15, *28*, 136, 140, 143, *147*
Francis, S., 303, *315*
Francois, J., 18, *29*
Frangeul, L., 335, *351*
Frank, C. G., 104, *112*, 155, *158*
Fransen, J., 307, *318*
Franzot, S. P., 67, *86*
Frasch, A. C., 20, *30*
Fraser-Reid, B., 165, *178*, 200, 204, *225*, 231, *243*
Fredrick, L. R., 363, *372*
Freeze, H. H., 104, *112*
Freiberg, N., 72, *88*, 103, *112*, 250, 258, *262*
Fremond, C., 105, *113*
Frerker, N., 297, 301, *313*
Frieman, M. B., 325, 328, 329, *349*
Fritz, B. A., 292, *311*
Fryauff, D. J., 235, *244*
Fuchimoto, J., 340, *354*
Fucini, R. V., 305, *316*
Fujioka, H., 3, *22*
Fujita, H., 307, *317*
Fujita, M., 2, 3, 6–8, 15, 16, 18, *21*, *22*, *28*, *29*, 32, *44*, 77–79, 82, *89*, *90*, 125, *131*, 258, *266*, 270, 271, *284*, 297, 299, *314*, 332, 333, *350*, 358–360, *370*
Fujita, T., 2, 12, *21*, *26*, 33, 42, *44*, 47, 92, 104, *109*, *113*, 124, *131*, 254, *264*, 362, *371*
Fujita, Y., 341, *354*
Fujiwara, T., 293, 308, *318*, *319*
Fujiwara, Y., 366, *373*
Fukamizu, A., 18, *29*, 271, *284*
Fukasawa, M., 17, *29*, 279, 281, *287*, *288*

Fukuchi, S., 134, *146*
Fukuda, H., 123, *130*, 338–341, *353*, *354*
Fukuda, T., 338, 339, 342, *353*
Fukui, T., 6, *23*
Fukuma, T., 3, *22*, 248, 250, 256, *261*, *262*
Fukushi, M., 79, *89*, *90*
Fukushima, K., 6, *23*
Fullekrug, J., 292, 297, *311*, *313*
Fuller, R. S., 323, 335, 336, *347*
Fülöp, V., 142, *149*
Funato, K., 7, 17, *24*, *29*, 76, *89*, 156, *158*, 278, 281, *287*
Funk, M., 93, *110*
Furge, K., 366, *372*
Furukawa, M., 70, *310*
Furukawa, Y., 134, *146*
Furusawa, M., 366, *372*
Futai, E., 271, *284*
Futerman, A. H., 17, *29*, 66, *83*, 278, *287*
Futter, C. E., 302, *315*

G

Gabius, H. J., 297, *313*
Gaertig, J., 304, *316*
Gagnon-Arsenault, I., 323, 335, 336, *347*, *351*
Gaillardin, C., 125, *131*, 332, *350*
Galli, T., 307, *317*, *318*
Gallop, J. L., 295, *312*
Gallup, J. L., 230, *242*
Gallwitz, D., 272, *285*
Galperin, M. Y., 120, 121, *130*
Galvan, C., 302, *315*
Gamarro, F., 137, *148*
Gan, Y., *318*, 366, *373*
Gancedo, J. M., 40, *46*
Gao, B., 339, 341, *354*
Gao, L., 307, *317*
Gao, X. D., 15, *28*, 42, *46*, 136, 142, 144, *148*
Garcia, S., 223, *227*
Garegg, P. J., 218, *226*, *227*
Garen, C., 52, *63*
Garg, N., 140, *149*
Garraud, O., 235, *244*
Garrett, S., 16, *29*
Garvik, B., 342, *355*
Gast, C., 328, 329, *349*
Gatti, E., 322, 329, 334, *346*, *349*
Gautier, C., 35, *45*

Gaynor, E. C., 12, *27*, 92, *109*, 120–122, *130*, 153, *158*, 332, *350*
Gazzinelli, R. T., 173, *176*, 231, 241, *243*, *245*, 259, *267*
Geerts, W. J., 291, *318*
Gelli, A., 322, *346*
Gemmill, T. R., 328, *349*
Gentles, A. J., 42, 43, *47*
Gentzsch, M., 153, *158*, 338, 339, *353*
Georgescu, S. P., 14, *27*, 137, *147*
Gerard-Vincent, M., 328, *349*
Gerber, L. D., 134, 136, *145–147*, 259, *267*
Gerdes, H. H., 301, *314*
Geremia, R. A., 35, *45*
Geren, I. N., 344, *356*
Gerisch, G., 66, *84*
Gerke, V., 297, 301, *313*
Gerl, M. J., 306, *317*
Gerold, P., 3, 12, 22, *27*, 71, 72, *87*, *88*, 92, 103, *109*, *112*, 123, *131*, 162, 163, 166, 167, 169, 170, 173, *176–180*, 231, 232, *242*, *243*, 248, 250, 252–255, 257, 258, *261–265*
Gerwig, G. J., 165, *179*
Gething, M. J., 276, *286*
Geyer, H., 165, 170, *176*, *180*, 241, *245*
Geyer, R., 165, 170, *176*, *180*, 241, *245*
Ghugtyal, V., 15, 18, *28*, 78, *89*, 271, *284*
Giannoni, E., 276, *286*
Gibbs, J., 276, *286*
Gibson, T. J., 74, *90*
Gicquel, B., 106, 109, *114*, *115*
Giga-Hama, Y., 322, 331, *346*
Gil, C., 322, 329, *346*
Gill, G. N., 293, *319*
Gilleron, M., 105–108, *113–115*
Gillmor, C. S., 3, *21*
Gillooly, D., 304, *316*
Gilmore, R., 137, *148*, 162, *178*
Gilson, P. R., 165, *177*
Gimeno, C. J., 40, *46*
Gingerich, G., 35, *45*
Giocondi, M. C., 293, *318*
Gish, W., 50, 55, *63*
Gissi,C., 330, 334, *350*
Glimcher, L. H., 276, *286*
Gliozzi, A., 342, *355*
Gluckman, E., 362, *371*
Goder, V., 277, *286*
Godi, A., 297, *313*, *314*
Goffard, N., 335, *351*

Gojobori, T., 335, *351*
Gokel, G. W., 20, *30*, 258, *266*
Goldberg, J., 271, *285*
Goldenring, J. R., 303, *315*
Golenbock, D. T., 173, *176*, 231, *243*
Golenbock, R. R. D., 259, *267*
Gomi, F., 276, *286*
Gomis, R. R., 35, *45*
Gonzalez, M., 322, 326, 328, 331, *346*
Gordon, J. I., 20, *30*, 258, *266*
Gordon, V. M., 248, 255, *261*
Gou, D., 307, *317*
Goud, B., 308, *318*
Gousset, K., 293, *314*, 361, *371*
Gouveia, C., 342, *355*
Gouyer, V., 297, *313*
Gowda, D. C., 3, *21*, 71, *88*, 162, 163, 168, 173, *176–178*, *180*, 209, *226*, 231, 232, 235, *243*, *244*, 255, 265, 322, *345*
Graff, C. P., 343, *355*
Grandori, R., 324, 335, *348*
Grant, B. D., 306, *317*
Grard, G., 297, *313*
Grau, C., 336, *352*
Grau, G. E., 231, *242*
Gray, J. J., 139, *148*
Gray, S. G., 366, *372*
Gray, W., 12, *26*, 35, 36, *45*, 250, *262*
Green, L. G., 190, *224*, *225*
Green, R. F., 295, *313*
Green, S., 230, *242*
Greenblatt, J., 366, *372*
Greenblatt, J. F., 16, *29*
Greenwood, B. M., 230, *242*
Greiner, R. A., 277, *286*
Grimme, J. M., 97, *111*
Grimme, S. J., 12, 15, *27*, *28*, 92, 96, 97, *109*, *111*, 118–122, 124–127, *129*, *130*, *132*, 136, 142, 144, *148*, 153, *158*, 259, *267*, 332, *350*
Grindstaff, K. K., 303, *315*
Grobe, H., 104, *112*
Groos, E. M., 325, 330, *348*
Gross, U., 173, *178*, 322, *345*
Grossniklaus, U., 3, *21*
Grubenmann, C. E., 104, *112*
Gruber, C., 324, *348*
Grunewald, S., 104, *112*
Guerriero, C. J., 301, 303, *314*, *315*
Guether, M. L. S., 103, *112*
Guha-Niyogi, A., 241, *245*

Guidotti, G., 124, 125, *131*
Guilliams, M., 176, *179*
Guinovart, J. J., 35, *45*
Gumkowski, F. D., 304, *316*
Guo, X. J., 308, *318*, 366, *372*
Guo, Z., 144, *149*, *226*
Gupta, D. K., 66, *84*, 126, *131*
Gupta, N., 338, 340, *353*
Gupta, P., 163, *177*, 231, *243*
Gupta, S., 230, *242*
Gurcha, S. S., 107, *114*
Gurkan, C., 271, *284*, *285*
Guse, A. H., 297, *313*
Gustafsson, C. M., 331, *350*
Güther, M. L., 122, *130*, 248, 252, 253, 255, 256, 258, 260, *261*, *263–265*, *267*
Guther, M. L. S., 10, 14, 19, *24*, *27*, *29*, 50, 62, *63–64*, 66, 69, 70, 79, *85–87*, *90*, 102, *111*
Guyot, B., 366, *373*

H

Habermann, B., 304, *316*
Hackett, F., 170, *179*, 231, *242*
Hahn, M. G., 190, *225*
Haites, R., 107, *114*
Haites, R. E., 107, *114*
Hajjar, A. M., 173, *178*, 231, *243*
Halachmi, S., 144, *149*
Hall, A., 305, *316*
Hall, B. F., 238, *245*
Hallaq, Y., 126, *131*, 255, 257, *265*, *266*
Hamada, K., 134, *146*, 322, 328, 329, 335, *346*, *349*
Hamakawa, N., 6, *23*
Hamamoto, S., 271, 272, *284*, *285*
Hamana, H., 52, *63*
Hamann, T., 3, *21*
Hamburger, D., 15, *28*, 137, 140, *147*
Hammersen, G., 104, *112*
Hammett, M., 366, *373*
Han, L., 366, *373*
Han, S., 366, *373*
Hanada, K., 17, *29*, 279, 281, *287*, *288*
Hancock, J. F., 295, 306, *312*, *317*
Handa, N., 52, *63*
Hannan, L. A., 295, *313*
Hannedouche, O., 297, *313*
Hannun, Y. A., 297, *314*
Hansen, G. H., 293, *318*

Hanzal-Bayer, M. F., 295, *312*
Hao, M., 293, *318*
Hara, T., 3, *22*, 248, 250, 256, *261*, *262*
Harada, A., 306, *317*
Harada, R., 306, *317*
Harada Nishida, M., 6, *23*
Harder, T., 293, *319*
Harding, H. P., 276, *286*
Harms, E., 104, *112*
Haro, D., 366, *372*
Harris, C. L., 11, *25*, 76, *89*, 94, *111*, 254, *264*, 365, *372*
Harrison, J., 254, *264*
Hart, G. W., 2–4, 7, 10, 20, *21*, *23*, *24*, *30*, 32, *44*, 50, 59, 62, *62*, *64*, 66, 69, 70, *85–87*, 119, *129*, 135, *145*, 258, *266*
Hartland, R. P., 322, 334, *346*
Harvey, D. J., 3, 6, *22*, 66, *84*, 360, *371*
Hasegawa, T., 366, *372*
Haselbeck, A., 167, *177*
Hasilik, M., 104, *112*
Haslam, S. M., 137, *148*
Hata, K., 12, *26*, *27*, 67, 75, *86*, *88*
Hata, Y., 338, *353*
Hauri, H. P., 272, 273, *285*
Hauser, K., 324, *348*
Hay, J. C., 275, *286*
Hayakawa, J., 8, *24*, 32, *44*, 67, *85*
He, H. T., 308, *318*
He, S., 365, 366, *372*
Hederos, M., 221, *227*
Heese-Peck, A., 279, *287*
Hegde, R. S., 270, 271, *284*
Heine, M., 297, 301, 303, 305, 306, *313*, *315*, *317*
Heise, N., 71, *87*
Heitman, J., 40, *46*
Helenius, A., 295, *312*
Helenius, J., 153, *158*
Heller, P., 338, *353*
Hellingwerf, K. J., 3, *21*, 134, *146*, 322, *346*, *347*
Helms, M. J., 14, *28*
Hemming, F. W., 254, *264*
Henderson, R. M., 293, *318*, *319*
Hendrickson, T. L., 139, *148*
Hendrix, G. M., *319*
Henneberry, A. L., 282, *288*
Hennebicq-Reig, S., 295, *312*
Hennet, T., 104, *112*

Henrissat, B., 11, 12, *25*, 35, *44*, *45*, 323, 333, 334, *347*
Hensel, A., 334, *351*
Heringa, J., 50–52, *63*
Hernez-Deviez, D., 307, *317*
Herold, M., 105, *113*
Herrmann, A., 152, *157*, 214, *226*
Herrmann, J. M., 271, *285*
Herscovics, A., 167, *177*
Hese, K., 137, *148*
Hess, D., 335, *351*
Hetz, C., 276, *286*
Heudier, P., 362, *371*
Heuser, J. E., 238, *245*
Hewitt, M. C., 175, *179*, 204, *225*, 230, 233, 234, 240, *242*, *243*
Higgins, D. G., 50–52, *63*, 74, *90*, 168, *180*
High, S., 139, *148*
Hille-Rehfeld, A., 302, *315*
Hilley, J. D., 138–140, *148*, *149*, 258, *267*
Hillman, P., 42, *47*
Hillmen, P., 2, *21*, 104, *113*, 248, 254, *261*, 362, 363, 365, 369, *371*
Hino, J., 11, 15, *25*, *28*, 34, 36, 38, 39, *44*, *46*, 93, *110*, 136–140, 143, *147*, 250, *262*
Hippe, D., 174, 175, *177*
Hirata, A., 327, *349*
Hiroi, Y., 14, *27*, 137, 141, *147*, *149*
Hirokawa, N., 304, *315*, *316*
Hirose, S., 8, *24*, 67, *85*, 119, 126, 127, *129*, 257, *267*
Hirotsune, S., 43, *47*
Hirschberg, C. B., 151, 154, *157*, *158*
Hirschberg, K., 305, *315*
Hitchen, P. G., 107, *114*, 137, *148*
Hodson, C., 291, *309*
Hoepfner, S., 304, *316*
Hoessli, D. C., 162, *177*
Hoflack, B., 167, *177*
Hofmann, K., 152, *157*
Hofsteenge, J., 93, *110*, 153, *158*
Holder, A. A., 3, *22*, 87, 163, *177*, *178*, 231, 232, *242*
Holguin, M. H., 363, *372*
Holst, B., 17, *29*
Holzmann, K., 366, *373*
Homans, S. W., 3, 6, 7, *22*, 97, *111*, 118, 119, 127, *128–129*, 182, *224*, 238, *245*, 260, *267*, 270, 271, *284*

Hong, Y., 8, 11, 12, 14, 15, *24–28*, 37, 38, 41, 42, *46*, *47*, 70, 72, 76, *87–89*, 92, 94, 97, 102, *109*, *111*, 118–127, *128–130*, *132*, 136–140, 142, 143, *147*, *148*, 152, *157*, 168, *178*, 254, *264*, 365, *372*
Honys, D., 3, *21*
Hooper, N. M., 3, *22*, 295, *312*
Hopkins, C. R., 302, *315*
Hoque, M. O., 144, *149*
Horber, J. K., 293, *319*
Hori, H., 203, *225*
Horii, Y., 18, *29*
Horvath, A., 12, *26*, 35, *45*, 123, *131*, 134, *146*, 250, 257, *262*, *265*, 278, *287*
Hoshi, K., 79, *90*
Hosoda, T., 14, *27*, 137, *147*
Hosomi, A., 322, 331, *346*
Houjou, T., 8, 15, 16, *24*, *28*, 32, *44*, 67, 77, *85*, *89*, 258, *266*, 270, *284*, 299, *314*, 360, *370*
Howes, M. T., 306, *317*
Howland, S. W., 344, *356*
Hu, R., 33, *44*
Hua, W., 302, 303, *314*
Huang, B., 366, *373*
Huang, D., 339, 340, *353*
Huang, X., 144, *149*
Hubbard, A. L., 307, *317*
Hube, B., 322, 335, *346*, *351*
Hudson Keenihan, S. N., 235, *244*
Huet, G., 295, *312*
Hulsmeier, A., 7, *23*
Hülsmeier, A., 66, *85*
Humphrey, D. R., 6, 7, *22*, 66, 67, *84*, *85*, 118, *128*
Hunter, C. A., 173, *178*
Hunter, S. W., 104, *113*
Hurvitz, H., 104, *112*
Huse, J. T., 335, *351*
Hussain, S., 337, 338, 341, 342, *352*
Hutchinson, D. W., 202, *225*
Huttner, W. B., 295, *312*
Hyman, R., 66, *85*, 126, *131*

I

Ibam, E., 235, *244*
Ibata-Ombetta, S., 332, *350*
Ielpi, L., 35, *45*
Igarashi, A., 79, *89*
Igarashi, K., 123, 124, *130*, *131*

Iguchi-Ariga, S. M., 366, *372*
Ihara, K., 7, *24*, 76, *89*, 155, 156, *158*, 278, 281, *287*
Ihmels, J., 16, *29*
Ihrke, G., 295, 303, *312*, *315*
Iida, Y., 2, 11, *21*, *24*, 33, 35, *44*, 104, *113*, 362, *371*
Iino, R., 308, *318*
Ikawa, M., 33, *44*, 77, *89*, 361, *371*
Ike, H., 308, *318*, *319*
Ikehara, Y., 6, *23*, 67, 79, *86*, *89*, *90*, 250, 258, *262*, *266*
Ikezawa, H., 6, 10, *23*, *24*, 134, *146*, 322, *345*
Ikonen, E., 278, *287*, 292, 293, *311*
Ilg, T., 238, *245*
Illarionov, P. A., 107, *114*
Imaizumi, K., 276, *286*
Imanaka, T., 341, *354*
Imbach, T., 104, *112*
Imhof, I., 6–8, 14, 15, *23*, *28*, 118, 119, 122, *128*, *129*, 136, 138, 140, *147*, *148*, 250, 257, *262*, *266*
Immerdal, L., 293, *318*
Inamine, J. M., 107, *114*
Ince, S. J., 190, *225*
Ingles, C. J., 366, *372*
Inoue, N., 11, 12, 14, 15, *24–26*, *28*, 32, 33, 35, 37, 38, 42, *44*, *47*, 50, 58, *63*, 67, 76, *86*, *88*, *89*, 92, 93, 97, *109–111*, 119, 121, 124–127, *129*, *131*, 136–143, *147*, *148*, 152, *157*, 250, 254, *262–264*, 322, *345*
Inoue, S. B., 330, *350*
Intersberger, E. W., 366, *372*
Inuoe, N., 34–36, 38, 39, *44–45*, *47*
Ira, K., *265*, *314*, 361, *371*
Irie, K., 330, *350*
Ishiguro, M., 343, *355*
Ishihara, S., 11, 14, *25*, 72, 76, *88*, 102, *111*, 127, *132*, 152, *157*, 327, *349*
Ishitani, H., 17, *29*, 279, *287*
Isobe, K., 366, *372*
Itami, S., 33, *44*
Ito, J., 338, 341, *353*, *354*
Ito, K., 333, *350*, *351*
Iturriaga, G., 342, *354*
Iuzzolino, L., 293, *319*
Ivanov, I. E., 303, *315*
Iwakoshi, N. N., 276, *286*
Izquierdo, L., 20, *30*

J

Jaafar, L., 339, *354*
Jackson, M., 104–106, 108, 109, *113–115*
Jacob, R., 295, 297, 301, 303–306, *313–317*
Jacobs, M., 105, *113*
Jacobs, P. P., 343, *355*
Jacobs, W. R. Jr., 108, *115*
Jacobson, K., 293, *318*, *319*
Jadallah, S., 144, *149*
Jaeken, J., 104, *112*, 370, *373*
Jager, D., 338, *352*
Jager, E., 338, *352*
Jambou, R., 235, *244*
James, M. N. G., 52, *63*, 335, *351*
Jamieson, J. D., 304, *316*
Jankowski, A. W., 254, *264*
Jansch, L., 322, *345*
Januschke, E., 323, *347*
Jaquenoud, M., 17, 19, *29*, *30*, 271, *284*, 299, *314*
Jaspars, M., 260, *267*
Jaulin, F., 303, 304, *315*
Jayaprakash, K. N., 167, *178*, 200, 204, *225*, 231, *243*
Jedrzejas, M. J., 120, 121, *130*
Jelk, J., 19, *30*, 140, *149*, 299, *314*
Jeno, P., 334, *351*
Jentoft, N., 324, *348*
Jiang, S., 342, *355*
Jiang, Z. B., 338–341, *353*, *354*
Jigami, Y., 6, 7, 12, 15, 16, 18, *26–29*, 67, 75–79, 82, *86*, *88–90*, 125, *131*, 258, *266*, 270, 271, 276, *284*, *286*, 299, *314*, 332, 333, 338–340, 342, *350*, *353*, *354*, 360, *370*
Jih, G. T., 304, *316*
Jin, M., 343, *355*
Jin Lee, Y., 322, *346*
Johnson, A. E., 3, *21*, 139, 141, *148*
Jones, E. W., 95, *111*
Jostock, T., 337, *352*
Jouault, T., 332, *350*
Julius, M., 11, *25*, 34, 36, 38, 39, *44*, 250, *262*
Jung, N., 72, *88*, 103, *112*, 250, 255, 258, *262*, *265*
Jungeblut, C., 104, *112*
Jurka, J., 42, 43, *47*

K

Kaether, C., 301, *314*
Kai, J., 12, *26*, 75, *88*
Kaiser, C. A., 78, *89*, 273, *285*
Kajiwara, K., 7, *24*, 76, *89*, 155, 156, *158*, 278, 281, *287*
Kaksonen, M., 295, *312*
Kalb, S. R., 121, *130*
Kamada, Y., 330, *350*
Kamasawa, N., 339–341, *354*
Kamena, F., 209, *226*, 235, 238, 240, *244*, *245*
Kamewari, Y., 52, *63*
Kaminska, J., 273, *285*
Kamitani, T., 11, *25*, 36, *46*, 66, *84*, 126, *131*, 134, *146*, 255, 257, *265*, *266*
Kanai, M., 6, *23*
Kaneko, A., 124, *131*
Kanemitsu, T., 204, 206, *226*, 234, *243*
Kang, J. H., 342, *354*
Kang, J. Y., 8, 11, 14, *24*, *25*, *28*, 72, 76, *88*, 102, *111*, 123, 127, *130*, *132*, 137, 142, 143, *148*, 152, *157*
Kang, M. S., 330, *350*
Kang, R. S., 307, *318*
Kang, X., 14, *28*, 137, 139, 140, *147*
Kangawa, K., 11, 15, *25*, *28*, 34, 36, 38, 39, *44*, *46*, 93, *110*, 136–140, 143, *147*, 250, *262*
Kappe, S. H., 230, *242*
Kapteyn, J. C., 3, *21*, 125, *131*, 248, 256, *261*, 325, 327, 328, 338, 339, *348*, *353*
Kaptinov, D., 35, *45*
Karadimitris, A., 104, *113*, 362, 366, 367, *371*, *373*
Karnes, P. S., 104, *112*
Kasai, R. S., 293, 308, *318*, *319*
Kasper, M., 297, *313*
Kassel, R. L., 230, *242*
Katayama, T., 276, *286*
Kato, K., 343, *355*
Kato, M., 339, 341, *353*, *354*
Kato, Y., 306, *317*
Katzin, A. M., 254, *264*
Kaur, D., 105, 108, 109, *113*, *115*
Kaur, R., 335, *351*
Kawada, T., 200, *225*, 256, *265*
Kawagoe, K., 2, *21*, 33, *44*, 104, *113*, 361, 362, *371*
Kawakami, M., 339, 343, *353*, *355*
Kawanishi, Y., 335, *351*

Kaya, M., 338, *353*
Kedees, M. H., 250, *263*
Keenan, T. W., 66, *84*
Kehl, H. G., 104, *112*
Kelleher, D. J., 137, *148*, 162, *178*
Keller, G. A., 270, *284*
Keller, P., 292, 293, 300–303, 305, 306, *310*, *314*, *315*, *317*, *319*, 361, *371*
Kellogg, B. A., 344, *356*
Kenworthy, A. K., *319*
Keranen, S., 17, *29*
Kerwin, J. L., 6, *23*
Kessels, M. M., 305, *316*
Khan, A. R., 335, *351*
Khan, R., 202, *225*
Khazanovich-Bernstein, N., 335, *351*
Khiar, N., 223, *227*
Khoo, K. H., 105, 107, 108, *113–115*
Kieke, M. C., 343, *356*
Kielland-Brandt, M. C., 17, *29*
Kienz, P., 104, *112*
Kim, I., 295, *312*
Kim, J. S., 343, *356*
Kim, M., 333, *351*
Kim, S., 104, *112*
Kim, S. Y., 338, *352*
Kim, W., 78, *89*, 291, *309*
Kim, Y. S., 343, 344, *355*, *356*
Kim, Y. U., 11, 12, *25*, 94, *111*, 168, *178*
Kimmel, J., 71, 72, *88*, 163, 167, 169, 170, *178*, *179*, 250, 254, 255, 258, *262*, *265*
Kimura, E. A., 254, *264*
King, M. D., 104, *112*
Kinoshita, K., 11, *25*, 52, *63*, 76, *88*, 92–94, 97, 102, 104, 107, *109–111*, *113*, *114*, 254, *264*, 365, *372*
Kinoshita, T., 2–4, 6–8, 11, 12, 14–16, 18, *21*, 22, *24–29*, 32–39, 41, *47*, 50, 55, 58, *63*, 67, 70, 72, 76, 77, *85–89*, 118–143, *145*, *147*, *148*, 152, *157*, 168, *178*, 248, 250, 254, 256, 258, *261–264*, *266*, 270–272, *284*, *285*, 291, 299, *314*, *318*, 358, 365–367, *370–373*
Kirchhausen, T., 271, *285*
Kitada, K., 134, *146*, 322, 328, 329, *346*, *349*
Kitagaki, H., 333, *350*, *351*
Kitagawa, H., 17, *29*, 279, *287*
Kitaguchi, T., 338, 339, 342, *353*
Kitamura, D., 361, *371*

Kitani, T., 2, 11, *21*, *24*, 33, 35, *44*, 104, *113*, 362, *371*
Kitt, K. N., 307, *317*
Klaassen, Z., 278, *287*
Klein, C., 66, *84*
Klis, F. M., 3, 12, 18, *21*, *26*, *29*, 125, *131*, 134, *146*, 248, 256, *261*, 322–329, 335, 337–340, *345–350*, *352–354*
Klohn, P. C., 277, *286*
Klopfenstein, D. R., 304, *316*
Knez, J. J., 14, *27*, 66, *84*, 119, 127, *129*, 136, 137, 139, *147*
Knijff, R., 52, 54, *63*
Knuth, A., 338, *352*
Kobayashi, S., 17, *29*, 279, 281, *287*, *288*
Kobayashi, T., 292, *311*, 322, *345*
Kobe, S., 72, *88*, 103, *112*, 250, 258, *262*
Koch, A., 304, *316*
Koch, H. G., 104, *112*
Kochibe, N., 6, *23*
Kociuba, K., 277, *287*
Kodukula, K., 134, *145*, *146*, *149*
Koessler, J. L., 223, *227*
Koide, Y., 292, *310*
Koivisto, U. M., 306, *317*
Kollar, R., 3, *21*, 248, 256, *261*, 325, 328, 330, *348*, *350*
Komano, H., 323, 336, *347*
Komuro, I., 14, *27*, 137, *147*
Komuves, L. G., 307, *317*, *318*
Kondo, A., 123, *130*, 337–341, *352–354*
Kondo, J., 293, *319*
Kondo, K., 322, *345*
Konradsson, P., 200, 218, 221, *225–227*
Konstantakopoulos, A., 297, *313*
Koolen, M., 165, *179*
Kopecka, M., 323, *349*
Koppel, D. E., 248, 253, *261*
Koranda, M., 366, *372*
Kordulakova, J., 105, 106, *114*
Korf, J., 176, *179*
Korner, C., 104, *112*
Kornfeld, S., 167, *177*
Kornitzer, D., 322, *346*
Korn-Lubetzki, I., 104, *112*
Korsmeyer, S. J., 276, *286*
Korting, H. C., 323, *347*
Kosik, O., 323, 336, *347*
Koster, A. J., 291, *318*

Kostova, Z., 11, 12, *25*, *26*, 35, 36, *45*, 152, *157*, 250, *262*
Kotaka, A., 338, *353*
Kotsianidis, I., 2, *21*, 104, *113*, 248, 254, *261*, 362, 363, 365, 369, *371*
Kovacevic, S., 107, *114*
Kovacs, P., 255, *265*
Kowalsky, K., 238, *245*
Krafft, G. A., 323, 336, *347*
Krakow, J. L., 66, *85*
Krall, J. A., 66, *84*
Kranz, C., 104, *112*
Kranz, D. M., 343, *356*
Kratzer, B., 195, 215, *225*
Krauland, E. M., 342, *355*
Krautter, M., 292, *311*
Kreger, D. R., 323, *349*
Kreissel, G., 104, *112*
Kreitzer, G., 290–292, 303, 305–307, *315*, *316*
Kremer, L., 105, 107, *113*, *114*, 297, *313*
Krishnamoorthy, G., 293, *314*, 361, *371*
Krishnegowda, G., 71, *88*, 168, 173, *176*, *178*, *180*, 231, 235, *243*, *244*
Krisin, H., 235, *244*
Krogan, N. J., 16, *29*
Krol, S., 342, *355*
Kroos, L., 323, 335, 336, *347*
Kroschewski, R., 305, *316*
Krul, E. S., 238, *245*
Krumbach, K., 107, 109, *114*, *115*
Krysan, D. J., 323, 335, 336, *347*
Kubo, Y., 306, *317*
Kudoh, S., 14, *27*, 137, *147*
Kuehn, M. J., 271, *285*
Kuksis, A., 6, *23*, 66, *84*
Kulaksiz, H., 292, *311*
Kular, G. S., 297, *313*
Kumagai, K., 281, *288*
Kumar, M., *319*
Kumar, S., 235, *244*
Kunihiro, S., 335, *351*
Kunz-Renggli, C., 3, *22*, 93, *110*
Kuramitsu, S., 52, *63*
Kurjan, J., 125, *131*, 324, 325, 328, 338, 339, *348*, *349*, *353*
Kurniawan, H., 136, 140, 141, *147*
Kuroda, K., 338, 339, 342, 343, *353*, *355*
Kurokawa, K., 107, *114*

390

Kuroki, M., 6, *23*
Kuromitsu, J., 12, *26*, 75, *88*
Kurtev, V., 366, *372*
Kurzchalia, T. V., 200, *225*, 292, *311*
Kusumi, A., 293, *318, 319*
Kuurzchalia, T., 256, *265*
Kuzmin, P. I., 295, *312*
Kwiatkowski, D., 235, *244*
Kwon, M. H., 343, *356*
Kwon, Y. U., 165, *178*, 204, 206, 208, 209, *226*, 231, 234, 235, 238, *243–245*
Kyle, D. E., 230, *242*

L

Labrijn, A. F., 337, 338, *352*
Lacana, E., 324, 329, 335, *348, 349*
Lachmann, P. J., 66
Lacomis, L., 305, *316*
Lafont, F., 297, 303, 307, *313, 315, 317*
Lafontaine, I., 335, *351*
Lagace, T. A., 281, *288*
Lages, F., 17, *29*
Lagger, G., 366, *372*
Lagog, M., 235, *244*
Lai, Y., 301, *314*
Lal, A. A., 71, *88*, 163, *178*, 231, 232, 235, *243*
Lalanne, E., 3, *21*
Lallem-Breitenbach, V., 308, *318*
Lamb, R. A., 295, *312*
Lamikanra, A. A., 238, *244*
Landsberger, F. R., 238, *244*
Lange, Y., 282, *288*
Langhorne, J., 238, *244*
Langhorst, M. F., 297, *313*
Lapointe, P., 271, *284, 285*
La Porte, S. L., 344, *356*
Lapurga, J. P., 67, *86*, 250, 258, *262, 266*
Larsen, M., 323, 331, 332, *347*
Latek, R. R., 12, *26*, 35, 36, *45*, 250, *262*
Latge, J. P., 322, 323, 327, 330–332, 334, *346, 347, 349, 350*
Lawrence, J. C., 293, *318, 319*
Layton, D. M., 2, *21*, 104, *113*, 248, 254, *261*, 362, 363, 365–367, 369, *371, 373*
Lazaro-Dieguez, F., 305, 306, *316*
Lazarow, P. B., 258, *266*
Leal, S., 14, *27*
Lea-Smith, D. J., 107, *114*
Le Bert, M., 105, *113*

Le Bivic, A., 280, *288*, 295, 297, 302, 304, 307, *313, 315, 317, 318*
Le Bowitz, J. H., 154, *158*
Lebreton, S., 295, 299, 301, 305, *312, 314*
Lecat, S., 297, *313*
Lechenne, B., 323, 334, *347*
Lederkremer, G. Z., 271, *285*
Ledesma, M. D., 302, *315*
Lee, A. H., 276, *286*
Lee, E. K., 342, *354*
Lee, H. W., 343, *356*
Lee, J. P., 35, *45*
Lee, M. C., 271, 272, *285*, 290, *309*
Lee, S. H., 343, *356*
Lee, V. M.
Lee, Y. C., 104, *113*
Le Gall, A. H., 295, *313*
Legler, D. F., 293, 297, 299, *313, 314*
Le Grimellec, C., 293, *318*
Lehle, L., 92, 93, *110*, 137, *148*, 326, 328, 330, 339, *348*
Lehrman, M. A., 7, *24*, 67, *85, 86*, 93, 94, 103, 104, *110–112*, 153, *158*
Leidich, S. D., 12, *26*, 35–38, *45, 46*, 248, 250, 255, *261, 262*, 361, *371*
Leigh, M. B., 337, 338, 341, 342, *352*
Leitinger, B., 302, *315*
Leke, R. G., 235, *244*
Le Lay, S., 297, *313*
Lellouch, A. C., 35, *45*
Lemansky, P., 66, *84*, 126, *131*
Lemberg, M. K., 136, *146*
Lennarz, W. J., 254, *264*
Lenne, P. F., 308, *318*
Leporrier, M., 362, *371*
Leppla, S. H., 248, 255, *262*
Lesage, G., 322, 323, 325, 327, 328, 336, *346*
Leslie, G. J., 335, *351*
Lester, R. L., 7, *23*
Lesuffleur, T., 295, *312*
Leteurtre, E., 297, *313*
Levin, D. E., 12, *26*, 34, 36–38, 40, 41, *44, 46*, 330, *350*
Levy, R., 344, *356*
Lewis, S. E., 12, *26*, 36, *46*
Lewis, S. M., 362, *371*
Ley, S. V., 190, *224, 225*
Li, B., 342, *355*
Li, G., 322, *345*
Li, J., 338, *352*

Li, L., 338, *352*, 365, 366, *372*
Li, W., 270, *284*
Li, X., 291, 307, *317*, *318*
Lifely, M. R., 66, *83*
Lilley, K. S., 2, 3, *21*
Lillico, S., 3, *22*, 139, 140, *148*
Lindberg, J., 218, *226*, *227*
Linder, T., 331, *350*
Linehan, J., 277, *286*
Lingelbach, K., 71, *88*, 169, *178*
Lingwood, D., 294, *319*
Linn, J. F., 144, *149*
Linton, D., 137, *148*
Lipardi, C., 280, *288*, 292, 295, 297, 301, *311*
Lipke, P. N., 125, *131*, 322–328, 330, 331, 336, 338, *345–349*
Lipman, D. J., 50, 55, *63*
Lippincott-Schwartz, J., 302, 305, *315*, *319*
Lippow, S. M., 343, *355*
Lisanti, M. P., 136, *147*, 248, 252, 255, *261*, *265*, 280, *288*, 292, 295, *311*, *313*
Liu, B., 343, *355*
Liu, J., 11, *25*, 94, *111*, 295, *312*, 338, *352*
Liu, L., 307, *317*
Liu, X. Y., 165, *178*, 204, 209, *226*, 231, 235, 238, 240, 241, *243–245*
Ljungdahl, P. O., 40, *46*
Llobell, A., 125, *131*, 325, 330, *348*
Lo, R. S., 36, 38, 41, 42, *46*
Lo, W. S., 322, *346*
Locco, L., 307, *317*
Lock, J. G., 303, *315*
Lockhart, D. J., 276, *286*
Loh, Y. P., 323, *347*
London, E., 292, 293, 295, *311*, *318*, *319*
Longacre, S., 235, *244*
Longo, L., 42, *47*
Lopez, A. D., 238, *244*
Lopez-Prados, J., 223, *227*
Lorenz, M. C., 40, *46*
Lorenzi, M. V., 18, *29*
Lorry, K., 235, *244*
Louvard, D., 307, *317*
Low, M. G., 6, *23*, 66, *83*, *84*
Low, P., 167, *178*
Low, S. H., 291, 307, *317*, *318*
Lowary, T. L., 109, *115*
Lowe, A. W., 292, *311*
Lu, C. F., 125, *131*, 324, 325, 338, 339, *348*

Lu, J., 167, *178*, 200, 204, *225*, 231, *243*, 366, *373*
Lu, T., 20, *30*, 258, 266
Lucas, C., 17, *29*
Lucas, J. J., 254, *264*
Lucet, I. S., 108, *114*
Lucking, U., 190, *224*
Lucocq, J. M., 297, *313*
Luder, A. S., 104, *112*
Luder, C. G., 173–175, *177*, *178*
Luini, A., 301, *314*
Lukowitz, W., 3, *21*
Lunde, C., 17, *29*
Luo, W., 302, *315*
Lupetti, P., 291, *318*
Luzzatto, L., 2, *21*, 42, *47*, 362, *371*

M

Ma, B., 335, *351*
Ma, L. X., 338, 340, *353*
Ma, W., 276, *286*
Ma, Y., 339, 341, *354*
Maccallum, D., 322, 335, *346*
Macedo, C. S., 255, *265*
Machado, M. D., 342, *355*
Machida, M., 335, *351*
Machray, G., 20, *30*
Mack, J. C., 93, *110*
Madsen, R., 200, *225*
Madzak, C., 338, *352*
Maeda, H., 339, *353*
Maeda, K., 11, *24*, 33, 35, *44*
Maeda, Y., 2, 3, 6–8, 11, 12, 14–16, 18, *21*, *22*, *24–29*, 32, 34, 36–39, 41, *44*, *46*, 55, *63*, 67, 70, 72, 76, 77, *85*, *87–89*, 92–94, 97, 102, 104, 107, *109–111*, *113*, *114*, 118–127, *128–132*, 136–143, *147*, *148*, 152, *157*, 168, *178*, 248, 250, 254, 256, 258, *261*, *262*, *264*, *266*, 270–272, *284*, *285*, 291, 299, *314*, *318*, 358–363, 365–367, 369, *370–373*
Maes, E., 297, *313*
Magez, S., 176, *179*
Magnelli, P., 162, *178*
Magrassi, R., 342, *355*
Maguire, G. P., 235, *244*
Maisch, B., 175, *180*
Makarow, M., 280, *288*, 338, 339, *353*
Malherbe, T. A. T. I., 361, *371*
Malhotra, R., 124, 125, *131*

Malkus, P. N., 272, *285*
Mallet, A., 191, *225*
Mallet, J. M., 191, *225*
Mallucci, G. R., 277, *286, 287*
Mambo, E., 144, *149*
Manabe, S., 343, *355*
Mancini, P., 338, *352*
Maneesri, J., 123, *130*
Mann, K. J., 120, 122, *130*, 257, *266*
Manninen, A., 297, 306, *313, 314, 317*
Manning-Krieg, U., 278, *287*
Mansur, D. S., 173, *176*
Mansur, M. G., 259, *267*
Mao, Y., 328, 329, *349*
Maraskovsky, E., 344, *356*
Marches, F., 173, *180*
Marck, C., 335, *351*
Marguet, D., 308, *318*
Mariadason, J. M., 366, *372*
Mariuzza, R. A., 343, *356*
Marks, J. D., 344, *356*
Marland, Z., 108, *114*
Marmor, M. D., 11, *25*, 34, 36, 38, 39, *44*, 250, *262*
Marmorstein, A., 303, *315*
Marotta, A., 304, *316*
Marra, P., 297, *313*
Marrama, L., 235, *244*
Marrero, P. F., 366, *372*
Marsh, K., 230, *242*
Martin, D. W., 134, *145*
Martin, K. L., 107, *114*, 250, *263*
Martin, P. S., 15, *28*, 136, 142, 144, *148*
Martin-Belmonte, F., 290, 297, *313*
Martinez, S. E., 304, *316*
Martinez-Duncker, I., 94, *111*
Martinez-Lopez, R., 322, 329, *346*
Martin-Lomas, M., 223, *227*
Martin-Yken, H., 18, *29*
Martoglio, B., 136, *146*
Martone, M. E., 291, *318*
Marwoto, H., 235, *244*
Mary, J. Y., 362, *371*
Marykwas, D. L., 324, 339, *348*
Masaki, R., 11, *25*, 35, 36, 38, *45*, 76, *88*, 152, *157*
Mason, P. J., 42, *47*
Masterson, W. J., 7, 10, 19, *23, 24*, 32, 35, *44, 45*, 50, 55, 56, 58, 59, 62, *62–64*, 66, 69, 70, *85–87*, 119, 122, *129, 130*, 250, 253, 258, *264, 266*

Matera, K. M., 260, *267*
Mathias, A., 339, 340, *354*
Matsui, Y., 332, 338, *350*
Matsumoto, K., 330, *350*
Matsumoto, T., 123, *130*, 338–341, *353*
Matsumura, K., 338, *353*
Matsuura, Y., 335, *351*
Matter, K., 290, 292, *309*
Matthies, H. J., 282, *288*
Matthijs, G., 104, *112*, 370, *373*
Maurer-Stroh, S., 11, *25*, 34–37, *45*, 134, 142, *146*
Maxfield, F. R., 293, *318*, 361, *371*
Maxwell, S. E., 136, *147*, 259, *267*
Mayer, T. G., 195, 200, 215, *225*, 256, *265*
Maynes, J. T., 52, *63*
Mayor, S., 2, *21*, 66, 69, 70, *84–87*, 119, *130*, 136, *147*, 167, *178*, 267, 278, *287*, 290, 291, 293, 306, *309, 314, 317, 319*, 361, *371*
Mays, R. E., 292, *311*
Mazhari-Tabrizi, R., 93, *110*
Mazon, M. J., 12, *27*, 92, *109*, 123, *131*, 254, *264*
Mazza, D., 342, *355*
Mazzucchelli, R., 366, *372*
Mc Bride, H., 306, *317*
Mc Carthy, A. A., 52, 54, *63*
Mc Conville, M. J., 3, 7, 20, *22, 30*, 67, 72, *85, 88*, 107, *114*, 140, *149*, 151, *157*, 170, 173, *179, 180*, 182, *224*, 231, 238, *243, 245*, 248, 253, 258, *261, 266*, 270, 271, *284*
Mc Cutchan, T. F., 168, *178, 180*
Mc Dowell, M. A., 270, *284*, 291, *309*
Mc Gwire, B. S., 154, *158*
Mc Mahon, H. T., 295, *312*
Mc Millan, B. N., 107, *114*
Mc Neil, M. R., 104, 108, 109, *113, 115*
Mc Phie, P., 323, 336, *347*
Meder, D., *319*
Medina-Acosta, E., 255, *265*
Medof, M. E., 8, 14, *24, 27*, 66, 67, *85, 86*, 119, 120, 122, 126, 127, *129, 130*, 136, 137, 139, 141, *147, 149*, 250, 257, 258, *262, 266, 267*
Meguro, H., 203, *225*
Mehlert, A., 3, 7, 20, *22, 23, 30*, 67, *86*, 241, *245*, 259, *267*
Mehta, A., 152, *157*, 214, *226*
Mehta, D. P., 104, *112*
Meiss, H. K., 295, *313*

Meitzler, J. L., 139, *148*
Melgers, P. A., 165, *179*
Mellman, I., 290, 292, 302, 303, 305, 306, *310, 314–317*
Melo, J. S., 342, *355*
Menon, A. K., 2–4, 6–11, 14, 15, *21, 24, 25,* 32, 34, 35, *44, 45,* 66, 67, 69, 70, 76, *85–88,* 119, 120, 126, *129–132,* 135–137, 139–141, 145, *147–149,* 151–156, *157–158,* 167, *178,* 182, 214, *224, 226,* 248, 254, 255, 257–259, *261, 262, 264–267,* 270, 278, 282, *283, 287,* 358–360, *370*
Mensa-Wilmot, K., 154, *158*
Mercier, C., 165, *176*
Mergler, M., 320, 339, *354*
Merika, M., 366, *372*
Metenou, S., 235, *244*
Meury, A., 270, *284*
Meyale, S., 66, *83, 84,* 126, *131*
Meyer, U., 14, 15, *28,* 119, *129,* 136, 138, 140, *147, 148*
Meyhofer, E., 304, *316*
Mgone, C. S., 235, *244*
Micanovic, R., 134, *145*
Michaelson, D. M., 66
Michell, R. H., 66
Micksche, M., 366, *373*
Miki, K., 306, *317*
Miki, T., 18, *29*
Mikusova, K., 106, *114*
Milgram, S. L., 291, *309*
Milhiet, P. E., 293, *318*
Millan, J. L., 67, *86,* 250, 258, *262, 266,* 297, *313*
Miller, C. M., 305, *315*
Miller, E. A., 271, 272, *285*
Miller, G., 366, *373*
Miller, J. P., 36, 38, 41, 42, *46*
Miller, L. H., 168, *178*
Miller, W., 50, 55, *63*
Milne, F. C., 72, *88,* 249, 255, *262*
Milne, K. G., 14, 19, *27, 29,* 35, *45,* 50, 55, 56, 58, 62, *63,* 69, *87,* 103, *112,* 248, 250, 252, 255, 260, *263, 265, 267*
Mineo, C., 293, *319*
Mirabile, I., 277, *287*
Mironov, A. A., 291, *318*
Mirre, C., 307, *317*
Mischo, A., 338, *352*
Mishkind, M., 11, *26,* 118–123, 127, *128*
Mishkind, M., 92, *109*

Mishra, A., 307, *317*
Mishra, A. K., 108, *115*
Misra, S., 307, *317*
Misumi, Y., 79, *89, 90*
Mitchell, A. P., 333, *351*
Miura, M., 307, *317*
Miyata, T., 2, 11, *21, 24,* 33, 35, *44,* 104, *113,* 362, *371*
Mizuno, T., 14, *27,* 137, *147*
Mneimneh, S., 327, 330, 336, *349*
Mohney, R. P., 119, 127, *129*
Mol, P., 322, *347*
Mollica, L., 2, *21,* 104, *113,* 248, 254, *261,* 362, 363, 365, 369, *371*
Mollicone, R., 94, *111*
Mologni, D., 366, *372*
Mondesert, G., 12, *27,* 92, *109,* 120–122, *130,* 153, *158,* 332, *350*
Monlauzeur, L., 295, *313*
Monod, M., 322, 323, 334, 335, *346, 347, 351*
Montagna, G., 20, *30*
Montijn, R. C., 125, *131,* 324, 325, 327, 330, 338, *348, 353*
Moon, C., 144, *149*
Mooren, A. T., 337, *352*
Mootoo, D. R., 200, *225*
Mora, R., 297, *311*
Morales, E. Q., 223, *227*
Moran, P., 134, *146,* 270, *284*
Moreau-Hannedouche, O., 297, *313*
Morehouse, C. B., 107, *114*
Moreno, M. J., *319*
Morgan, B. P., 3, 6, *22,* 66, *84,* 238, *245,* 360, *371*
Mori, K., 168, *178,* 276, *286*
Morii, E., 77, *89*
Moriondo, V., 366, *372*
Morita, T., 322, 331, *346*
Morita, Y. S., 8, 14, 19, *24, 27, 30,* 70, *87,* 102, 107, *111, 114,* 123, *130,* 140, *148,* 248, 252, *261,* 270, *284*
Moritz, R. L., 165, *177*
Morone, N., 308, *318*
Morrice, N. A., 14, *27,* 87
Morris, H. R., 105, 107, *113, 114,* 137, *148*
Morris, J. C., 154, *158*
Morsomme, P., 78, *89,* 271, 272, 274, *285,* 291, *318*
Mosch, H. U., 333, *351*
Mosimann, S. C., 35, *45*
Mostov, K. E., 290–292, 307, *317, 318*

Mottram, J. C., 3, 14, *22*, *28*, 138–140, *148*, *149*, 258, *267*
Moukadiri, I., 339, *354*
Mouritsen, O. G., 293, *319*
Mouyna, I., 322, 331, 332, 334, *346, 348, 350*
Movva, N. R., 278, *287*
Mowatt, M. R., 66, *84*
Mrsa, V., 324, 326, 328, 330, 338, 339, *348*
Msaki, A., 260, *267*
Mueller, R., 93, *110*
Muhlschlegel, F. A., 334, *351*
Mukherjee, S., 293, *318*, 361, *371*
Mulchandani, A., 337–342, *352, 353*
Mullins, C., 292, *310*
Mumberg, D., 93, *110*
Muniz, M., 271–273, *285*, 291, *318*
Muñiz, M., 78, *89*
Munnik, T., 327, 339, *348, 354*
Munro, S., 361, *371*
Munstermann, F., 200, *225*, 256, *265*
Murai, K., 366, *373*
Murai, T., 339–341, *354*
Murakami, S., 7, *24*, 76, *89*, 155, 156, *158*, 278, 281, *287*
Murakami, Y., 2, 8, 11, 12, 14, *21*, *24*, *25*, 33, 34, 36–39, 41, *44*, *46*, 72, 76, *88*, 94, 102, 104, *111*, *113*, 123, 127, *130*, *132*, 152, *157*, 248, 250, 254, *261*, *262*, 362, 363, 365–367, 369, *373*
Murakata, C., 183, *224*
Murakoshi, H., 293, 308, *318, 319*
Murase, K., 293, 308, *318, 319*
Murata, C., 18, *29*
Murray, C. J., 238, *244*
Murrells, L. J., 303, *315*
Musch, A., 290–292, 303, 305, 306, *315, 316*
Mushegian, A., 11, *25*, 94, *111*
Mushiake, S., 306, *317*
Muthusamy, A., 168, *176*
Myers, C. L., 322, 329, *346*
Myers, E. W., 50, 55, *63*
Myher, J. J., 6, *23*, 66

N

Nadal, A., 366, *372*
Nagamune, K., 3, 14, 15, *22*, *27*, *28*, 70, *87*, 93, *110*, 136–140, 143, *147*, *148*, 248, 250, 256, *261*, *262*
Nagarajan, S., 14, *27*, 137, 139, *147*

Nagasu, T., 12, *26*, 75, *88*
Naglik, J. R., 322, 335, *346*
Nagpal, J. K., 144, *149*
Nagy, A., 2, *21*, 33, *44*, 248, 254, *261*
Nahlen, B. L., 71, *88*, 163, *178*, 231, 232, 235, *243*
Naik, R. R., 95, *111*
Naik, R. S., 71, *88*, 163, *178*, 231, 232, 235, *243*, *244*, 255, *265*
Naim, H. Y., 295, 297, 301, 303, 305, 306, *313–315*, *317*
Naito, K., 335, *351*
Nakada, C., 293, *319*
Nakagawa, T., 304, *316*
Nakakuma, H., 11, 14, *25*, 72, 76, *88*, 102, *111*, 127, *132*, 152, *157*
Nakamoto, K., 12, *26*, 75, *88*
Nakamura, C., 70, *310*
Nakamura, H., 52, *63*
Nakamura, N., 11, 12, *25*, *26*, 50, 55, 58, *63*, 76, *89*, 92, *109*, 254, *264*
Nakamura, Y., 339, *353*
Nakanishi, M., 20, *30*
Nakatani, F., 14, *27*, 70, *87*, 107, *114*
Nakatsu, F., 292, *310*
Nakayama, K., 12, *27*, 67, 76, *86*, *88*, 276, *286*
Nasab, F. P., 137, *148*
Nasir, R. T., 162, *177*
Navarrete, A., 305, *316*
Navarro, V. M., 342, *355*
Nebl, T., 165, *177*, 231, *242*
Necochea, R., 342, *354*
Nelson, K. L., 248, 255, *261*
Nelson, W. J., 290, 292, 303, *311, 315*
Netto, G. J., 144, *149*
Neuveglise, C., 335, *351*
Neville, D. M., 339, 340, *354*
Newbold, C., 230, *242*
Newman, H. A., 12, *26*, 36, *46*
Newton, G. L., 52, *63*
Ng, D. T., 16, *29*, 78, *89*, 276, *286*, 291, *309*
Ngo, L. L., 342, *355*
Nichan, C., 34–37, *45*
Nichols, B. J., *319*
Nicolle, D., 105, *113*
Niehues, R., 104, *112*
Niehus, S., 165, 167, *176*, *179*
Niels-Christiansen, L. L., 293, *318*
Nierman, W., 331, 332, *350*
Nieto, P. M., 223, *227*

Nigou, J., 105, 107, 108, *113–115*
Nikolaev, A. V., 3, *22*, 61, *64*, 102, *112*, 215, 223, *226*, *227*, 241, *245*, 252, 253, 256, *263*
Nischan, C., 11, *25*
Nishida, Y., 203, *225*
Nishijima, M., 17, *29*, 279, 281, *287*, *288*
Nishikawa, A., 42, *46*
Nishimura, J., 2, 14, *21*, *28*, 137, 142, 143, *148*
Nishimura, T., 292, *310*
Nishizaki, R., 322, *345*
Nisogi, H., 332, 338, *350*
Nissan, A., 144, *149*
Nita-Lazar, M., 137, *148*
Nitsch, L., 280, *288*, 292, 295, 297, 299–302, 304, *311*, *315*
Noble, W. S., 36, 38, 41, 42, *46*
Noda, M., 248, 253, *261*
Noda, Y., 12, *27*, 96, *111*, 304, *315*
Nogami, S., 327, *349*
Noguchi, C. T., 366, *372*
Nojima, K., 6, *23*
Nolte, M., 342, *355*
Nombela, C., 336, *352*
Norbury, C. C., 276, *286*
North, S. J., 137, *148*
Nosjean, O., 292, *311*
Notredame, C., 50–52, *63*
Novakovic, S., 170, 173, *179*, *180*, 231, *243*
Novatchkova, M., 11, *25*, 34–37, *45*, 134, 142, *146*
Nozaki, M., 2, *21*, 248, 254, *261*
Nozaki, T., 3, *22*, 248, 250, 256, *261*, *262*
Nuoffer, C., 134, *146*, 272, 273, *285*, 334, *351*
Nussenzweig, V., 3, *22*, 134, *145*

O

Ockenhouse, C. F., 71, *88*, 163, *178*, 231, 232, 235, *243*, *244*
Oda, K., 79, *89*, *90*
Oda, N., 366, *372*
Odds, F. C., 322, 335, *346*
Odenthal-Schnittler, M., 162, *179*, 254, *264*
Ogasawara, K., 304, *316*
Ogata, S., 6, *23*
Ogawa, T., 183, *224*
Ogino, C., 338, *353*
Oguchi, T., 119, 120, 125, *130*, *131*, 332, 338, *350*

Ogun, S. A., 163, *177*, *178*
Ohashi, Y., 79, *89*
Ohba, F., 12, *26*, 75, *88*
Ohberg, L., 218, *227*
Ohishi, K., 2, 3, 11, 12, 14, 15, *21*, *22*, *25*, *26*, *28*, 33, 37, 38, 42, 43, *44*, *46*, *47*, 55, *63*, 76, *89*, 92–94, 97, *109–111*, 118–127, *128–130*, 136–143, *147*, *148*, 248, 250, 254, 256, *261*, *262*, *264*, 365, *372*
Ohno, H., 292, *310*
Ohrui, H., 203, *225*
Ohtsuka, M., 343, *355*
Ohvo-Rekila, H., 278, *287*
Ohya, Y., 327, 328, 330, *349*, *350*
Okabe, M., 33, *44*, 77, *89*, 361, *371*
Okada, H., 66, *83*
Okada, Y., 304, *315*
Okami, K., 144, *149*
Okamoto, M., 12, *27*, 67, 76, *88*, 125, *131*, 276, *286*, 332, 333, *350*
Okamoto, P., 303, *315*
Okemoto, K., 281, *288*
Old, L. J., 230, *242*, 338, 344, *352*, *356*
Olfo, A., 293, *319*
Oliveira, R., 17, *29*
Oliver, G. J., 254, *264*
Olivo, J. C., 303, 305, 306, *315*
Olsen, M. J., 344, *356*
Olsen, V., 323, *347*
Omine, M., 2, *21*
Omura, S., 79, *90*
Oono, K., 276, *286*
Oosawa, F., 339, *354*
Ooshima, H., 123, 124, *130*, *131*
Oppenheimer, S. B., 342, *355*
Opperdoes, F. R., 20, *30*
Orci, L., 271, 272, *284*, *285*
Orii, T., 12, *26*, 124, *131*
Oriol, R., 94, *111*
Orkin, S. H., 366, 369, *372*, *373*
Orlean, P., 2–4, 6–8, 10–12, 14, 15, *21*, *23*, *25–28*, 32, 34–38, 40, 41, *44–46*, 66, 67, *84–86*, 92, 93, 95–97, *109–111*, 118–127, *129*, *130*, *132*, 135–137, 142, 144, *145*, *148*, 151–153, 155, *157*, *158*, 167, *176*, *177*, 182, 224, 248, 250, 255, 258, 259, *261–263*, *267*, 270, *283*, 324, 332, *348*, *350*, 358–361, *370*, *371*
Orr, G., 330, *350*
Osada, M., 144, *149*

Oscarson, S., 218, *226*
Oster, G., 295, *312*
Osumi, M., 339–341, *354*
Ota, M., 52, *63*
Otto, C., 137, *148*
Ovalle, R., 322, 326, 327, *347*
Overath, P., 238, *245*
Oxley, D., 3, *22*
Ozawa, H., 79, *90*
Ozono, K., 306, *317*
Oztan, A., 303, *315*

P

Paces, J., 42, 43, *47*
Paces, V., 42, 43, *47*
Pagac, M., 19, *30*, 299, *314*
Paidhungat, M., 16, *29*
Paladino, S., 280, *288*, 292, 295, 297, 299–302, 304, 305, *311–314*
Pallavi, B., 307, *317*
Palleschi, C., 338, *352*
Pan, L., 366, *373*
Pan, Y. T., 255, *265*
Pandolfi, P. P., 361, *371*
Pang, S., 295, *312*
Panico, M., 137, *148*
Pankuweit, S., 175, *180*
Parida, S., 105, *113*
Paris, G., 20, *30*
Parise, L., 323, 335, 336, *347*
Park, H., 322, 329, *346*
Park, K. J., 343, *356*
Park, S. Y., 52, *63*
Parker, C. J., 2, *21*, 167, *177*, 363, *372*
Parker, G., 67, *86*, 250, 258, *262*, *266*
Parodi, A. J., 137, *148*
Parthasarathy, R., 342, *355*
Parton, R. G., 295, 297, 306, *312*, *313*, *317*
Pasquali, C., 295, *312*
Paterson, M. J., 58, 60–62, *64*, 72, *88*, 102, *112*, 167, *179*, 250, 252, 255, 256, *263*, *265*
Paterson, N. A., 52, 54, *63*
Pathak, S., 230, *242*
Patil, C. K., 276, *286*
Patterson, J. H., 107, *114*
Patterson, M. C., 104, *112*
Patterson, M. J., 58, 61, *64*
Patterson, S., 2, *21*, 104, *113*, 248, 254, *261*, 362, 363, 365, 369, *371*

Paul, K. S., 19, *30*, 270, *284*
Paulick, M. G., 358, *370*
Pavlicek, A., 42, 43, *47*
Pearce, E. J., 238, *245*
Peelle, B. R., 342, *355*
Pekari, K., 166, *178*, 200, 204, 209, *225*, 241, *245*
Pekosz, A., 295, *312*
Pelham, H. R., 279, *287*
Pelkmans, L., 295, *312*
Pellegrini, G., 323, *347*
Peng, R., 272, *285*
Pepper, L. R., 337, *352*
Pepperkok, R., 272, *285*, 291, *318*
Peranen, J., 292, *310*
Peranovich, T. M., 71, *87*
Peres, V. J., 254, *264*
Perez, F., 308, *318*
Perkins, D. J., 71, *88*, 163, *178*, 231, 232, 235, *243*
Perraut, R., 235, *244*
Pessin, M. S., 70, *87*
Peterson, N. A., 52, *63*
Peterson, S., 104, *112*
Petrakova, E., 248, 256, *261*, 325, 328, *348*
Petráková, E., 3, *21*
Petre, B. M., 271, *285*
Pfreundschuh, M., 338, *352*
Phair, R. D., 305, *315*
Philipsen, S., 365, 366, *372*
Phinney, B. S., 322, *346*
Piatesi, A., 344, *356*
Pichler, H., 7, *24*, 76, *89*, 155, 156, *158*, 278, 279, 281, *287*
Pichova, I., 322, 335, *346*
Pijak, D. S., 335, *351*
Pike, L. J., 293, *319*, 339, *353*
Pillich, R., 280, *288*, 292, 295, 299–301, *311*
Pirker, C., 366, *373*
Pirofski, L. A., 322, 323, *345*
Pittet, M., 2, 3, 6–8, 10, *21*, 67, *86*, 94, *111*, 137, *147*, 248, 255, *261*, 270, *283*
Piyawattanasakul, N., 33, *44*
Plaine, A., 322, 323, *356*
Platt, F. M., 297, *314*
Plattner, H., 297, *313*
Playfair, J. H. L., 230, 231, *242*
Plebanski, M., 162, *180*
Pleshak, E. N., 248, 255, *261*
Ploegh, H. L., 136, *146*

Plowman, S. J., 306, *317*
Pluschke, G., 209, *226*, 235, 240, *244*
Pocard, T., 295, 302, 304, 307, *313*, *314*, *318*
Poindexter, P., 3, *21*
Poirier, F., 304, *316*
Polakoski, K., 238, *245*
Polgár, L., 142, *149*
Polishchuck, R. S., 301, *314*
Polishchuk, E., 297, *314*
Polishchuk, R., 302, *315*
Polokoff, M. A., 152, *157*
Pomorski, T., 151–154, 156, *157*, 214, *226*
Pons, A., 297, *313*
Popolo, L., 322, 324, 329, 330, 334, 335, *346*, *348–350*
Porcelli, S. A., 105, *113*
Porcher, C., 366, *373*
Porello, S., 322, 334, *346*
Potocnik, A., 238, *244*
Pottekat, A., 152, *157*, 278, *287*
Potter, B. A., 295, 303, *312*, *315*
Potter, C. S., 271, *285*
Poulain, D., 125, *131*
Pous, C., 308, *318*
Pouwels, P. H., 340, 343, *354*
Pralle, A., 293, *319*
Prescianotto-Baschong, C., 78, *89*, 274, *285*, 291, *318*
Prescott, A. R., 7, 14, *23*, *27*, 66, 79, *85*, *90*, 260, *267*
Presley, J. F., 305, *315*
Preuss, U., 295, *313*
Price, M. S., 52, *63*
Priem, J., 327, 339, *348*, *354*
Prince, G. M., 8, *24*, 66, 67, *85*, 119, 126, 127, *129*, 136, *147*, 257, *267*
Prinz, W. A., 282, *288*
Procopio, D. O., 173, *176*, 231, 241, *243*, *245*
Prohaska, R., 297, *313*
Proux-Gillardeaux, V., 307, *318*
Prusiner, S. B., 79, *89*, 277, *287*
Puertollano, R., 297, *313*
Punna, T., 16, *29*
Puoti, A., 7, *23*, 67, *85*, *86*, 126, *131*, 250, *262*
Puzo, G., 105, 106, *113*, *114*
Pyke, J. S., 107, *114*
Pypaert, M., 303, 306, 307, *315*, *317*, *318*
Python, C. P., 330, *350*
Pyun, Y. R., 338, *352*

Q

Qadota, H., 330, *350*
Qi, Q., 338, *352*
Qian, C. N., 366, *372*
Qiu, W. G., 327, 330, 336, *349*
Qualmann, B., 305, *316*
Quarles, T., 342, *355*
Quaroni, A., 302, 304, *315*
Quesniaux, V., 105, *113*
Quesnoit, M., 308, *318*
Quilici, D., 170, *180*, 231, *243*, 248, 253, *261*

R

Raas-Rotschild, A., 104, *112*
Raben, D. M., 70, *87*
Rachel, R., 338, 339, *353*
Rademacher, T. W., 3, 6, *22*, 66, *85*, 97, *111*, 118, 119, 127, *128–129*, 182, 223, *224*, *227*
Radke, J., 173, *180*
Raetz, C. R., 32, *44*, 121, *130*
Rafalski, J. A., 322, *345*
Ragni, E., 330, 334, *350*
Ragnini-Wilson, A., 301, *314*
Raja, S. M., 248, 255, *261*
Rajendran, L., 294, *319*
Rakestraw, J. A., 344, *356*
Ralton, J. E., 20, *30*, 72, *88*, 103, *112*, 255, 258, *265*, *266*
Ram, A. F., 12, 18, *26*, *29*, 134, *146*, 322, 324, 325, 327, 329, 330, 332, *345*, *348*, *349*
Ramalingam, S., 136, *147*, 259, *267*
Ramponi, G., 276, *286*
Rancour, D. M., 11, *25*, 35, *45*, 152, *157*
Ransom, D. M., 270, *284*, 291, *309*
Rao, B., 343, *355*
Rao, M., 293, *314*, *319*, 361, *371*
Raper, J., 7, 20, *23*, *30*, 70, 71, *87*, 258, *266*
Rapoport, T. A., 277, *286*
Ratiwayanto, S., 235, *244*
Ratner, D. M., 235, *244*
Ratnoff, W. D., 66, *84*
Ratovitski, E. A., 144, *149*
Rau, S., 338, 340, 341, *352*
Rauceo, J. M., 322, 323, 325, *345*
Ravi, L., 8, *24*, 67, *85*, 119, 126, 127, *129*, 257, *267*
Rawer, M., 14, *28*, 137, 139, 140, *147*
Ray, S., 93, 104, *110*, *112*, 153, *158*

Raymond, G. V., 104, *112*
Rea, D., 142, *149*
Real, F. X., 295, *312*
Redman, C. A., 6, *23*
Reed, N. A., 304, *316*
Reed, S. I., 12, *27*, 92, *109*, 120–122, *130*, 153, *158*, 332, *350*
Reggiori, F., 6–8, 11, 12, 18, *23*, *26*, 119, 120, 122, 124, 125, *129*, 250, 257, 258, *262*, *266*, *267*, 270, *283*, 332, *350*
Regitz, E., 338, *352*
Reichardt, H. C., 223, *227*
Reichardt, N. C., 223, *227*
Reichel, S., 104, *112*
Reinhold, B. B., 3, *21*, 248, 256, *261*, 325, 328, *348*
Reinhold, V. N., 6, 7, *22*, 66, 67, *84*, 118, *128*
Reiter, K., 104, *112*
Remington, J. S., 165, 173, *177*, *178*
Remmert, C. L., *319*
Ren, R., 339, 341, *354*
Ren, X., 293, *319*
Renner, C., 344, *356*
Renner, V., 142, *149*
Rescher, U., 297, 301, *313*
Resende, M. G., 173, *176*
Resende, U., 259, *267*
Reynolds, C. M., 121, *130*
Ribas, J. C., 328, *349*
Richard, M. L., 125, *131*, 322, 323, 332, *345*, *350*
Richards, A. A., 295, *312*
Richards, M. K., 66, *84*
Richards, S., 2, *21*, 104, *113*, 248, 254, *261*, 362, 363, 365, 369, *371*
Richardson, J. M., 7, *23*, 67, *86*, 259, 260, *267*
Richie, T. L., 235, *244*
Richier, P., 66, *84*
Richmond, G. S., 258, *267*
Ridgway, N. D., 281, *288*
Rietveld, A., 82, *90*
Rietveld, A. G., 270, 280, *284*, *288*, 295, 299, *312*
Riezman, H., 2, 3, 7, 11, 12, 14, 15, 17, *21*, *22*, *24*, *26–29*, 34–38, 40, 41, *44*, *45*, 76, 78, *89*, 92, *109*, 118–123, 127, *128*, *131*, 134, 137–140, *146–148*, 155, 156, *158*, 248, 250, 254, 256, 257, *261*, *262*, *264*, *265*, 270–274, 278, 279, 281, 282, *284*, *285*, *287*, 291, *318*, 324, 327, 329, 330, 334, *347*, *349*, *351*

Rifkin, M. R., 238, *244*
Rigneault, H., 308, *318*
Riley, E. M., 235, *244*
Rindler, M. J., 303, *315*
Rio, B., 362, *371*
Ritchie, K., 293, 308, *318*, *319*
Ritter, G., 338, 344, *352*, *356*
Rittmann, D., 108, 109, *115*
Rittner, C. M., 107, *114*
Riva, A., 366, *372*
Robbins, P. W., 93, *110*, 162, *178*, 325, 328, *348*
Roberts, C., 200, *225*
Roberts, D. J., 238, *244*
Roberts, I. A., 104, *113*, 366, 367, *373*
Roberts, W. L., 6, *23*, 66, 67, *84*, 118, *128*
Robson, N., 344, *356*
Rochant, H., 362, *371*
Rockwell, N., 323, 336, *347*
Rodan, G. A., 248, 253, *261*
Rodgers, G., 165, *177*
Rodgers, W., 308, *318*
Roditi, I., 140, *149*, 361, *371*
Rodriguez-Boulan, E. J., 136, *147*, 248, 252, 255, *261*, *265*, 280, *288*, 291, 292, 295, 297, 302–306, *311*, *313*, *315*, *316*, *318*
Rodriguez-Pena, J. M., 336, *352*
Rodriguez-Suarez, R. J., 328, *349*
Roemer, T., 334, *351*
Roger, A. J., 322, *345*
Roh, D. H., 327, 329, 330, *349*
Roitbak, T., 290, *309*
Rollins, C., 11, *25*, 36, *46*
Romeo, M. J., 12, *26*, 34, 36–38, 40, 41, *44*, *46*
Romeo, M. R., 37, 40, *46*
Ron, D., 276, *286*
Rooney, I. A., 238, *245*
Roper, J. R., 260, *267*
Rose, C., 362, *371*
Rose, J. K., 17, *29*, 278, 280, *287*, *288*, 292, *311*
Rosenberry, T. L., 6–8, *22–24*, 66, 67, *84*, *85*, 103, *112*, 118, 119, 126–127, *128–129*, 255, 257, *265*, *267*
Ross, A. J., 223, *227*
Rosse, W. F., 2, *21*, 362, *371*
Rossjohn, J., 108, *114*
Rosti, V., 361, *371*
Roth, M. G., 292, *310*
Rotheneder, H., 366, *372*
Rotoli, B., 362, *371*

Roubaty, C., 6, 7, 15, 18, *23*, *28*, 78, *89*, 118, 122, *128*, 271, *284*
Routier, F. H., 137, *148*
Roux, B., 292, *311*
Roy, A., 324, 339, *348*
Rubin, E., 144, *149*
Ruda, K., 218, *226*
Rudd, P. M., 3, 6, *22*, 66, *84*, 360, *371*
Ruhela, D., 214, *226*
Runge, A., 304, *316*
Ruppert, V., 175, *180*
Rusconi, S., 15, *28*, 137–139, *147*, 154, *158*
Rush, J. S., 93, 103, *110*, *112*, 153, *157*, *158*
Rustom, A., 301, *314*
Rvel-Vik, S., 104, *112*
Ryals, P. E., 255, *265*
Ryazanov, A. G., 306, *317*
Ryckaert, S., 343, *355*
Ryu, K., 342, *354*

S

Sabatini, D. D., 303, *315*
Sabharanjak, S., 306, *317*
Sachchidan, S., 292, *311*
Sachs, J. D., 230, *242*
Sadeghi, H., 66, *84*
Safar, J. G., 277, *287*
Saffitz, J. E., 238, *245*
Sagane, K., 12, *26*, *27*, 67, 75, *86*, *88*
Sagi, D., 104, *112*
Sahara, H., 338, *353*
Sailer, A., 277, *286*
Saito, F., 339, *354*
Saito, M., 308, *318*
Saito, N., 304, *315*
Saito, R., 43, *47*
Sajid, M., 14, *28*
Sakaguchi, K., 18, *29*
Sakai, Y., 123, 124, *130*, *131*
Sakuragi, N., 322, *345*
Sakurai, T., 322, *345*
Saleem, M., 337, 338, 341, 342, *352*
Salkar, R., 292, *311*
Saltiel, A. R., 66, *83*, 280, *288*
Salvarezza, S. B., 305, *316*
Sambrook, J. F., 66, *84*, 167, *177*, 276, *286*
Samuelson, J., 162, *178*
Sanada, M., 343, *355*
Sanglard, D., 335, *351*

Sano, M., 335, *351*
Santiago, D., 338, *352*
Santikarn, S., 66, 67, *84*, 118, *128*
Santoro, M., 297, *314*
Santos, M. S., 342, *355*
Santos de Macedo, C., 255, *265*
Santovito, M. L., 293, *319*
Sanyal, S., 155, 156, *158*
Sarasij, R. C., 293, *314*, 361, *371*
Sargeant, T., 165, *177*
Sarnataro, D., 280, *288*, 292, 295, 299–301, *311*, *312*
Saslowsky, D. E., 293, *318*, *319*
Sato, K., 12, *27*, 96, *111*, 306, *317*
Sato, M., 306, *317*
Sato, T., 306, *317*
Sayeed, A., 276, *286*
Schackert, C., 323, *347*
Schaeffer, M. L., 107, *114*
Schafer, W., 322, 335, *346*
Schaller, M., 322, 323, 335, *346*, *347*
Schares, G., 166, *179*
Scheiffele, P., 292, *310*
Schekman, R., 270–272, *284*, *285*
Schenk, B., 104, *112*, 153, 154, *158*
Schiebe-Sukumar, M., 104, *112*
Schimmoller, F., 273, *285*
Schinzel, A., 276, *286*
Schleinkofer, K., 334, *351*
Schlueter, U., 167, *178*, 200, 204, *225*, 231, *243*
Schmidt, A., 163, *179*
Schmidt, J., 165, 166, 170, 173, *176*, *179*, 241, *245*
Schmidt, N., 259, *267*
Schmidt, R. E., 11, *25*, 34–37, *45*, 259, *267*, 332, *350*
Schmidt, R. R., 166, 170, 173, *176*, *178*, 195, 200, 204, 209, 215, *225*, 241, *245*
Schmitt, M. J., 334, 338, 340, 341, *351*, *352*
Schmoranzer, J., 307, *310*
Schneeberger, E. E., 297, *313*
Schneider, G., 11, *25*, 34–37, *45*, 134, 142, *146*
Schneider, P., 66, *83*, 293, 299, *314*
Schofield, D. A., 344, *356*
Schofield, L., 3, *22*, 69, *87*, *128*, *130*, 163, 165, 170, 173, 175, *176–177*, *179*, *180*, 204, *225*, 230, 231, 233, 234, 240, *242*, *243*, 248, 253, *261*
Schollen, E., 104, *112*
Scholler, N., 342, *355*

Schönbächler, M., 12, *26*, 35, *45*, 250, *262*
Schonfeld, G., 238, *245*
Schreiner, R., 305, *316*
Schreuder, M. P., 337–343, *352–354*
Schroder, S., 273, *285*
Schroeder, R. J., 293, *319*
Schubert, J., 11, *25*, 34–37, *45*, 332, *350*
Schubert, U., 276, *286*
Schuck, S., 295, 306, *312*, *317*, 361, *371*
Schuldiner, M., 16, *29*
Schulz, B. L., 137, *148*
Schulz, T. A., 282, *288*
Schuppert, A. A., 162, *179*
Schutzbach, J. S., 93, *110*
Schwab, H., 338, 340, 341, *352*
Schwartz, R., 35, *45*
Schwarz, R. T., 3, 12, *22*, *27*, 70–72, *87*, *88*, 92, 93, 103, *109*, *110*, *112*, 119, 123, *128*, *130*, *131*, 162, 163, 165–167, 169, 170, 173–175, *176–180*, *231*, *232*, *242*, *243*, *245*, 250, 252, *261–265*
Schwarz, R. T. T., 248, 259, *267*
Scmidt, R. R., 256, *265*
Sedbrook, J. C., 3, *21*
Seeber, F., 71, *88*, 169, *178*
Seeberger, P. H., 165, 175, *178*, *179*, 204, 206, 208, 209, *225*, 230, 233, 234, 238, 240, 241, *243–245*
Seidl, T., 338, 339, *353*
Seiser, C., 366, *372*
Sellick, G. S., 2, *21*, 104, *113*, 248, 254, *261*, 362, 363, 365, 369, *371*
Sena, C. B., 107, *114*
Seog, D., 304, *316*
Sernee, M. F., 107, *114*
Setou, M., 304, *315*, *316*
Severin, F., 304, *316*
Sevlever, D., 8, *24*, 67, *85*, 103, *112*, 119, 120, 122, 126, 127, *129*, *130*, 255, 257, *265–266*
Sexton, A., *128*, *130*
Shakespeare, W., 322, 345, *345*
Shams-Eldin, H., 71, 72, *88*, 165, 167, 170, *176*, *179*, 250, 254, 258, *262*, *263*
Shams-Elsin, H., *128*, *130*
Shamu, C. E., 276, *286*
Shao, N., *226*
Shao, Y., 366, *373*
Sharma, D. K., 10, 19, *24*, *29*, 60–62, *64*, 67, 69, 70, 72, *87*, *88*, 102, *112*, 136, 138, 139, *147*, *148*, 248–250, 252–256, 258, 259, *263*, *267*

Sharma, N., 307, *317*
Sharma, P., 293, 306, *314*, *317*, 361, *371*
Sharma, S., 230, *242*
Sharon, N., 4, *30*
Shaz, D., 93, *110*, 270, *284*
Shedd, D., 366, *373*
Sheehan, J., 7, *23*, 66, *85*
Sheff, D., 302, 303, *314*
Shen, K., 366, *372*
Shen, Y. H., 366, *373*
Sher, A., 238, *245*
Sherman, D., 335, *351*
Sherrill, C., 52, *63*
Shevchenko, A., 17, *29*
Shi, W., 361, *371*
Shibasaki, S., 338, 339, *353*
Shibasaki, Y., 339–341, *354*
Shibata, H., 79, *89*
Shiel, M. J., 302, *315*
Shimaoka, M., 343, *355*
Shimma, Y., 338–340, 342, *353*, *354*
Shimoi, H., 125, *131*, 333, *350*, *351*
Shiraga, S., 339, 343, *353*, *355*
Shirouzu, M., 52, *63*
Shishioh, N., 8, 11, 12, *24–26*, 94, 102, *111*, 121, 123–127, *130*
Shneider, B. L., 292, *310*
Shteyn, E., 307, *318*
Shukla, S. D., 66, *83*
Shusta, E. V., 337, 339, 340, 343, *352*, *353*, *356*
Siafakas, A. R., 323, 331, 332, *347*
Sidenius, N., 293, *319*
Sidransky, D., 144, *149*
Siemers, K. A., 292, *311*
Sievi, E., 280, *288*, 338, 339, *353*
Siggia, E. D., 305, *315*
Signorell, A., 19, *30*, 299, *314*
Silman, I., 66, *83*
Silva, L. S., 241, *245*
Simon, S. M., 291, *318*
Simonot, C., 278, *287*
Simons, K., 17, *29*, 82, *90*, 270, 278–280, *284*, *287*, *288*, 292–295, 297, 299–303, 305–307, *310–319*, 361, *371*
Sinay, P., 191, *225*
Singh, A., 109, *115*
Singh, N., 66, *83*, 93, *110*, 258, *266*
Singh, S. P., 167, *177*
Singleton, D., 66, *83*
Sinha, A., 152, *157*, 214, *226*

Siomos, M. A., 175, *179*, 204, *225*, 230, 233, 234, 240, *242*
Sipos, G., 7, 11, 12, 18, *23*, *26*, 67, *86*, 119, 120, 122, 124, 125, *129*, 258, *266*, 270, *283*, 332, *350*
Siripanyaphino, U., 33, *44*
Siripanyaphinyo, U., 11, *25*, 37, 38, 41, *46*
Siripanyapinyo, U., 11, 14, *25*, 72, 76, *88*, 102, *111*, 127, *132*, 152, *157*
Smith, D. F., 82, *90*
Smith, J. A., 173, *176*, 231, *243*
Smith, J. D., 291, *318*
Smith, L. A., 344, *356*
Smith, R. F., 72, *88*
Smith, T. K., 3, 10, 14, 19, *22*, *24*, *27*, *29*, 50, 52–54, 56, 58, 60–62, *63–64*, 67, 69–72, *86–88*, 102, *112*, 165, 167, 169, 170, *176*, *178*, *179*, 248–250, 252–258, *262–265*, 267
Snider, M. D., 154, *158*
Snow, R. W., 230, *242*
Snyder, D. A., 165, *178*, 204, 206, 208, *225*, 234, 235, *243*, *245*
Snyers, L., 297, *313*
Soares, E. V., 342, *355*
Soares, H. M., 342, *355*
Sobering, A. K., 12, *26*, 34, 37, 38, 40, 41, *44*, *46*
Socie, G., 2, *21*, 362, *371*
Soebianto, S., 235, *244*
Sohlbach, C., 104, *112*
Sohn, J. H., 338, *352*
Soler, M. N., 105, *113*
Somerville, C., 3, *21*
Sommer, S. S., 334, *351*
Song, G., 343, *355*
Song, H. T., 338, 340, *353*
Song, S., 322, *345*
Sook Kim, M., 144, *149*
Soroka, C. J., 292, *310*
Sorrell, T. C., 323, 329, 331, 332, *347*
Sosinska, G. J., 322, *347*
Soucy, R. L., 165, *178*, 204, 206, 208, *226*, 234, 235, *243*, *244*
Spang, A., 273, *285*
Sparla, R., 292, *311*
Spear, E. D., 16, *29*, 78, *89*, 276, *286*, 291, *309*
Specht, C. A., 12, *26*, 34, 37, 38, 40, 41, *44*, 93, *110*
Speed, T. P., *128*, *130*, 165, *177*
Spencer, N., 107, *114*

Spiess, M., 302, *315*
Spiliotis, E. T., 307, *317*
Spreghini, E., 333, *351*
Springer, S., 271, *285*
Springer, T. A., 343, *355*
Spurway, T. D., 139, *148*
Srikrishna, G., 104, *112*
Stadler, J., 66, *84*
Stafford, F. J., 93, *110*, 270, *284*
Stagg, S. M., 271, *284*, *285*
Stagljar, I., 36, 38, 41, 42, *46*
Stamnes, M., 305, *316*
Stavrou, S., 339, 340, *354*
Stefani, M., 276, *286*
Stein, M., 290, *309*
Steiner, E., 366, *373*
Steiner, G., 144, *149*
Stelzer, E. H., 292, *311*
Stenmark, H., 304, *316*
Stephens, D. J., 272, *285*, 291, *318*
Stevens, L., 76, *89*
Stevens, S. L., 67
Stevens, T. J., 2, 3, *21*
Stevens, V. L., 32, *44*, 50, 58, *63*, 69, *85*, 119, *129*, 248, 255, *261*
Stewart Campbell, A., 200, *225*
Stijlemans, B., 176, *179*
Stockert, E., 338, *352*
Storey, M. K., 124, 125, *131*, 281, *288*
Stow, J. L., 303, *315*
Strahl, S., 92, 93, *110*
Strahl-Bolsinger, S., 153, *158*
Stremmel, W., 292, *311*
Stricker, H. M., 307, *317*
Striepen, B., 162, 165, 169, *179*, *180*
Stringaro, A., 323, *347*
Strub, J. M., 15, *28*, 136, 140, *147*
Strynadka, N. C., 35, *45*
Stuermer, C. A., 297, *313*
Sturley, S. L., 282, *288*
Stutz, A., 104, *112*
Stuurman, N., 304, *316*
Styles, C. A., 40, *46*
Su, L., 307, *317*
Su, T., 290, *309*
Subaran, R., 333, *351*
Subauste, C. S., 173, *178*
Suchy, F. J., 292, *310*, *311*
Sugahara, M., 292, *310*
Sugimoto, C., 322, *345*

Sugimoto, H., 292, *310*
Sugimoto, N., 11, 12, *25*, 94, *111*
Sugiyama, E., 66, *84*, *85*, 126, *131*, 134, 146, 167, *177*
Suguitan, A. L., 235, *244*
Sullivan, D. P., 278, *287*
Sullivan, D. R., 241, *245*
Sumita, T., 339, *354*
Sun, A. Q., 292, *310*, *311*
Sun, J. M., 365, 366, *372*
Sun, X., 342, *355*
Sundberg, E., 343, *356*
Sung, C. H., 292, *311*
Suntio, T., 280, *288*
Suske, G., 365, 366, *372*
Sutterlin, C., 3, 12, *22*, *27*, 92, *109*
Sütterlin, C., 123, *131*, 248, 250, 254, 256, 257, *261*, *262*, *264*, *265*, 273, 278, *285*, *287*
Suye, S., 338, 339, 342, *353*
Suzuki, K., 293, 308, *318*, *319*
Suzuki, Y., 173, *178*
Swers, J. S., 344, *356*
Szallies, A., 14, *28*, 137, 139, 140, *147*
Szaniszlo, P. J., 330, *350*
Szeltner, Z., 142, *149*

T

Tabrizi, S. J., 278, *287*
Tabudravu, J. N., 260, *267*
Tachado, S. D., 170, 173, *179*, *180*, 231, *243*, 248, 253, *261*
Taddei, N., 276, *286*
Taguchi, R., 6, 8, 15, 16, 18, *23*, *24*, *28*, *29*, 32, *44*, 67, 77, *89*, 258, *266*, 270, *284*, 299, *314*, 360, *370*
Taguchi, T., 303, *315*
Tailler, D., 166, *178*, 200, 204, 209, *225*
Taira, T., 366, *372*
Takada, Y., 124, *131*
Takahashi, M., 2, 11, 12, *21*, *24*, 26, 33, 35, *44*, 50, 58, *63*, 76, *89*, 92, 104, *109*, *113*, 254, *264*, 362, *371*
Takai, Y., 330, *350*
Takayama, K., 108, *115*, 338, 339, 342, *353*
Takeda, J., 2, 11, 12, 14, *21*, *24*–*26*, *28*, 32, 33, 35–38, 42–43, *44*–*45*, *47*, 50, 58, *63*, 76, *88*, *89*, 92, 93, 104, *109*, *110*, *113*, 124, *131*, 137–139, *148*, 152, *157*, 248, 250, 254, *261*–*264*, 361, 362, *371*

Takeda, N., 306, *317*
Takegawa, K., 322, 331, *346*
Takeuchi, O., 173, *176*, 231, *243*
Takida, S., 272, *285*, 291, *318*
Takio, K., 304, *316*
Takvorian, P., 330, *350*
Talla, E., 335, *351*
Tamaru, Y., 343, *355*
Tamborrini, M., 209, *226*, 235, 240, *244*
Tame, J. R. H., 52, *63*
Tanaka, A., 337–343, *352*–*355*
Tanaka, K., 330, 332, 338, *350*
Tanaka, N., 292, *310*, 322, 331, *346*
Tanaka, S., 14, 15, *28*, 38, *46*, 77, *89*, 93, *110*, 137, 142, 143, *148*, 258, *266*, 271, *284*, 291, *309*, 360, *370*
Tanesaka, E. C., 322, *345*
Tanino, T., 340, *354*
Taniuchi, I., 361, *371*
Tanner, M. E., 35, *45*
Tanner, W., 92, 93, *110*, 153, *158*, 167, *177*, 324, 326, 328, 330, *348*, *353*
Tao, X., 339, 341, *354*
Taron, B. W., 6, *23*, 97, *111*, 118–120, 122, 123, *129*, *130*
Taron, C. H., 6, 11, 12, *23*, *25*, *27*, 32, 37, 38, *44*, 93, 95–97, *110*, *111*, 118–120, 122–127, *129*, *130*, *132*, 250, *262*
Tartakoff, A. M., 66, *83*, *84*, 93, *110*, 126, *131*, 258, *266*
Tarutani, M., 33, *44*
Tashima, Y., 8, 11, 15, 16, 18, *24*, *25*, *28*, *29*, 32, 37, 38, 41, *44*, *46*, 67, 77, *85*, *89*, 258, *266*, 270, 271, *284*, 291, 299, *314*, 360, *370*
Tateno, Y., 335, *351*
Tatituri, R. V., 107, 108, *114*, *115*
Taverne, J., 230, 231, *242*
Taylor, D. W., 235, *244*
Tcheperegine, S. E., 15, *28*, 136, 142, 144, *148*
te Biesebeke, R., 339, *353*
Teh, B. T., 366, *372*
Teh, H. J. C., 3, *21*
Tekaia, F., 331, 332, *350*
Tempre, R., 295, *312*
Tempst, P., 305, *316*
Terada, T., 52, *63*
Terashima, H., 322, 328, 329, 335, *346*, *349*
ter Beest, M. B., 290, 307, *317*
Terrance, K., 338, *353*
Tettelin, H., 134, *146*, 322, 329, *345*

Thevananther, S., 292, *310*
Thevelein, J. M., 40, *46*
Thiele, C., 297, *313*
Thomas, D. Y., 335, *351*
Thomas, J., 238, *245*
Thomas, L. J., 66, *84*, *85*, 126, *131*, 134, *146*, 167, *177*
Thomas-Oates, J. E., 3, 6, 7, *22*, *23*, 238, *245*, 270, 271, *284*
Thompson, F., 209, *226*, 235, 240, *244*
Thompson, J. D., 74, *90*
Thompson, N. J., 16, *29*
Thorsen, E., 293, *318*
Thuita, L., 235, *244*
Thurnheer, S., 295, *312*
Thurnher, M., 105, *113*
Tiede, A., 34–37, *45*, 332, *350*
Ting, E. L., 323, 335, 336, *347*
Tivodar, S., 280, *288*, 292, 295, 299–301, *311–313*
Todd, J., 235, *244*
Toh-e, A., 119, 120, 125, *130*, *131*, 332, 338, *350*
Tohyama, M., 276, *286*
Tomavo, S., 162, *179*, 250, 254, *262*, *264*
Tomishige, M., 304, *316*
Tomita, M., 43, *47*
Tomita, S., 11, *25*, 38, *46*, 93, *110*
Tong, M., 342, *355*
Toomre, D., 292, 293, 300–303, 305, 306, *310*, *314*, *315*, *319*
Torkko, J., 297, *313*
Toschka, H. Y., 337, 339, *352*, *353*
Toubaji, A., 144, *149*
Toutant, J. P., 66, *84*
Traister, A., 361, *371*
Traub, L. M., 292, *310*
Traumann, A., 259, *267*
Travassos, L. R., 173, *176*, 231, 241, *243*, *245*
Travers, K. J., 276, *286*
Tremblay, J., 323, 335, 336, *347*, *351*
Tremblay, P., 277, *287*
Tremml, G., 361, *371*
Treumann, A., 7, *23*, 66, 67, *83*, *85*, *86*
Triggs, V. P., 270, *284*
Trimble, R. B., 328, *349*
Trink, B., 144, *149*
Trucco, A., 291, *318*
Tsuchiya, M., 12, *26*, 75, *88*
Tsuchiyama, K., 338, 339, 342, *353*
Tsuji, A., 306, *317*

Tsukahara, K., 12, *26*, *27*, 67, 75, *86*, *88*
Tsukamoto, K., 134, *146*
Tu, K., 15, *28*, 136, 142, 144, *148*
Tucker, G., 66, *84*, 126, *131*
Tucker, S., 154, *158*
Tuininga, A. R., 6, *23*
Tull, D., 107, *114*
Tuma, P. L., 307, *317*
Turco, S. J., 241, *245*
Turner, A. J., 3, *22*
Turner, K. M., 323, 329, *347*
Twell, D., 3, *21*
Tweten, R. K., 248, 255, *261*
Tykocinski, M. L., 258, *266*

U

Uccelletti, D., 338, *352*
Udeinya, I. J., 255, *265*
Udenfriend, S., 14, *27*, 67, *86*, 134, 136, 137, 139, *145–147*, *149*, 250, 258, 259, *262*, *266*, *267*
Udodong, U. E., 200, *225*
Ueda, E., 8, *24*, 67, *85*, 119, 126, 127, *129*, 257, *267*
Ueda, M., 337–343, *352–353*
Ueda, Y., 77, *89*
Uematsu, S., 173, *178*, 231, *243*
Uemura, H., 3, *22*, 93, *110*
Ueno, M., 17, *29*, 279, *287*
Ulloa, F., 295, *312*
Umemura, M., 12, 16, 18, *27–29*, 67, 76, 82, *86*, *88*, *90*, 270, 271, 276, *284*, *286*, 297, *314*
Umemura, Y., 308, *318*
Umeyama, T., 124, *131*
Umlauf, E., 297, *313*
Urakaze, M., 66, *84*, *85*, 126, *131*, 134, *146*
Urano, F., 276, *286*
Urban, N., 342, *355*
Urbaniak, M. D., 3, 14, *22*, *27*, 52–56, 58, 60, 61, *63*, *64*, 102, *112*, 252, 253, 256, 258, 260, *263*, *267*
Urquhart, P., 295, *312*
Usukura, J., 308, *318*
Utama, B., 366, *373*

V

Vadakkan, C., 305, *316*
Vai, M., 322, 324, 329, 334, 335, *346*, *348*, *349*

Vaillancourt, P., 165, *177*
Vainauskas, S., 35, *45*, *132*, 136, 139–141, *147–149*, 152, *157*
Valderrama, B., 342, *354*
Valdez-Taubas, J., 279, *287*
Vale, R. D., 304, *316*
Valente, E. P., 173, *176*, 231, *243*
Valentijn, K. M., 304, *316*
Vallee, R., 303, *315*
Valverde-Garduno, V., 366, *373*
Van, L. C., 366, *372*
van Aalten, D. M., 14, *27*, 252, *263*
van Berkel, M. A., 338, *353*
Van Den Abbeele, J., 176, *179*
van den Berg, C. W., 3, 6, *22*, 66, 360, *371*
Van Den Ende, H., 134, *146*, 322, 325, 327–330, 339, *345*, *348*, *349*, *353*, *354*
van den Hondel, C. A., 331, 332, *350*
van Der Goot, F. G., 295, *312*
van der Spoel, A. C., 297, *314*
Van der Vaart, J. M., 339, *353*
van Dorsselaer, A., 15, *28*, 136, 140, *147*
Van Dyke, K., 255, *265*
Van Egmond, P., 338, 339, *353*
VanKuyk, P. A., 331, 332, *350*
van Meer, G., 292, *311*
van Rinsum, J., 324, *348*
Van Schaftingen, E., 104, *112*
van Schagen, F. A., 324, *348*
van Solinge, W. W., 370, *373*
van Wijk, R., 370, *373*
Varki, A., 4, *30*
Varma, R., 293, *314*, 361, *371*
Vashist, S., 78, *89*, 291, *309*
Vasioukhin, V., 290, *309*
Vassella, E., 140, *149*
Vay, H. A., 37, 40, *46*
Vaz, W. L., 293, *318*, *319*
Vazquez-Duhalt, R., 342, *354*
Vehring, S., 152, *157*, 214, *226*
Venkataraman, K., 17, *29*, 278, *287*
Verdin, E., 366, *372*
Verges, M., 291, 292, *309*
Verhey, K. J., 304, *316*
Verkade, P., 297, 307, *313*, *314*, *317*, *319*
Verrips, C. T., 337, 339, *352*, *353*
Verstrepen, K. J., 322, *347*
Vidugiriene, J., 7, 9, *24*, 32, *44*, 76, *88*, 120, *130*, 136, 139, 141, *147*, *148*, 152, 154, *157*, 254, 259, *264*, *267*

Vieira, O. V., 297, *314*
Vijaykumar, M., 71, *88*, 163, *178*, 231, 232, 235, *243*
Vilbois, F., 295, *312*
Vilella, M., 305, *316*
Vink, E., 125, *131*, 328, *349*
Vionnet, C., 6, 7, 11, 12, 15, 18, 19, *23*, *26*, *28*, 29, 78, *89*, 118–120, 122–125, *128*, *129*, *131*, 136, 140, 143, *147*, 258, *266*, 270, 271, *283*, *284*, 332, *350*
Vishwakarma, R. A., 7, *24*, 152, 156, *157–158*, 209, 214, *226*, 241, *245*
Vissa, V. D., 109, *115*
Vivas, L., 163, *177*, *178*
Vliegenthart, J. F., 165, *179*
Voelker, D. R., 124, 125, *131*
Vogan, E., 271, *285*
von Figura, K., 104, *112*
Von Hanwehr, S. H., 297, *313*
von Itzstein, M., 162, *180*
Vossen, J. H., 12, *26*, 134, *146*, 322, 324, 327, 329, *345*, *348*, *349*
Vreeke, T. M., 167, *177*
Vriend, G., 53, 54, *63*
Vukcevic, D., 165, *177*
Vyas, P., 366, *373*

W

Wacker, M., 137, *148*
Wadle, A., 338, *352*
Waechter, C. J., 93, 103, 107, *110*, *112*, *114*, 153–154, *157*, *158*, 254, *264*
Walker, S. L., 342, *355*
Waller, R. F., 107, *114*
Walter, E. I., 66, 67, *84*, *86*, 250, 258, *262*, *266*
Walter, P., 16, *29*, 276, 278, *286*, *287*
Walther, E., 190, *224*
Walz, T., 271, *285*
Wanachiwanawin, W., 33, *44*
Waneck, G. L., 11, *25*, 36, *46*, 66, *84*, 167, *177*
Wang, C., 108, *115*
Wang, E., 306, *317*
Wang, F., 330, *350*
Wang, G. T., 323, 336, *347*
Wang, J., 366, *373*
Wang, L., 338, 342, *352*, *354*
Wang, P., 307, *317*

Wang, P. G., 338, *352*
Wang, Q., 3, *22*, 338, *352*
Wang, X. L., 276, *286*, 338, *352*, 366, *373*
Wang, Y., 123, *131*, 257, *265*, 322, *346*
Wang, Z., 307, *317*, 339, 340, *354*
Wantabe, R., 55, *63*
Wantanabe, R., 50, 58, *63*
Ward, G. E., 168, *180*
Ware, F. E., 93, 94, *110*, *111*, 153, *158*
Ware, R., 2, *21*
Ware, R. E., 362, *371*
Warren, C. D., 66, *84*, *85*, 126, *131*, 134, *146*, 167, *177*, 255, 257, *265*, *266*
Warren, G., 303, *315*
Watanabe, N. A., 12, *26*, 75, *88*
Watanabe, R., 7, 8, 11, 12, 17, *24–26*, *29*, 32, 34–41, *44–46*, 67, 76, 78, *85*, *89*, 92–94, 97, *109–111*, 118–123, 125–*129*, 152, 156, *157–158*, 250, 254, *262–264*, 270, 278, 279, 281, *284*, *287*, 291, *318*, 365, *372*
Watanabe, T., 330, *350*, 361, *371*
Waters, A. P., 168, *180*
Watzele, G., 324, *348*
Watzele, M., 324, *348*
Watzig, K., 338, *352*
Wawrezinieck, L., 308, *318*
Weddell, G. N., 134, *145*
Weerasinghe, G., 342, *355*
Wei, D., 339, 341, *354*
Wei, S., 173, *180*
Weig, M., 322, *345*, *347*
Weihofen, A., 136, *146*
Weik, J., 93, *110*, 153, *158*
Weimbs, T., 291, 307, *317*, *318*
Weingart, D. S., 259, *267*
Weingart, R., 166, 170, 173, *176*, *178*, 200, 209, *225*, 241, *245*, 256, *265*
Weiss, L. M., 330, *350*
Weissman, J. S., 16, *29*, 276, *286*
Weissman, Z., 322, *346*
Weissmann, C., 277, *286*
Weisz, O. A., 295, 301, 303, *312*, *314*, *315*
Weixel, K. M., 295, 301, *312*, *314*
Weller, C. T., 60, 62, *64*, 250, 252, *263*
Wells, S., 366, *373*
Weng, T., 307, *317*
Wenk-Siefert, I., 322, *345*
Wennicke, K., 175, *180*
West, G., 297, *314*

Westfall, B. A., 11, 12, *25–27*, 32, 36–38, *44*, *46*, 96, *111*, 118, 119, 125, 127, *129*, 250, 259, *262*, *267*
Westphal, V., 104, *112*
Westwater, C., 344, *356*
Whisstock, J. C., 108, *114*
White, J., 292, 300–303, 305, 306, *310*, *315*
White, M. D., 173, *180*, 277, *287*
Whiteheart, S. W., 307, *318*
Wichmann, D., 175, *180*
Wichroski, M. J., 168, *180*
Wiechers, M. F., 297, *313*
Wiedman, J. M., 6, 12, *23*, *27*, 96, *111*, 118–120, 122–127, *129*, *130*, 259, *267*
Wierstra, I., 365, *372*
Wiese, B. A., 72, *88*
Wijnaendts-van-Rest, R. W., 292, *311*
Wilce, M. C., 108, *114*
Wilcox, L. A., 363, *372*
Wildpaner, M., 134, *146*
Williams, A. F., 3, *22*, 66, 67, *83*, 118, 127, *128*, 182, *224*, 358, *370*
Williamson, B., 230, *242*
Williamson, P. R., 323, 331, 332, *347*
Wilson, C., 323, 331, 332, *347*
Wilson, J. M., 307, *317*
Wimmer, C., 307, *317*, *318*
Winger-Ness, A., 290, 292, 300, *318*
Winkler, J., 254, *264*
Wittrup, D. K., 343, *355*
Wittrup, K. D., 338, 339, 342–344, *353*, *355*, *356*
Wittwer, C. T., 167, *177*
Wodicka, L., 276, *286*
Woerle, B., 323, *347*
Wojciechowicz, D., 324, 338, *348*
Wojzynski, M. K., 72, *88*
Wolf, K., 339, 342, *354*
Wolfson, W., 234, *243*
Wong, B., 328, 329, *349*
Wong, Y. W., 66, *84*
Woods, A. S., 71, *88*, 163, 173, *178*, 231, 232, 235, *243*
Worley, K. C., 72, *88*
Wormald, M. R., 3, 6, *22*, 66, *84*, 360, *371*
Wren, B. W., 137, *148*
Wright, L. C., 323, 329, 331, 332, *347*
Wu, C. H., 337, 340–342, *352*
Wu, F., 144, *149*
Wu, G., 144, *149*

Wu, H., 333, *350*
Wu, R., 104, *112*
Wu, W. I., 124, 125, *131*
Wu, Y. S., 338, *353*
Wu, Z. B., 338, 340, *353*
Wurtz, O., 308, *318*
Wynshaw-Boris, A., 43, *47*

X

Xiao, H., 366, *372*
Xu, D., 275, *286*
Xu, S., 292, *311*
Xu, Y., 304, *315*, 330, *350*
Xue, J., *226*
Xue, X., 303, 304, *315*

Y

Yabuki, N., 322, 329, 335, *346*, *349*
Yamada, N., 2, 11, *21*, *24*, 33, 35, *44*, 248, 254, *261*
Yamagata, Y., 335, *351*
Yamaguchi, R., 77, *84*, *89*
Yamamoto, A., 11, *25*, 35, 36, 38, *45*, 76, *88*, 152, *157*
Yamamura, M., 339, 340, *354*
Yamanaka, H., 343, *355*
Yamanaka, R., 18, *29*
Yamashita, H., 308, *318*
Yamashita, K., 6, *23*
Yan, B. C., 12, *26*, 34–38, 40, 41, *44–46*, 152, *157*
Yang, B., 293, *318*
Yang, S. H., 6, 7, *22*, 66, *84*, 118, *128*
Yang, X., 255, *265*
Yano, Y., 43, *47*
Yasa, I., 344, *356*
Yashar, B. M., 330, *350*
Yashunsky, D. V., 3, *22*, 61, *64*, 102, *112*, 215, *226*, 241, *245*, 252, 253, 256, *263*
Yasuda, S., 17, *29*, 279, 281, *287*, *288*
Yasunaga, S., 332, 338, *350*
Yazaki, Y., 14, *27*, 137, *147*
Ye, J., 7, *24*, 67, *84*, *85*, 366, *373*
Yeaman, C., 290, 292, 295, *313*
Yeh, E. T. H., 11, *25*, 36, *46*, 66, *85*, 126, *131*, 134, *146*, 255, 257, *265*, *266*
Yeh, H. J., 248, 256, *261*, 325, 328, *348*
Yeremeev, V., 105, *113*

Yewdell, J. W., 276, *286*
Yin, Q. Y., 322, 323, 335, *346*, *347*
Yoda, K., 12, *27*, 96, *111*
Yoko, O. T., 18, *29*, 125, *131*, 258, *266*, 270, 271, 276, *284*, *286*, 299, *314*, 332, 333, 339, *350*, *354*, 360, *370*
Yoko-o, T., 15, 16, *28*, 76–79, 82, *89*, *90*
Yokoyama, S., 52, *63*
Yoneda, T., 276, *286*
Yoon, K., 248, 253, *261*
Yoshida, M., 322, *345*
Yoshida, N., 43, *47*
Yoshikawa, K., 33, *44*
Yoshiki, A., 43, *47*
Yoshimatsu, K., 12, *26*, 75, *88*
Yu, G., 35, *45*
Yu, J., 14, *27*, 137, 139, *147*, 342, *355*
Yu, R. K., 35, *45*
Yuan, L. C., 93, *110*, 270, *284*
Yuasa, S., 308, *318*
Yue, L., 338, *352*
Yui, D., 276, *286*

Z

Zacchetti, D., 297, *313*
Zacks, M. A., 140, *149*
Zaker-Tabrizi, L., 108, *114*
Zanetta, J. P., 297, *313*
Zanolari, B., 279, *287*
Zanone, P. B., 366, *372*
Zawadzki, J. L., 140, *149*
Zem, G., 342, *355*
Zeng, L., 292, *311*
Zerial, M., 304, *316*, *317*
Zhang, H., 67, *85*, 307, *317*, 330, *350*
Zhang, J., 295, *312*
Zhang, Y., 276, *286*
Zhang, Z., 304, *315*, 328, 329, *349*, 361, *371*
Zheng, Y., 330, *350*
Zhong, X., 124, 125, *131*
Zhou, A., 235, *244*
Zhou, M. M., 292, *311*
Zhu, J. Z., 173, *178*, 231, *243*
Zhu, Y., 19, *29*, 123, 124, *131*, 136, 143, *147*, 293, *319*
Zia-ul, H., 337, 338, 341, 342, *352*
Zimmer, C., 338, 340, 341, *352*
Zimmer, K. P., 104, *112*, 295, 297, 301, *313*
Zimmerberg, J., 295, *312*, *319*

Zimmerman, J. W., 93, *110*
Zimmermann, M., 339, 342, *354*
Zinecker, C. F., 162, 165, 166, *179*, *180*
Zitzmann, N., 7, *23*, 66, 67, *85*, *86*, 259, *267*
Zlotnik, H., 328, *349*
Zoeller, R. A., 258, *266*
Zorn, M., 292, *311*

Zou, W., 173, *180*, 340, 342, 343, *354*, *355*
Zueco, J., 339, *354*
Zurdo, J., 276, *286*
Zurzolo, C., 280, *288*, 292, 295, 297, 299–302,
    304, 305, 307, *311–315*, *318*
Zweibaum, A., 295, *312*
Zwick, M. B., 337, 338, *352*

Index

A

Actin filaments
  GTPase cdc42, 305–306
  protein transport, 305
  sucrose isomaltase (SI), 306
Apicomplexan protozoa, GPIs
  glycosylation, 162
  human and agricultural diseases
    Cryptosporidium, 161
    malaria, 161–162
  immunological functions
    antidisease vaccine target, 175–176
    apoptosis, 173–175
    inflammation, 170, 173
    TLR signaling, 175
  life cycle, 160–161
  Plasmodium falciparum
    biosynthesis, 166–168
    structure, 162–165
  Toxoplasma gondii
    biosynthesis, 168–170
    structure, 165–166
  ultrastructure, 161f

B

Bovine serum albumin (BSA), 235
Butyrate responsive elements (BRE),
  366–367, 370

C

Calcofluor white (CFW), 16, 18, 19
Cell-free system, 59–60
Chinese hamster ovary (CHO), 8, 18, 50, 58,
  72, 77, 94, 95, 103, 173
Chitin
  CRH family, 336

Chitin—cont'd
  β1,3 glucan and, 329, 330
  glucans, 325–326
  reducing ends, 328
  yeasts, 327
Congenital disorders of glycosylation (CDG),
  103–104

D

Decay-accelerating factor (DAF), 66–67,
  134, 144
deNAc gene, 50
De novo biosyntheses, 342–343
Detergent-resistant membranes
  (DRMs), 16
  apical proteins, 292
  basolateral GPI-AP (GFP-PrP), 299
  GPI-APs, 16–17, 275, 283, 295–296, 301
  mammalian cells, 280
  rafts, 293–294
  yeast, 278–279
Diisopropylfluorophosphate (DFP), 70, 72,
  79, 169, 257–258
Dolichol phosphate (Dol-P), 167–168,
  170–172
Dolichol-phosphate mannose (Dol-P-Man),
  8, 9, 12, 39
  Dol-P-β-Man, 103
  GDP-Man, 167
  GPI biosynthesis, 103–104
  GPI-Man-Ts substrate, 92–98
  as mannose donor, 359–360
  Plasmodium species, 168
  subunit, 11, 38
  synthesis, 254
Dolichol phosphate mannose (DPM)
  synthase, 167, 170–172
DRMs. *See* Detergent-resistant membranes

E

Endoplasmic reticulum (ER)
  Etn-P moieties addition, 119
  exit and GPI-anchored proteins
    COPII-mediated, 271–272
    receptors, 272–273
    SEC13 genetic interactions, 273
    sorting signals and factors, 273–275
  export and GPI anchor structure
    biosynthesis, defect, 270–271
    lipid moiety, 270
ER-associated protein degradation (ERAD),
  276–278
Etn-P. *See* Phosphoethanolamine

F

Functional complementation assays, 53,
  58–59
Flippase, 152–153, 155–156

G

Genetically modified organisms (GMO), 342
GlcNAc-PI de-N-acetylase
  activity assays
    cell-free system, 59–60
    fluorescence, 61
    functional complementation, 58–59
    mass spectrometry, 60–61
    radiolabeled acetate, 60
    substrate obtainment, 58
  alanine mutants analysis, 53
  catalytic activity, 52–53
  enzyme substrate specificity
    lipid portion, 61–62
    mammalian and trypanosomal
      system, 62
  identification, 50
  mechanism, 54–55
  molecular model, 54f
  putative mechanism, 55f
  rat truncated forms, 57
  recombinant protein expression
    soluble protein, 55–56
    truncations and domain boundaries,
      56–58
  sequence alignment, 51f
  structural homology, 52
  zinc metalloenzymes, 52–53

GlcNAc-PI synthase
  GlcN-PI generation, 155f
  synthesis and de-N-acetylation, 151–152
Glycosylinositol phospholipids (GIPLs),
  140, 221
Glycosylphosphatidylinositol-anchored
  proteins (GPI-APs)
  deficiency, 2–3
  ER exit
    COPII-mediated, 271–272
    exit receptors, 272–273
    genetic interactions and *sec13*, 273
    GPI anchor structure and ER export,
      270–271
    in yeast, 275f
  eukaryotic cells, 358
  fatty acid (FA), 360
  functions, 361
  glycan and lipid structures, 3
  GPI and lipid remodeling
    mammalian cells, 16f
    protozoan parasites, 19–20
    yeast and mammalian cells, 15–19
  GPI glycan modification, enzymes, 20
  IGD, 363–364
  and lipid traffic, 278–282
  modification, 2
  proliferative stages, 3
  sorting and ER exit
    factors, 274–275
    signals, 273–274
  sphingolipids and sterols
    DRM, 278–281
    mammalian cells secretory pathway, 280f
    secretory pathway, yeast, 279f
    transport, yeast, 281–282
  structure
    mammalian cells, 4f, 6
    Trypanosoma brucei, 7
    yeast, 6–7
  trafficking defects and folding
    ERAD, 277
    protein response, unfolded, 276
    PrP prion, 277–278
    unfolded degradation, 275–276
  yeast and mammalian cells, 283
GPI anchor
  ceramide remodeling, yeast
    CWH43, 18–19
    in ER, 18

GPI anchor—cont'd
  endoplasmic reticulum (ER), 248
  human proteins, 2
  mannosylation, 254
  precursors, 1, 3, 6–8
  and rafts environment, 299–300
  Saccharomyces cerevisiae, proteins, 2–3
  saturated fatty acid transfer, 17–18
  unsaturated fatty acid and sn-2 position
    gene identification, 17
    PGAP3 and yeast Per1p, 16
GPI anchor attachment
  cancer cell lines, 144
  eukaryotic proteins, 144–145
  GAA1/Gaa1p
    Etn-P cap, 141
    glycosylation sites, 140
  GPI8/Gpi8p
    as catalytic center, 138–139
    cell surface expression, 140
    yeast structure, 139f
  GPIT
    OST and, 137
    subunits, 136–137
    T. brucei and human subunits, 138f
  PIG-S/Gpi17, 143
  PIG-T/Gpi16p, 141–142
  PIG-U/Gab1p
    long chain fatty acids, 143
    N-terminal 430 amino acids, 142f
  signal sequence, ω-minus and ω-plus,
    134–135
  transamidation reaction, 136
  TTA1 and TTA2, 143–144
GPI anchor synthesis
  CD52
    pseudodisaccharide, 218f
    synthetic design, 216, 217f
  Leishmania parasite, LPG
    benzylation, 219, 220f
    building blocks, 218
    ethyl-1-thio-β-galactopyranoside,
      219, 220f
    furanoside derivative, 221
    synthetic approach, 219
  membrane association, 182–183
  P. falciparum
    coupling, 206
    reagents and conditions, 207f, 208f
    Seeberger synthesis, 204, 207f, 208–209

GPI anchor synthesis—cont'd
  subunit assembly, 205
  synthetic approach, 206
  PIs and, 183
  rat brain Thy-1
    coupling, 204
    galacto-manno building block, 202f
    glucosamine inositol building block,
      202, 203f
    inositol building block, 203f
    synthetic approach, 200f
    trimannose building block, 201f
  Saccharomyces cerevisiae
    ceramide lipid intermediate, 195, 196f
    glucosamine-inositol fragment, 197
    hexasaccharide block assembly, 198f
    lipid anchor and deprotection, 199f
    2-*O*-acetyl group removal, 199
    phosphitylation, 199–200
    synthetic approach, 195f
    tetramannose fragment, 197f, 198f
  structure, 182
  Trypanosoma cruzi
    coupling, 213, 214
    Dundee group, design, 215f
    GIPL, 221–223
    glucosamine-inositol glycosylation, 212
    glycoinositol backbone, 216f
    mannobiose, 211
    myo-inositols, 209
    synthetic strategy, 211f
    tetramannose, 211, 214f
  VSG of *T. brucei*
    building blocks, 183
    chemical structures, 184f
    coupling, 188f, 193f
    galactobiosyl donor, 187f
    GlcN-inositol building block, 191f
    glycosidation, 186
    hydrogenation, 195
    myo-inositol building block, 185f, 186f
    phosphoramidite, 194
    reagents and conditions, 189f, 194f
    synthetic approach, 190f
    trisaccharide donor and
      mannobiose, 192f
GPI-APs polarized sorting, epithelial cells
  apical, 294f
  cytoskeleton
    actin filaments, 305–306

GPI-APs polarized sorting, epithelial
    cells—cont'd
    microtubules, 303–305
    direct and indirect delivery
        endocytic compartments, 302–303
        transport analysis, 302
    mechanism
        protein ectodomain, 295–296
        putative apical receptor, 296–298
        rafts environment and GPI anchor,
            299–300
    membrane traffic
        cargo vesicles and target membrane
            fusion, 307
        Rab8, 306
        Rab14, 306–307
    pathways, 298f
    and secretory pathway
        endoplasmic reticulum (ER), 291
        Golgi apparatus and, 291–295
        sorting signals, 290–291
    TGN
        apical and basolateral cargoes
            segregation, 300–301
        MDCK cells, 301–302
        raft dependent and independent
            mechanisms, 301
GPI biosynthesis
    anchored proteins (GPI-APs), 2–6
        posttranslational attachment, 10
        structure of, 6–7
    attachment to proteins
        gene comparison, 13–14
        mammalian enzymes, 11–12
        mammalian pathway, 7–8
        posttranslational, GPI-AP, 10
        T. brucei, pathway, 9–10
        transamidases, 14–15
        trypanosome enzymes, 12, 14
        yeast enzymes, 12
        yeast S. cerevisiae, pathway, 8
    biological characterization, 260
    cell-free system, 249–250
    in ER membrane, mammalian cells, 5f
    ethanolamine phosphatetransferases, 257
    GlcNAc-PIDe-N-acetylase, 250–253
    GlcNActransferase, 250
    inositol acyltransferase, 253–254
    inositol deacylation, 257–258
    in vitro, 254

GPI biosynthesis—cont'd
    lipid remodelling, 258
    mammalian cells, 248
    mannose analogs, 254–255
    MT-I, 255–256
    MT–II, 256
    MT–III, 256–257
    pathways, 70, 253–254
    and remodeling, genes, 9
    species-specific modifications, 259–260
    split topology
        assembly reactions, 153–154
        GlcNAc-PI, synthesis and de-N-
            acetylation, 151–152
        GlcN-PI flips, 154–156
        inositol acylation, 152–153
    structures and anchors, 249
    transamidase, 258–259
GPI-Etn-P-T-I
    moiety, 123–124
    PIG-N and Mcd4, 122–123
    yeast cells, ATP, 124
GPI-Etn-P-T-II
    PIG-G and Gpi7, 124–125
    yeast gpi7 null mutants, 125
GPI-Etn-P-T-III
    Gpi11 roles, 127
    pig-f mutant cell, 126
    PIG-O and Gpi13 proteins, 125–126
    S. cerevisiae PIG-F homolog, 126–127
GPI-Etn-P-T, transmembrane organization,
    121f
GPI-GlcNAc transferase
    biosynthetic steps, communication, 41–42
    complex modeling, 38
    mammalian enzyme regulation, DPM2
        Dol-P-Man synthase subunits, 38–39
        mechanism, 39
    PIG-A gene somatic mutation, 33
    PIG-A/Gpi3, 35
    PIG-C/Gpi2, PIG-H/Gpi15 and PIG-P/
        Gpi19, 36
    PIG-Q/Gpi1, 37
    PIG-Y/Eri1, 37–38
    pseudogenes
        generation, 42
        RNAs retrotransposition, 43
    subunit composition and topology, 34f
    UDP-GlcNAc concentration regulation, 41
    yeast and mammalian cells, 33–34

GPI-GlcNAc transferase—cont'd
 yeast regulation, RAS
  ERI1, 39–40
  functional significance, 40–41
  gpi1Δ mutant, 40
GPI-Man-Ts and Dol-P-Man substrate
 dolichol-phosphate-glucose (Dol-P-Glc),
  92–93
 GPI-Man-TI
  characteristics, 94
  PIG-X identification, 94–95
 GPI-Man-TII
  CAZy classification, 95–96
  PIG-V and gpi18p, 95
 GPI-Man-TIII, 96
 GPI-Man-TIV
  mRNA expression, 97
  Smp3p identification, 96–97
 synthase components, 93
GPI proteins and yeast cell walls
 cell wall assembly
  ordered addition, 329
  Rho1p, 330
 cellwall cross-link, biogenesis
  CRH family, 336–337
  Dcw1p and Dfg5p, 333–334
  GAS/GEL/PHR family, 334–335
  phosphoethanolamine transferases,
   332–333
  Yapsin/SAP family, 335–336
 fungal
  biogenesis model, 325f
  glycans and, 324
  types, 322
 linked proteins, 323
 structure, 326f
  mannoproteins, 328–329
  polysaccharides, 327–328
 transglycosylation
  bioinformatic analysis, 331
  Cryptococcus, 331–332
  fungal genomes, 330
  wall glucan, 331f
GPI transamidase (GPI-TA), 1, 9, 10, 13–15
GPI transamidase (GPIT)
 as catalyst, 135f
 GAA1/Gaa1p, 140–141
 GPI8/Gpi8p, 138–140
 PIG-T/Gpi16p, 141–142
 PIG-U, 143

GPI transamidase (GPIT)—cont'd
 subunits, 136–138, 144–145
 transamidation reaction, 133–134, 136
 TTA1 and TTA2, 143–144
Guanosine diphosphate mannose
 (GDP-Man), 93, 106, 154, 166–167, 172,
 250, 254, 255

H

Hematopoietic stem cell (HSC), 361–362
High molecular weight (HMW) complexes,
 292, 299, 300, 308

I

ICAM-1. *See* Intercellular adhesion
 molecule-1
IGD. *See* Inherited GPI deficiency
Induced nitric oxide synthase (iNOS), 162
Inherited GPI deficiency (IGD)
 acquired, PNH, 362
 cell and tissue-dependent variability, 369
 clinical spectrum, 362–363
 GPI anchor
  biological significance, 361
  biosynthesis, 358–360
  lipid modification and protein
   attachment, 360
 HDAC inhibition, 367, 370
 house-keeping gene, 370
 hypomorphic promoter mutation, 365
 mutations and mechanism, 369
 PIG-M
  identification, 363–365
  transcriptional control and Sp1, 367–368
 Sp1 and gene-specific histone
  hypoacetylation
   HDAC inhibition, 366
   PIG-M promoter, 367
   tentative model, 368f
 surface expression, 363
 TF Sp1, 365–366
Inositol acylation/deacylation
 Cryptococcus neoformans, 70–71
 enzymes and functions
  inositol acyltransferase, 72–77
  inositol deacylase, 77–82
 Leishmania, 72
 PI-PLC-resistant GPI molecules

Inositol acylation/deacylation—cont'd
  CD52, 66–67
  GPI anchor, 66
  Plasmodium falciparum and *Trypanosoma cruzi*, 71
  Toxoplasma gondii, 71–72
  Trypanosoma brucei
    enzymatic characteristics, 70
    and mammalian pathways, 69
  yeast and mammals
    enzymatic activity, 69
    glucosamine-phosphatidylinositol (GlcN-PI), 67
    steps, 68f
Inositol acyltransferase
  mammalian PIG-W
    cDNA expression, 72
    PI species, 72, 74–75
    multiple alignment, 73f–74f
  S. cerevisiae Gwt1p
    gene characterization, 75–76
    identification, 75
  in *T. brucei*, 76–77
  topology, 76
Inositol deacylase
  mammalian PGAP1, 77
  multiple alignment, 80f, 81f
  S. cerevisiae Bst1p
    ER and GPI-anchored proteins, 78–79
    identification, 78
  T. brucei GPIdeAc
    identification, 79–82
    multiple alignment, 80f
  T. brucei GPIdeAc2
    identification, 79
    knockdown, 80, 82
    production, 79–80
Intercellular adhesion molecule-1 (ICAM-1), 170, 174, 343

K

Keyhole limpet hemocyanin (KLH), 233–234

L

LAM. *See* Lipoarabinomannan
Lipases, 321–323, 331, 339, 341, 343
Lipoarabinomannan (LAM)
  biosynthetic pathway, 106f

Lipoarabinomannan (LAM)—cont'd
  immuno-suppressive activities, 104–105
  structure, 105f
Lipomannan (LM)
  biosynthetic pathway, 106f
  structure, 105f
Lipopeptidophosphoglycan (LPPG), 221–223
Lipophosphoglycans (LPGs), 140, 238
LM. *See* Lipomannan

M

Malarial vaccine
  anemia
    GPI involvement, 239f
    P. falciparum, 240
    RBCs, 238
  antibody response
    BSA and, 235
    GPI microarray, 236f
    Man$_3$ and Man$_4$-GPIs, 236–237
  GPI and
    homogeneity, 232
    parasites, 230
    P. falciparum, 231
    toxin theory, 230–231
  synthetic GPI, rodent model
    glycan–KLH conjugate, 233f
    immunization, 234
    P. falciparum, 234–235
Mammalian GPI-anchored proteins, 248
Mannoproteins
  design, 328
  transglycosylation, 328–329
Mannosylation
  biosynthetic pathway, 92
  congenital disorders of glycosylation (CDG), 103–104
  Dol-P-Man
    acceptor sugars structure, 97f
    GPI-Man-TI, 94–95
    GPI-Man-TII, 95–96
    GPI-Man-TIII, 96
    GPI-Man-TIV, 96–97
    synthase activity, 93
    utilization, 92–93
  glycosyltransferases and Dol-P-monosaccharides
    classification, 97–98
    extracellular loop phylogram tree, 100f

Mannosylation—cont'd
  mannosyltransferases extracellular loop
      alignment, 99f
  peptide sequences, 98–102
  protein accession numbers, 101
mannosyltransferases, mycobacteria, 104
  GDP-mannose-dependent, 106–107
  LAM, 104–105
  PIMs and LM/LAM, 105–106
mycobacteria genes and mammalian
      PIG-M
  MSMEG_4247, 109
  MSMEG_3120 and MSMEG_4241, 108
  PimE, 107–108
paroxysmal nocturnal hemoglobinuria,
      104
substrate specificities and inhibitors
  Dol-P-Man, 103
  T. brucei, 102–103
α1–2–Mannosyltransferase (MT–III), 171,
      256–257
α1–4–Mannosyltransferase (MT-I), 168,
      171, 255–256
α1–6–Mannosyltransferase (MT–II), 171,
      256
MDCK cells, 292, 296, 297, 301–303,
      305, 306
Membrane-bound O-acyltransferase
      (MBOAT), 152
Membraneform variant surface glycoprotein
      (mfVSG), 238
Microtubules (MTs)
  alignment, 303
  apical and basolateral transport, 303–304
  KIFC3, 304

N

N-acetylglucosamine-PI transfer
      reaction, 32

O

Oligosaccharyltransferase (OST)
  GPIT and, 138
  subunits, 137
Organophosphate hydrolases (OPH), 341,
      344
Oxysterol binding protein
      (OSBP), 281

P

Paroxysmal nocturnal hemoglobinuria
      (PNH), 33
  genetic lesions, 362
  GPI expression, 364
Phenylmethylsulfonyl fluoride (PMSF),
      70–72, 169, 251, 253, 254, 257, 258
Phosphatidylcholine (PC), 154–156
Phosphatidylethanolamine (PE), 153, 155
Phosphatidylinositol (PI)
  GlcNAc-PI, 151–152, 155
  GlcN-PI flips, 151
      generation, 152
      glycolipid synthesis, 154
      GPI assembly steps, 155
      membrane protein, 156
  inositol ring of, 32
Phosphatidylinositol mannosides (PIMs)
  biosynthetic pathway, 106f
  glycolipids biosynthesis, 91
  LM/LAM, 104–106
Phosphatidylinositol-specific phospholipase
      C (PI-PLC), 66
Phosphoethanolamine (Etn-P), 331–333
  addition and proteins
      GPI-Etn-P-T-I, 122–124
      GPI-Etn-P-T-II, 124–125
      GPI-Etn-P-T-III, 125–127
      GPI-Etn-P-transferases, 120–122
  donor, 119–120
  GPI7, 332–333
  GPI anchors, 117–118
  modification sites
      free GPIs, 119
      protein-bound GPIs, 118–119
PIG-A gene, 33
PIG-A/Gpi3, 35
PIG-C/Gpi2, 36
PIG-H/Gpi15, 36
PIG-L, sequence alignment, 52
PIG-P/Gpi19, 36
PIG-Q/Gpi1, 37
PIG-Y/Eri1
  deletion mutant, 37–38
  identification, 37
Plasmodium falciparum
  biosynthesis
      Dol-P-Man, 167
      GDP-Man, 166

Plasmodium falciparum—cont'd
  genes, 171–172
  GPI-MT-I, 168
  substrate specificity, 166–167
  GPI structure, 164f, 232f
    asexual, intraerythrocytic stage, 162
    chemical synthesis, 165
    malarial precursors, 163
PNH. *See* Paroxysmal nocturnal
  hemoglobinuria
Prion diseases, 277–278
Prion protein (PrP), 2, 270, 277–278,
  299–300, 302
Protozoan GPIs
  antidisease vaccine target, 175–176
  apoptosis
    morphological changes, 173
    T. gondii, 173–175
  inflammation
    cytokines, 170
    TLRs and, 173
  TLR signaling, 175

R

Red blood cells (RBCs), 238–240
rPIGL(EC-184), 56–57

S

SD. *See* Surface display
Severe malaria anemia (SMA), 238
Single-chain monoclonal antibody peptides
  (scFv), 342–343
Surface display (SD)
  anchorage modes, 338–340
  exogenous proteins, 323
  heterologous proteins, 337f, 340–344
  yeast, 321–322, 337–344

T

Tetrabutylammonium fluoride (TBAF), 186,
  191, 194, 197, 198, 201, 212, 216
Toll-like receptor 2 (TLR2), 160, 173–175
Toxoplasma gondii
  biosynthesis
    GalNAc, 168
    genes, 171–172
    glucosylation, 169

Toxoplasma gondii—cont'd
  serine protease inhibitors, 169–170
  GPI structure, 164f
  glycoforms, 165
  inositol structure, 166
Transferrin receptor (TfR), 302
trans-Golgi network (TGN)
  GPI-APs polarized sorting
    apical surface pathways, 301
    in nonpolarized cells, 301–302
    segregation, 300–301
    surface sorting, 291–292
Transmembrane helix (TMH), 167
Trypanosomatid transamidase 1 (TTA1)
  PIG-S, 14–15
  TTA2 and, 137–138, 143–144
Trypanosome cell-free system, 52
TT1534, sequence alignment, 52

V

Variant surface glycoproteins
  (VSGs)
  GPI
    attachment, 134
    inositol moiety, 69
  T. brucei, GPI anchor synthesis, 182–195
  Trypanosoma brucei, 134, 144
Vascular cell adhesion molecule-1
  (VCAM-1), 170
Vesicular stomatitis virus glycoprotein
  (VSV-G), 302
VSGs. *See* Variant surface glycoproteins

W

WHAT IF programme, 53

Y

Yapsins, 335–336
Yeast
  cell wall
    GPI-APs, 323
    model, 326f
    structure, 326–329
  surface display
    anchorage modes, 338–340
    vs. bacterial expression systems, 338
    heterologous proteins, 340–344

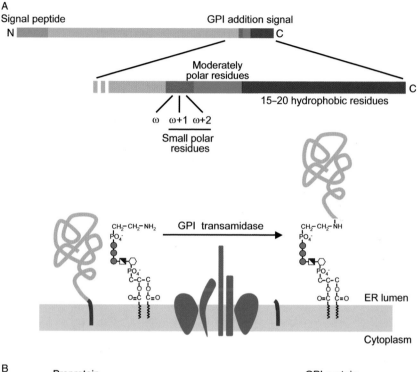

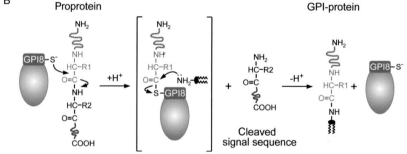

Aita Signorell and Anant K. Menon, Fig. 7.1, page 135. The GPI signal sequence and attachment of a GPI anchor to proteins. (A) GPI-anchored proteins contain an N-terminal ER import signal (orange) and a C-terminal GPI addition sequence; the C-terminal signal is replaced by GPI in a reaction catalyzed by GPIT (modified from Ref. [2]). See text for details. (B) The GPI-anchoring reaction. Proproteins are recognized by GPIT on the luminal face of the ER. The C-terminal signal sequence is cleaved between ω (in red) and ω + 1 residues by the GPIT subunit GPI8, thereby activating the carbonyl group of the ω amino acid. An amide bond is formed by nucleophilic attack on the activated carbonyl by the amino group of the ethanolamine-phosphate cap of a GPI [3].

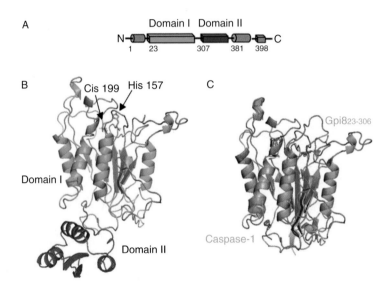

A

Domain I Domain II

N—◼▯▭—◼◼—▯—◼—C

1   23   307   381 398

B

Cis 199 His 157

C

Gpi8₂₃₋₃₀₆

Domain I

Caspase-1

Domain II

Aɪᴛᴀ Sɪɢɴᴏʀᴇʟʟ ᴀɴᴅ Aɴᴀɴᴛ K. Mᴇɴᴏɴ, Fɪɢ. 7.3, ᴘᴀɢᴇ 139. Predicted structure for yeast Gpi8p. The protein sequence predicts an N-terminal signal sequence and a C-terminal TM domain (gray tubes, panel (A)), and two soluble domains (domain I in turquoise and domain II in purple). Panel (B) depicts the predicted structure of yeast Gpi8p, showing the two domains in the same color as in the top panel. The putative catalytic dyad, consisting of Cys199 and His157, is highlighted by arrows. (C) Overlay of a model of the caspase-like domain of yeast Gpi8p (turquoise) with the resolved crystal structure of Caspase-1 from *Spodoptera frugiperda* (PDB: 1M72, orange) (adapted from Ref. [49]).

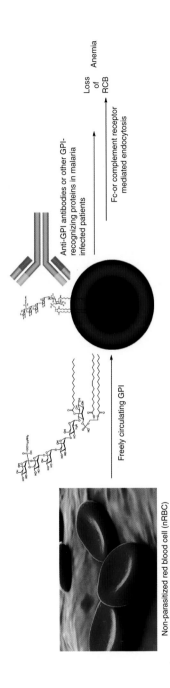

XINYU LIU *ET AL.*, FIG. 11.5, PAGE 239. GPI's involvement in malaria-associated anemia. Freely circulating or released *P. falciparum* GPI inserts into nRBCs and results in the recognition by anti-GPI antibodies that may contribute subsequent RBC eliminations.

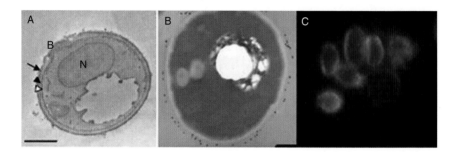

Marlyn Gonzalez et al., Fig. 15.1, page 323. GPI-anchored proteins in yeast cell walls. (A) Transmission electron micrograph of a permanganate-fixed cell showing outer wall proteins, many of which are GPI linked (arrow); glucan layer (filled arrowhead); plasma membrane/periplasmic region (open arrowhead); B bud scar; N, nucleus. (B) Micrograph of a yeast cell stained after sectioning with gold-labeled antibody to a GFP-GPI fusion protein. (C) Direct fluorescence of GFP-GPI protein in isolated, membrane-free washed cell walls. Note the retention of cell shape in the isolated walls.